ENVIRONMENTAL REMOTE SENSING
and
SYSTEMS ANALYSIS

EDITED BY NI-BIN CHANG

CRC Press
Taylor & Francis Group
Boca Raton   London   New York

CRC Press is an imprint of the
Taylor & Francis Group, an **informa** business

CRC Press
Taylor & Francis Group
6000 Broken Sound Parkway NW, Suite 300
Boca Raton, FL 33487-2742

First issued in paperback 2019

© 2012 by Taylor & Francis Group, LLC
CRC Press is an imprint of Taylor & Francis Group, an Informa business

No claim to original U.S. Government works

ISBN-13: 978-1-4398-7743-2 (hbk)
ISBN-13: 978-0-367-38155-4 (pbk)

Visit the Taylor & Francis Web site at
http://www.taylorandfrancis.com

and the CRC Press Web site at
http://www.crcpress.com

Contents

PART II Sensing and Monitoring for Land Use Patterns, Reclamation, and Degradation

PART III   *Air Quality Monitoring, Land Use/Land Cover Changes, and Environmental Health Concern*

Preface

In the last few decades, rapid urbanization and industrialization have altered the priority of environmental protection and restoration of air, soil, and water quality many times. Yet it is recognized that the sustainable management of human society is necessary at all phases of impact from the interactions among energy, environment, ecology, public health, and socioeconomic paradigms. The multidisciplinary nature of this concern for sustainability is truly a challenging task that requires employing a systems analysis approach. Such a systems analysis approach links several disciplinary areas with each other to promote the concept of sustainable management. Just as a sophisticated piece of music involves many different instruments played in unison, systems analysis requires a holistic viewpoint and a plethora of tools in sensing, monitoring, and modeling that have to be woven together to explore the state and function of air, water, and land resources at all levels.

With the aid of systems analysis, this comprehensive collection includes a variety of research work that results from years of experience and that reflects the contemporary advances of remote sensing technologies. This unique publication presents and applies the most recent synergy of remote sensing technologies that will advance the overall understanding of the sensitivity of key environmental quality issues in relation to human perturbations. These perturbations can be caused by collective or individual impacts of economic development and globalization, population growth and migration, and climate change on atmospheric, terrestrial, and aquatic environmental systems.

Specifically, this book aims to address the following intertwined research topics in the nexus of the environmental remote sensing and systems analysis:

- What are the potential impacts on water quality when the management of the nitrogen cycle in a watershed changes, affecting ecosystem health in marine and fresh waters?
- What are the regional impacts of an oil spill in coastal environments?
- How will water quality in coastal bay and estuary regions be affected by changing salinity concentrations, turbidity levels, and sediment transport processes?
- How will landslide and land subsidence in association with the changing hydrologic cycle influence human society?
- How will the effects of urbanization affect the rate of water infiltration at urban–rural interfaces?
- How can the impact of air pollution on meteorology, climatology, and public health be evaluated in association with the changing land use and land cover patterns from urban to global scales?

The presentations in this book uniquely elaborate on the intrinsic links of the above questions that capture important interactions among three thematic areas.

They include (1) water quality monitoring, watershed development, and coastal management; (2) sensing and monitoring for land use patterns, reclamation, and degradation; and (3) air quality monitoring, land use/land cover changes, and environmental health concerns.

On this foundation, many new techniques and methods developed for spaceborne, airborne, and ground-based measurements, mathematical modeling, and remote sensing image-processing tools may be realized across these three distinctive thematic areas. This book will be a useful source of reference for undergraduate and graduate students and working professionals who are involved in the study of environmental science, environmental management, sustainability science, environmental informatics, and agricultural and forest sciences. It will also benefit scientists in related research fields, as well as professors, policy makers, and the general public.

As the editor of this book, I wish to express my great appreciation for the contributions of many individuals who helped write, proofread, and review these book chapters. I am indebted to the 58 authors and coauthors within the scientific community who have shared their expertise and contributed much time and effort in the preparation of this book. I also wish to give credit to the numerous funding agencies promoting scientific research in environmental remote sensing that have led to the generation of invaluable findings presented here. I acknowledge the management and editorial assistance of Irma Shagla and Kari Budyk.

Dr. Ni-Bin Chang
Director, Stormwater Management Academy
University of Central Florida
Orlando, Florida

MATLAB® is a registered trademark of The MathWorks, Inc. For product information, please contact:

The MathWorks, Inc.
3 Apple Hill Drive
Natick, MA 01760-2098 USA
Tel: 508 647 7000
Fax: 508-647-7001
E-mail: info@mathworks.com
Web: www.mathworks.com

About the Editor

 **Dr. Ni-Bin Chang** is currently a professor with the Civil, Environmental, and Construction Engineering Department, University of Central Florida (UCF). He is also a senior member of the Institute of Electrical and Electronics Engineers (IEEE) affiliated with the IEEE Geoscience and Remote Sensing Society and the IEEE Computational Intelligence Society. He has earned the selectively awarded titles of Certificate of Leadership in Energy and Environment Design (LEED) in 2004, Board Certified Environmental Engineer (BCEE) in 2006, Diplomat of Water Resources Engineer (DWRE) in 2007, elected member (Academician) of the European Academy of Sciences (MEAS) in 2008, and elected Fellow of American Society of Civil Engineers (ASCE) in 2009. He was one of the founders of the International Society of Environmental Information Management and the former editor-in-chief of the *Journal of Environmental Informatics*. He is currently an editor, associated editor, or editorial board member of 20+ international journals.

Contributors

Robert F. Adler
NASA Goddard Space Flight Center
Greenbelt, Maryland

and

University of Maryland
College Park, Maryland

Fahad A. M. Alawadi
Regional Organization for the
    Protection of the Marine
    Environment (ROPME)
Safat, Kuwait

Pasquale Avino
Chemical Laboratory DIPIA, INAIL
    (ex-ISPESL)
Rome, Italy

Lakshimi Madhavan Bomidi
City College
City University of New York
New York, New York

Vittorio Brando
CSIRO Land & Water
Canberra, Australia

Jon Brodie
James Cook University
Townsville, Australia

Halil I. Cakir
U.S. Environmental Protection Agency
Research Triangle Park, North Carolina

Ni-Bin Chang
University of Central Florida
Orlando, Florida

Liang-Chien Chen
National Central University
Jhongli, Taiwan

Heather M. Cheshire
North Carolina State University
Raleigh, North Carolina

Sundar A. Christopher
The University of Alabama
    in Huntsville
Huntsville, Alabama

Maryam Dehghani
Shiraz University
Shiraz, Iran

and

K. N. Toosi University of Technology
Tehran, Iran

Arnold Dekker
CSIRO Land & Water
Canberra, Australia

Michelle Devlin
James Cook University
Townsville, Australia

Paul Elsner
Birkbeck College
University of London
London, United Kingdom

Iman Entezam
Engineering Geology Group
Geological Survey of Iran (GSI)
Tehran, Iran

Wei Gao
Colorado State University
Fort Collins, Colorado

Zhiqiang Gao
Colorado State University
Fort Collins, Colorado

Barry M. Gross
City College
City University of New York
New York, New York

Ernst F. Hain
North Carolina State University
Raleigh, North Carolina

Min Han
Dalian University of Technology
Dalian, China

Xianjun Hao
College of Science
George Mason University
Fairfax, Virginia

Jerry L. Hatfield
National Laboratory for Agriculture
    and the Environment
USDA-ARS
Ames, Iowa

Bin He
Kyoto University
Kyoto, Japan

D. Barry Hester
Bryan Cave LLP
Atlanta, Georgia

Scott Hetrick
Indiana University
Bloomington, Indiana

Yang Hong
University of Oklahoma
Norman, Oklahoma

N. Christina Hsu
NASA Goddard Space Flight Center
Greenbelt, Maryland

Shih-Jen Huang
National Taiwan Ocean University
Keelung, Taiwan

Siamak Khorram
North Carolina State University
Raleigh, North Carolina

Guiying Li
Anthropological Center for
    Training and Research on Global
    Environmental Change
Indiana University
Bloomington, Indiana

Zonghu Liao
University of Oklahoma
Norman, Oklahoma

Tang-Huang Lin
National Central University
Jhongli, Taiwan

Chun Liu
Tongji University
Shanghai, China

Gin-Rong Liu
National Central University
Jhongli, Taiwan

Dengsheng Lu
Indiana University
Bloomington, IN

Maurizio Manigrasso
Chemical Laboratory DIPIA, INAIL
    (ex-ISPESL)
Rome, Italy

Randal S. Martin
Utah State University
Logan, Utah

Lachlan McKinna
Curtin University
Perth, Australia

Iris Möller
University of Cambridge
Cambridge, United Kingdom

Emilio Moran
Indiana University
Bloomington, Indiana

Kunal Nayee
University of Central Florida
Orlando, Florida

Stacy A. C. Nelson
North Carolina State University
Raleigh, North Carolina

Kazuo Oki
Institute of Industrial Science
University of Tokyo
Tokyo, Japan

Taikan Oki
Institute of Industrial Science
University of Tokyo
Tokyo, Japan

Min M. Oo
University of Wisconsin–Madison
Madison, Wisconsin

John J. Qu
College of Science
George Mason University
Fairfax, Virginia

Sassan Saatchi
UCLA Center for Tropical Research
Los Angeles, California

Thomas Schroeder
CSIRO Land & Water
Brisbane, Australia

Mohammad Sharifikia
Tarbiat Moddaress University
Tehran, Iran

Geoff Smith
Specto Natura Ltd.
Cambridge, United Kingdom

Tom Spencer
University of Cambridge
Cambridge, United Kingdom

Si-Chee Tsay
NASA Goddard Space Flight Center
Greenbelt, Maryland

Fugui Wang
LSU Agricultural Center
Louisiana State University
Baton Rouge, Louisiana

and

University of Wisconsin–Madison
Madison, Wisconsin

Michael D. Wojcik
Utah State University Research
    Foundation
Logan, Utah

Y. Jun Xu
LSU Agricultural Center
Louisiana State University
Baton Rouge, Louisiana

Zhemin Xuan
University of Central Florida
Orlando, Florida

Wei Yao
Dalian University of Technology
Dalian, China

Mohammad Javad Valadan Zoej
K. N. Toosi University of Technology
Tehran, Iran

1 Linkages between Environmental Remote Sensing and Systems Analysis

Ni-Bin Chang

CONTENTS

1.1 INTRODUCTION

The interactions of physical, chemical, and biological processes in coupled natural systems and the built environment have given rise to the intertwined complexity, diversity, and persistence of various types of environmental problems. Environmental protection and restoration of air, soil, and water quality in relation to land use and regional planning are deeply rooted in spatiotemporal evolution at different scales. To achieve sound environmental resources management, there is often a need to investigate pollutant storage, transport, and transformation in both natural systems and the built environment. However, it is recognized that the sustainable management of human society is necessary at all phases of impact from the interactions among energy, environment, ecology, public health, and socioeconomic paradigms. Such a multidisciplinary nature of sustainability concern is truly a challenging task that requires employing a systems analysis approach.

Environmental sensing and monitoring networks are deemed an integral part of environmental cyberinfrastructures and may produce comprehensive and accurate spatial information over time, providing the basis for sustainable development. To properly respond to natural and human-induced stresses to the environment, however, environmental resource managers often consider the functions and values of systems analysis that may be geared toward synergistic integration among remote sensing technologies, data/image processing tools,

1

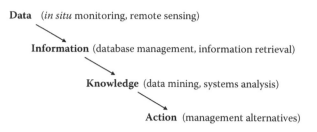

FIGURE 1.1   System complexity to be tackled by large-scale systems analysis. (From Chang, N. B., *Systems Analysis for Sustainable Engineering*, McGraw Hill, New York, 2010.)

and supportive environmental cyberinfrastructures. Systems analysis can provide a coordinated, multidisciplinary effort to identify and understand these needs. As a consequence, systems analysis has become an important task for essential environmental resources management throughout the world. Major momentum to improve systems analysis emerged as a pressing priority during the late 1990s when the Internet became a norm in information exchange. Rapid advances in the integration of remote sensing (RS), global positioning system (GPS), and geographical information system (GIS) technologies motivate more integrative sensing, monitoring, and modeling with system thinking for sound decision making. Such understanding leads to the proper integration of sensing, monitoring, and modeling technologies in order to aid in the decision making involved in the preservation or remediation of the environment.

For sound decision making, a holistic approach is thus required that encapsulates the technical, institutional, social, economic, and environmental dimensions with systems thinking and provides an environmental basis for addressing cultural needs, social evolution, economic reality, and national policies. This movement requires expertise in acquisition, storage and warehousing, quality assurance, and presentation of environmental data from which the information can be retrieved and knowledge can be developed for decision making (Figure 1.1). To fulfill such a synergistic integration, it requires the following: collecting and maintaining environmental data; analyzing environmental data; using data for environmental protection actions; engaging the community to promote policies and to improve the sustainable management with environmental information; evaluating the effectiveness of environmental management processes, programs, and efforts with environmental knowledge; and implementing total quality management through integrated environmental sensing, monitoring, modeling, and decision making.

1.2 CURRENT CHALLENGES

Due to global climate change, economic development and globalization, increased frequency of natural hazards, rapid urbanization, and population growth and migration, an integrated, quantitative, systems-level method of remote sensing is essential to track the dynamics of coupled natural systems and the built environment. However, existing environmental systems that have been degraded and even contaminated face a reduced solution space spatially and temporally, due to competing and conflicting stakeholders' interests over demand for water supply, industrial production, recreation, land development, air quality management, and environmental flow requirements. This reduced solution space also magnifies hydroclimatic variability, leading toward more vulnerability to seemingly unbalanced economic development and ecosystem conservation. The increasing hydroclimatic variability could further translate the pollution impact into aggravation of resources scarcity, land degradation, environmental health and safety, and insufficient agricultural production at different scales. As a consequence, rapid change detection using remote sensing becomes an indispensable tool for future sustainable management. This entails an acute need to integrate environmental remote sensing and systems analysis in complex sociotechnical systems (Laracy 2007). Catastrophic failures are associated with ignoring social, political, economic, and institutional elements when determining the system boundaries in concert with the temporal scales of the environmental issues that need to be sensed, monitored, and investigated (Laracy 2007).

Remote sensing, one of the core technologies in environmental informatics, is not a panacea or an anecdote but may become powerful when fundamental physical, chemical, and biological processes in environmental systems can be sensed, monitored, and analyzed by a systematic approach. Yet how to optimize the synergistic effects of sensors, platforms, and sensor networks to provide decision makers and stakeholders timely decision support tools with respect to species diversity, spectral heterogeneity, spatial variability, and temporal scaling issues is deemed a critical challenge (NCAR 2002; NSF 2003; Chang et al. 2009, 2010, 2011).

Further, the identification of comparative advantages across data mining, image processing, machine learning, and information retrieval techniques applied to exhibit information and knowledge in support of synergistic integration of sensing, monitoring, and modeling creates an ever-growing challenge for a sound systems analysis (Back et al. 1997; Zilioli and Brivio 1997; Volpe et al. 2007; Chang et al. 2009, 2010, 2011). In order to apply the systems thinking archetypes to the environmental problem being observed, the problem comes down to methodology with regard to how the techniques of systems analysis can be connected with a modern understanding of environmental remote sensing. It leads to the creation of case-based remote sensing practices by developing operable systems that meet requirements within imposed constraints (Pyschkewitsch et al. 2009). This may be illuminated by some ways through assessing four dimensions of novelty, complexity, technology, and scale simultaneously (i.e., the NCTS framework) when a new environmental remote sensing project has to be launched (Figure 1.2). The demonstrated selection across the four dimensions in Figure 1.2 entails how the different sensors, images, and spectral analysis skills can be integrated for scale-dependent sensing, monitoring, and modeling in case-based remote sensing practices.

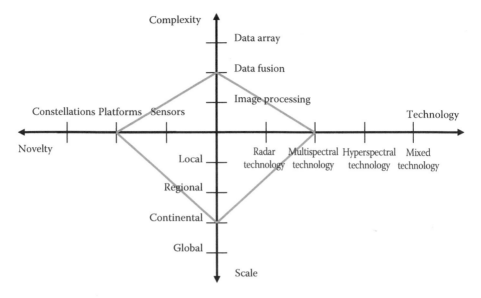

FIGURE 1.2   The NCTS framework.

1.3   FEATURED AREAS

This book is designed to address the grand challenges in the nexus of environmental remote sensing and systems analysis under global changes. Recent advances in environmental remote sensing with various data-mining, machine-learning, and image-processing techniques provide us with a reliable and lucid means to explore the changing environmental quality via a temporally and spatially sensitive approach. It leads to the improvement of our understanding of the sensitivity of key factors in environmental resources management. Due to space limitations, the main focus of current research in the context of environmental remote sensing and systems analysis may be classified into three topical areas as follows:

- *Topical area I: Water quality monitoring, watershed development, and coastal management.* The interactions among aquatic environments, such as lakes, bays, and estuaries, and associated watersheds are emphasized to monitor the human-induced changes in the regional nutrient cycle. Addressing these interactions is as critical as coping with the impacts of land degradation through sea–land interactions, energy and transportation, and natural hazards on water quality management. A few applications and case studies at different scales worldwide in Chapters 2 through 8 demonstrate a contemporary coverage of these issues in association with both point and nonpoint sources.
- *Topical area II: Sensing and monitoring for land use patterns, reclamation, and degradation.* The environmental consequences of urbanization effect in association with land use and land cover change include, but are not limited to, changing pervious areas and altering the hydrological cycle,

land subsidence, landslide and mud flows, reclamation of land from aquatic environments, and associated complexity of land management policies. A few applications and case studies at different scales worldwide in Chapters 9 through 13 demonstrate a contemporary coverage of these issues.

- *Topical area III: Air quality monitoring, land use/land cover changes, and environmental health concerns.* From local, to regional, to continental scale, urbanization effect and desertification seriously contribute to a number of environmental problems in air quality management. As an example, municipal and agricultural activities require intensive long-term air quality monitoring. Human-induced dust storms reacting with desertification result in rising global particulate matter, with unintended social and health impacts. Global changes such as ozone depletion and the resulting ultraviolet impact on human society and ecosystems trigger a holistic view of environmental management. A few applications and case studies at different scales worldwide in Chapters 14 through 20 demonstrate a contemporary coverage of these issues in association with both point and nonpoint sources.

1.4   DISTINCTIVE ASPECTS

Macroenvironment (e.g., social, political, economic, technological, and legal) and market demand will certainly shape the most appropriate synergistic efforts between environmental remote sensing and systems analysis techniques. Complementing this emerging focus with respect to the actual need of coordination and exchange of data for improved understanding of environmental informatics, this book brings together forward-looking scholars with the requisite experience for showing coordinated interdisciplinary approaches between environmental remote sensing and system analysis. Compared to previous publications, this book uniquely emphasizes the following distinctive aspects:

- *Comparative approach for information retrieval.* Throughout the book, comparisons between data-mining, machine-learning, and image-processing methods are presented to help managers and researchers make the optimal selection when dealing with satellite images.
- *Integration with ground-based monitoring network.* To incorporate the strength of environmental cyberinfrastructures, a few case studies emphasize the inclusion of ground-based monitoring networks in dealing with air quality and water quality management issues.
- *Emphasis on environmental information management.* The book also focuses on environmental information management with proper integration of Global Positioning System (GPS), Geographical Information System (GIS) and existing environmental databases of remote sensing images and *in situ* measurements in several case studies.
- *Modeling for decision making.* Implications in environmental resources management and policy using integrated simulation and optimization processes are demonstrated throughout some case studies.

- *Remote sensing and environmental health.* Emphasis has been placed on the linkages between remote sensing and environmental health implications.
- *Remote sensing and environmental management.* Emphasis has also been placed on the linkages between remote sensing and environmental management policy.
- *Policy analysis for decision making.* Scenario- or index-based systems analysis is demonstrated throughout some case studies to aid in environmental policy analysis and decision making.
- *Enhancement of environmental education.* Multidisciplinary education and research are demonstrated explicitly to indicate opportunities for integrated field and laboratory studies in environmental remote sensing education.

REFERENCES

Back, T., Hammel, U., and Schwefel, H. P. (1997). Evolutionary computation: Comments on the history and current state. *IEEE Transactions on Evolutionary Computation*, 1(1), 3–17.

Chang, N. B., Daranpob, A., Yang, J., and Jin, K. R. (2009). A comparative data mining analysis for information retrieval of MODIS images: Monitoring lake turbidity changes at Lake Okeechobee, Florida. *Journal of Applied Remote Sensing*, 3: 033549.

Chang, N. B. (2010). *Systems Analysis for Sustainable Engineering*, McGraw Hill, New York.

Chang, N. B., Han, M., Yao, W., Xu, S. G., and Chen, L. C. (2010). Change detection of land use and land cover in a fast growing urban region with SPOT-5 images and partial Lanczos extreme learning machine. *Journal of Applied Remote Sensing*, 4, 043551.

Chang, N. B., Yang, J., Daranpob, A., Jin, K. R., and James, T. (2011). Spatiotemporal pattern validation of chlorophyll a concentrations in Lake Okeechobee, Florida using a comparative MODIS image mining approach. *International Journal of Remote Sensing*, doi: 10.1080/01431161.2011.608089.

Laracy, J. R. (2007). Addressing system boundary issues in complex socio-technical systems, *Proceedings of Systems Engineering Research Forum*, 2(1), 19–26, Hoboken, NJ.

National Science Foundation (NSF) (2003). *Complex Environmental Systems: Synthesis for Earth, Life, and Society in the 21st Century.* NSF Environmental Cyberinfrastructure Report, Washington, DC.

National Center for Atmospheric Research (NCAR) (2002). *Cyberinfrastructure for Environmental Research and Education*, Boulder, CO.

Pyschkewitsch, M., Schaible, D., and Larson, W. (2009). The art and science of systems engineering, *Proceedings of Systems Engineering Research Forum*, 3(2), 81–100, Loughborough University, Leicestershire, UK.

Volpe, G., Santoleri, R., Vellucci, V., d'Alcalà, M. R., Marullo, S., and D'Ortenzio, F. (2007). The colour of the Mediterranean Sea: Global versus regional bio-optical algorithms evaluation and implication for satellite chlorophyll estimates. *Remote Sensing of Environment*, 107, 625–638.

Zilioli, E. and Brivio, P. A. (1997). The satellite derived optical information for the comparative assessment of lacustrine water quality, *Science of the Total Environment*, 196, 229–245.

Part I

*Water Quality Monitoring,
Watershed Development,
and Coastal Management*

2 Using Remote Sensing– Based Carlson Index Mapping to Assess Hurricane and Drought Effects on Lake Trophic State

Ni-Bin Chang and Zhemin Xuan

CONTENTS

2.1   INTRODUCTION

Lake Okeechobee, the second largest freshwater lake in the United States, is the source of fresh water to the Everglades. To the north, in the Kissimmee River Basin, major land uses are ranching and dairy farms, and as a result, excessive nutrient loads of phosphorus have entered the lake for more than three decades, resulting in cultural eutrophication. About 40% of the entire lake bed is covered with black, carbonate, organic phosphorus-enriched mud (Mehta et al. 1989). This phosphorus-laden sediment can be resuspended into the water column by wind and wave action (Maceina and Soballe 1990) and can be a primary source of phosphorus to the water column (Evans 1994) through the diffusion and desorption processes, which is highly related to the shear stress of the sediment bed.

Excessive phosphorus loads from the Lake Okeechobee watershed over the last few decades have led to increased eutrophication and water quality deterioration in the lake. According to long-term monitoring records, average annual surface water chlorophyll-a (Chl-a) concentrations from 1974 to 2010 were 14 to 28 mg/m^3, and average annual loading of total nitrogen and TP were 59 to 206 and 58 to 155 mg/m^3, respectively. Hence, the lake has long been regarded as a shallow (average depth 2.7 m), large (1990 km^2), frequently turbid eutrophic lake in south Florida. Mud sediment resuspension and transportation can extensively impact the water quality and environment of Lake Okeechobee (Jin and Ji 2004). Higher concentrations of suspended sediments also change light attenuation and affect the cycling of nutrients, organic micropollutants, and heavy metals in the water column and sediment bed (Blom et al. 1992; Van Duin et al. 1992; Jin et al. 2002). Overall, nutrient management for improving the water quality of this lake through nonpoint source pollution control in the lake watershed is a long-term issue in the relevant Total Maximum Daily Load (TMDL) programs. Yet the contribution of phosphorus to the lake's water column from internal loading was about equal to the contribution from external loading in late 1990s (Moore and Reddy 1994). Steinman et al. (1999) indicated that the lake may not respond to reductions in external phosphorus inputs in the TMDL efforts due to this high internal loading.

Lake Okeechobee has been threatened in recent decades by excessive phosphorus loading, harmful extreme high and low water levels, intermittent hurricanes, and rapid expansion of exotic plants. Four major hurricanes in the past decade, including Irene in 1999, Frances and Jeanne in 2004, and Wilma in 2005, made landfalls in this area and impacted the lake's hydrodynamic pathways and ecosystem (James et al. 2008). Some hurricanes affected the water quality condition in Lake Okeechobee through persistent, sustained wind speed (Table 2.1). Resuspension of sediments due to hurricanes and local wind gusts during the drought period later on also contributed to nutrient release, regardless of the changing nutrient loads in the Kissimmee River Basin. Erosion induced by wind promotes the sediment resuspension and diffusion process. In contrast, a long-term drought from 2000 to 2001 followed by another from 2007 to 2008 brought about salient ecosystem impacts due to the lower water depth. Coupling effects of these continuous natural hazards resulted in the resuspension of a large quantity of sediment, lower light transparency, and the release of a large amount of nutrients into the water column. Nevertheless, a long-term drought

TABLE 2.1
Hurricane Winds and Persistent Time in Lake Okeechobee

Hurricane Name	Landfall Location (Florida)	Peak Date	Sustained Wind Speed (m/s)	Persistent Time (Days) (=8 m/s)
Irene	Cape Sable	10/15/99	23	2.3
Frances	Cat Island	9/5/04	31	4.7
Jeanne	Hutchinson Island	9/26/04	33	2.5
Wilma	Cape Romano	10/24/05	41	1.5

Source: Chang, N. B., Makkeasorn, A., and Shah, T., *Technical Report for Contract 4100000079*, submitted to South Florida Water Management District, West Palm Beach, FL, 2008.

from 2007 to 2008 triggered a trajectory of ecosystem recovery in mud, littoral, and transition zones due to higher light penetration after the hurricane impacts in 2004 and 2005 (Figure 2.1). The wet and dry seasons (Figure 2.1) are designated from May to October and from November to April, respectively, according to the weather pattern in south Florida. The dual impact on submerged aquatic vegetation (SAV) after the landfalls of several hurricanes in 2004 and 2005 followed by the drought is apparent (Figure 2.1). As a consequence, algal blooms have been occasionally observed in Lake Okeechobee since these events (Figure 2.2).

We investigated whether the drought impact was more influential than the hurricane impact on lake eutrophication, a question that can be analyzed more thoroughly with the aid of remote sensing technology. Lake trophic states that move from oligotrophic (lakes with low production associated with low nitrogen and phosphorus) to eutrophic (lakes type with high production, associated with high nitrogen and

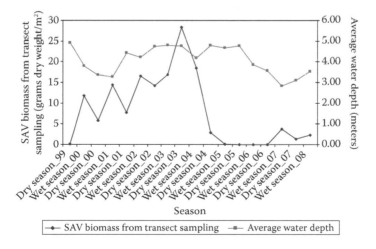

FIGURE 2.1　The decadal interactions between hurricane, drought, and submerged aquatic vegetation (SAV) on a seasonal scale.

<table>
<tr><td>(a) Incipient stage</td><td>(c) Patch formation stage</td></tr>
<tr><td>(b) Fostering stage</td><td>(d) Bloom spreading stage</td></tr>
</table>

FIGURE 2.2 Algal bloom event observed by the authors from September 18 through September 22, 2008. (From Chang, N. B., Makkeasorn, A., and Shah, T., *Technical Report for Contract 4100000079*, submitted to South Florida Water Management District, West Palm Beach, FL, 2008.)

phosphorus) are related to various water quality factors, mainly Chl-a, Secchi disk depth (SDD), and TP (TP) concentrations in this study. The monitoring of Chl-a, SDD, and TP concentrations have not been fully developed spatially and temporally to assess the trophic states of a lake using remote sensing. Collecting and analyzing water quality data is costly and time consuming, and whether limited numbers of field data can truly characterize the spatial variation of trophic state within a vast water body is still unknown. Within this study, we estimated spatiotemporal patterns of Chl-a concentrations on the lake in four prespecified time periods to holistically compare the variations of eutrophication potential of the lake using both remote sensing and *in situ* measurements. Our assessment was based on the derived Carlson's trophic state index maps translating Chl-a concentrations to trophic state.

2.2 MATERIALS AND METHODS

2.2.1 FIELD MEASUREMENTS, DATA COLLECTION, AND ANALYSIS

Lake water quality is traditionally monitored and evaluated based on field data. Since 1972, 23 water quality monitoring stations have been deployed, maintained, and operated monthly by the South Florida Water Management District (SFWMD) (Table 2.2 and Figure 2.3). DBHYDRO is SFWMD's corporate environmental database that stores hydrologic, meteorological, hydrogeologic, and water quality data. This database is the source of historical and up-to-date environmental data for the 16-county region covered by the SFWMD and includes all data collected by the Lake Okeechobee water monitoring stations. The DBHYDRO browser allows end users to search DBHYDRO (http://www.sfwmd.gov/dbhydroplsql/show_dbkey_info .main_menu) using one or more criteria to generate a summary of the data from the available period of record (i.e., since DBHYDRO only has information of TP instead of total phosphorus, the limitation results in some inconvenience when evaluating the trophic state according to the literature). We can then select datasets of interest and have the time series data dynamically displayed as tables or graphs. All the ground truth data used in this study such as water depth, Chl-*a*, SDD, and TP concentrations were acquired from the DBHYDRO Web database.

TABLE 2.2

Site Identification and Coordinates of the Monitoring Stations for *In Situ* Chlorophyll-*a* and Environmental Data

Site	Latitude (Degree)	Longitude (Degree)
KBAROUT	27.12	−80.84
L001	27.13	−80.78
L002	27.08	−80.79
L003	27.04	−80.7
L004	26.97	−80.7
L006	26.82	−80.78
L007	26.77	−80.78
L008	26.95	−80.89
LZ2	27.18	−80.82
LZ25	26.74	−80.75
LZ30	26.79	−80.86
LZ40	26.9	−80.78
LZ42	26.88	−80.91
LZ42N	27.04	−80.89
PALMOUT	26.83	−80.94
PELMID	26.76	−80.7
PLN2OUT	26.86	−80.94
POLE3S	26.73	−80.84
POLESOUT	27.03	−80.91
RITAEAST	26.71	−80.78
RITAWEST	26.72	−80.82
STAKEOUT	27.01	−80.93
TREEOUT	26.9	−80.96

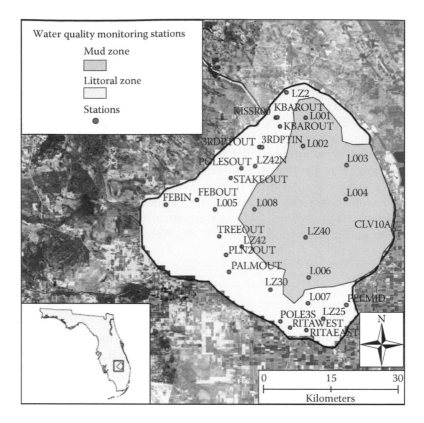

FIGURE 2.3   Location of water quality monitoring stations.

2.2.2   EUTROPHICATION ASSESSMENT

2.2.2.1   Definition of Trophic State Index (TSI)

The concept of a trophic state is based on changes in nutrient levels (measured by TP) that cause changes in algal biomass (measured by Chl-*a*), which in turn cause changes in lake clarity (measured by Secchi disk transparency). Carlson (1977) proposed a trophic state index (TSI) that retains the expression of the diverse aspects of the trophic state found in multiparameter indices, which can be computed from any of three interrelated water quality parameters: SDD, Chl-*a*, and TP according to the literature (see Equations 2.1 through 2.3). Thus, TSI may be defined as (Carlson and Simpson 1996):

$$TSI = 60 - 14.41 \ln SDD \ (m) \tag{2.1}$$

$$TSI = 9.81 \ln Chl\text{-}a \ (\mu g/L) + 30.6 \tag{2.2}$$

$$TSI = 14.42 \ln (total\ phosphorus) \ (\mu g/L) + 4.15 \tag{2.3}$$

where ln is natural logarithm.

2.2.2.2 Classification Methods

Several combinatorial sets of single- and multiparameter indices were developed for trophic classification of lakes (Brezonik and Shannon 1971; Beeton and Edmonson 1972). Secchi disk transparency, Chl-*a* (an indirect measure of phytoplankton), and TP (an important nutrient and potential pollutant) are often chosen to define the degree of eutrophication, or trophic state, of a lake. A list of possible changes that might be expected in a north temperate lake as the amount of algae changes along the trophic state gradient can be summarized with respect to varying ranges of SDD, Chl-*a*, and TP according to the literature (Table 2.3) in which a more delicate classification method for water quality management can be shown in Table 2.3b.

2.2.2.3 The Role of Remote Sensing

Remote sensing images can be used directly or indirectly to derive lake water quality information from SDD, Chl-*a*, and TP for eutrophication assessment. For example, Landsat TM data was used to depict trophic status of lake water (Lillesand et al. 1983; Baban 1996; Zilioli and Briviou 1997; Cheng and Lai 2001; Fuller and Minnerick 2007). In particular, based on the Landsat TM imageries, Baban (1996) found that the TSI for TP in some lakes in the United Kingdom is much higher than the TSI for other parameters, which affects eutrophication assessment. Cheng and Lai (2001) used Landsat TM images to derive a more detailed, locally calibrated version of the original Carlson's model for a reservoir in Taiwan. Fuller and Minnerick (2007) predicted water quality by relating SDD and Chl-*a* measurements to Landsat satellite imagery in support of the TSI assessment in the Lake Michigan area.

For the purpose of demonstration, this study used Moderate Resolution Imaging Spectroradiometer (MODIS) imagery to depict trophic state of Lake Okeechobee based on remote sensing–derived Chl-*a* concentrations retrieved by a genetic programming (GP) model. This implies that only Equation 2.2 was applied to generate the TSI maps for the classification of the trophic state with respect to the designated criteria (Table 2.3a). No previous effort has been made to develop a locally calibrated version of the original Carlson's model for the Lake Okeechobee study based on the eutrophication assessment that may be carried out by a more detailed classification method of the trophic state (Table 2.3).

TABLE 2.3a

Classification of the Trophic State from Oligotrophic to Mesotrophic, to Eutrophic, and to Hypereutrophic Conditions

TSI	Chl-*a*	Total Phosphorus	SDD	Trophic Class
<30–40	0–2.6	0–12	>8–4	Oligotrophic
40–50	2.6–20	12–24	4–2	Mesotrophic
50–70	20–56	24–96	2–0.5	Eutrophic
70–100+	56–155+	96–384+	0.5–<0.25	Hypereutrophic

Source: Carlson, R. E. and Simpson, J., *A Coordinator's Guide to Volunteer Lake Monitoring Methods*, North American Lake Management Society, Madison, WI, 1996.

TABLE 2.3b

Implications of the Trophic State in Environmental Management Entailing Detailed Attributes of Each Condition and Relevant Implications of Water Resources and Ecosystem Management

TSI	Chl-a (µg/L)	SDD (m)	Total Phosphorus (µg/L)	Attributes	Water Supply	Fisheries and Recreation
<30	<0.95	>8	<6	*Oligotrophy:* Clear water, oxygen throughout the year in the hypolimnion.	Water may be suitable for an unfiltered water supply.	Salmonid fisheries dominate.
30–40	0.95–2.6	8–4	6–12	Hypolimnia of shallower lakes may become anoxic.		Salmonid fisheries in deep lakes only.
40–50	2.6–7.3	4–2	12–24	*Mesotrophy:* Water moderately clear; increasing probability of hypolimnetic anoxia during summer.	Iron, manganese, taste, and odor problems worsen. Raw water turbidity requires filtration.	Hypolimnetic anoxia results in loss of salmonids. Walleye may predominate.
50–60	7.3–20	2–1	24–48	*Eutrophy:* Anoxic hypolimnia, macrophyte problems possible.		Warm-water fisheries only. Bass may dominate.
60–70	20–56	0.5–1	48–96	Blue-green algae dominate, algal scums and macrophyte problems.	Episodes of severe taste and odor possible.	Nuisance macrophytes, algal scums, and low transparency may discourage swimming and boating.
70–80	56–155	0.25–0.5	96–192	*Hypereutrophy:* (light limited productivity). Dense algae and macrophytes.		
>80	>155	<0.25	192–384	Algal scums, few macrophytes.		Rough fish dominate; summer fish kills possible.

Source: Carlson, R. E. and Simpson, J., *A Coordinator's Guide to Volunteer Lake Monitoring Methods,* North American Lake Management Society, Madison, WI, 1996.

2.2.3 Remote Sensing for the Estimation of Chl-*a* Concentrations

Remotely sensed Chl-*a* concentrations were estimated in some early studies with the aid of Coastal Zone Color Scanner (CZCS) (Gordon et al. 1988) and the Landsat Thematic Mapper (TM) (Tassan 1987; Pattiaratchi et al. 1994). A neural network model was derived for estimating sea surface chlorophyll and sediments from Landsat TM imagery (Keiner and Yan 1998). Multitemporal datasets from an LISS-III sensor mounted on an Indian Remote Sensing Satellite IRS-IC and field reflectance spectra were evaluated for estimating Chl-*a* content in Mecklenburg Lake, Germany (Thiemann and Kaufmann 2000). For vast water bodies, Chl-*a* concentrations correlate well with the ratio of Landsat TM green to red reflectance in the eutrophic East River and Long Island Sound in New York (Hellweger et al. 2004). Sea-viewing wide field-of-view sensor (SeaWiFS) and Landsat TM and enhanced thematic mapper ETM+ satellite images might be suitable for a general assessment purpose in vast water bodies such as the Mediterranean Sea (Volpe et al. 2007) and the northern Baltic Sea (Erkkila and Kalliola 2004). In particular, SeaWiFS was used to evaluate bio-optical algorithms for the Laurentian Great Lakes in comparison with 10 published marine bio-optical algorithms (nine empirical and one semianalytical algorithms) (Li et al. 2004). With MODIS data available, Chang et al. (2011) used a highly nonlinear GP model as an inverse modeling tool for Chl-*a* estimation; this study follows the same track to conduct comparative eutrophication assessment.

2.2.3.1 Remote Sensing Data Collection

MODIS is an advanced multipurpose National Aeronautics and Space Administration (NASA) sensor and a key instrument aboard the Terra (EOS AM) and Aqua (EOS PM) satellites. With MODIS images, Chl-*a* concentrations can be derived several ways, including statistical regression, neural network model, and other evolutionary computing methods such as the GP model, which is deemed as an extension of evolutionary computation (EC) (Chang et al. 2011). In Phase I of this study, all data of the seven bands of MODIS Aqua (MYD09GA) was downloaded associated with selected dates (Tables 2.4 and 2.5). The criterion for the selection of appropriate dates in both phases is mainly based on the absence of cloud contamination and whether it is closely associated with an event or episode. In Phase II of this study,

---

TABLE 2.4
Dates of the MODIS Images Synchronized with Ground Truth Data Used in Phase I

	Jan.	Feb.	March	April	May	June	July	Aug.	Sept.	Oct.	Nov.
2003			13	1, 16	6	10	15	26	8	13	12
2004	6, 22	3	16	19	17	8, 09	7, 20	09		11	17

Source: Chang, N. B., Yang, J., Daranpob, A., Jin, K. R., and James, T., *International Journal of Remote Sensing*, doi:10.1080/01431161.2011.608089.

TABLE 2.5

Corresponding Bandwidths of MODIS Bands 1 to 7

MODIS Band	Bandwidth (nm)
1	620–670
2	841–876
3	459–479
4	545–565
5	1230–1250
6	1628–1652
7	2105–2155

Source: National Aeronautics and Space Administration (NASA),
           http://modis.gsfc.nasa.gov (accessed by June 2011).

all relevant water quality data across all *in situ* monitoring stations was downloaded from DBHYDRO according to the four preselected dates (focused scenarios hereafter), including December 20, 2003 (before Hurricane Jeanne landfall), November 29, 2004 (after Hurricane Jeanne landfall), August 22, 2006 (before drought in 2007), and June 11, 2007 (during drought in 2007). These focused scenarios were selected to enhance the comparison of eutrophication assessment with the aid of either remote sensing data or *in situ* measurements, or both.

2.2.3.2 Machine Learning for Remote Sensing: The GP Model

The principle of EC is rooted in genetic algorithms (GA) first developed by Holland (1975), which is similar to evolution strategies (ES) (Rechenberg 1965; Back et al. 1997) and evolutionary programming (EP) developed by Fogel (1966). All three were eventually combined into one entity called evolutionary computation (Gagne and Parizeau 2004). Under the EC framework, the well-known GP approach was invented by Koza (1992), which became the best advancement to create best-selective nonlinear regression models in terms of later multiple independent variables.

For model development in this study, model screening among the neural network model, the statistical linear regression model, and GP models was conducted. We used the Discipulus software package, developed by Francone (1998), to perform GP modeling analysis for the estimation of Chl-*a* concentrations over the entire lake. We used the Statistica software package to derive the neural network model and statistical linear regression models simply for comparison. Rigorous calibration and verification of these three types of models were carried out to ensure the best fit (Chang et al. 2011). To overcome the misinterpretation of using r-square values, three more indices, including the root-mean-square error (RMSE), mean of percent error (PE), and ratio of standard deviation of predicted to observed value (CO), were also introduced as additional indicators for holistic assessment (Chang et al. 2011). Overall, the GP model obtained the highest r-square values and lowest RMSE in both calibration and verification stages (Chang et al. 2011); hence, the GP approach is favored relative to other methods (Chang et al. 2011).

The GP model is designed to produce a Chl-*a* estimation algorithm by linking seven MODIS bands to ground measured Chl-*a* concentrations. Based on the regression relationships, Chl-*a* maps at 1 km resolution can be developed over the study area (Chang et al. 2011). The calibration and verification were carried out in one of our previous studies to prove the credibility of this GP model (Chang et al. 2011). With the same GP model, Chl-*a* concentrations (µg/L) can be estimated with respective to these focused scenarios in relation to hurricane and drought events as described below (Chang et al. 2011):

$$C_{Chlorophyll-a} = [-3.78226 \sin X_1] - 2X_2$$

$$X_1 = 2\left[X_3^2 - 10585\right]$$

$$X_2 = \frac{(X_{10} - X_{11})}{\sqrt[4]{X_7}\sqrt[2]{X_5}}$$

$$X_3 = -1.9244 in\left(\frac{X_4}{-0.8277\, Band5} - 0.7233\right)$$

$$X_4 = 2\left\{2 \sin\left(4\sqrt{X_5}\, Band7\right) - X_2 + 1.6426\right\}$$

$$X_5 = \left|\sin(X_6 - Band7)\right| + 0.42816$$

$$X_6 = 2\sqrt[4]{X_7} + \frac{(X_{10} - X_{11})}{\sqrt[4]{X_7}}$$

$$X_7 = \left|\sin(X_8^2 - 0.72339)\right| + 0.42816$$

$$X_8 = \left[-1.96667 \sin(X_9 - X_{10} + X_{11})\right](X_{10} - X_{11})$$

$$X_9 = \left|\sin(X_{10} - X_{11})\right|$$

$$X_{10} = X_{15} + \left[\frac{(X_{12} + X_{15} - 0.634207)}{Band6}\right]^2 - X_{11}$$

$$X_{11} = \sin\left[\frac{(X_{12} + X_{15} - 0.634207)}{Band6}\right]^2$$

$$X_{12} = 2\cdot\left[\sin(X_{13} - Band5)\right]^2$$

$$X_{13} = \left[\cos\left(\frac{X_{14}}{Band6}\right)\right]^2$$

$$X_{14} = \frac{2\left[\sin X_{16}\right]^2 - 4\, Band7}{Band3}$$

$$(2.4)$$

$$X_{15} = X_{17} - \left(\sin X_{16}\right)^2$$

$$X_{16} = 2\cos\left(\frac{\sin\left(\frac{X_{18}}{Band5} - 0.723392\right)}{-0.236008}\right) + X_{20}$$

$$X_{17} = \cos\left(\frac{\sin\left(\frac{X_{18}}{Band5} - 0.723392\right)}{-0.236008}\right) + X_{20}$$

$$X_{18} = \left|X_{19} - X_{20}\right| - 0.827745$$

$$X_{19} = 8\sin X_{21} + 2Band7 - 2.432348$$

$$X_{20} = X_{23} - X_{24} + X_{21}$$

$$X_{21} = 1.530830\sin\left(\frac{2 \cdot Band6\left[\sin\left(X_{22}\right) + 1.6426456\right]}{-0.236008}\right)$$

$$X_{22} = 2\left\{\left[\left(\frac{X_{24}}{Band6}\right)^2 \left(X_{23} - X_{24}\right)^2 + 1.0868337\right]\right\}^2 \qquad (2.4)$$

$$X_{23} = X_{25} - \left(\frac{X_{26}}{Band5}\right) - X_{24}$$

$$X_{24} = \frac{0.061677\left(\frac{X_{26}}{Band5} + 1.259177\right)^2}{Band6\left(X_{25} - \frac{X_{26}}{Band5}\right)}$$

$$X_{25} = X_{28} + 2X_{27}$$

$$X_{26} = -2\left(X_{28} + X_{27}\right)$$

$$X_{27} = \sin\left(0.00421512 \cdot Band7 + 0.3275442\right)$$

$$X_{28} = -6.1905594 + 0.00421512\,Band7$$

in which X_is are intermediate variables; $C_{\_Chlorophyll-a}$ is the estimation of Chl-a concentrations (µg/L); and $Band3$, $Band5$, $Band6$, and $Band7$ are MODIS bands 3, 5, 6, and 7, respectively. The GP model does exhibit highly nonlinear structures to infer the correlations between input and output variables.

2.3 RESULTS AND DISCUSSION

2.3.1 LAKE OKEECHOBEE WATER QUALITY ANALYSIS

Factors that influence algal blooms mainly include the availability of nutrients (e.g., nitrogen and phosphorus), solar energy, and stable water column with transparency. Thus, Chl-*a* concentration provides a linkage among algae standing stocks, water quality changes, and hydrodynamic factors in Lake Okeechobee. The recorded trends (Figure 2.4a and b) confirm our observations. Both turbidity levels and TP concentrations went up abruptly immediately following landfalls of these three hurricanes in 2004 and 2005. The subsequent drought in and following 2006 kept the turbidity levels high because a small wind gust can easily disturb the sediment bed and trigger turbid water columns (i.e., DBHYDRO only has TP instead of total phosphorus data, limiting the eutrophication assessment). Both time series datasets of spatially averaged TP and turbidity levels reveal a consistent pattern over the period

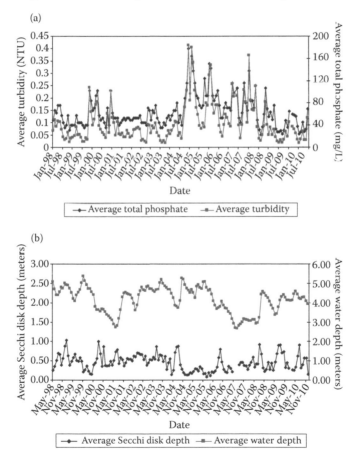

FIGURE 2.4 Covariations between (a) turbidity, TP, (b) Secchi disk depth, and water depth during hurricane and drought events. Average water depth was measured based on stage data NVGD 1929, relative to sea level.

of time (Figure 2.4a), yet the variations of SDD did not coincide with the changing water levels (Figure 2.4b). This deeply complicates the presence or absence of an algal bloom event and implies that factors critical to the determination of SDD are highly intricate.

Further quantitative insight into interactions of SDD, turbidity, and TP (Figure 2.5a and b) may cohesively illustrate the embedded patterns in between. For example, the relationship between SDD and turbidity may be delineated based on the spatially averaged monthly datasets collecetd at the *in situ* monitoring stations in 2003 and 2004 (Figure 2.5a), clarifying that the higher the turbidity level, the lower the SDD via a linear structure betweem the natural logarithms of SDD and the reciprocal of turbidity. Such a relationship also indicates that the higher the turbidity level, the less light is available for phytoplanton photosynthesis, thereby resulting in lower Chl-*a* concentrations. Further, the linear trend discovered in the regression analysis (Figure 2.5b) indicates a strong correlation between turbidity levels and TP concentrations, implying that situations of either higher wind speed or lower water level or both may promote an increased TP concentration supporting the growth of algal blooms. However, situations of either higher wind speed or lower water level or both may lead to higher turbidity and lower SDD, resulting in relatively lower light penetration, slowing down fast growth of algal blooms due to insufficient light. Such a counterbalance effect ends up a highly nonlinear dynamic process of the final Chl-*a* concentration, limiting the presence of aglal blooms on a long-term basis. We envisioned that any subtle disturbance through interactions among physical, chemical, and biological systems may incidentally disturb the delicate equilibrium, leading to chaotic algal bloom events.

The scale-dependent interactions among physical, chemical, and biological systems in the lake, however, are often strongly nonlinear, making it difficult to address the possible causes of algal bloom events. As a consequence, the use of traditional discrete reduction-based experiments, *in situ* sampling, or small-scale modeling cannot result in an adequate understanding of the system ecology issues such as algal bloom events. Estimation of Chl-*a* concentrations is deemed challenging due to such embedded complexity. With the aid of remote sensing derived maps of Chl-*a*

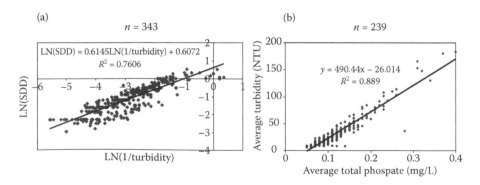

FIGURE 2.5 Interrelationships between spatially averaged (a) SDD, (b) TP, and turbidity from 1998 to 2010.

concentrations, it is possible to holistically know how the Chl-*a* concentrations on the lake respond spatially and temporally to natural hazards such as hurricanes and droughts. Remote sensing efforts with a highly nonlinear GP model in this study actually lead to a better understanding of how higher Chl-*a* concentration events could be triggered by higher or lower SDD, turbidity, TP, and water levels, collectively resulting in different trophic states.

2.3.2 Lake Okeechobee Eutrophication Assessment

2.3.2.1 Remote Sensing–Based Carlson Index Mapping

MODIS images collected from 2003 to 2004 were analyzed to retrieve the spatio-temporal patterns of Chl-*a* concentrations in Lake Okeechobee, Florida. The proposed GP model can successfully generate the Chl-*a* maps of focused scenarios (Figure 2.6). With visual interpretation, the spatial distribution of Chl-*a* concentrations before drought and hurricane seem to be spread around the lake. After the hurricane landfall, turbid water hinders the photosynthesis of phytoplanktons, thereby reducing the Chl-*a* concentrations and altering the spatiotemporal patterns of TSI (Figure 2.7).

2.3.2.2 Remote Sensing–Based Eutrophication Assessment

Based on the remote sensing maps (Figures 2.6 and 2.7), statistics of these focused scenarios can be summarized (Table 2.6) for eutrophication assessment. The TSI estimates across the preselected focused scenarios show that the trophic state during the middrought time period is relatively lower than the other three scenarios. Overall, the trophic state right after the hurricane landfall is higher than the other cases, evidenced by the mean values of TSI (Table 2.6). Only the middrought time period can be classified as mesotrophy (mean of TSI = 41.77), which implies a gradual change of trophic state from hurricane landfall to predrought. Because the predrought scenario has a relatively lower mean TSI value (50.44), which is at the brink of the eutrophy, the drought scenarios would holistically entail a slightly different situation relative to that during the hurricane scenarios. We can conclude that the lower the water depth, the more turbid the water quality, mainly due to wind-induced effect lowering of Chl-*a* concentrations. With the mean values of TSI, both focused scenarios of pre- and middrought may be classified as mesotrophic as opposed to eutrophic, generally indicating that the drought time period can actually temper the eutrophication impact in the lake.

2.3.2.3 Eutrophication Assessment Based on *in Situ* Measurements

For a broader basis for comparison, we can also choose four 3-month time periods (e.g., seasonal time periods) to analyze the turbidity, SDD, TP, and Chl-*a* concentrations based on the point measurements collected at all 23 *in situ* monitoring stations in Lake Okeechobee, Florida (Table 2.7). These four generalized scenarios include (1) prehurricane (06/01/2004–08/31/2004), (2) post-hurricane (09/27/2004–12/26/2004), (3) predrought (05/01/2006–07/31/2006), and (4) middrought (05/01/2007–07/31/2007).

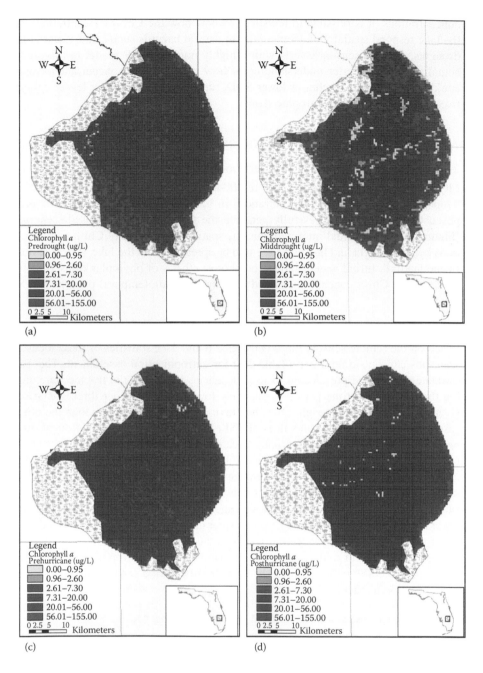

FIGURE 2.6 Chl-*a* concentrations of the selected four scenarios based on the derived GP model. (a) Predrought event (Aug. 22, 2006), (b) middrought event (June 11, 2007), (c) prehurricane event (Dec. 20, 2003), and (d) posthurricane event (Nov. 29, 2004).

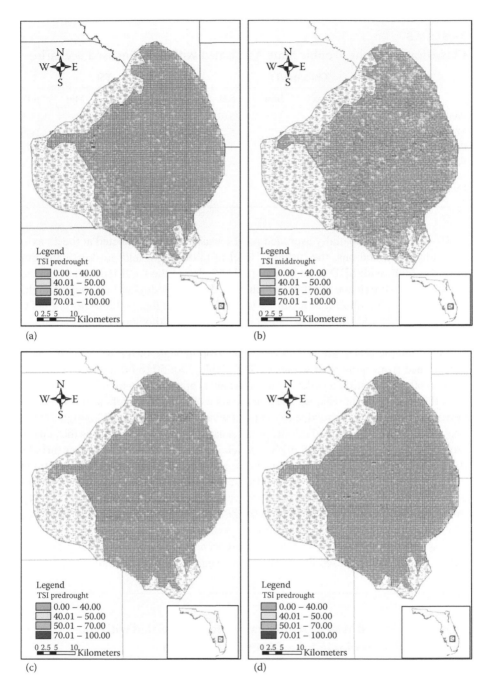

FIGURE 2.7 TSI estimates associated with the selected four scenarios based on the derived GP model. (a) Predrought event (Aug. 22, 2006), (b) middrought event (June 11, 2007), (c) prehurricane event (Dec. 20, 2003), and (d) posthurricane event (Nov. 29, 2004).

TABLE 2.6

Classification of the Trophic State Associated with Four Focused Scenarios

	Chl-a (μg/L)				TSI			
	Mean	Max	Min	S.D.	Mean	Max	Min	S.D.
Predrought	11.54	54	0	5.03	50.44	69.00	0	13.38
Middrought	10.53	49	0	5.28	41.77	68.00	0	21.70
Prehurricane	11.00	31	0	4.74	52.77	64.28	0	6.18
Posthurricane	11.86	33	0	4.57	53.53	64.9	0	6.38

Note: S.D. = standard deviation.

If we follow the spatially averaged *in situ* measurements collected at the 23 existing monitoring stations, the turbidity level in middrought and posthurricane time period is high, with SDD at its lowest, probably due to the wind-induced turbulence caused by the hurricane landfall, which stirred up the sediment bed causing highly turbid water. This observation can be verified by the higher turbidity (66.79 NTU) and lower SDD (0.14 m) in the predrought scenario, although TP concentration is relatively high (0.18 mg/L) (Table 2.7).

Other unique observations can be made. Trophic state of the water body can be determined from numerical values of TSI (Table 2.8); 50 to 60 indicates eutrophic waters; 60 to 70 is still considered eutrophic but is approaching hypereutrophic, such as is the case for prehurricane. The trophic state of the middrought scenario is more eutrophic than mesotrophic due to the local wind-induced effects. Although TP concentration of the generalized middrought scenario has the same value as the generalized predrought scenario (0.18 mg/L), which is possibly induced by higher turbidity (66.79 NTU), the SDD is limited to 14 cm, and therefore the Chl-*a* concentration is constrained to 9.2 mg/m^3 on average. In fact, the mean Chl-*a* concentratstion in the generalized scenario of predrought is lowest in the predrought time period, whereas the mean Chl-*a* concentration is highest in the generalized scenario of prehurricane (Table 2.8 and Figure 2.8) across all four generalized scenarios. Yet this is not the case when looking into these four focused scenarios when the spatially averaged values are calcualted based on the remote sensing images (Table 2.6).

TABLE 2.7

Spatially Averaged Water Quality Parameters Associated with Four Generalized Scenarios in Lake Okeechobee, Florida

	Turbidity (NTU)	Secchi Disk Depth (m)	Total Phosphate (mg/L)
Predrought	47.89	0.25	0.18
Middrought	66.79	0.14	0.18
Prehurricane	11.73	0.70	0.07
Posthurricane	69.24	0.17	0.21

TABLE 2.8

Eutrophication Assessment with Four Generalized Scenarios Based on Spatially Averaged Chl-*a* Concentration in Lake Okeechobee, Florida

Monitoring Period	Mean (Chl-a) (mg/m³)	S.D. (Chl-a) (mg/m³)	Mean TSI	Trophic State
Predrought	7.02	9.01	49.7	Mesotrophy
Middrought	9.20	9.74	52.4	Eutrophy
Prehurricane	18.40	17.50	59.2	Eutrophy
Posthurricane	10.53	9.34	53.7	Eutrophy

Note: S.D. = standard deviation.

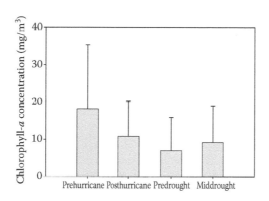

FIGURE 2.8 Chl-*a* measurements associated with the selected four generalized scenarios.

2.3.2.4   Comparative Eutrophication Assessment

In comparison, after the hurricanes hit in August 2004, levels of turbidity and TP increased substantially, and the SDD decreased to less than 0.2 m. Thus, less sunlight was available for photosynthesis in the lake, and the Chl-*a* concentrations decreased during the posthurricane and predrought periods. Local fluctuations of Chl-*a* concentrations at the 23 stations in pre- and posthurricane scenarios partially confirm our justification of eutrophic status (Table 2.9), although intermittent trends among stations are difficult to determine, leading to a mesotrophic status in the predrought period. The levels of turbidity and TP started to decrease during the predrought period, which may explain the higher level of Chl-*a* concentration during the following middrought period. Because turbidity also increased during the middrought period, however, we might expect the Chl-*a* concentrations to decrease in subsequent time periods. When examining the changing trophic state over the entire drought period from August 2006 to June 2007 (Table 2.10), the mesotrophic state can be observed intermittently in the predrought period except in October 2006.

TABLE 2.9

Chl-a Distribution across 23 Stations in Pre- and Posthurricane Scenarios

ID	Site	Latitude	Longitude	Date	Chl-a (mg/m³)	Date	Chl-a (mg/m³)
1	KBAROUT	27.12	−80.84	6/3/2004	16	11/3/2004	15
2	L001	27.13	−80.78	6/9/2004	24	11/18/2004	10
3	L002	27.08	−80.79	6/9/2004	20	11/18/2004	8
4	L003	27.04	−80.70	6/9/2004	22	10/11/2004	6
5	L004	26.97	−80.70	8/10/2004	32	10/11/2004	5
6	L006	26.82	−80.78	6/8/2004	7	12/29/2004	19
7	L007	26.77	−80.78	6/8/2004	9	11/17/2004	10
8	L008	26.95	−80.89	6/9/2004	20	1/10/2005	13
9	LZ2	27.18	−80.82	6/9/2004	16	11/18/2004	8
10	LZ25	26.74	−80.75	6/8/2004	15	12/26/2004	11
11	LZ30	26.79	−80.86	6/8/2004	10	11/17/2004	12
12	LZ40	26.90	−80.78	6/9/2004	13	11/18/2004	12
13	LZ42	26.88	−80.91	6/8/2004	10	11/17/2004	14
14	LZ42N	27.04	−80.89	6/9/2004	59	12/29/2004	11
15	PALMOUT	26.83	−80.94	6/8/2004	9	11/17/2004	15
16	PELMID	26.76	−80.70	6/8/2004	6	12/29/2004	10
17	PLN2OUT	26.86	−80.94	6/8/2004	12	11/17/2004	20
18	POLE3S	26.73	−80.84	6/8/2004	13	12/29/2004	18
19	POLESOUT	27.03	−80.91	6/9/2004	66	11/18/2004	10
20	RITAEAST	26.71	−80.78	6/8/2004	8	11/17/2004	30
21	RITAWEST	26.82	−80.82	6/8/2004	4	11/17/2004	21
22	STAKEOUT	27.01	−80.93	6/9/2004	78	11/18/2004	9
23	TREEOUT	26.90	−80.96	6/8/2006	10	11/17/2004	32

TABLE 2.10

The Changing Trophic State over the Entire Drought Period from August 2006 to June 2007

Month	Chlorophyll-a Average (µg/L)	Standard Deviation	TSI	Trophic State
August 2006	6.71	5.67	49.27	Mesotrophic
September 2006	7.04	5.37	49.75	Mesotrophic
October 2006	8.26	5.59	51.31	Eutrophic
November 2006	7.17	3.76	49.92	Mesotrophic
December 2006	7.14	2.18	49.89	Mesotrophic
January 2007	11.42	6.46	54.45	Eutrophic
February 2007	9	5.71	52.15	Eutrophic
March 2007	7.73	4.60	50.28	Eutrophic
April 2007	14.8	9.22	57.02	Eutrophic
June 2007	14.4	10.97	56.77	Eutrophic

During this time period, local flucutations of Chl-*a* concentrations at the 23 stations between pre- and postdrought scenarios showed a general increasing trend across most stations. This evidence further confirms our justification of the changing status of eutrophication from predrought to middrought (Table 2.11). Yet that is not the case when looking into possible discrepancies of eutrophication assessment (Figure 2.6b). A comparison of the middrought scenario (Figure 2.7b) (June 11, 2007) and trophic state shifts over the entire drought period (Table 2.10) implies that possible discrepancies of eutrophication assessment between remote sensing and *in situ* measurements also exist. Whereas the mean TSI value based on the remote sensing estimation is 41.77, the corresponding value based on the two different sets of *in situ* point measurements is 52.4 (Table 2.8) and 56.77 (Table 2.10), respectively, in generalized scenarios. While the former is recognized as mesotrophy (Table 2.6), the latter is classified as eutrophy (Tables 2.8 and 2.10) in the eutrophication assessment, mainly because most grids in the central lake area did not appear eutrophic in the middrought scenario (June 11, 2007). However, that area has no local monitoring

TABLE 2.11
Chl-*a* Distribution across 23 Stations in Pre- and Middrought Scenarios

ID	Site	Latitude	Longitude	Date	Chl-*a* (mg/m^3)	Date	Chl-*a* (mg/m^3)
1	KBAROUT	27.12	−80.84	8/3/2006	9	N/A	0
2	L001	27.13	−80.78	8/14/2006	8	6/7/2007	16
3	L002	27.08	−80.79	8/14/2006	18	8/18/2007	32
4	L003	27.04	−80.70	8/14/2006	5	6/7/2007	5
5	L004	26.97	−80.70	8/14/2006	2	8/13/2007	9
6	L006	26.82	−80.78	8/15/2006	11	6/7/2007	28
7	L007	26.77	−80.78	8/15/2006	4	6/7/2007	4
8	L008	26.95	−80.89	8/14/2006	8	6/7/2007	13
9	LZ2	27.18	−80.82	8/14/2006	4	N/A	0
10	LZ25	26.74	−80.75	8/15/2006	1	N/A	0
11	LZ30	26.79	−80.86	8/15/2006	4	8/13/2007	15
12	LZ40	26.90	−80.78	8/14/2006	1	8/13/2007	14
13	LZ42	26.88	−80.91	8/15/2006	21	8/13/2007	13
14	LZ42N	27.04	−80.89	8/14/2006	10	8/13/2007	81
15	PALMOUT	26.83	−80.94	8/15/2006	2	N/A	0
16	PELMID	26.76	−80.70	8/15/2006	4	8/14/2007	6
17	PLN2OUT	26.86	−80.94	8/15/2006	2	8/7/2007	3
18	POLE3S	26.73	−80.84	8/15/2006	3	4/23/2007	3
19	POLESOUT	27.03	−80.91	8/14/2006	20	8/7/2007	20
20	RITAEAST	26.71	−80.78	8/15/2006	1	8/14/2007	5
21	RITAWEST	26.82	−80.82	8/15/2006	3	N/A	0
22	STAKEOUT	27.01	−80.93	8/14/2006	31	7/5/2010	4
23	TREEOUT	26.90	−80.96	8/15/2006	12		

Note: N/A = date was too far out of range.

station to detect such changes (Figures 2.2 and 2.6b, comparatively). A similar comparison is possible for the predrought scenario (August 22, 2006) based on the remote sensing estimation and the *in situ* measurements. Whereas the mean TSI value based on the remote sensing estimation is 50.44, the corresponding value based on the *in situ* point measurements is 49.7 (Table 2.8) and 49.27 in generalized scenarios (Table 2.10). While the former is recognized as eutrophy (Table 2.6), the latter is classified as mesotrophy (Tables 2.8 and 2.10) in the eutrophication assessment. Again, this is due to the lack of monitoring stations in the central lake area, causing a discrepancy in the eutrophication assessment. Both comparisons from pre- and middrought scenarios confirm the comparative advantages of using a remote sensing approach.

2.3.3   FINAL REMARKS: COMPLEXITY IN THE ESTIMATION OF CHL-*a* CONCENTRATIONS IN LAKE OKEECHOBEE

For *in situ* measurements, pairwise statistical nonlinear regression analyses were performed between the Chl-*a* concentrations (Table 2.8) and the corresponding parameters associated with four prescribed scenarios (Table 2.7) on a long-term basis. A polynomial regression of turbidity versus Chl-*a* concentration (Figure 2.9) explained the most variation (as indicated by the R^2 value of 0.998). Thus, Chl-*a* concentration appears to decrease in a nonlinear fashion as the turbidity concentration increases. There was a negative, nonlinear relationship between turbidity and Chl-*a* concentration. In comparison, SDD showed a positive nonlinear relationship with Chl-*a* concentration (Figure 2.10). A polynomial regression again explained the most

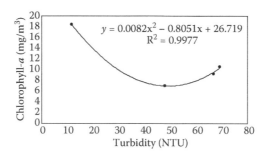

FIGURE 2.9   The nonlinear relationship between turbidity and Chl-*a* concentration.

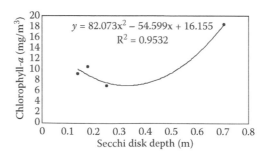

FIGURE 2.10   The nonlinear relationship between SDD and Chl-*a* concentration.

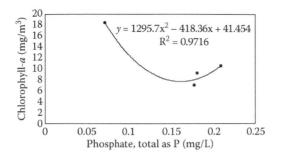

FIGURE 2.11 The nonlinear relationship between TP and Chl-*a* concentration.

variation in the data ($R^2 = 0.953$). Finally, there was a negative, nonlinear relationship between TP and Chl-*a* concentration (Figure 2.11). Once again, a polynomial regression explained the greatest amount of variation ($R^2 = 0.972$). Such highly nonlinear relationships intertwined among these relevant parameters further compound the challenges of modeling and prediction of Chl-*a* concentrations in the shallow lake. Yet it also justifies the need of using the GP model to address internal complexity embedded within the highly nonlinear functional structures of the production of Chl-*a* in the lake aquatic environments.

2.4 CONCLUSIONS

This study explored the estimation of Chl-*a* concentrations on Lake Okeechobee using the MODIS-based GP model that accounts for highly nonlinear structure among several MODIS band data. Chl-*a* concentrations in the lake can eventually be examined at high spatial resolution (1000 m pixel size) MODIS images and be translated into the corresponding TSI values. Based on the mean values of TSI, both focused scanarios of pre- and middrought fall into the category of mesotrophic as opposed to the eutrophic condition, which includes both focused scenarios of pre- and posthurricane. Yet a local effect causes a different eutrophication assessment outcome of the middrought scenario if only *in situ* measurements are taken into account, which makes the TSI value of middrought higher. After a series of hurricane landfalls in 2004 and 2005, the gradual drop of water depth in 2006 actually brought the water quality condition from eutrophy to mesotrophy. This implies that some coupling effect between hurricanes in 2004 and 2005 and drought during and after 2006 jointly affected the Chl-*a* concentrations in lake water body. Overall, the trophic state of posthurricane, predrought, and middrought scenarios is actually lower than that of the prehurricane scenario, which was approaching a hypereutrophic condition. Yet it is indicative that the drought episode in 2007 had a relatively lower eutrophication potential than that of the hurricane landfall in 2004.

With the aid of remote sensing technology, a series of TSI and Chl-*a* maps can lead to a more holistic and objective identification of possible hot spots (algal blooms) and cold spots (clean areas) on the lake, which is helpful for long-term water quality management in the lake. Such comparative analysis between remote sensing and *in situ* measurements implies that possible discrepancies of eutrophication assessment might be salient if we only use *in situ* point measurements for environmental management decision making. In real-world applications, we highly recommend using spatiotemporal analysis based on remote sensing images in concert with time series *in situ* measurements.

REFERENCES

Baban, S. M. J. (1996). Trophic classification and ecosystem checking of lakes using remotely sensed information. *Hydrological Sciences Journal*, 41(6), 939–957.

Back, T., Hammel, U., and Schwefel, H. P. (1997). Evolutionary computation: Comments on the history and current state. *IEEE Transactions on Evolutionary Computation,* 1(1), 3–17.

Beeton, A. M. and Edmonson, W. T. (1972). The eutrophication problem. *Journal of the Fisheries Research Board of Canada*, 29, 673–682.

Blom, G., Van Duin, E. H. S., Aalderink, R. H., Lijklema, L., and Toet, C. (1992). Modeling Sediment Transport in Shallow Lakes—Interactions between Sediment Transport and Sediment Composition. *Hydrobiologia,* 235/236, 153–166.

Brezonik, P. L. and Shannon, E. E. (1971). Trophic State of Lakes in North Central Florida. *Water Resources Research Center Publication*, 13, 102.

Carlson, R. E. and Simpson, J. (1996). *A Coordinator's Guide to Volunteer Lake Monitoring Methods.* North American Lake Management Society, Gainesville, FL.

Chang, N. B., Makkeasorn, A., and Shah, T. (2008). *Lake Okeechobee Sediment Profile Characterization.* Technical Report for Contract 4100000079, submitted to South Florida Water Management District, West Palm Beach, FL.

Chang, N. B., Yang, J., Daranpob, A., Jin, K. R., and James, T. (In press). Spatiotemporal pattern validation of chlorophyll-*a* concentrations in Lake Okeechobee, Florida using a comparative MODIS image mining approach. *International Journal of Remote Sensing*, doi:10.1080/01431161.2011.608089.

Cheng, K. S. and Lei, T. C. (2001). Reservoir trophic state evaluation using Landsat TM images. *Journal of the American Water Resources Association*, 37(5), 1321–1334.

Erkkila, A. and Kalliola, R. (2004). Patterns and dynamics of coastal waters in multi-temporal satellite images: Support to water quality monitoring in the Archipelago Sea, Finland. *Estuarine, Coastal and Shelf Science*, 60, 165–177.

Evans, R. D. (1994). Empirical evidence of the importance of sediment resuspension in lakes. *Hydrobiologia*, 284, 5–12.

Fogel, L. J., Owens, A. J., and Walsh, M. J. (1966). *Artificial Intelligence through Simulated Evolution.* John Wiley & Sons, New York.

Francone, F. D. (1998). *Discipulus™ Software Owner's Manual, version 3.0 DRAFT.* Machine Learning Technologies, Inc., Colorado.

Fuller L. M. and Minnerick, R. J. (2007). *Predicting Water Quality by Relating Secchi-Disk Transparency and Chlorophyll a Measurements to Landsat Satellite Imagery for Michigan Inland Lakes*, 2001–2006. http://pubs.usgs.gov/fs/2007/3022/pdf/FS2007-3022.pdf.

Gagne, C. and Parizeau, M. (2004). *Genericity in Evolutionary Computation Software Tools: Principles and Case-Study. Technical Report RT-LVSN-2004-01.* Laboratoire de Vision et Systemes Numerique (LVSN), Universite Laval, Quebec, Canada.

Gordon, H. R., Brown, O. B., Evans, R. H., Brown, J.W., Smith, R. C., Baker, K. S., and Clark, D. K. (1988). A semianalytic radiance model of ocean color. *Journal of Geophysical Research.* 93, 10909–10924.

Hellweger, F. L., Schlosser, P., Lall, U., and Weissel, J. K. (2004). Use of satellite imagery for water quality studies in New York Harbor. *Estuarine, Coastal and Shelf Science*, 61, 437–448.

Holland, J. M. (1975). *Adaptation in Natural and Artificial Systems*. University of Michigan Press, Ann Arbor, MI.

James, R. T., Chimney, M. J., Sharfstein1, B., Engstrom, D. R., Schottler, S. P., East1, T., and Jin, K. R. (2008). Hurricane effects on a shallow lake ecosystem, Lake Okeechobee, Florida (USA). *Fundamental and Applied Limnology, Archiv für Hydrobiologie*, 172(4), 73–287.

Jin, K. R. and Ji, Z. G. (2004). Case Study: Modeling of sediment transport and wind-wave impact in Lake Okeechobee. *Journal of Hydraulic Engineering*, ASCE, 130(11), 1055–1067.

Jin, K. R., Ji, Z. G., and Hamrick, J. H. (2002). Modeling winter circulation in Lake Okeechobee, Florida. *Journal of Waterway, Port, Coastal, and Ocean Engineering*, ASCE, 128(3), 114–125.

Keiner, L. E. and Yan, X. H. (1998). A neural network model for estimating sea surface chlorophyll and sediments from Thematic Mapper Imagery. *Remote Sensing of Environment*, 66, 153–165.

Koza, J. R. (1992). *Genetic Programming: On the Programming of Computers by Means of Natural Selection*. MIT Press, Cambridge, MA.

Li, H., Budd, J. W., and Green, S. (2004). Evaluation and regional optimization of bio-optical algorithms for central Lake Superior. *Journal of Great Lakes Research*, 30(1), 443–458.

Lillesand, T. M., Johnson, W. L., Deuell, R. L., Lindstorm, O. M., and Meisner, D. E. (1983). Use of Landsat data to predict the trophic state of Minnesota Lakes. *Photogrammetric Engineering and Remote Sensing*, 49(2), 219–229.

Maceina, M. J. and Soballe, D. M. (1990). Wind-related limnological variation in Lake Okeechobee. *Lake and Reservoir Management*, 6, 93–100.

Mehta, A. J., Hayter, E. J., Parker, W. R., Krone, R. B., and Teeter, A. M. (1989). Cohesive sediment transport. I: Process description. *Journal of Hydraulic Engineering*, 115, 1076–1093.

Moore, P. A., Jr., and Reddy, K. R. (1994). Role of Eh and pH on phosphorus biogeochemistry in sediments of Lake Okeechobee, Florida. *Journal of Environmental Quality*, 23, 955–964.

National Aeronautics and Space Administration (NASA). http://modis.gsfc.nasa.gov (accessed September 12, 2008).

Pattiaratchi, C., Lavery, P., Wyllie, A., and Hick, P. (1994). Estimates of water quality in coastal waters using multidate Landsat Thematic Mapper data. *International Journal of Remote Sensing*, 15, 1571–1584.

Rechenberg, I. (1965). *Cybernetic Solution Path of an Experimental Problem*. Royal Aircraft Establishment, Farnborough, Library Translation 1122.

Steinman, A. D., Havens, K. E., Aumen, N. G., James, R. T., Jin, K. R., Zhang, J., and Rosen, B. H. (1999). *Phosphorus in Lake Okeechobee: Sources, Sinks, and Strategies*. In K. R. Reddy, G. A. O'Conner, and C. L. Schelske (eds.), *Phosphorus Biogeochemistry in Subtropical Ecosystems*, Lewis Publishers, Cherry Hill, NJ.

Tassan, S. (1987). Evaluation of the potential of the Thematic Mapper for marine applications. *International Journal of Remote Sensing*, 8, 1455–1478.

Thiemann, S. and Kaufmann, H. (2000). Determination of chlorophyll content and trophic state of Lakes using field spectrometer and IRS-1C satellite data in the Mecklenburg Lake District, Germany. *Remote Sensing of Environment*, 73, 227–235.

Van Duin, E. H. S., Blom, G., Lijklema, L., and Scholten, M. J. M. (1992). Aspects of modeling sediment transport and light conditions in Lake Marken. *Hydrobiologia*, 235/236, 167–176.

Volpe, G., Santoleri, R., Vellucci, V., d'Alcalà, M. R., Marullo, S., and D'Ortenzio, F. (2007). The colour of the Mediterranean Sea: Global versus regional bio-optical algorithms evaluation and implication for satellite chlorophyll estimates. *Remote Sensing of Environment*, 107, 625–638.

Zilioli, E. and Brivio, P. A. (1997). The satellite derived optical information for the comparative assessment of lacustrine water quality. *Science of the Total Environment*, 196, 229–245.

3 Mapping Potential Annual Pollutant Loads in River Basins Using Remotely Sensed Imagery

Kazuo Oki, Bin He, and Taikan Oki

CONTENTS

3.1   INTRODUCTION

Water quality is an important factor for human health and quality of life (Sivertun and Prange 2003). In the last several decades, rapid land use and land cover changes (LUCCs) across river basins have oftentimes caused serious environmental degradation. The increase of pollutant loads such as nitrogen and phosphorus to the river due to LUCC at the basin scale has become one of the major sources of water pollution (Krenkel and Novotny 1980; Lewis et al. 1984; He et al. 2011). The eutrophication of coastal waters as a result of the delivery of excess nutrients by rivers is a critical environmental issue on local, regional, and global scales (Smith et al. 2005;

Seitzinger and Kroeze 1998; Turner and Rabalais 2003). The United Nations has estimated that the economic cost of environmental deterioration due to water pollution is very severe in the Asia–Pacific region in terms of restoring the quality of life and installing controls (UN 1998). In many European rivers, including the major streams of the Rhine and Elbe basins, the nutrient load exceeds target levels, which cause eutrophication in fresh waters and coastal zones (Owens 1993; Oslo and Paris Commission 1993; Stanners and Bourdeau 1995; Van Dijk et al. 1995). In the context of the South Asian region, specifically in Nepal, India, and Bangladesh, pollution of surface waters has become more severe and critical near the urban areas due to high pollution loads discharged within short stretches of rivers from urban activities.

Nutrient pollution has several undesirable effects, most of which are related to the increased growth of phytoplankton and other aquatic plants. This so-called eutrophication leads to a shift in the biological structure, and in severe cases, even to oxygen depletion, production of toxins, and the collapse of entire aquatic ecosystems (OECD 1982). Failure to provide adequate urban sanitary infrastructures and the lack of sophisticated planning and effective implementation of necessary pollution control measures are leading to a worsening situation (Karn and Harada 2001). In order to reduce river nutrient loads, policy makers often need a rapid analysis of nutrient fluxes from all sources (e.g., agriculture, atmospheric deposition, households, and industry) to help them grasp the most effective measures for improving water quality.

In Japan, the water quality has been enhanced remarkably during recent decades. A classic example would be the Yodo River, which is the principal river in Osaka Prefecture on Honshu, where the average annual concentration of biochemical oxygen demand (BOD) was 3.4 mg·L^{-1}, 2.5 mg·L^{-1}, 2.3 mg·L^{-1}, and 1.5 mg·L^{-1} for the years 1985, 1990, 1995, and 2004, respectively, with a decreasing trend. In the Tone River, deemed one of the three greatest rivers in Japan, the average annual concentration of BOD was 2.6 mg·L^{-1}, 2.3 mg·L^{-1}, 1.9 mg·L^{-1}, and 1.4 mg·L^{-1} for the years 1985, 1990, 1995, and 2004, respectively, with a decreasing trend as well. Despite these observed improvements, rivers in Japan are still heavily impacted by many human-induced stresses and natural hazards. They include the canalization effect (i.e., tunneling and straightening of rivers or change of the bottom infrastructure by reinforced concrete slabs without transverse reinforcement), loss of most dynamic buffer zones in flood plains, flow regulation, invasion of exotic species, and intensive urbanization (Yoshimura et al. 2005; Kimura and Hatano 2007). In addition to the urbanization effect, Japanese agriculture has also created runoff with high nitrogen concentration across the agricultural fields due to the increasing application rate of chemical fertilizer and livestock husbandry (Mishima 2001; Kimura 2005). The water quality impact is still a big concern with regard to water-related issues in Japan.

The direct monitoring of water quality is inherently difficult since it requires spatial and temporal measurement tools for tackling rapid LUCCs in river basins; yet simulation models relating water quality changes in association with LUCCs can be performed to overcome this barrier (Oki and Yasuoka 1997; Omasa et al. 2006). Prediction of pollutant loads is therefore crucial to environmental management under the impact of intensified human activities. In this chapter, we first review recent research with regard to the estimation methods of pollutant loads in river basins. This review is followed by the mapping of several potential annual total pollutant loads (PATLs) in river basins across Japan using integrated remotely sensed images and a simulation model with the aid of a geographic information system (GIS).

3.2 LITERATURE REVIEW

Estimating the impact of nonpoint source (NPS) pollution in river basins is a complex problem. Models that can describe nutrient fluxes from pollution sources to river outlets can help policy makers to choose the most effective source-control measures to achieve a further reduction of nutrient levels in rivers and coastal waters. There are several existing models that describe the major processes involved in the transport of nutrients through soil, groundwater, and rivers. They include ANIMO (Groenendijk and Kroes 1997), RIVERSTRAHLER (Billen et al. 1994), DAISY (Hansen et al. 1991), MT3D (Zheng 1990), CREAMS (Knisel 1980), AGNPS (Young et al. 1989), ANSWERS (Beasley et al. 1980), CASC2D (Ogden and Julien 2002), DWSM (Borah et al. 2002), KINEROS (Woolhiser et al. 1990), MIKE SHE (Refsgaard and Storm 1995), and SWAT (Arnold et al. 1998). These models have been designed for different purposes with varying spatial and temporal scales. Still, they require data that may or may not always be available at the river basin scale to support these models (de Wit 2001). The simple models are sometimes incapable of giving detailed results whereas the sophisticated models are computationally inefficient and could be prohibitive for applications in large river basins. This means that finding an appropriate model for a broad application across a variety of watersheds is quite a challenging task. Since the behavioral patterns (transport, retention, and loss) of nutrients in the soil, groundwater, and river networks is a very complex function in terms of numerous biological, physical, and chemical processes, models (simplified representations of the real system) are needed to analyze nutrient fluxes (de Wit 2001).

In recent decades, the distributed hydrologic models (e.g., watershed models) has led to improved forecasting capability at the expense of requiring more detailed spatial information. These simulation models can assist in environmental management for alleviating the water quality problems that have been shown to be effective tools for specific areas of interest (Neill 1989; Fernandez-Santos et al. 1993; He et al. 1993; Johnes 1996; Gardi 2001; Bhuyan et al. 2003; Tadamus and Bergman 1995). Incorporation of watershed models into a GIS environment has improved modeling capability by streamlining data input and providing better interpretation of model outputs (Pullar and Springer 2000). With this advancement, GIS has been coupled with three general categories of NPS pollution control models, including regression models (Corwin et al. 1989), index models (Rundquist 1991), and transient-state solute transport models (Corwin et al. 1993). More recently, numerous water quality–based hydrologic models linking runoff with soil erosion impact have been used within a GIS environment to determine the surface water quality impact of NPS pollutants in different watersheds (Pelletier 1985; Warwick and Haness 1994), agricultural areas (Hopkins and Clausen 1985; Tim and Jolly 1994), and urban regions (Smith and Brilly 1992; Smith 1993). Most of the commonly used models were formulated in the 1970s through the 1980s, and since the early 1990s, most modeling research has focused on development of dedicated graphic user interfaces (GUIs) and integration with GIS and remote sensing data.

For estimating the influent pollutant load to rivers, Corwin et al. (1988), Corwin and Rhoades (1988), and Corwin et al. (1989) applied GIS techniques to delineate areas of salinity accumulation in the vadose zone by coupling a GIS of the Wellton-Mohawk Irrigation District to a phenomenological model of salinity development. Petach et al. (1991) and Corwin et al. (1993) used transient state solute transport models coupled to a GIS to assess the leaching potential of some common NPS agricultural chemicals under nonequilibrium conditions. Tim and Jolly (1992) applied an integrated approach, coupling water quality simulation models with GIS to delineate critical NPS pollution control issues at the watershed level. Burrough (1996) identified three components to constitute a GIS-based environmental modeling platform. These three components were identified as input data, GIS modules, and spatial modeling outputs. Burrough stated that first and foremost, the solute transport model must be developed; second, data for the model must be obtained; and finally, the solute transport model must be coupled to a GIS containing the spatial input data. Basnyat et al. (2000) examined a methodology to determine nitrate pollution "contributing zones" within a given basin based on basin characteristics. In their study, the types of land use and land cover (LULC) were classified, and basins along with contributing zones were delineated using GIS and remote sensing analysis tools. Oki and Yasuoka (1997) proposed a method for evaluating the total nitrogen load to a lake from each neighboring river basin. They investigated a remote sensing method using the Landsat Multispectral Scanner (MSS) and Thematic Mapper (TM) to monitor LUCC in lake basins and analyzed the relation between changes in LULC and changes in the total amount of nitrogen loaded to the lake. As a result, it was indicative that the runoff load factors of total nitrogen of urban areas are greater than those of forested areas and paddy fields. It can be concluded that the extension of urban areas affects the eutrophication of lakes through an increase in nitrogen load to the lakes. Oki and Yasuoka (2008) analyzed the relation between the LULC types estimated from monthly maximum normalized difference vegetation index (NDVI) and the annual total nitrogen load discharged from river basins. The monthly NDVI values used were produced by the National Oceanic and Atmospheric Administration (NOAA) Advanced Very High Resolution Radiometer (AVHRR) images. By considering such a relation that is basinwide in Japan, the annual total nitrogen load discharged from river basins may be generated based on two advanced maps including the potential annual total nitrogen load (PTNL) index and the potential annual total nitrogen load for each river basin area (PTNL/area) index. Additional efforts were proposed to pursue an integrated approach to assess daily nitrogen loads from anthropogenic and natural sources in Japan by using a process-based modeling framework with the aid of a high resolution environmental database (He et al. 2009a,b). More delicate descriptions of other models may be found in Singh (1995) and Singh and Frevert (2002a,b).

As stated above, most studies involved handling the runoff impact from agricultural fields rather than other types of LULC. The impact of LUCC has not yet been fully investigated by these models in Japan. In addition, these simulation models are all data-intensive, in which the input data normally covers soil type, precipitation, and slope in support of water quality assessment. Several methods have been proposed for estimating the annual total nitrogen load to rivers, lakes, and the sea (e.g., Ibaraki Prefecture of Japan 1983). Most of the methods estimated the annual total nitrogen load of inflow based on the use of export coefficients associated with differing types of LULC. It is important to note, however, that export coefficients vary from time to time, depending on the estimation methods, site-specific conditions, and seasonal effects. It can be deduced that these export coefficients methods can only be used consistently in small areas where the discharge rate of total nitrogen load can be measured for verification. In any circumstance, simplified models are needed to analyze nutrient fluxes at the national level.

3.3 MATERIALS AND METHODS

3.3.1 An Efficient Method to Estimating Pollutant Loads

In this study, a model for estimating each type of the annual pollutant load discharged from a river basin was developed using the LULC data derived from remotely sensed images and simulation models. This is based on an extension of the conventional model that was formulated from the relationship between the load and the river flow in a river basin as follows (Gunnerson 1967; Yamaguchi et al. 1980; Ebise 1984; Lewis et al. 1984; Neill 1989):

$$l = a \cdot q^b \tag{3.1}$$

where l is the pollutant load discharged from a river basin such as total nitrogen, q is the river flow, and a and b are coefficients. The empirical model in Equation 3.1 can be modified to suit new applications by considering an additional variable, s, which is defined as the pollutant load accumulated in a river basin as follows (Yamaguchi et al. 1980):

$$l = k \cdot q^n \cdot s^m \tag{3.2}$$

$$ds/dt = s_0 - l \tag{3.3}$$

where k, n, and m are coefficients, t is time, and s_0 is the initial value of the pollutant load accumulated in a basin at time zero. With the empirical model in Equations 3.2 and 3.3, we can estimate a suite of pollutant loads discharged from a river basin after considering the load accumulation in a river basin over time. Oki and Yasuoka (1997) modified the model described in Equations 3.2 and 3.3 to include basinwide LULC information leading to the generation of the annual total nitrogen load discharged from each river basin. The modified model for each river basin i is shown as follows:

$$\frac{L_i}{A_i} = k \cdot \left(\frac{Q_i}{A_i}\right)^n \cdot \left(\frac{S_i}{A_i}\right)^m \tag{3.4}$$

where L_i is the annual total nitrogen load discharged from a river basin i (kg·yr⁻¹), A_i is the area of the river basin I (km²), Q_i is annual river flow of a river in a river basin I (m³/yr), S_i is the annual total nitrogen load accumulated in a river basin i (kg·yr⁻¹), and k, n, and m are coefficients. In Equation 3.4, the aggregate effect associated with each type of LULC in a basin can be calculated with the following empirical model in Equation 3.5. Such a total pollutant load can be estimated as the sum of the individual pollutant load divided by the area of the corresponding type of LULC across a river basin. The annual total pollutant load accumulated in a river basin i, S_i (kg·yr⁻¹) is expressed by

$$S_i = \sum_{j=1}^{LC} \left(U_j \cdot A'_{ij} \right)$$
(3.5)

where the subscript j (=1 to LC) represents each category of LULC from class 1 to class LC. U_j is the export coefficient defined as a kind of runoff-driven load factor associated with each category j (kg·km⁻²·yr⁻¹), and A'_{ij} is the area of each category j of LULC in a river basin i (km²). In our study, five classes were included. These five land use categories of LULC from class 1 to 5 include urbanized area, beech land 1 (natural vegetation), beech land 2 (secondary vegetation), community, plantation, and grassland, respectively. Due to the temporal variability of runoff in a 1-year time period, we redefined that the runoff-driven load factor associated with each category j of LULC, which is U_j (kg·km⁻²·yr⁻¹), can be expressed as an "average value" of the runoff-driven load factor across 12 months to come up with the annual pollutant load in each category of LULC j.

To estimate k, n, m, and U_j (j = 1 to LC) in Equations 3.4 and 3.5, minimization of the sum of deviations between predicted and observed values was carried out and constrained by the nonnegativity requirements of k, n, m, and U_j (j = 1 to 5) (see Equations 3.6 and 3.7), which represents an integrated simulation and optimization model for parameter identification:

$$Min \sum_{j=i}^{5} \left(\frac{L_i}{A_i} - k \cdot \left(\frac{Q_i}{A_i} \right)^n \cdot \left(\frac{S_i}{A_i} \right)^m \right)^2$$
(3.6)

$$subject\ to: k.n.m,\ U_j \geq 0 \quad j = 1,\ldots,5$$
(3.7)

3.3.2 GENERATION OF THE LULC MAP WITH REMOTE SENSING IMAGES

Remotely sensed images have been widely used to classify the global vegetation cover from which the consistent LULC maps may be derived (DeFries and Townshend 1994a,b). Recent work in classifying vegetation cover with regional, continental, and global scales has been extended to use multitemporal remotely sensed data sets, which describe vegetation dynamics by viewing their phenological variation throughout the course of a year (Verhoef et al. 1996; Tucker et al. 1985; Townshend et al. 1987; Stone et al. 1990). Loveland et al. (1999) produced a global LULC map layer with a 1-km resolution in which each continent is classified separately and then unified together. The work was carried out with the aid of the DISCover product, produced by the International Geosphere–Biosphere Programme (IGBP). They used a 12-month NDVI as the data source in an unsupervised clustering algorithm to achieve the goal. Oki and Yasuoka (1997) used Landsat MSS and TM images, with spatial resolutions of about 80 and 30 m, respectively, for retrieving the thematic information of LULC for only part of Japan. This was due to the limitation of the cloud-free Landsat MSS and TM images. Still, the repeat cycle of 16 days for these remote sensing NDVI products can hinder the assessment of vegetation dynamics even in relatively cloud-free areas (Hall et al. 1991). This underlines the difficulties in obtaining the consistent categories of LULC maps from images taken at different space and time intervals with variability of atmospheric effects. This may impede rapid and efficient sharing of information for many applications though. Besides this factor, the NOAA AVHRR images were also used to produce multitemporal profiles of NDVI in order to measure and assess changes in vegetation phenology and conditions (Maselli et al. 1998). It can also support the LULC classification at the global scale (Tucker et al. 1985). Moreover, the frequent repeat cycle of the satellite providing fine temporal data of at least one sample per day (depending on the latitude) is capable of monitoring the vegetation dynamics. We chose to use the monthly maximum NDVI calculated from NOAA AVHRR imagery in this study to cover the whole area of Japan, which is deemed appropriate for the holistic assessment.

3.3.3 MODEL CALIBRATION FOR DIFFERENT POLLUTANTS

Parameters in Equation 3.6 are determined by an optimization analysis using the simplex method throughout the solution space to minimize the root-mean-squared error between the actual and observed annual mean pollutant loads. We followed the linear approximation of the multiplicative parameters in relation to the root-mean-squared error until the convergence criterion can be met (Nelder and Mead 1965; Press et al. 1988). Still, there may be cases of multiple near-optimal solutions in which the same objective function value in relation to the root-mean-squared error can be reached with different values of k, n, m, and U_j ($j = 1$ to 5). In conclusion, we determined the values of m and U_j ($j = 1$ to 5) when the values of k and n were similar to the historical values so as to pin down the final near-optimal solution.

3.4 RESULTS AND DISCUSSION

3.4.1 DATA COLLECTION AND ANALYSIS

The annual mean concentration of total nitrogen (TN), TP, suspended solid (SS), BOD, chemical oxygen demand (COD), and dissolved oxygen (DO) were routinely measured by the Ministry of Land, Infrastructure and Transport of the Government of Japan (MLIT). The hydrochemical data applied in this analysis, including TN, TP, SS, BOD, COD, and DO, were collected at the outfalls of 30 selected rivers in Japan

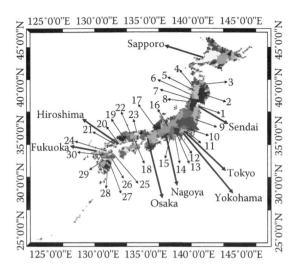

FIGURE 3.1 Individual river basins in Japan determined using the basin database of the digital national land information and the selected 30 test river basins in Japan.

(Figure 3.1). With river discharge information available in MLIT, the annual mean of pollutant concentrations may be calculated using a set of pollutant concentrations within the prescribed time windows for either monthly or seasonal variations. A summary of the all river basins is listed in Table 3.1.

The *in situ* quantitative and qualitative measurements at the mouth of the river allow us to calculate the particular type of annual pollutant load, L (kg·yr^{-1}) discharged from the whole river basin as follows:

$$L = P_{con} \cdot 10^{-3} \cdot Q \qquad (3.8)$$

TABLE 3.1

List of River Basins in Japan

ID	River	ID	River	ID	River
1	Hirose	11	Tsurumi	21	Saba
2	Kitakami	12	Fuji	22	Hino
3	Mabechi	13	Abe	23	Sendai
4	Yoneshiro	14	Oui	24	Yamakuni
5	Omono	15	Tenryu	25	Oita
6	Koyoshi	16	Kiso	26	Ono
7	Mogami	17	Nagara	27	Omaru
8	Aka	18	Yoshii	28	Oyodo
9	Kuji	19	Asida	29	Shira
10	Tama	20	Oze	30	Kikuchi

where P_{con} is pollutant concentration (mg·L^{-1}) and Q is river discharge (m^3·yr^{-1}). The pollutant loads were estimated for the 1996 study since 1996 is the year in which the LULC map was generated to feature the land use categories based on the AVHRR NDVI products.

3.4.2 CHARACTERISTICS OF RIVER BASINS IN JAPAN

With the aid of national river basin databases created and maintained by the MLIT, we can show the geographical distribution of all 30 river basins along with a few major cities (i.e., Tokyo, Yokohama, Osaka, Nagoya, Fukuoka, Sapporo, Sendai, and Hiroshima) in Japan (Figure 3.1). Most river basins are situated within rugged terrain that is comprised of steep landscapes and shorter river reaches. As a consequence, NPS pollution is the nation's largest water quality problem. This type of pollution is commonly scattered around through the overland flows via various types of LULC such as urbanized areas, agricultural land, and forested areas. Given that the spatial information is readily available, it enables us to estimate each of the particular annual pollutant loads with respect to the LULC and long-term average river discharge across those river basins using Equations 3.4 and 3.5. Simulation analysis was carried out based on the data sets collected from these 30 river basins to generate the annual mean value of pollutant loads representing regional differences across Japan in this study.

3.4.3 LAND COVER CLASSIFICATION

To achieve the classification of LULC, the monthly maximum NDVI dataset of 1996 (after removing the cloud effects) were used to produce the LULC maps. To create the 12 monthly maximum NDVI datasets, daily NDVI was calculated using Band 1 (0.58–0.68 μm) and Band 2 (0.725–1.10 μm) images, which were measured by AVHRR mounted on NOAA-12 and -14. The maximum NDVI value was then screened within a series of daily NDVI values for each pixel on a monthly basis from January to December to select the monthly maximum NDVI values (Cihlar 1996; Dye and Tucker 2003). Such an implementation took temporal changes into consideration with a spatial resolution of 1.1 km^2 produced by the National Institute for Environmental Studies of Japan (Matsushita and Tamura 2002).

With these monthly maximum NDVI values associated with each pixel, the ISODATA classification method was used to produce a series LULC map of Japan. ISODATA, which is an unsupervised classification method, leads to the calculation of classes evenly distributed in the data space that is produced by clustering iteratively the pixels using minimum distance techniques. During the iteration, ISODATA was reapplied to break each confused cluster into additional clusters. These clusters were then reevaluated using the previous process. The iteration is carried out to recalculate the means and reclassify pixels using the new means so as to minimize the discrepancies. The final step can be eventually reached with splitting and merging of clusters until the number of pixels in each class changes by less than the prescribed threshold. Finally, the 12 monthly maximum NDVI datasets were classified into five land use categories of LULC using such an ISODATA method. With the coarse

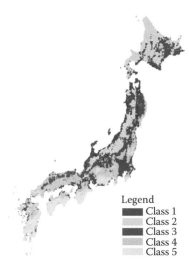

FIGURE 3.2   Generated land cover from remote sensing imagery. (From Oki, K. and Yasuoka, Y., *Remote Sensing of Environment*, 112, 6, 3091–3098, 2008. With permission.)

spatial resolution (1.1 km²), the classification was still made possible to produce the much-needed LULC maps that are similar to the existing vegetation maps produced by the Ministry of the Environment of Japan. They are actually carried out based on the same types of classes that the Ministry of the Environment's vegetation maps employed. The generated LULC map is shown in Figure 3.2.

3.4.4   Statistical Analysis

A statistical study was conducted for the collected data sets to further summarize the environmental features across these river basins. Statistical parameters used in this study included maximum, median, minimum, mean, standard deviation (SD), and coefficient of variation (CV) (Table 3.2). For all 30 river basins in Japan, large variations of annual concentration of TN, TP, and BOD were observed among river basins. The range of CV varies from the smallest one of 10.3% associated with DO to the highest value of 162.6% associated with TP. In addition to water quality variations, large variations were also found in terms of river discharge and LULC classification over time (Table 3.2). The maximum variation of LULC was observed in class 2 (235.4%) and the minimum variation of LULC in class 4 (117.7%). These large spatial variations correspond to the large differences in regional characteristics including basin size, population, topography, geology, agricultural activities, and policy.

To detect the relationships between pollutant load and basin characteristics, the correlation analysis was also conducted with the collected data sets. A significant correlation was not found between the pollutant load per area and five LULC classes (Tables 3.3 and 3.4). Only class 2 exhibited relatively higher correlation coefficients associated with TN, SS, COD, and DO, although significant correlation between pollutant loads and classes 1 and 3 was noted. The largest one was identified between SS and class 2, which reflects the high impact of beech-type natural vegetation on SS

TABLE 3.2
Statistical Parameters of the Studied Variables

Variables	Maximum	Median	Minimum	Mean	SD	CV (%)
TN (mg/l)	11.3	1.0	0.5	1.7	2.2	129.7
TP (mg/l)	0.8	<0.01	<0.01	0.1	0.2	162.6
SS (mg/l)	24.0	5.0	1.0	7.4	5.0	66.6
BOD (mg/l)	14.2	1.2	0.2	1.8	2.5	137.9
COD (mg/l)	9.5	2.5	0.8	2.9	1.9	63.5
DO (mg/l)	11.7	10.4	6.7	10.1	1.0	10.3
Q (m^3/s)	387.1	37.6	3.3	69.7	90.3	129.5
LC_1 (km^2)	1363.0	107.0	4.0	222.3	299.8	134.8
LC_2 (km^2)	4721.0	<0.01	<0.01	450.2	1059.7	235.4
LC_3 (km^2)	5365.0	51.5	<0.01	579.7	1091.1	188.2
LC_4 (km^2)	2696.0	402.0	<0.01	646.0	760.6	117.7
LC_5 (km^2)	1653.0	75.5	<0.01	311.0	464.9	149.5

Source: He, B., Oki, K., Wang, Y., and Oki, T., *Water Science and Technology*, 60, 8, 2009–2015, 2009.
Note: LC_1 = class 1 of LULC; SD = standard deviation; CV = coefficient variation.

generation. We also found that the urbanized areas are highly related to the export of TN, TP, BOD, COD, and DO. Finally, we analyzed the relationship between pollutant loads and river discharges. It is indicative that all of the pollutant loads are highly flow-proportional except for TP, which has a correlation coefficient of 0.69 with river discharge. This statistical analysis provides us with a base to develop an empirical approach to assess annual pollutant load with respect to river basin characteristics.

TABLE 3.3
Correlation Coefficients between Pollutant Load per Unit Area (kg·km^{-2}) and Types of LULC (km^2)

	LC_1	LC_2	LC_3	LC_4	LC_5
TN	0.10	0.44	0.08	−0.34	−0.33
TP	0.08	−0.07	−0.14	−0.16	−0.20
SS	0.13	0.53	0.17	−0.35	−0.23
BOD	−0.01	0.10	−0.11	−0.37	−0.30
COD	0.10	0.44	0.08	−0.34	−0.33
DO	0.01	0.41	0.08	−0.26	−0.13

Source: He, B., Oki, K., Wang, Y., and Oki, T., *Water Science and Technology*, 60, 8, 2009–2015, 2009.
Note: LC_1 = class 1 of LULC.

TABLE 3.4

Correlation Coefficients between Pollutant Load (kg) and Land Covers (km²)

	LC₁	LC₂	LC₃	LC₄	LC₅
TN	0.71	0.81	0.58	0.24	0.14
TP	0.81	0.64	0.66	0.37	0.02
SS	0.57	0.93	0.49	−0.09	−0.10
BOD	0.68	0.91	0.61	0.03	−0.10
COD	0.67	0.92	0.60	0.06	−0.06
DO	0.60	0.88	0.56	0.09	0.04

Source: He, B., Oki, K., Wang, Y., and Oki, T., *Water Science and Technology*, 60, 8, 2009–2015, 2009.

Note: LC_1 = class 1 of LULC.

3.4.5 ANNUAL POLLUTANT LOAD ASSESSMENT

The pollutant loads of TN, TP, BOD, and COD in urban runoff are higher than those of other LULC types (classes 2 to 5) in Japan. In urbanized areas, the nitrogen and phosphorus concentrations are higher than those of other types of LULC categories due to various factors such as fertilizer application and atmospheric pollution. Unlike the case in forested land, where the total pollutant load cannot be discharged easily due to the uptake by the forest and the attachment of the soil, urban runoff carries direct impact of nutrients. In addition, the BOD and COD concentrations of river water in urbanized areas are higher than those of other types of LULC due to human activities. The factor of river discharge (i.e., the coefficient n in Equation 3.1) is the smallest for TP, the largest for BOD and COD, and almost the same for TN and SS. This does not mean that the correlation between TP load and river discharge is weak, however. The factor n in fact is heavily attributed to the order of magnitude of data since the value of BOD load ranges from several hundreds to over tens of thousands, but TP load ranges are only from zero to several hundred.

All coefficients estimated by the simplex method based on the 1996 data sets can be derived based on Equations 3.6 and 3.7 (Table 3.5). The potential pollutant load discharged from each river basin in Japan can then be estimated by using river flow data, the basin area, LULC map, and runoff load factors as shown in Table 3.5. For nutrient management, it is possible to estimate the annual total nitrogen load discharged from these river basins and compare the outcome with those obtained by using conventional methods. Still, conventional methods involve too many ramifications across different methods, sites, and seasons (Oki and Yasuoka 1997).

Model verification based on the observed and predicted pollutant loads for the year 1996 in 30 river basins in Japan is presented in Figure 3.3. To summarize, the regression of predicted versus observed pollutant loads data resulted in a set of high correlation coefficients across all water quality categories. Some of these values were 0.98, 0.87, 0.96, 0.98, 0.98, and 0.99 for TN, TP, SS, BOD, COD, and

TABLE 3.5

Estimated Coefficients in This Study

	k	n	U_1	U_2	U_3	U_4	U_5	m	R-Square Value
TN	0.91	0.65	1.00	0.28	0.25	0.24	0.48	1.90	0.96
TP	0.98	0.12	9.95	2.71	1.76	3.05	2.79	1.71	0.76
SS	0.91	0.63	1.45	2.31	0.99	0.86	1.56	1.92	0.92
BOD	0.85	1.06	0.06	0.01	0.02	0.01	0.02	1.90	0.96
COD	0.87	1.06	0.06	0.01	0.02	0.01	0.02	1.90	0.96
DO	0.91	1.00	0.08	0.10	0.10	0.09	0.09	1.90	0.98

Source: He, B., Oki, K., Wang, Y., and Oki, T., *Water Science and Technology*, 60, 8, 2009–2015, 2009.

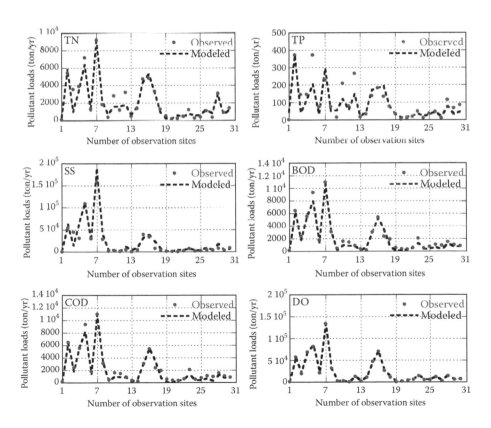

FIGURE 3.3   Model verification based on observed and predicted pollutant loads in 30 river basins in Japan (unit: metric ton/yr at vertical axes).

TABLE 3.6
Estimated S_i and PTNL for Each River Basin in This Study

ID	S_i	PTNL	ID	S_i	PTNL	ID	S_i	PTNL
1	468	0.12	11	235	0.06	21	200	0.05
2	3745	1.00	12	1340	0.36	22	261	0.07
3	613	0.16	13	280	0.07	23	349	0.09
4	1271	0.34	14	506	0.14	24	198	0.05
5	1683	0.45	15	1747	0.47	25	234	0.06
6	345	0.09	16	1691	0.45	26	488	0.13
7	2425	0.65	17	1476	0.39	27	235	0.06
8	340	0.09	18	604	0.16	28	974	0.26
9	592	0.16	19	269	0.07	29	202	0.05
10	714	0.19	20	116	0.03	30	395	0.11

DO, respectively. It shows that the estimation of pollutant loads can be considered reproducible.

The annual total pollutant load in each river basin can then be estimated based on Equation 3.5, which considers the runoff load factors and LULC of river basin simultaneously. It is worthwhile to note that the estimated values of U_j ($j = 1$ to 5) may not be accurate, which might disturb the estimation of annual total nitrogen load in river basins as calculated by Equation 3.5. To make the evaluation of PTNL in river basins comparable, the PTNL index was proposed. The PTNL index, ($=S_i/S_{max}$), is defined by using each annual total pollutant load, S_i, divided by the maximum value of annual total pollutant load, S_{max}, in 30 river basins. S_{max} is 3745 at the Kitakami river basin of ID No. 2. The estimated S_i and PTNL for each river basin in this study can be summarized in Table 3.6.

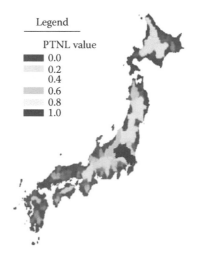

Legend

PTNL value
- 0.0
- 0.2
- 0.4
- 0.6
- 0.8
- 1.0

FIGURE 3.4 PTNL index in each river basin over the whole area of Japan. (From Oki, K. and Yasuoka, Y., *Remote Sensing of Environment*, 112, 6, 3091–3098, 2008. With permission.)

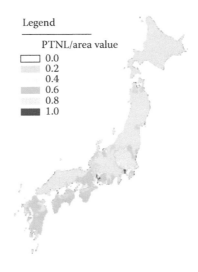

FIGURE 3.5 PTNL/area index associated with each river basin over the whole area of Japan. (From Oki, K. and Yasuoka, Y., *Remote Sensing of Environment*, 112, 6, 3091–3098, 2008. With permission.)

Furthermore, the potential annual total nitrogen load for each river basin area (PTNL/area), which is considered to reflect individual basin characteristics, was also proposed as $(S_i/A_i)/(S_i/A_i)_{max}$ because the PTNL index varies with basin area. The PTNL/area index is calculated as a ratio between (S_i/A_i) and $(S_i/A_i)_{max}$ for a given river basin. Both the PTNL index and PTNL/area index can be shown for each river basin in Japan, respectively (Figures 3.4 and 3.5). Within Figure 3.4, the Tokyo metropolitan region has a relatively high potential annual total nitrogen load. The PTNL map may be used for risk assessment or environmental impact assessment of total nitrogen load to lakes and coastal waters from each basin. In general, additional information is required in detail because the PTNL index strongly depends on the

TABLE 3.7
Estimated S_i/Area and PTNL/Area for Each River Basin in This Study

ID	S_i/Area	PTNL/Area	ID	S_i/Area	PTNL/Area	ID	S_i/Area	PTNL/Area
1	0.43	0.44	11	0.99	1.00	21	0.39	0.40
2	0.35	0.36	12	0.35	0.35	22	0.28	0.28
3	0.27	0.27	13	0.48	0.49	23	0.29	0.29
4	0.30	0.31	14	0.40	0.40	24	0.36	0.36
5	0.37	0.37	15	0.36	0.36	25	0.36	0.36
6	0.29	0.29	16	0.35	0.36	26	0.36	0.36
7	0.35	0.36	17	0.40	0.40	27	0.46	0.47
8	0.40	0.40	18	0.29	0.29	28	0.45	0.45
9	0.38	0.39	19	0.32	0.33	29	0.40	0.40
10	0.59	0.59	20	0.32	0.32	30	0.43	0.43

basin area. By using the PTNL/area index, which takes into account the effects of the basin area, the status of river basins can be further evaluated as shown in Figure 3.5. It shows that there are several small river basins with a large PTNL/area index value. These two PTNL related maps were produced based on the estimated S_i/Area and PTNL/Area for each river basin as shown in Table 3.7.

3.5 CONCLUSIONS

This chapter reviews a plethora of methods in regard to the gross estimation of the potential annual pollutant load within river basins in Japan. The monthly maximum NDVI products yielded from the NOAA AVHRR images were used to generate an LULC map after removing the cloud effects. The LULC is directed to support the estimation of potential annual pollutant loads of river basins in Japan via a semiempirical formula. With the aid of GIS and remote sensing technologies, the assessment of watershed NPS pollution may be improved. In our application across Japanese river basins, we have successfully analyzed the relationship between the types of LULC based on monthly NDVI and the annual total pollutant load discharged from river basins. Within this context, a suite of typical water quality indicators were investigated, including TN, TP, SS, BOD, COD, and DO. Statistical analysis shows that large spatial variations in terms of pollutant concentrations of TN, TP, BOD, river discharge, and LULC were observed. A correlation analysis was conducted for pollutant load and basin characteristics. Our findings reveal that significant correlation between pollutant loads and the types of LULC can be confirmed. Urbanized areas are highly correlated with the export of TN, TP, BOD, COD, and DO. All the pollutants are highly flow-proportional except for TP.

The modified model that we developed is deemed as a very efficient method in dealing with the environmental impact assessment at the national level. It can be used not only for within basins with respect to different LULC categories, but also between basins with different runoff patterns, size, landscape, and topography across the nation. With our optimization analysis to characterize the essential parameters in support of simulation analysis, the empirical approach in our modified model can smoothly assess the annual pollutant loads with respect to two types of index categories. Both of these classifications can be useful as a set of effective tools for environmental planning and management of river basins. Such integration enables us to quickly estimate the annual pollutant loads directly from satellite images and hydrological data. Results clearly show that the proposed simulation technique for pollutant loading estimation can effectively produce the potential annual pollutant load maps that may be useful for risk assessment of water quality impact at the basin scale.

ACKNOWLEDGMENTS

This study was sponsored by the Japanese Society for the Promotion of Science (JSPS) Grants-in-Aid for Scientific Research and the Kyoto University Global COE program "Sustainability/Survivability Science for a Resilient Society Adaptable to Extreme Weather Conditions." The authors are grateful for their financial support.

REFERENCES

Arnold, J. G., Srinivasan, R., Muttiah, R. S., and Williams, J. R. (1998). Large-area hydrologic modeling and assessment: Part I. Model development. *Journal of American Water Resources Association*, 34(1), 73–89.

Basnyat, P. (2000). The use of remote sensing and GIS in watershed level analyses of nonpoint source pollution problems. *Forest Ecology and Management*, 128(1–2), 65–73.

Beasley, D. B., Huggins, L. F., and Monke, E. J. (1980). ANSWERS: A model for watershed planning. *Transactions of American Society of Agricultural Engineering (ASAE)*, 23(4), 938–944.

Bhuyan, S. J., Koelliker, J. K., Marzen, L. J., and Harrington, J. A., Jr. (2003). An integrated approach for water quality assessment of a Kansas watershed. *Environmental Modelling & Software*, 18, 473–484.

Billen, G., Garnier, J., and Hanset, P. (1994). Modelling phytoplankton development in whole drainage networks: The RIVERSTRAHLER model applied to the Seine river system. *Hydrobiologia*, 289, 119–137.

Borah, D. K., Xia R., and Bera M. (2002b). DWSM—A dynamic watershed simulation model. Chapter 5 in *Mathematical Models of Small Watershed Hydrology and Applications*, V. P. Singh and D. K. Frevert (eds.), Water Resources Publications, Highlands Ranch, CO, pp. 113–166.

Burrough, P. A. (1996). Opportunities and limitations of GIS-based modeling of solute transport at the regional scale. In *Applications of GIS to the Modeling of Non-Point Source Pollutants in the Vadose Zone*, D. L. Corwin and K. Loague (eds.), SSSA Special Publication, Madison, WI.

Corwin, D. L., Werle, J. W., and Rhoades, J. D. (1988). The use of computer-assisted mapping techniques to delineate potential areas of salinity development in soils: I. A conceptual introduction. *Hilgardia*, 56(2), 1–17.

Corwin, D. L. and Rhoades, D. J. (1988). The use of computer-assisted mapping techniques to delineate potential areas of salinity development in soils: II. Field verification of the threshold model approach. *Hilgardia*, 56(2), 8–32.

Corwin, D. L., Sorensen, M., and Rhoades, J. D. (1989). Field-testing of models which identify soils susceptible to salinity development. *Geoderma*, 45, 31–64.

Corwin, D. L., Vaughan, P. J., Wang, H., Rhoades, J. D., and Cone, D. G. (1993). Coupling a solute transport model to a GIS to predict solute loading to the groundwater for a non-point source pollutant. *Proceedings of ASAE Applications of Advanced Information Technology for Management of Natural Resources Conference*, Spokane, WA, June 17–19, 1993, pp. 485–492.

Cihlar, J. (1996). Identification of contaminated pixels in AVHRR composite images for studies of land biosphere. *Remote Sensing of Environment*, 35, 149–163.

de Wit M. J. M. (2001). Nutrient fluxes at the river basin scale. I: The PolFlow model, *Hydrological Processes*, 15, 743–759.

DeFries, R. S. and Townshend, J. R. G. (1994a). Global land cover: comparison of ground based data sets to classifications with AVHRR data. In *Environmental Remote Sensing from Regional to Global Scales*, G. Foody and P. Curran (eds.), Wiley-Interscience, Chichester, West Sussex, UK.

DeFries, R. S. and Townshend, J. R. G. (1994b). NDVI-derived land cover classifications at a global scale. *International Journal of Remote Sensing*, 15, 3567–3586.

Dye, D. G. and Tucker, C. J. (2003). Seasonality and trends of snow-cover, vegetation index, and temperature in northern Eurasia. *Geophysical Research Letters*, 30(7), 3–6.

Ebise, S. (1984). *Comprehensive Studies on the Eutrophication Control of Freshwaters— Estimation of Input Loading of Lake Kasumigaura*. Research Report from the National Institute for Environmental Studies, Japan, p. 50 (in Japanese).

Fernandez-Santos, J., Zekri, S., and Herruzo, A. C. (1993). On-farm costs of reducing nitrogen pollution through BMP. *Agriculture, Ecosystems and Environment*, 45, 1–11.

Gardi, C. (2001). Land use, agronomic management and water quality in a small northern Italian watershed. *Agriculture, Ecosystems and Environment*, 87, 1–12.

Gunnerson, C. G. (1967). Streamflow and quality in the Columbia River Basin, Journal Sanitary Engineering Division. *Proceeding American Society Civil Engineers*, 95(SA6), 1–16.

Groenendijk, P. and Kroes, J. G. (1997). *Modelling the Nitrogen and Phosphorus Leaching to Groundwater and Surface Water: Animo 3.5*. Report 144, DLO Winand Staring Centre, Wageningen, the Netherlands.

Hall, F. G., Sellers, P. J., Strevel, D. E., Kanemasu, E. T., Kelly, R. D., Blad, B. L., Markham, B. J., Wang, J. R., and Huemmrich, F. (1991). Satellite remote sensing of surface energy and mass balance: results from FIFE. *Remote Sensing of Environment*, 35, 187–199.

Hansen, S., Jensen, H. E., Nielsen, N. E., and Svendsen, H. (1991). Simulation of nitrogen dynamics and biomass production in winter wheat using the Danish simulation Daisy. *Fertilizer Research*, 27, 245–259.

He, C., Riggs, J. F., and Kang, Y. T. (1993). Integration of geographic information systems and a computer model to evaluate impacts of agricultural runoff on water quality. *Water Resources Bulletin*, 29(6), 891–900.

He, B., Oki, T., Kanae, S., Mouri, G., Kodama, K., Komori, D., and Seto, S. (2009a). Integrated biogeochemical modeling of nitrogen load from anthropogenic and natural sources in Japan. *Ecological Modeling*, 220, 2325–2334.

He, B., Oki, K., Wang, Y., and Oki, T. (2009b). Using remotely sensed imagery to estimate potential annual pollutant loads in river basins. *Water Science and Technology*, 60(8), 2009–2015.

He, B., Oki, T., Sun, F., Komori, D., Kanae, S., Wang, Y., Kim, H., and Yamazaki, D. (2011). Estimating monthly total nitrogen concentration in Japanese streams by using artificial neural networks. *Journal of Environment Management*, 92(1), 172–177.

Hopkins, R. and Clausen, J. (1985). Monitoring and assessment for non-point source pollution control. In *Perspective on Non-point Source Pollution*. U.S. EPA, Kansas City, MO, pp. 25–29.

Ibaraki Prefecture of Japan. (1983). *The Report of Eutrophication at the Lake Kasumigaura, Ibaraki Prefecture, Japan*.

Johnes, P. J. (1996). Evaluation and management of the impact of change on the nitrogen and phosphorus load delivered to surface waters: The export coefficient modeling approach. *Journal of Hydrology*, 183, 323–349.

Karn, K. S. and Harada, H. (2001). Surface water pollution in three urban territories of Nepal, India, and Bangladesh. *Environmental Management*, 28(4), 483–496.

Kimura, S. D. (2005). *Creation of an Eco-balance Model to Assess Environmental Risks Caused by Nitrogen Load in a Basin Agroecosystem*. Ph.D. thesis, Hokkaido University Graduate School of Agriculture, Japan.

Kimura, S. D. and Hatano, R. (2007). An eco-balance approach to the evaluation of historical changes in nitrogen loads at a regional scale. *Agricultural Systems*, 94, 165–176.

Knisel, W. G. (1980). *CREAMS. A field Scale Model for Chemical/Runoff and Erosion from Agricultural Management Systems*. U.S. Department. Agriculture Conservation Research Report No. 26, Washington, DC.

Krenkel, P. A. and Novotny, V. (1980). *Water Quality Management*. Academic Press, London.

Laporte, M., Justice, C., and Kendall, J. (1995). Mapping the dense humid forest of Cameroon and Zaire using AVHRR satellite data. *International Journal of Remote Sensing*, 16, 1127–1145.

Lewis, W. M., Jr., Saunders, J. F., Crumpacker, D. W., Sr., and Brendecke, C. (1984). *Eutrophication and Land Use: Lake Dillon, Colorado*. Springer-Verlag, NY.

Loveland, T. R., Reed, B. C., Brown, J. F., Ohlen, D. O., Zhu, Z., Yang, L., and Merchant, J. W. (2000). Development of a global land cover characteristics database and IGBP discover from 1 km AVHRR data. *International Journal of Remote Sensing*, 21, 1303–1330.

Maselli, F., Gilabert, M. A., and Conesec, C. (1998). Integration of high and low resolution NDVI data for monitoring vegetation in Mediterranean environments. *Remote Sensing of Environment*, 63, 208–218.

Matsushita, B. and Tamura, M. (2002). Integrating remotely sensed data with an ecosystem model to estimate net primary productivity in East Asia. *Remote Sensing of Environment*, 81, 58–66.

Mishima, S. (2001). Recent trend of nitrogen flow associated with agricultural production in Japan. *Soil Science* and *Plant Nutrition*, 47(1), 157–166.

Neill, M. (1989). Nitrate concentrations in river water in the South–East of Ireland and their relationship with agricultural practice. *Water Research*, 23(11), 1339–1355.

Nelder, J. A. and Mead, R. (1965). A Simplex method for function minimization. *The Computer Journal*, 7, 308–313.

Organisation for Economic Co-operation and Development (OECD). (1982). *Eutrophication of Waters: Monitoring, Assessment and Control*. OECD, Paris, France.

Oki, K. and Yasuoka, Y. (1997). Estimation of annual total nitrogen load in Lake Basin with remote sensing—Case study at Lake Kasumigaura. *Journal of the Remote Sensing Society of Japan*, 17(1), 22–35 (in Japanese).

Oki, K. and Yasuoka, Y. (2008). Mapping the potential annual total nitrogen load in the river basins of Japan with remotely sensed imagery. *Remote Sensing of Environment*, 112(6), 3091–3098.

Oslo and Paris Commission. (1993). *Nutrients in the Convention Area—1993*. The Chameleon Press Limited, London.

Omasa, K., Oki, K., and Suhama, T. (2006). Remote sensing from satellite and aircraft. In *Handbook of Agricultural Engineering, Information Technology*, VI, A. Munack (ed.), pp. 231–244.

Ogden, F. L. and Julien, P. Y. (2002). CASC2D: A two-dimensional, physically based, Hortonian hydrologic model. In *Mathematical Models of Small Watershed Hydrology and Applications*, V. P. Singh and D. K. Frevert (eds.). Water Resources Publications, Highlands Ranch, CO, pp. 69–112.

Owens, N. J. P. (1993). Nitrate cycling in marine waters. In *Nitrate*, T. P. Burt, A. L. Heathwaite, and A. L. Trudgill (eds.). Wiley-Interscience, Chichester, West Sussex, UK, pp. 169–209.

Petach, M., Wagenet, R., and DeGloria, S. (1991). Regional water flow and pesticide leaching using simulations with spatially distributed data. *Geoderma*, 48, 245–269.

Pelletier, R. (1985). Evaluating nonpoint pollution using remotely sensed data in soil erosion models. *Journal of Soil* and *Water Conservation*, 40(4), 332–335.

Press, W. H., Flannery, B. P., Teukoisky, S. A., and Vetterling, W. T. (1988). *Numerical Recipes in C: The Art of Scientific Computing*. Cambridge University Press, Cambridge, UK.

Pullar, D. and Springer, D. (2000). Towards integrating GIS and catchment models. *Environmental Modeling and Software*, 15(5), 451–459.

Rundquist, D. (1991). Statewide groundwater-vulnerability assessment in Nebraska using the DRASTIC/GIS model. *Geocarto International*, 2, 51–58.

Refsgaard, J. C. and Storm, B. (1995). MIKE SHE. In *Computer Models of Watershed Hydrology*, V. P. Singh (ed.). Water Resources Publications, Highlands Ranch, CO, pp. 809–846.

Smith, M. (1993). A GIS-based distributed parameter hydrologic model for urban areas. *Hydrological Processes*, 7, 45–61.

Smith, M. and Brilly, M. (1992). Automated grid element ordering for GIS-based overland flow modeling. *Photogrammetric Engineering & Remote Sensing*, 58(5), 579–585.

Smith, S. V., Swaney, D. P., Buddemeier, R. W., Scarsbrook, M. R., Weatherhead, M. A., Humborg, C., Eriksson, H., and Hannerz, F. (2005). River nutrient loads and catchment size. *Biogeochemistry*, 75, 83–107.

Sivertun, A. and Prange, L. (2003). Non-point source critical area analysis in the Gisselo Watershed using GIS. *Environmental Modelling & Software*, 18, 887–898.

Singh, V. P., ed. (1995). *Computer Models of Watershed Hydrology*. Water Resources Publications, Highlands Ranch, CO.

Singh, V. P. and Frevert, D. K. eds. (2002a). *Mathematical Models of Large Watershed Hydrology*. Water Resources Publications, Highlands Ranch, CO.

Singh, V. P. and Frevert, D. K. eds. (2002b). *Mathematical Models of Small Watershed Hydrology and Applications*. Water Resources Publications, Highlands Ranch, CO.

Seitzinger, S. P. and Kroeze, C. (1998). Global distribution of nitrous oxide production and N inputs in freshwater and coastal marine ecosystems. *Global Biogeochemical Cycles*, 12, 93–113.

Stanners, D. and Bourdeau, P. (1995). *Europe's Environment*. The Dobris Assessment. European Environment Agency (EEA). Copenhagen, Denmark.

Stone, T. A., Schlesinger, P., Houghton, R. A., and Woodwell, G. M. (1994). A map of the vegetation of South America based on satellite imagery. *Photogrammetric Engineering and Remote Sensing*, 60, 541–551.

Tadamus, C. and Bergman, M. (1995). Estimating nonpoint source pollution loads with a GIS screening model. *Journal of the American Water Resources Association*, T31 (4), 647–655.

Tim, U. and Jolly, R. (1994). Evaluating agricultural nonpoint source pollution using integrated geographic information systems and hydrologic/water quality model. *Journal of Environmental Quality*, 23, 25–35.

Townshend, J. R. G., Justice, C. O., and Kalb, V. T. (1987). Characterization and classification of South American land cover types using satellite data. *International Journal of Remote Sensing*, 8, 1189–1207.

Tucker, C. J., Townshend, J. R. G., and Goff, T. E. (1985). African land-cover classification using satellite data. *Science*, 227, 369–375.

Turner, R. E. and Rabalais, N. N. (2003). Linking landscape and water quality in the Mississippi river basin for 200 years. *Bioscience*, 53, 563–572.

United Nations (UN) (1998). *Sources and Nature of Water Quality Problems in Asia and the Pacific*, Environment and Development Division (EDD), ST/ESCAP/1875, New York, USA.

Van Dijk, G. M., Marteijn, E. C., and Schulte-Wülwer-Leidig A. (1995). Ecological rehabilitation of the river Rhine: Plans, progress and perspectives. *Regulatory Review*, 11, 377–388.

Verhoef, W., Meneti, M., and Azzali, S. (1996). A colour composite of NOAA-AVHRR NDVI based on time series analysis (1981–1992). *International Journal of Remote Sensing*, 17, 231–235.

Warwick, J. and Haness, S. (1994). Efficacy of ARC/INFO GIS application to hydrologic modeling. *Journal of Water Resources Planning and Management*, 120(3), 366–381.

Woolhiser, D. A., Smith, R. E., and Goodrich, D. C. (1990). *KINEROS, A Kinematic Runoff and Erosion Model: Documentation and User Manual*. ARS-77. Fort Collins, CO: USDA Agricultural Research Service.

Yamaguchi, T., Yoshikawa, K., and Koshiishi, H. (1980). A hydrologic approach to characteristics of water quality and pollutant load of the river. *Journal of the Japan Society of Civil Engineers*, 49–63 (in Japanese).

Yoshimura, C., Omura, T., Furumai, H., and Tockner, K. (2005). Present status of rivers and streams in Japan. *River Research Application*, 21, 93–112.

Young, C. A., Onstad, C. A., Bosch, D. D., and Anderson, W. P. (1989). 'AGNPS': A non point source pollution model for evaluating agricultural watersheds. *Journal of Soil and Water Conservation*, 44(2), 168–173.

Zheng, C. (1990). *A Modular Three-dimensional Transport Model for Simulation of Advection, Dispersion, and Chemical Reactions of Contaminants in Groundwater Systems*. Report to the Kerr Environmental Research Laboratory, US Environmental Protection Agency, Washington, DC.

4 Identifying the Impact of Oil Spills with the Remote Sensing–Based Oil Spread Index Method

Fahad A. M. Alawadi

CONTENTS

4.1 INTRODUCTION

Oil spills resulting from accidents or hostilities are probably the most detrimental man-made disaster impacting the marine ecosystem. The direct and indirect adverse effects caused by large oil spills on the natural flora and fauna may take several years to achieve satisfactory environmental mitigation through natural processes (Hawkins and Southward 1992; Shigenaka et al. 1997; Hayes et al. 2005). The 1991 Gulf War was a salient example of a massive oil spill incident during which the release from the Kuwaiti oil wells into the Gulf was about 4 to 6 million gallons of crude oil and 732 oil wells were set on fire (Abdali and Al-Yakoob 1994). Recently, the Gulf of Mexico experienced the world's worst accidental oil spill from the Deepwater Horizon oil well on April 20, 2010, where scientists estimated that 205 million gallons (4.9 million barrels) of oil had spewed from the leaking well 5000 feet below sea surface. This event affected an area equivalent to the size of Oklahoma and fouled over 960 km (600 miles) of beaches and wetlands spreading across five states (Norse and Amos 2010). A wealth of supporting data provided by satellite imagery during these two major accidents confirmed that remote sensing for oil spill disaster management is considered an indispensable tool to combat such an impact. The role of remote sensing in relation to oil spills starts with the early warnings issued upon their occurrence and then continues with the monitoring of their evolution in space and time, providing quantitative mapping information such as thickness estimates during the event.

Oil spills vary greatly in appearance depending on the remote sensing imaging system used for detection and on the conditions surrounding their viewing geometry. When monitoring, oil spills can appear with different contrasts depending on the sensor used and/or the viewing geometry. Passive optical sensors are not as commonly used for oil spill detection as their counterpart of microwave synthetic aperture radar (SAR) due to their lack of coverage in cloudy and night conditions. Under the active SAR sensors, oil spills appear as dark surface formations relative to their surrounding waters. This occurs because they dampen the capillary and short water gravity waves that cause the incident microwaves to reflect off their mirrorlike surface away from the sensor. Researchers therefore have focused their attention in extracting dark patterns from SAR data as proxies to oil spills.

Passive satellite optical sensors have also been used for monitoring oil spills, though not as extensively as the active SAR images due to limited visibility in the nighttime and possible cloudy conditions. Under such sensors, oil spills can appear dark (negative contrast) or bright (positive contrast) relative to the water, depending on the properties of the bidirectional reflectance distribution function (BRDF) that governs the sun-target-sensor configuration. In the event of turbulent surface currents, the rate of shear diffusion may be greatly enhanced and could lead to the breakup of oil slicks, thus lowering the detection limit of most available sensors.

Several papers reviewed the performance of optical sensors such as the Sea-viewing Wide Field-of-view Sensor (SeaWiFS)* and the Moderate Resolution Imaging Spectroradiometer (MODIS)[†] for oil spill detection applications. A good example would be when Friedman et al. (2002) compared the RADARSAT-1 SAR images with the corresponding SeaWiFS and concluded that multiple data sets can be used to discriminate between algal blooms and man-made slicks. The drawback of SeaWiFS, however, is its coarse spatial resolution of ~1 km (Brekke and Solberg 2005).

Hu et al. (2003) investigated the usage of MODIS in delineating historical oil spill incidents in and around Lake Maracaibo in Venezuela. Shi et al. (2007) used a fuzzy

---

* SeaWiFS is on board the SeaStar satellite launched in 1997.
[†] MODIS is on board the Terra satellite (launched in 2000) and the Aqua satellite (launched in 2002).

C-means (FCM) cluster algorithm to delineate an oil spill in MODIS images. Shaban et al. (2007) studied the spills in MODIS data that resulted from the bombing of oil tanks at Jiyeh power station in Lebanon. Alawadi et al. (2008) developed a spectral method to discriminate oil spills using MODIS's highest 250 m/pixel resolution red and near-infrared (NIR) band group, and Alawadi (2010) further developed the surface algae bloom index (SABI) algorithm, which discriminates surface floating algae from oil spills. Hu et al. (2009) used MODIS imagery to estimate the surface area of natural oil slicks in the Gulf of Mexico. Still, new interactive and automatic methods for extracting oil spills have been continually developing based on the spatial and texture features of oil spills acquired by optical sensors. This chapter presents a new method: the oil spread index (OSI), developed to interactively identify and discriminate oil spills from look-alikes based on their shape and texture via using both optical sensor images and SAR data.

4.2 PHYSICAL PROPERTIES OF OIL

Due to their chemical composition, oils are usually characterized with a set of physical properties that distinguishes each type from another (Payne 1994). These properties will also determine how they will behave when crude oils are spilled on the sea. The specific gravity (SG) or relative density of an oil type is its density in relation to pure (fresh) water. It determines whether the oil will float. Most oils have an SG below 1 and are lighter than seawater, which has an SG of about 1.025. The American Petroleum Institute (API) gravity scale is commonly used to describe the SG of crude oils and petroleum products, and is calculated as follows:

$$API = \frac{141.5}{s.g.60/60F} - 131.5 \tag{4.1}$$

where $s.g.60/60F$ is the SG of the oil at 60°F (15.6°C) relative to the water at the same temperature (ITOPF 1986). To put into perspective, high API (low SG) tends to be of low viscosity and contains a high proportion of volatile components. Viscosity* of an oil type is its resistance to flow or the ratio of stress to shear per unit time, which is a temperature-dependent property. Petroleum fluids are generally complex fluids, which normally require using simulation models to predict their dynamics. In general terms though, oils become more viscous (i.e., flow less readily) as their temperature falls, with some proving to be more viscous than others depending on their composition. Oils of lighter viscosity, such as lighter refined "fuel oils," may spread at a higher rate than heavy "crude oils" because viscosity is temperature dependent. In high-temperature waters, oil viscosity decreases, making it more susceptible to spread than cold waters (Tsukihara 1995). There are several viscosity–temperature equations, some of which are purely empirical, whereas others are derived from

---

* It is usually measured in millipascal-seconds (mPa s) or in centipoises, with $1cP = 1mPa\ s = 10^{-3}$ $kgm^{-1}s^{-1}$ being the viscosity of water at 20°C. Centistoke (cSt) is a unit of measurement used in defining the kinematic viscosity of a fluid. $cSt = 10^{-2}$ St. Water at 20°C has a kinematic viscosity of about 1 cSt.

TABLE 4.1
Different Characteristics for Different Crude Oil Categories

Very Light Oils (Jet Fuels, Gasoline)	Light Oils (Diesel, No. 2 Fuel Oil, Light Crude Oils)	Medium Oils (Most Crude Oils)	Heavy Oils (Heavy Crude Oils, No. 6 Fuel Oil, Bunker C)
Should all evaporate within 1–2 days	Moderately volatile; will leave residue (up to one-third of spill amount) after a few days	About one-third will evaporate within 24 hours	Little or no evaporation or dissolution
High concentrations of toxic (soluble) compounds	Moderate concentrations of toxic (soluble) compounds, especially distilled products	Maximum water-soluble fraction 10–100 ppm	Water-soluble fraction is less than 10 ppm
Result: Localized, severe impacts to water column and intertidal areas	Has potential for subtidal impacts (dissolution, mixing, absorption onto suspended sediments)	Oil contamination of intertidal areas can be severe and long-term	Heavy contamination of intertidal areas likely
No dispersion necessary	No dispersion necessary	Oil impacts to waterfowl and fur-bearing mammals can be severe	Severe impacts to waterfowl and fur-bearing mammals (coating and ingestion)
No cleanup necessary	Cleanup can be very effective	Chemical dispersion is an option within 1–2 days	Long-term contamination of sediments possible
		Cleanup most effective if conducted quickly	Weathers very slowly
			Chemical dispersion seldom effective

Source: Michel, J., *Report HMRAD 92-4*, Hazardous Materials Response and Assessment Division, NOAA, Seattle, WA, 1992.

theoretical models. The pour point of oil is the lowest temperature at which the oil will flow. It is a function of the wax and asphaltene content of the oil.

4.3 CLASSIFICATION OF CRUDE OIL

Crude oils are often divided into very light, light, medium, and heavy categories. On the basis of API gravity measurements (Killops and Killops 1993), each category has its own distinct properties (Table 4.1).

4.4 FATE OF OIL OVER THE WATER SURFACE

A number of mathematical models were suggested to explain the spreading and transport mechanism of oil slicks in waters. A good example would be the spreading of oil in the open ocean that was investigated by Fannelop and Waldman (1972), where they took into consideration the effects of wind, waves, and currents. Fay (1969), Fay and Hoult (1971), and Buckmaster (1973) proposed a three-stage model to explain the various forces that influence oil dispersion and movement in calm seawater conditions. According to these theoretical models, the spill's gravitational forces caused by the sizable thickness of the slick at its early phase may control the spreading of the oil across the surface, during which inertia acts as the main retarding force. In the second phase of this evolution, the spill's inertial force would become negligible due to its weight. At the same time though, it may be dominated by a decelerating force attributed to the viscous drag along the slick bottom. In its third and final phase, when the slick has become sufficiently thin, the interfacial (surface tension) forces may dominate to further its spread. Spreading will continue as long as the surface and interfacial tensions are unchanged when the oils experience changing physical, chemical, and biological properties. This is collectively called weathering. This is a complex process where oil may typically undergo eight different progressions that may occur simultaneously and in different degrees. Some of these processes include spreading and advection, evaporation, natural dispersion, emulsification, photo-oxidation, sedimentation, shoreline stranding, and biodegradation (Ducruex et al. 1986).

4.4.1 Environmental Parameters That Impact Oil at Sea

4.4.1.1 Temperature

The apparent temperature difference between the oil spill and its surrounding water background is a complex function of incident solar radiation, thermal properties of the oil, oil thickness, and water temperature (Goodman 1989; Lunel et al. 1997). Thick oil may generally be warmer during the day due to thermal isolation from underlying water and the high absorption of solar energy since it is acting as a black body (Hurford 1989; Samberg 2005). At night, thick oil may become cooler than water due to lower emissivity (Tseng and Chiu 1994). Thin layers of oil or sheen may not be thermally detected (Svejkovsky and Muskat 2006). Despite this factor, Cross (1992) studied thermal responses in the Advanced Very High Resolution Radiometer

(AVHRR) datasets over Kuwaiti coastal oil slicks recorded in 1991 and found that the slick was warmer than the sea at night and cooler by day in contrast to the observations made by others. Such temperature variations were attributed to a variety of factors including the properties of the oil itself, degree of emulsification, the air and sea temperatures, intensity of incident light, and the thickness of the oil (Goodman 1989; Lunel et al. 1997).

4.4.1.2 Wind and Sea Currents

Ambient conditions like wind speed, sea state, and sea currents work collectively to transport oil at the water surface. Generally, an oil slick moves at 3.5% of the wind speed (Lewis and Aurand 1997) (Figure 4.1).

Less viscous oils may spread at higher rates than heavy oils. The mixing energy of sea waves plays a substantial role in the emulsification of the oil (i.e., the intake of seawater by crude oil), making it more viscous than the original oil. This is often referred to as "chocolate mousse" (Król et al. 2006). High frequency conditions, short-term wind patterns, and turbulent surface currents may greatly enhance the rate of spread and ultimately the breakup of the slick in a process called "shear diffusion" (Elliott et al. 1986; Spaulding 1988).

Sea currents also play a significant role in the movement of oil at the water surface. Windrows or roll vortices are organized in counterrotating secondary circulations (Langmuir circulations) embedded in the mean flow (Langmuir 1938). Such circulations result from instabilities in the convective marine boundary layer. This is caused by the movement of the wind that will act on the spill to break it (Figure 4.2) into streaks along the direction of the wind (Langmuir 1938; Brown 1980; Thurman 1983; Pavlakis et al. 2001).

The oil spill trajectory model (OSTM) is often used as a decision support tool to predict the behavior of various oils in the water column based on wind and tidal data. Data of local current movements and tidal effects near the shore and reef regions are more difficult to predict due to the complex dynamics involved. High inaccuracies are therefore expected when using OSTM in such regions. For this reason, surveillance by visual observation always remains the best means of estimating a slick's movement.

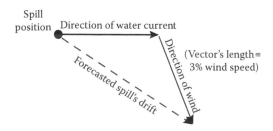

FIGURE 4.1 A slick's drift can be calculated by means of a vector diagram using 3% of the wind speed combined with 100% of the current movement. (From Alawadi, F., Detection and classification of oil spills in MODIS satellite imagery, PhD thesis, University of Southampton, England, 2011.)

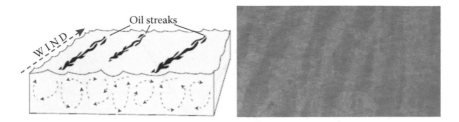

FIGURE 4.2   Langmuir cells in the mixed layer depth (left). Oil in Windrows or Langmuir cells (right). (From NOAA, *Trajectory Analysis Handbook*, Ocean Service, Hazardous Materials Response Division, and Office of Response and Restoration, NOAA, Seattle, WA, 2002. With permission.)

4.5   REMOTE SENSING OF OIL SPILLS

SAR sensors remain the most preferred spaceborne devices used for the detection of oil spills (Brekke and Solberg 2005). This is because microwave radar bands are used in active remote sensing in all weather conditions, which enables us to perform full-day imaging regardless of prevailing cloud cover. Oil has a distinctively different spectral signature in the thermal infrared region. Passive hyper- or multispectral optical sensors encompassing the visible, near- and middle-infrared wavelength regions also proved their effectiveness for the detection of oil spills under certain meteorological conditions (Malthus and Karpouzli 2007). Ultraviolet sensors can be used to map sheens of oil with the appearance of oil slicks with a broad range of fluorescent intensities and spectral signatures. Currently, there are no satellite systems suitable for detecting oil spills via the use of ultraviolet (UV) wavelength range (Malthus and Karpouzli 2007). Still, infrared, visible, and UV sensors will not be able to detect oil in inclement weather such as heavy rain.

MODIS is an example of an optical sensor on board two satellites, Terra and Aqua. MODIS, with its short repeat cycle (i.e., within 24 hours), has demonstrated the capability of detecting relatively large oil spills (~2500 pixels at its maximum 250 m/pixel spatial resolution), particularly in regions characterized with ~80% cloud-free conditions.

4.5.1   Detecting Oil Spills Using SAR

Real aperture radars (RARs) as well as SARs usually operate at incident angles (i.e., the angle between nadir* and the look direction of the antenna) that range between 20° and 70° (Valenzuela 1978). It is at such incidence angles that radar backscattering can be described by Bragg's scattering theory (Crombie 1955). In this case, radar microwaves are in resonance with ocean waves of similar scale, and can be related to the radar wavelength (Figure 4.3):

$$\sin\theta = \frac{\lambda_r}{2\lambda_w},\tag{4.2}$$

---

* Normal to the sea surface pointed down.

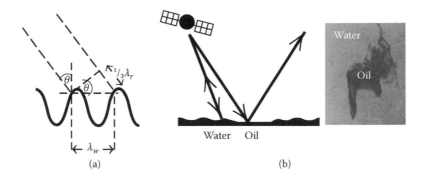

FIGURE 4.3 Bragg scattering waves in relation to (a) the sensor waves and (b) the SAR scattering mechanisms of an oil slick. (From Tahvonen, K., The use of remote sensing, drifting forecasts and GIS data in oil response and pollution monitoring, *Remote Sensing & Hydrosphere*, Helsinki, Finland, 2008; Alawadi, F., Detection and classification of oil spills in MODIS satellite imagery, PhD thesis, University of Southampton, England, 2011.)

where λ_w and λ_r are the ocean and radar wavelengths, respectively, and θ is the incidence angle (Valenzuela 1978). The capillary and short gravity waves* are called Bragg waves and thus obey the Bragg resonance condition for wind speeds that lie between a range of 2 to 3 m/s and 12 to 15 m/s (Hühnerfuss et al. 1978), and the SAR wavelengths should relatively match the size of surface waves. As an example, a range of SAR incident angles between 20° to 26° will have Bragg waves at the following common SAR frequencies and possible wavelengths: X-band (3 cm), 3.9 ± 0.5 cm; for C-band (5.5 cm), 7 ± 1 cm; and L-band (23 cm), 30 ± 4 cm (Holt 2004).

High-density surface films such as oil may appear as dark patches in the SAR image data due to the dampening effect of water roughness, and hence the reduction of Bragg's backscattering coefficient (Guinard 1971). There is also an upper threshold of wind speed beyond which the water surface becomes so rough (diffused surface) that the oil film may be fragmented by the breaking waves, depending on the oil type and age, and therefore undetectable by radar (Demin et al. 1985; Brekke and Solberg 2005). Figure 4.3 shows the concept of Equation 4.2 and the scattering of radar waves off an oil film and an oil-free water surface.

According to Solberg (2005), the general framework for allocating dark patches (oil spills and their look-alikes) in SAR images can be summarized by the following three steps (Figure 4.4): (1) isolation of dark spots in the image through an appropriate adaptive threshold and segmentation process, (2) the extraction and analysis of the main dark features, and (3) the classification of extracted features as either oil slicks or look-alikes, based on prior probability confidence values. Table 4.2 summarizes the most common methods used on SAR data for the extraction of dark features as oil spills.

The polarimetric properties for both transmission and reception of the SAR signal off an oil spill have also been investigated from the results acquired during the

---

* A wave occurring at the interface between two fluids, such as the interface between air and water on oceans, is called a capillary wave if the principal restoring force is controlled by surface tension, or a gravity wave if the restoring force is gravity.

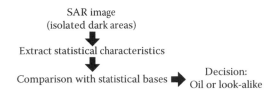

SAR image
(isolated dark areas)

Extract statistical characteristics

Comparison with statistical bases ➡ Decision:
Oil or look-alike

FIGURE 4.4 The three-step procedure for extracting dark features from SAR images. (From Tahvonen, K., The use of remote sensing, drifting forecasts and GIS data in oil response and pollution monitoring, *Remote Sensing & Hydrosphere*, Helsinki, Finland, 2008; Alawadi, F., Detection and classification of oil spills in MODIS satellite imagery, PhD thesis, University of Southampton, England, 2011.)

two spaceborne imaging radar-C/X band SAR (SIR-C/X-SAR) missions on board the space shuttle Endeavour in 1994. The results revealed that multifrequency and multipolarization SAR imagery is advantageous for the detection of oceanic surface films including oil (Melshelmer et al. 1998).

4.5.2 DETECTING OIL SPILLS USING MODIS

Hu et al. (2003) investigated the usage of MODIS in delineating historically verified oil spill incidents in and around Lake Maracaibo in Venezuela. The negative-contrasted spills (i.e., darker relative to surrounding waters) were visible even when significant sun glint contamination was present. The likelihood of the oil spills being phytoplankton blooms was dismissed first by using prior knowledge of the environment and knowing that the patterns did not fall within the seasonal bloom period. Second, it was determined that the spills were not phytoplankton blooms because their spectral characteristics did not match that of algal blooms. Hu et al. also pointed out that the degree of turbidity underneath and around the oil spills played a role in deriving accurate results in such empirical approaches. They concluded their findings by stating that there was no good reason to continue minimizing the potential of optical satellite sensors at detecting oil spill events and highlighted the need to develop good empirical algorithms to derive oil film thickness supported by *in situ* measurements.

Shi et al. (2007) used a fuzzy C-means (FCM) cluster algorithm with a texture feature analysis to delineate an oil spill using MODIS images. They estimated the degree of disorganization in the pixels' direction by applying a mathematical process based on the gray level co-occurrence matrix (GLCM) in pixels. In doing so, they managed to delineate the oil spill area as well as the shoreline regions because oil, like the shorelines, exhibits the most rapid changes in pixel values in relation to the moderately uniform water background. They concluded that their method was effective when it was implemented locally around the spill area only rather than on the image as a whole since other parts of the image, such as high sediment loads, may have similar reflectances as the oil spill. Their research also found no distinct temperature variations between the spill and its surrounding water. Shaban et al. (2007) studied the spills in MODIS data that resulted from the bombing of oil tanks at the Jiyeh power station in Lebanon and showed their delineation in the images. Hu et al. (2009) used MODIS imagery to estimate the surface area of natural oil slicks

TABLE 4.2
Most Common Methods Used in Oil Spill Feature Extraction Processing

#	Feature
1	Slick area (A)
2	Slick perimeter (P)
3	P/A
4	Slick complexity
5	Spreading (low for long thin slicks, high for circular shape)
6	Slick width
7	First invariant planar moment
8	Dispersion of slick pixels from longitudinal axis
9	Object/dark area standard deviation
10	Background/outside dark area standard deviation
11	Max contrast (between object and background)
12	Mean contrast (between object and background)
13	Max border gradient
14	Mean border gradient
15	Gradient standard deviation
16	Local area contrast ratio
17	Power-to-mean ratio of the slick
18	Homogeneity of surroundings
19	Average NRCS inside dark area
20	Average NRCS in limited area outside dark area
21	Gradient of the NRCS across the dark area perimeter
22	Ratio #9 to #10
23	Ratio #19 to #9
24	Ratio #20 to #10
25	Ratio #23 to #24
26	Ratio #19 to #20
27	Distance to a point source
28	Number of detected spots in the scene
29	Number of neighboring spots

Source: Brekke, C. and Solberg, A. H., *Remote Sensing of Environment*, 95, 1, 1–13, 2005.

in the northwest Gulf of Mexico. In some of the images, the spills have shown both negative and positive slick contrasts. They attributed the negative contrast to a reduction in glint due to dampening of the surface roughness, while the positive contrast was observed due to higher specular reflection within the oil film patch. It is clear from the aforementioned papers that MODIS data was only utilized to describe and delineate already-documented and known oil spills and that none of them have demonstrated or argued whether it is possible to identify unknown patterns as positive oil spills using the MODIS data.

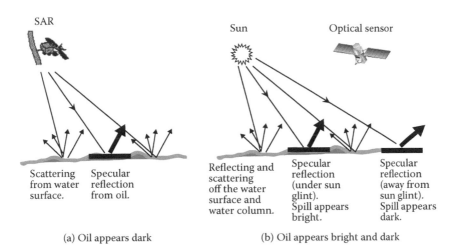

Scattering   Specular
from water   reflection
surface.   from oil.

Reflecting and   Specular
scattering   reflection
off the water   (under sun
surface and   glint).
water column.   Spill appears
                 bright.

Specular
reflection
(away from
sun glint).
Spill appears
dark.

(a) Oil appears dark   (b) Oil appears bright and dark

FIGURE 4.5 An observation comparison between SAR and optical sensors of an oil spill in moderate sea state roughness. (From ESA/EDUSPACE, *Galathea 3 EMU Radar Look at Ocean Features and Oil Spills—Background*, http://galathea3.emu.dk/satelliteeye/casestudies/radar1/back_uk.html, 2010. With permission.)

4.5.3 COMPARISON OF IMAGERY BETWEEN MODIS AND SAR

Under SAR data, oil spills will always appear with negative contrast (i.e., dark with respect to water) independently from the sun position in relation to the sensor. In optical data however, oil spills can appear with either contrast due to the geometry of the sensor-target-sun configuration and the spectral band used (Figure 4.5). The key factors that appear to determine the degree of this contrast in, for example, MODIS are the optical properties of the oil, the film's thickness, the coefficients of light absorption and scattering in sea water, the overhead light conditions, sea surface state and depth, the type of sea bottom, the angle of illumination, and the angle of observation (Otremba 1999; Otremba and Król 2001).

Simulation studies (Figure 4.6) by Otremba and Piskozub (2001) showed that oil spills appeared dark in optical sensors when the oil film dampened the short water waves, and hence less scattering occurred, resulting in a smoother surface. Still, when an observer sees the oil slick close to the sun's specular reflection (sun glint), the contrast becomes positive (brighter) (Otremba and Piskozub 2001; Chust and Sagarminaga 2006). Furthermore, SAR has only one "side-looking" viewing geometry, while optical sensors can be either "down-looking" or "side-looking." The side view and bottom view make optical sensors more flexible.

4.5.4 OIL SPILL LOOK-ALIKES

Oil look-alikes are oceanographic and atmospheric phenomena that can be mistaken for oil slicks and may lead to false-positive sightings of oil spills (Solberg 2005). These issues are common to all satellite sensors. Look-alikes can include natural surface films (biogenic), low wind speed areas (wind speed less than 3 m/s),

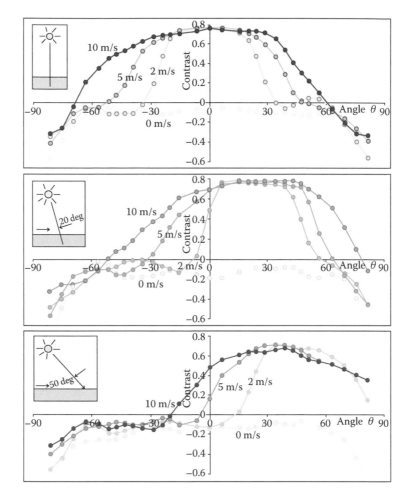

FIGURE 4.6 Contrast of an oil film as a function of angle of observation at various light incidence angles and at various wind speeds. (From Otremba, Z. and Piskozub, J., *Optics Express*, 9, 8, 411–416, 2001. With permission.)

wind front areas, wind sheltering by land, internal waves, rain cells, grease ice, cold upwelling zones, kelp beds,* silt or suspended sediment, jellyfish, and algal mat formations (Pavia and Payton 1983; Elliott 1986; Otremba and Piskozub 2001; Alpers and Hühnerfuss 1989; Espedal 1999; Hovland et al. 1994; Chust and Sagarminaga 2006).

MODIS has an advantage over SAR sensors in that if weather conditions permit, it is able to retrieve the spectral signatures for some of the observed biological and physical targets that appear at the water surface. Some of these life forms or objects that resemble oil may include macro algae and cloud shadows. A good example would be particular cloud shapes and cloud shadows that can sometimes be

---

* Kelp are large seaweeds (algae), belonging to the brown algae.

misclassified as oil spills, which can be masked off using their spectral absorption properties in the thermal infrared band. Prior knowledge of local conditions such as winds, ocean currents, and vessel traffic (Karathanassi et al. 2006) can also play a role in improving the accuracy of detection and the reduction of false-positives.

4.6 THE DEVELOPMENT AND APPLICATION OF THE OSI

The OSI is a new empirical technique proposed to distinguish oil spills in terms of their shape and texture based on how they appear within SAR images and optical-sensor data. Its aim is to provide environmental scientists with a rapid and robust method to discriminate between oil spills and their look-alikes with the least available environmental information, such as wind and current observations or forecasts. Although wind and tidal current histories have recently been added to images in order to determine slick age (Espedal 1999), their near-real-time (NRT) estimate remains unreliable. It can therefore be deduced that the proposed OSI method seems more appropriate for NRT estimates.

Generally, less viscous oil may spread at a higher rate than a heavy crude oil if both spills were experiencing the same environmental conditions. Because viscosity is a temperature-dependent property, the viscosity of oil may decrease in higher temperature waters, prompting its lateral spread to be higher than it would be in colder waters. Only certain criteria of oil spills can thus be examined by the OSI method:

1. The spill must be relatively large (no less than ~50 × 50 pixels)
2. The spreading of the spill should be solely influenced by its inertial and interfacial forces
3. The oil is assumed to have maintained its original viscosity and therefore undergone negligible weathering conditions

The first criterion was necessary to compensate for the coarse spatial resolution with which MODIS is characterized (i.e., maximum 250 m/pixel at nadir, and cell area of 62,500 m²). Still, the maximum spatial resolution of an ENVISAT ASAR image is 30 m/pixel (i.e., cell area of 900 m²), which might be a complementary tool for the detection of oil spills.

A spill area made up of 10 pixels in ASAR may therefore be represented by only one pixel in MODIS, and thus cannot be detected. In order to have a recognizable homogeneous feature in an image, however, its size has to be greater than the resolution cell of the sensor; otherwise, it becomes indistinguishable. Similarly, if the spill is large enough to be detected yet its gray level pixel values lack any contrast with its neighboring pixels, it will not be distinguishable even if it was spatially large. Also, the details of larger oil spills may appear scattered or inhomogeneous when they are viewed in ASAR, but may appear integrated and averaged when the same scene is viewed in MODIS.

Oil spills that violate criterion (2) above may include those that originate from a moving object that creates a distinctive linear-shaped spill (Figure 4.7). If, however, the source of the spill was stationary, then the OSI analysis can be smoothly carried out at a distance from the point of release. This is aimed to eliminate the additional

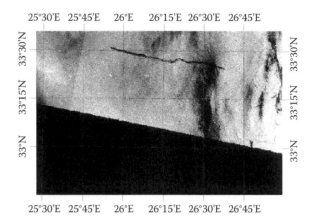

FIGURE 4.7 A threadlike oil slick near Crete detected by ENVISAT ASAR on September 4, 2004. (From Adamo, M., De Carolis, G., De Pasquale, V., and Pasquariello, G., Exploiting sunglint signatures from MERIS and MODIS imagery in combination to SAR data to detect oil slicks, *Proc. 'Envisat Symposium 2007'*, Montreux, Switzerland, 2007. With permission.)

forces that may influence its lateral spread other than its inertial and surface tension forces. Also, to satisfy this condition, the OSI method must confine its investigation to offshore spills in order to avoid the additional forces involved in the spread of near-shore oil spills. These added forces may include river discharges along the coastlines, baroclinic currents, tidal currents, and stokes drift (Carracedo et al. 2006).

Violation of criterion (3) can include old spills that have been dispersed and broken down into "feathered" streaks (Figure 4.8) caused by winds pushing the heavy

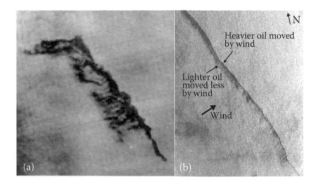

FIGURE 4.8 (a) A 250 m/pixel spatial resolution MODIS Aqua image in false RGB color composite bands ($\lambda_2 = 859$, $\lambda_1 = 645$, $\lambda_1 = 645$ nm) on April 28, 2010, at 09:27 UTC showing an aging oil spill (~119 km2) in the Arabian sea. (Generated from MODIS by the author through ROPME. With permission.) (b) A section of an ERS-2 (C-band, VV) SAR image from the Indian Ocean acquired April 6, 1999, at 04:58 UTC showing a "feathered" structure of an oil trail. The wind direction arrow in the image was obtained from the modeled surface wind field provided by the European Center for Medium-range Weather Forecasts (ECMWF). (© ESA.)

components of the mineral oil film accumulated at the downwind side (e.g., dark line in the image), while the feathered side is always located upwind.

The false red, green, and blue (RGB) color composite corresponding to bands ($\lambda_2 = 859$, $\lambda_1 = 645$, $\lambda_1 = 645$ nm) of the 250 m/pixel band group in MODIS may be used for field applications (Figure 4.8). First, this color band composite is useful because it uses the only two available bands in the highest spatial resolution group of MODIS (250 m/pixel), which are red ($\lambda_1 = 645$ nm) and NIR ($\lambda_2 = 859$). Second, this particular band combination, which was assigned the red color to the NIR reflectance, can immediately highlight the red-edge effect.* Such capability makes it very efficient and quick to discriminate between surface algae (i.e., appearing reddish in color) and oil.

Freshness of oil spills is defined by assuming that the "age" of the spill (i.e., the elapsed time between its release and detection) does not go beyond 12 to 24 hours in order to minimize the duration of exposure to different weathering processes. Suspected patterns positively classified as oil spills by the OSI model can have their viscosity group estimated by the same model. Needless to say, an oil spill treated with a dispersant renders the algorithm inapplicable because it dissipates the oil by altering its natural viscous properties.

4.6.1 OSI THEORETICAL BACKGROUND

The OSI method can be summarized in the following three steps by (1) delineating the suspected spill patterns using an edge enhancement filter on a gray scale image produced from a single band of the 250 m/pixel band group, (2) conducting an analysis of the spill's edge features and texture distribution relative to the surrounding water texture, and (3) estimating its relative viscosity type using the OSI algorithm once it can be classified as oil. An additional contrast stretching may be needed in order to highlight the granularity level of the inspected patch. The delineation process is currently performed interactively by applying a 3×3 convolution mask in the x and y directions, respectively. This operator was chosen over other edge detection operators because it does not deteriorate the fine detailed structures on the spill's surface area and its surrounding sea area. This is deemed necessary to allow for the evaluation of the spill's body texture in terms of its homogeneity to that of water. Such a process is illustrated in Figure 4.9. The edge gradient of the pattern's border can be assessed in terms of the uniformity of its border's width (i.e., mean gradient with the least standard deviation value) (Figure 4.10), its continuity, and the granularity of the patch's surface structure relative to its surroundings. It can therefore be deduced that relatively fresh oil spills detected within the first ~12 to 24 hours from the time of their occurrence should have a relatively homogeneous border width free from sharp points or geometric discontinuities (apart from the apparent point source) as defined in Figure 4.11.

---

* Red-edge is the highest reflectance slope between the maximum absorption in the red and the maximum reflectance in the IR, due to the presence of chlorophyll pigments (Tucker 1979; Jackson et al. 1983).

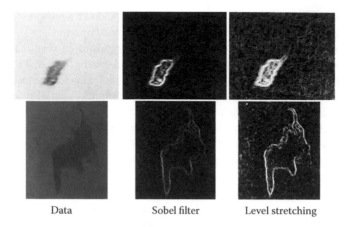

| Data | Sobel filter | Level stretching |

FIGURE 4.9 Two grayscale images of different oil spills. (Top) MODIS image from 250 m/pixel NIR band, generated from MODIS by the author through ROPME, March 26, 2009. (Bottom) RADARSAT image taken on March 22, 2007. The three sets of each image represent the original data, the image after applying the Sobel filter, and contrast stretching to highlight texture differences, respectively. (From Macdonald Dettwiler and Associates Ltd. [MDA]. With permission.)

After applying a Sobel filter, the texture of the oil spill surface area may generally appear with less noisy structure to that of the surrounding water. This is probably due to the dampening effect exerted by the viscous oil at the water surface and the suppressing of its capillary waves. Such features may be enhanced considerably in MODIS when the spill is viewed under the sun glint zone (i.e., the specular reflection of sun illumination off the water surface). According to the criteria set by the OSI model, an offshore oil spill involving a heavy crude oil at moderate wind speeds between 2 to 3 m/s and 12–15 m/s (Hühnerfuss et al. 1978) would appear with a smoothly curved, continuous, and equal-width boundary whose surface area may be characterized with less granular structure than the surrounding waters. Failing

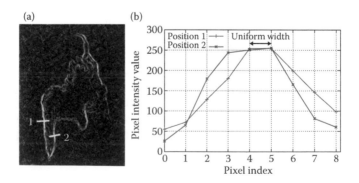

FIGURE 4.10 (a) A RADARSAT image on March 22, 2007, shown after applying a Sobel filter with two line transects across its border. (From MDA. With permission.) (b) The results of the transects are shown. The gradient border at the two transects are almost equal.

(a) (b) (c) (d)

FIGURE 4.11 Different levels of continuity to demonstrate the OSI shape compatibility criteria. (a) Discontinuity (incompatible with the OSI); (b) zero curvature level (due to the angle), but continuous (incompatible with the OSI); (c) inflection point curvature (first derivative is a constant but not zero, second derivative changes signs at the point, compatible with the OSI); and (d) curvature (first derivative is zero, second derivative can be either positive or negative, compatible with the OSI).

any of these conditions may imply several situations: (1) the original viscosity of the oil was changed due to natural weathering or artificially due to a dispersant, (2) the oil type involved in the spill was of a light viscous type (i.e., its viscosity is close to water), (3) it is of low concentration or thickness enough to be classified as sheen, and/or (4) it could be an oil spill look-alike.

4.6.1.1 Estimating Oil Viscosity Group

When an investigated pattern is classified as an oil spill according to the shape and texture prerequisites set by the OSI, it becomes possible to estimate the oil's viscosity group (η). This is done by approximating the spill shape with a series of successive circles (Figure 4.12). The circles' dimension should change as the spill's surface area progresses away from its presumed point source. If, for example, two circles are to be drawn inside the spill's inner surface area, each with radii r_1 and r_2 ($r_1 \neq r_2$, for lateral spread), respectively, then the following should be considered while drawing: the circle drawn should cover as much of the region as possible for the spill's area to the extent that it acts as a tangent for at least two points of the inner border of the spill, and the next adjacent circle with radius r_2 should be drawn in contact with the first circle with radius r_1 at a tangential point common to both of them. The sum of the circles' diameters may represent that part of the spill's length.

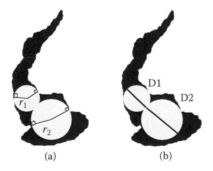

(a) (b)

FIGURE 4.12 Demonstrating the methodology of the OSI. (a) Two circles, each with radius r_1 and r_2 ($r_1 \neq r_2$), respectively, drawn over the spill's area. (b) The sum of the two circles' diameters D1 and D2, respectively, is approximately equal to the spill's length for the part that is covered by the circles.

The OSI and viscosity (η) can both then be evaluated using Equations 4.3 and 4.4, respectively:

$$OSI = \sin(\theta/2) = \frac{D_{i+1} - D_i}{D_{i+1} + D_i},\qquad(4.3)$$

where OSI is the ratio of measure of lateral spread and part of the spill length, D_i and D_{i+1} are the diameters of two consecutive circles i and $i + 1$, respectively. The sum $(D_{i+1} + D_i)$ represents the part of the spill length, which is covered by the two consecutive circles. The angle θ is a measure of the lateral spread and is defined as it is shown in Figure 4.13.

$$\eta \propto \frac{1}{OSI}.\qquad(4.4)$$

If two spills, S_1 and S_2, are caused by two different oil types, each with viscosity η_1 and η_2, respectively, and both are experiencing the same environmental conditions (Figure 4.14), then the lateral spread of S_1 becomes greater than that of S_2 ($OSI_1 > OSI_2$) and hence, in view of Equation 4.3, $\eta_1 < \eta_2$.

Linear-shaped spills (i.e., lacking lateral spread) are not compatible with the criteria proposed by the OSI criteria. This is because their size is not large enough to apply the method and also because the circles' diameters would be equal and thus yield a zero OSI value. Following the calibration process carried out on 20 preconfirmed oil spills, crude oil can be classified into three main categories according to their SG values (or viscosity): light (SG < 0.82), medium (SG 0.82–0.97), and heavy (SG > 0.97) (Concawe 1998). A classification table was derived (Table 4.3) to estimate the viscosity group of oil involved in unidentified spills based on the OSI values estimated.

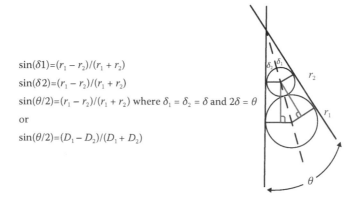

$\sin(\delta 1)=(r_1 - r_2)/(r_1 + r_2)$

$\sin(\delta 2)=(r_1 - r_2)/(r_1 + r_2)$

$\sin(\theta/2)=(r_1 - r_2)/(r_1 + r_2)$ where $\delta_1 = \delta_2 = \delta$ and $2\delta = \theta$

or

$\sin(\theta/2)=(D_1 - D_2)/(D_1 + D_2)$

FIGURE 4.13 A schematic diagram showing two consecutive circles, each with radii r_1 (diameter D_1) and r_2 (diameter D_2), respectively, representing an assumed spill area. The lateral spread of the spill is represented by $\sin(\theta/2)$.

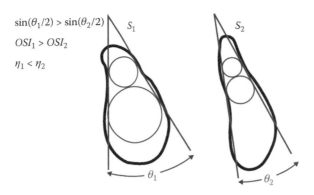

$\sin(\theta_1/2) > \sin(\theta_2/2)$

$OSI_1 > OSI_2$

$\eta_1 < \eta_2$

FIGURE 4.14 Two spills, S_1 and S_2, under the same environmental conditions caused by two different oil types, each with viscosity η_1 and η_2, respectively, showing that the lateral spread of S_1 is greater than that of S_2 ($OSI_1 > OSI_2$), indicating that $\eta_1 < \eta_2$.

TABLE 4.3

Different OSI Values and Their Corresponding Oil Viscosity Classification (η)

Oil Viscosity Group (η)	OSI
Light to medium	OSI > 0.02
Heavy	OSI < 0.02

4.6.2 OSI Application on Surface Algae

Floating cyanobacteria (*Trichodesmium erythraeum*) blooms similar to the blooms that appeared inside Qatari waters (seen in Figure 4.15 on April 24, 2009) can sometimes be misclassified as oil spills in satellite imagery. The same blooms seen in Figure 4.15 were viewed 5 days earlier in MODIS Aqua on April 19, 2009 (Figure 4.16) and ENVISAT ASAR (Figure 4.17), respectively, but at different timings. In both data sets, the border gradient of algae appears with no solid uniformity and is full of discontinuities after applying the Sobel filter. The border gradients for low wind patterns may sometimes appear with or without multiple discontinuities. Textually, however, their surface areas always appear virtually dark and with no clear patterns even after performing contrast stretching. This is because their surface water currents are almost nonexistent, and thus reflect light off their mirrorlike surface away from the sensor in a manner similar to the appearance of oil as dark patterns suppress water surface roughness.

4.6.3 OSI Application on Oil Spill

The OSI was applied to the Gulf of Mexico oil spill (~1000 km²) viewed in RADARSAT-2 on April 23, 2010 (Figure 4.18), 2 days after the release of light sweet

FIGURE 4.15 *In situ* pictures of the *Trichodesmium erythraeum* species, taken on April 24, 2009, 5 days after its occurrence in MODIS Aqua and ENVISTAR ASAR on April 19, 2009. (© Ministry of Environment Qatar.)

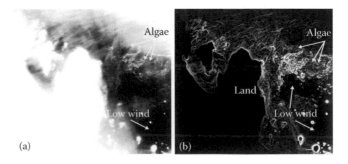

FIGURE 4.16 Floating algae as it appears in MODIS Aqua on April 19, 2009 at 10:06 UTC in (a) with a 250 m/pixel resolution false RGB color composite bands ($\lambda_2 = 859$, $\lambda_1 = 645$, $\lambda_1 = 645$ nm), and in (b) after applying the Sobel filter. (Generated from MODIS by the author through ROPME.)

FIGURE 4.17 (a) Floating algae as it appears in ENVISAT ASAR (C-band, HH polarization, 1 km/pixel resolution) data image on April 19, 2009, at 18:35 UTC, and (b) how it looks after applying the Sobel filter. (From ESA, http://earth.esa.int/EOLi/EOLi.html [accessed May 12, 2009]. With permission.)

crude oil (SG = 0.850). The mean wind speed estimated during the spill was 7 m/s (NOAA Magazine Online 2010). The computed OSI value (0.11) reflected a light-type viscous oil that matches the original oil type. It was, however, expected that the original viscosity of the oil would change since it had been flowing from 1524 m below the water surface for 3 days (since April 20, 2010) and parts of it had been sprayed with a dispersant.

The same spill (~1500 km^2) appeared in MODIS Aqua and SPOT 5, respectively, on April 25, 2010 (Figure 4.19), where the estimated wind speed and sea state, according to NOAA (2010), were 15–20 knots (7.7–10.3 m/s) and 4–6 feet (1.21–1.8 m), respectively. Figure 4.19 shows the spill in the 500 m/pixel MODIS RGB color composite corresponding to (λ_1 = 645, λ_4 = 555, λ_3 = 469 nm) and an unknown RGB color composite in SPOT 5. The oil slick in these images is likely to include areas of thicker oil surrounded by areas of relatively thin surface films. The SPOT 5 image in particular indicates a relatively thick oil layer in the center of the slick. The central stripe of red in this image is the thickest area. The darker edges are thinner oil regions—probably sheen. The MODIS image (from the sun glint zone) has a

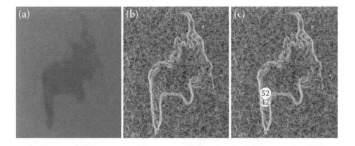

FIGURE 4.18 (a) The Gulf of Mexico oil spill (SG = 0.85) in RADARSAT-2 (scanSAR narrow, VV polarization) on April 23, 2010, (b) after applying the Sobel filter, and (c) after applying the OSI (0.11) method. (From Center for Southeastern Tropical Advanced Remote Sensing, http://www.cstars.miami.edu/cstars-projects/deepwaterhorizon/image-gallery/deep waterhorizon/radarsat2-2010-04-23-234908-utc-scansar-narrow-111 [accessed July 5, 2010]. With permission.)

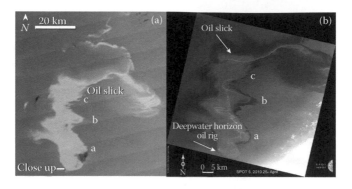

FIGURE 4.19 The Gulf of Mexico oil spill on April 25, 2010 (a) at 18:55 UTC in the 500 m/ pixel MODIS Aqua with RGB color composite bands ($\lambda_1 = 645$, $\lambda_4 = 555$, $\lambda_3 = 469$ nm) and in the corresponding. (Courtesy of National Aeronautics and Space Administration, http://lance .nasa.gov/imagery/rapid-response.) (b) SPOT 5 image (unknown color composite) at 10 m/ pixel resolution captured on the same day at 16:31 UTC. (From ASTRIUM, http://www .spot.com/web/SICORP/3198-oil-spill-in-the-gulf-of-mexico-spot-5-mobilized.php [accessed July 5, 2010]. With permission.) Points a, b, and c are shown on both images.

positive contrast for the whole area, including the fringes, although the thicker areas appear relatively brighter. The Sobel-filtered image of the spill indicates that the spill (Figure 4.20) began to depart from the criterion proposed by the OSI for a fresh oil spill. Figure 4.20 also reveals several discontinuities and lack of uniformity in the gradient border. This would classify the spill as an aging dispersed oil spill or a look-alike (i.e., low wind area or surface algae).

The low wind look-alike was ruled out, however, because its surface area is not completely pattern-free, which is a proxy for low wind feature according to the OSI. The discontinuities appearing at the borders of the spill were attributed to the extreme weathering conditions described earlier. The estimated viscosity was carried out anyway for testing purposes and yielded a value of 0.18, and thus matched that of the original viscosity group to which the oil belonged.

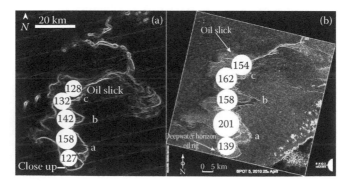

FIGURE 4.20 The Gulf of Mexico oil spill (April 25, 2010) after converting the image shown in Figure 4.19 into gray scale image, in (a) MODIS after applying the Sobel filter on them, and (b) in SPOT after applying the OSI method. Both OSI values were equal to 0.18.

4.7 CONCLUSIONS

Shape recognition methods are already in use in SAR data for classifying dark features as proxy to oil spills. The same shape methods can also be applied to data acquired from optical sensors to complement their existing spectral capabilities. The MODIS sensor provides free optical data with maximum resolution of 250 m (in nadir) from two satellites. These characteristics qualify the MODIS sensor and enable us to conduct NRT monitoring operations of relatively large oil spills, particularly over less cloudy locations.

The OSI requires a number of prerequisites without which the proposed technique becomes either invalid or inconclusive. In this chapter, the OSI method has proven to be a robust and interactive "shape-based" method developed to aid environmental scientists in discriminating between oil spills and their look-alikes with the minimal available ancillary data. Its application is restricted to spatially large and relatively fresh oil spills (i.e., those that have undergone minimal viscosity changes) originating from offshore stationary sources observed almost in NRT. Such assumptions were hypothesized to ensure that the oil's original viscosity had experienced negligible changes and to simplify the analysis of its lateral spread and advection. The Sobel filter was selected over other edge filters in its operation because it emphasizes the spill's edge gradient without compromising the details of its surface texture, including that of the surrounding water. Oil spills observed with this method will exhibit uniform continuous boundary gradient and a smoother textural granularity than the water surface due to the dampening effect of oil. Fresh spills caused by heavy to medium oil types with a viscosity greater than water should have well-defined Sobel-filtered edges. Less viscous oil types (close to water) may, however, produce inconclusive results using this method. Observed dark patterns positively classified by the OSI as oil spills are able to have their viscosity estimated by the same method. Such information might be useful in identifying the sources of spills, particularly when the oil is released illegally. Further research is needed to verify the relationship between OSI-derived viscosity values and sea surface temperatures to examine the possibilities of incorporating the concepts introduced by the OSI into existing automatic methods.

ACKNOWLEDGMENTS

The author acknowledges the valuable contributions of Dr. Peter Petrov and Ms. Noorah Riyadh in the production of this chapter.

REFERENCES

Abdali, F. and Al-Yakoob, S. (1994). Environmental dimensions of the Gulf War: Potential health impacts. In *The Gulf War and the Environment*, F. El-Baz, and R. M. Makharita (eds.), Gordon and Breach Science Publishers, Amsterdam, the Netherlands, p. 85.

Alawadi, F. (2010). Detection of surface algal blooms using the newly developed algorithm surface algal bloom index (SABI). In *Remote Sensing of the Ocean, Sea Ice, and Large Water Regions 2010*, C. R. Bostater, Jr., S. P. Mertikas, X. Neyt, and M. Velez-Reyes (eds.), *Proceedings of SPIE*, Toulouse, France: pp. 782506–14. http://link.aip.org/link/?PSI/7825/782506/1 (accessed January 4, 2011).

Alawadi, F. (2011). Detection and classification of oil spills in MODIS satellite imagery. PhD thesis. University of Southampton, England.

Alawadi, F., Amos, C., Byfield, V., and Petrov, P. (2008). The application of hyperspectral image techniques on MODIS data for the detection of oil spills in the RSA. *Proceedings of SPIE*, Cardiff, Wales, United Kingdom, 7110, pp. 71100Q–12.

Alpers, W. and Hühnerfuss, H. (1989). The damping of ocean waves by surface films: A new look at an old problem. *Journal of Geophysical Res*earch, 94, 6251–6265.

Brekke, C. and Solberg, A. H. (2005). Oil spill detection by satellite remote sensing. *Remote Sensing of Environment*, 95(1), 1–13.

Brown, R. A. (1980). Longitudinal instabilities and secondary flows in the planetary boundary layer: A review. *Reviews of Geophysics*, 18(3), 683–697.

Buckmaster, J. (1973). Viscous-gravity spreading of an oil slick. *Journal of Fluid Mechanics*, 59(3), 481–491.

Carracedo, P., Torres-López, S., Barreiro, M., Montero, P., Balseiro, C., Penabad, E., Leitao, P. and Pérez-Muñuzuri, V. (2006). Improvement of pollutant drift forecast system applied to the Prestige oil spills in Galicia Coast (NW of Spain): Development of an operational system. *Marine Pollution Bulletin*, 53(5–7), 350–360.

Chust, G. and Sagarminaga, Y. (2006). The multi-angle view of MISR detects oil slicks under sun glitter conditions. *Remote Sensing of Environment*, 107(1–2), 232–239.

Concawe. (1998). *Heavy Fuel Oils*. Product dossier no. 98/109. CONCAWE's Petroleum Products and Health Management Groups, Brussels, Belgium.

Crombie, D. D. (1955). Doppler spectrum of sea echo at 13.56 Mc./s. *Nature*, 175, 681–682.

Cross, A. M. (1992). Monitoring marine oil pollution using AVHRR data: Observations off the coast of Kuwait and Saudi Arabia during January 1991. *International Journal of Remote Sensing*, 13(4), 781–788.

Demin, B. T., Yermakov, S. A., Pelinovskkiy, Y., Talipova, T. G. and Sheremet'yeva, A. (1985). Study of the elastic properties of sea surface-active films. *Izvestiia*, 21(4), 312.

Ducruex, J., Berthou, F., and Bodennec, G. (1986). Study of the weathering of a petroleum product spilled on the ocean surface under natural conditions (Etude du Vieillissement d'un Petrole Brut Repandu a la Surface de l'eau de mer dans des Conditions Naturelles). *International Journal of Environmental Analytical Chemistry*, 24(2), 85–111.

Elliott, A. J. (1986). Shear diffusion and the spread of oil in the surface layers of the North Sea. *Ocean Dynamics*, 39(3), 113–137.

Elliott, A. J., Hurford, N., and Penn, C. J. (1986). Shear diffusion and the spreading of oil slicks. *Marine Pollution Bulletin*, 17(7), 308–313.

ESA/EDUSPACE (2010). Galathea 3 EMU radar look at ocean features and oil spills— Background. http://galathea3.emu.dk/satelliteeye/casestudies/radar1/back_uk.html (accessed January 5, 2011).

Espedal, H. (1999). Detection of oil spill and natural film in the marine environment by spaceborne SAR. *IGARSS'99 Proceedings of IEEE 1999 International Geoscience and Remote Sensing Symposium*, Hamburg, Germany, pp. 1478–1480.

Fannelop, T. K. and Waldman, G. D. (1972). Dynamics of oil slicks. *AIAA Journal*, 10, 506–510.

Fay, J. A. (1969). *The Spread of Oil Slicks on a Calm Sea*. Cambridge Fluid Mechanics Laboratory, Massachusetts Institute of Technology, Cambridge, MA.

Fay, J. A. and Hoult, D. P. (1971). *Physical Processes in the Spread of Oil on a Water Surface*. Department of Mechanical Engineering, Massachusetts Institute of Technology, Cambridge, MA.

Friedman, K. S., Pichel, W. G., Clemente-Colon, P., and Li, X. (2002). An experimental GIS system for the Gulf of Mexico region using SAR and additional satellite and ancillary data. *IGARSS'02 Proceedings of 2002 IEEE International Geoscience and Remote Sensing Symposium*, Toronto, Canada, pp. 3343–3345.

Goodman, R. H. (1989). *Application of the Technology in the Remote Sensing of Oil Slicks.* John Wiley and Sons Ltd., NY.

Guinard, N. W. (1971). The remote sensing of oil slicks. *Proceedings of the 7th International Symposium on Remote Sensing of Environment*, Ann Arbor, MI, p. 1005.

Hawkins, S. J. and Southward, A. J. (1992). The Torrey Canyon oil spill: Recovery of rocky shore communities. In *Restoring the Nations Marine Environment*, G. W. Thorpe (ed.), University of Maryland Sea Grant Publications, College Park, MD, pp. 583–631.

Hayes, M. O., Hoff, R., Michel, J., Scholz, D., and Shigenaka, G. (2005). *Introduction to Coastal Habitats and Biological Resources for Oil-Spill Response.* Office of Response and Restoration, NOAA Training Manual. First published on June 9, 2005, and last revised on July 22, 2010.

Holt, B. (2004). SAR imaging of the ocean surface. In *Synthetic Aperture Radar Marine User's Manual*, Department of Commerce, Washington, DC, pp. 25–81.

Hovland, H. A., Johannessen, J. A., and Digranes, G. (1994). Slick Detection in SAR Images. *IGARSS'94 Proceedings of 1994 IEEE Geoscience and Remote Sensing Symposium—Surface and Atmospheric Remote Sensing: Technologies, Data Analysis and Interpretation*, Pasadena, CA, pp. 2038–2040.

Hu, C., Muller-Karger, F. E., Taylor, C., Myhre, D., Murch, B., Odriozola, A. I.., and Godoy, G. (2003). MODIS detects oil spills in Lake Maracaibo, Venezuela. *EOS AGU Transactions*, 84(33), 313–319.

Hu, C., Li, X., Pichel, W. G., and Muller-Karger, F. E. (2009). Detection of natural oil slicks in the NW Gulf of Mexico using MODIS imagery. *Geophysical Research Letters*, 36, 2–5.

Hühnerfuss, H., Alpers, W., and Jones, W. L. (1978). Measurements at 13. 9 GHz of the radar backscattering cross section of the North Sea covered with an artificial surface film. *Radio Science*, 13, 979–983.

Hurford, N. (1989). Review of remote sensing technology: The remote sensing of oil slicks, *Proceedings of an International Meeting.* Organized by the Institute of Petroleum, May 1988, London.

ITOPF. (1986). *Fate of Marine Oil Spills.* The International Tanker Owners Pollution Federation Limited, London.

Jackson, R., Slater, P., and Pinter Jr., P. (1983). Discrimination of growth and water stress in wheat by various vegetation indices through clear and turbid atmospheres. *Remote Sensing of Environment*, 13(3), 187–208.

Karathanassi, V., Topouzelis, K., Pavlakis, P., and Rokos, D. (2006). An object-oriented methodology to detect oil spills. *International Journal of Remote Sensing*, 27, 5235–5251.

Malthus, T. J. and Karpouzli, E. (2007). *Passive Remote Sensing of Oil Slicks: A Review of the State-of-the-Art.* Report Number 001, for the EU MAPRES project, co-financed by the European Commission under the Community framework for cooperation in the field of accidental or deliberate marine pollution (No. 07.030900/2006/448578/SUB/A3).

Killops, S. D. and Killops, V. J. (1993). *An Introduction to Organic Geochemistry.* Addison-Wesley, Boston.

Król, T., Stelmaszewski, A., and Freda, W. (2006). Variability in the optical properties of a crude oil–seawater emulsion. *Oceanologia*, 48, 203–211.

Langmuir, I. (1938). Surface motion of water induced by wind. *Science*, 87(2250), 119–123.

Lewis, A. and Aurand, D. (1997). Putting dispersants to work: Overcoming obstacles. In *International Oil Spill Conference,* American Petroleum Institute Technical Report, IOSC-004, p. 78.

Lunel, T., Davies, L., Shimwell, S., Byfield, V., Boxall, S., and Gurney, C. (1997). Review of aerial/satellite remote sensing carried out at the sea empress incident. *The 3rd International Airborne Remote Sensing Conference and Exhibition—Development, Integration, Applications and Operations,* Copenhagen, Denmark, pp. I.731–I.732.

Melshelmer, C., Alpers, W., and Gade, M. (1998). Investigation of multifrequency/ multipolarization radar signatures of rain cells, derived from SIR-C/X-SAR data. *Journal of Geophysical Research*, 103, 18867–18884.

Michel, J. (1992). *Oil Behavior and Toxicity. Introduction to Coastal Habitats and Biological Resources for Spill Response*. Report HMRAD 92-4, Hazardous Materials Response and Assessment Division, NOAA, Seattle, WA.

NOAA. (2002). *Trajectory Analysis Handbook*, Ocean Service, Hazardous Materials Response Division, and Office of Response and Restoration, NOAA, Seattle, WA.

NOAA Magazine Online. (2010). Story 91. http://www.magazine.noaa.gov/stories/mag91.htm (accessed April 26, 2010).

NOAA Web Update. (2010). *DeepH2Oweb_25Apr.doc*. http://www.incidentnews.gov/ attachments/8220/526199/DeepH2Oweb_25Apr.doc (accessed July 22, 2010).

Norse, E. A. and Amos, J. (2010). *Impacts, Perception, and Policy Implications of the Deepwater Horizon Oil and Gas Disaster*. Environmental Law Institute, Washington, DC, p. 16.

Otremba, Z. (1999). Selected results of light field modelling above the sea surface covered by thin oil film. 41st Issue of Series: Computer Simulation and Boundary Field Problems, *Environmental Simulations*. Riga Technical University, 41, 5–13 (full paper: http:// www.rtu.lv/www_emc/issue_41.pdf/z_otremb.pdf).

Otremba, Z. and Król, T. (2001). Light attenuation parameters of polydisperse oil-in-water emulsion. *Optica Applicata*, 31(3), 599–609.

Otremba, Z. and Piskozub, J. (2001). Modelling of the optical contrast of an oil film on a sea surface. *Optics Express*, 9(8), 411–416.

Pavia, R. and Payton, D. (1983). An approach to observing oil at sea. *Proceedings of the 1983 Oil Spill Conference*, American Petroleum Institute, San Antonio, TX, pp. 345–349.

Pavlakis, P., Tarchi, D., and Sieber, A. J. (2001). On the monitoring of illicit vessel discharges using space-borne SAR remote sensing—A reconnaissance study in the Mediterranean Sea. *Annals of Telecommunications*, 56 (11), 700–718.

Payne, J. R. (1994). Section 4.0. Use of oil spill weathering data in toxicity studies for chemically and naturally dispersed oil slicks. *Proceedings of the First Meeting of the Chemical Response to Oil Spills: Ecological Effects Research Forum. Marine Spill Response Corporation*, Washington, DC, MSRC Technical Report Series, pp. 94–017.

Samberg, A. (2005). Advanced oil pollution detection using an airborne hyperspectral lidar technology. *Proceedings of the International Society for Optical Engineering, SPIE*, 5791, 308–317.

Shaban, A., Ghoneim, E., Hamzé, M., and El-Baz, F. (2007). A post-conflict assessment to interpret the distribution of oil spill off-shore Lebanon using remote sensing. *Lebanese Science Journal*, 8(2), 75.

Shi, L., Zhang, X., Seielstad, G., Zhao, C., and He, M. (2007). Oil spill detection by MODIS images using fuzzy cluster and texture feature extraction. 6 2007, Aberdeen, Scotland, pp. 1–5.

Shigenaka, G., Hayes, M. O., Michel, J., Henry Jr., C. B., Roberts, P., Houghton, J. P., and Lees, D. C. (1997). Integrating physical and biological studies of recovery from the Exxon Valdez oil spill. *NOAA Technical Memorandum NOS ORCA*, 114, 206.

Solberg, A. S. (2005). Automatic detection and estimating confidence for oil spill detection in SAR images. *Proceedings of the International Society of Photogrammetry and Remote Sensing*, Mestre-Venice, Italy.

Spaulding, M. L. (1988). A state-of-the-art review of oil spill trajectory and fate modeling. *Oil and Chemical Pollution*, 4(1), 39–55.

Svejkovsky, J. and Muskat, J. (2006). *Real-Time Detection of Oil Slick Thickness Patterns with a Portable Multispectral Sensor*. Minerals Management Service, Department of the Interior, United States.

Tahvonen, K. (2008). The use of remote sensing, drifting forecasts and GIS data in oil response and pollution monitoring. *Remote Sensing & Hydrosphere*, Helsinki, Finland.

Thurman, H. V. (1983). *Essentials of Oceanography*. Merrill Publishing, Columbus, OH.

Topouzelis, K. N. (2008). Oil spill detection by SAR images: Dark formation detection, feature extraction, and classification algorithms. *Sensors*, 8(10), 6642–6659.

Tseng, W. Y. and Chiu, L. S. (1994). AVHRR observations of Persian Gulf oil spills. *NOAA Geoscience and Remote Sensing Symposium, IGARSS'94 Proceedings of 1994 IEEE Geoscience and Remote Sensing Symposium—Surface and Atmospheric Remote Sensing: Technologies, Data Analysis and Interpretation*, Pasadena, CA, pp. 779–782.

Tsukihara, T. (1995). Weathering experiment on spilled crude oils using a circulating water channel. *Proceedings of Prevention, Behavior, Control, and Cleanup*. American Petroleum Institute, Atlanta, Seattle, WA, p. 435.

Tucker, C. J. (1979). Red and photographic infrared linear combinations for monitoring vegetation. *Remote Sensing of Environment*, 8(2), 127–150.

Valenzuela, G. R. (1978). Theories for the interaction of electromagnetic and oceanic waves— A review. *Boundary-Layer Meteorology*, 13(1), 61–85.

5 Remote Sensing to Predict Estuarine Water Salinity

Fugui Wang and Y. Jun Xu

CONTENTS

5.1 INTRODUCTION

5.1.1 BACKGROUND

"Estuary" has been defined in a multitude of ways, and each definition varies based on the emphasis of particular estuarine characteristics and processes (Dyer 1997). Based on salinity, for example, an estuary can be defined as a semienclosed body of water located along coastal regions where freshwater rivers and streams are mixed with saltwater from the ocean. Based on position, an estuary can be defined as a transitional zone between rivers and open oceans. Estuary has also been termed as a bay, harbor, sound, or lagoon (such as the Chesapeake Bay and the Long Island Sound). It is often difficult to define exactly where an estuary begins and ends.

Because of the large nutrient inflow from river water and the resuspension of decomposed organic matter in the bottom sediments, estuaries and fringe wetlands are considered among the world's most productive ecosystems, making them critical habitats for both terrestrial and aquatic wildlife to live, feed, and reproduce. It has been estimated that around 60% of the world's human population lives along estuaries and near the coast (Lindeboom 2002), two areas that comprise only about 5% of the earth's surface (Wolanski 2007). Although estuary regions constitute approximately 12.6% of the land area of the United States' contiguous 48 states, 43% of the country's population resides within them. This 43% make up 40% of the workforce, and produce 49% of the GDP of the U.S. economy (Colgan 2009). With the ascendance of late-industrial society, recreational aspects of the coastal zone have increased in relevance (Wilson and Farber 2008) and have served as a center for both shipping and commerce.

During the past century, rapid population increase and destructive human activity have caused significant changes in freshwater inflow to, and the water quality of, many of the world's estuaries. For instance, excessive input of nutrients into estuaries has been reported for the Chesapeake Bay basin (Kemp et al. 2005), the Mississippi–Atchafalaya Rivers (Rabalais et al. 2002; Xu 2006), and many other estuaries around the world (Diaz and Rosenberg 2008). These human-induced stresses and changes already have and continue to affect the structure and function of estuarine ecosystems.

5.1.2 ESTUARY FORMATION

Based on physical processes, Pritchard (1952) grouped estuaries into four categories: drowned river valleys (coastal plain estuaries), bar-built estuaries, fjords, and others (e.g., deltaic and tectonic estuaries). Most drowned river valleys were formed during the last glacial period (i.e., from about 110,000 years ago to approximately 9600–9700 BC). In the colder climates, sea level was tens to hundreds of meters lower than it is today, and rivers entrenched their downstream reaches. As glaciers receded during warmer climate periods, coastal plain estuaries were formed by a rise in sea level; the rising coastal waters filled in the valleys and inundated other portions of the coastal region to form the estuaries. Bar-built estuaries were created by the formation and movement of sandbars, which enclosed bodies of brackish water along the coastline, forming such estuaries as Vermilion and Wet Cote Blanche Bays in coastal Louisiana in the United States. Fjords, which are most commonly found along the Norwegian west coast and the coastal margins of British Columbia, Canada, are drowned glacier valleys that are formed when glaciers cut deep, u-shaped valleys through pre-existing river valleys advancing into sea. They can generally be identified by a steep ridge that has been formed at the farthest reach of the glacier where the cutting stopped (called sills), and by a steep-sided narrow channel. The other group includes estuaries such as San Francisco Bay in California, which were formed by earthquakes and faulting that caused the rapid sinking of coastal areas to below sea level.

5.1.3 TIDES AND WATER CIRCULATION

Water movement within estuaries is regulated by tidal cycles, differences in the density of water, fresh water discharge, wind, and topography of the estuary. Tides are the regular rise and fall of sea levels due to the gravitational attraction of the moon and the sun, and the rotation effects of the earth. The vertical rise and fall of the tides also creates a horizontal motion of the water's flood and ebb currents. In general, flood currents occur as rising tides push the water toward the shore or into the bays, harbors, and estuaries. Ebb currents are the tidal currents that flow out to the sea, thereby lowering water levels as the tides fall. The tide at the beach bordering open oceans is a symmetrical tide (i.e., roughly a 12-hour cycle with 6 hours of high tide and 6 hours of low tide). However, the rhythm of tides within a neighboring estuary and the strength of the tidal currents vary within an estuary by the interaction of the tides with topography, bottom friction, meteorology, and river discharge (Godin 1999; Garel et al. 2009; Reis et al. 2009). The result is that the symmetrical tide becomes asymmetrical. For instance, in the offshore waters of China's Pearl River Estuary, the average tidal range is low, but gradually increases as currents flow toward the estuary, reaching its maximum upstream (Mao et al. 2004). In the rockbound Guadiana Estuary, tidal asymmetry produces faster currents on the ebb than on the flood, especially at the lower estuary (Garel et al. 2009).

The lengths of flood and ebb tides have implications on fixed volumes of water moving in and out of an estuary with each tidal cycle. If the flood tide is shorter, then water flows faster to fill in an estuary, resulting in a net input of sediments. In addition to tidal asymmetry, estuaries receive greater volumes of water and sediment during spring and neap tide cycles (French 2001). Spring tides, which occur when the sun, earth, and moon are all in a straight line, cause higher high tides and lower low tides, thereby resulting in stronger bottom stirring and nitrate entrainment from the deep sea (24 mmol·m^{-2}) as compared to neap tides (17 mmol·m^{-2}) (Yin et al. 1995). Neap tides, which occur when the moon is at a right angle to the earth–sun line (i.e., during the first and last quarter of the lunar cycle), cause lower than average tides.

5.1.4 FRESHWATER AND SEDIMENT

Sedimentation in estuaries is normally viewed as a continuous process controlled by inputs of fluvial and marine materials, which migrate in response to tidal circulation (McManus 1998). Inputs of sediments from river discharge also vary seasonally with high concentrations during wet season and low concentrations during dry seasons (Kitheka et al. 2005). During a flood event, TSS concentration along an entire estuary is controlled by the sediment load from river discharge (Garel et al. 2009).

In addition to direct sediment input from river discharge, resuspension of bottom sediments is another important source of estuary suspended solids. Resuspension is mainly driven by wind-generated waves, resulting in highly turbid water that is pumped back into the estuary during flood tide (Kitheka et al. 2005). Due to wind resuspension, TSS concentrations at the Gulf end of the Breton Sound Estuary, for example, fluctuate highly during winter and spring (Lane et al. 2007). In addition to wind-generated waves, tidal currents are another force periodically driving resuspension (Siegle et al. 2009). Due to differences in tidal current velocity patterns during flood and ebb, resuspension of the deposited fluidized mud contributes to peak TSS concentration during flood tide (Kitheka et al. 2005). Spatially, TSS concentration varies in the estuary with local current magnitude and the proximity to patches of easily resuspendable sediment (Cook et al. 2007). In the Hudson River Estuary, regional sediment distribution is controlled by changes in morphology, bedrock, and tributary input, as well as by anthropogenic modifications of the estuary (Nitsche et al. 2007). The dominant forces that affect the transport process during spring or neap tides are tidal correlation term, gravitational circulation, and Stokes drift during spring tide (Siegle et al. 2009).

5.1.5 SALINITY

Salinity, usually expressed in parts per thousand (ppt), is a measure of the concentration of dissolved salts in water. In estuaries, salinity exhibits a gradual change throughout its length, from 0.5 ppt or less in the upper freshwater sector to more than 30.0 ppt near the connection with the open sea. Within the estuary, salinity levels are referred to as oligohaline (0.5–5.0 ppt), mesohaline (5.0–18.0 ppt), or polyhaline (18.0–30.0 ppt) (Mitsch and Gosselink 1986).

Based on salinity distribution and flow characteristics, estuaries are classified into three types including highly stratified, moderately stratified, and vertically homogeneous (well mixed) (Dyer 1997). In highly stratified estuaries (such as fjords), river output greatly exceeds marine input with tidal effects having minor importance. Rapidly flowing freshwater discharges push back a bottom layer of dense seawater, forming a wedge-shaped salty bottom layer. Strong velocity shear at the interface produces internal wave motion at the transitional zone between the two layers. Each layer, with the exception of the tip of the wedge where the interface meets the surface, has stable salinity. In moderately stratified estuaries, which are widespread in temperate and tropical climates, tidal volume exceeds river volume, causing a partial mix of seawater and freshwater, concurrently increasing the salinity of the top freshwater layer and decreasing the salinity of the seawater layer. However, as the layers advance seaward, the salinity of both layers increase, with the bottom layer always remaining more saline. In vertically homogeneous estuaries, tidal flows overwhelm river flows, causing the layers to mix, thereby erasing any stratification resulting in homogeneous salinity throughout the water column. This type of estuary is found in regions of particularly strong tides.

In addition to controlling water stratification in estuaries, salinity is a primary factor that affects community composition and structure and the habitats of estuarine plants and animals. For instance, salinity gradient has been identified as the most dominant factor influencing the abundance and size structure of meiobenthos (Yamamuro 2000), nematode density and community composition (Adao et al. 2009), and mollusk communities (Montagna et al. 2008). On the island of Minorca, growth rates (in length) of *Liza ramado* and *Mugil cephalus* are highest in oligo-mesohaline water (Cardona 2006); in Uruguay's Rio de la Plata Estuary, *Neomysis americana* occurs preferentially in oligomesohaline conditions (Calliari et al. 2007); in the estuarine portion of the Alafia River in Florida, changes in nekton community structure happen rapidly at each end of the estuarine salinity gradient (i.e., 0–0.5, 0.5–5, 5–18, 18–30, 30–40, >40) (Greenwood et al. 2007); and in the Gambia and Saloum estuaries of West Africa, the black-chinned tilapia is able to withstand saltier environments by limiting its growth and changing its fecundity, with the most profound changes visible in hypersaline conditions (Panfili et al. 2004). In the southeastern United States, river-dominated small estuaries with a lower salinity may represent a vital habitat for the maintenance of local blue crab populations (Posey et al. 2005).

5.1.6 IMPACTS OF SEA LEVEL RISE ON ESTUARIES

In its latest global climate assessment, the Intergovernmental Panel on Climate Change (IPCC) reported that global mean temperature of a 5-year period ending in 2005 has increased 0.76°C (0.57°C–0.95°C) compared to the global mean temperature of a 50-year period ending in 1899. In reference to the global mean temperature observed between 1980 and 1999, the IPCC also predicted that the temperature will continue to increase by 0.64°C to 0.69°C between the years 2011 and 2030, and 1.79°C to 3.13°C between the years 2080 and 2099 (Meehl et al. 2007). The increase in global temperature may have caused a rise in sea level as evidenced by an average increase of the world's oceans of 1.8 mm per year from 1961 to 2003. The rate of increase accelerated between 1993 and 2003 by approximately 3.1 mm per year. As projected under the A1B greenhouse gas emission scenario (IPCC Special Report on Emissions Scenarios), sea level will rise 0.21 to 0.48 m by the end of the century (2090 to 2099) (Meehl et al. 2007). For example, in the A1B climate scenario, the mean sea level is projected to rise 55 cm and a projected change in tidal range of 30 cm (high water +15 cm, low water –15 cm) of the upstream boundary of the brackish-water zone of Germany's Weser Estuary will move approximately 2 km inland (Grabemann et al. 2001). Sea level rise will certainly inundate wetlands and other low-lying land, erode beaches, intensify flooding, and increase the salinity of estuaries. In Chesapeake Bay, sea-level rise during the second half of the 20th century has already led to detectable increases of estuary salinity (Hilton et al. 2008). Long-term salinity increases may cause irreversible damage to estuarine ecosystems (Hilton et al. 2008). As sea and salinity levels increase in the low marsh along the San Francisco Estuary, halophytic species were found to be replacing salt-intolerant taxa (Watson and Byrne 2009). In the United Kingdom's Humber Estuary, a sea-level rise of 0.3 m would result in a 63% loss of intertidal area and a potential loss of macrobenthic biomass of up to 22.8% (Fujii and Raffaelli 2008).

5.2 ESTUARINE WATER OPTICS

5.2.1 ESTUARINE WATER COLOR

The color of estuarine water bodies is mainly controlled by scattering and absorption processes of the water itself, in conjunction with optically active substances that predominantly include color dissolved organic matter (CDOM), photosynthetic biota (i.e., phytoplankton and macrophytes where present), and suspended matter in the water (Arst 2003; Bowers et al. 2004; Bukata 2005). Estuarine water color can be highly variable due to sources, sinks, and transformations of these substances (Davies-Colley et al. 2003), which has been widely used to describe estuarine optics and for modeling of light attenuation in estuaries (Christian and Sheng 2003). CDOM plays a key role as an optical tracer to investigate processes related to carbon dynamics, nutrient availability, phytoplankton activity, microbial growth, and ecosystem productivity in estuarine and coastal zones (Nieke et al. 1997). Mainly originating from terrestrial ecosystems, CDOM's (commonly called yellow substance, Gelbstoff, or humic material), quantity and distribution are strongly affected by changes in the amount of freshwater discharge, precipitation, evaporation, biological activity, and photo-oxidation (Lubben et al. 2009). For instance, the freshwater input in Huon Estuary is intensely colored due to high levels of humic material and algal blooms (Clementson et al. 2004). Along with suspended matter, CDOM is a main colorant of estuary water. While CDOM does not significantly impact scattering within the water column, it does considerably increase absorption (Pozdnyakov and Grassl 2003). Its absorption spectrum declines exponentially from the ultraviolet (UV) to the visible. The spectrum varies in response to the nature and origin of CDOM as well as the transportation and mixing processes in estuaries.

Found in great abundance in estuaries, phytoplanktons provide food for a tremendous variety of organisms via photosynthesis and serve as the base of the aquatic food web. The coloration of phytoplankton cells is dependent on their pigment content and composition including chlorophyll, carotenoid, and phycobilin (Pozdnyakov and Grassl 2003). As phytoplankton bloom occurs, estuarine water color intensifies. Spatial distribution and productivity of phytoplankton in estuaries often vary by water temperature, salinity, and nutrient loadings (Mallin 1994), as well as water column depth, light availability, benthic grazing, topographic gradients, turbidity, and large-scale horizontal transport processes (Lucas et al. 1999a; Lucas et al. 1999b). Impacts of phytoplankton on optical properties of water vary with different phytoplankton chlorophyll pigments (e.g., a, b, or c) and properties such as, size, particle shape, and "package effect." The effect is caused by the fact that pigments are not in solution, but rather, packed inside the cells (and in cells inside chloroplasts) (Martinez-Guijarro et al. 2009).

Natural water bodies contain two types of suspended matters including organic and inorganic (Pozdnyakov and Grassl 2003). Suspended organic matter contains planktonic organisms and detritus particles (i.e., fragments of dead plankton and fecal pellets of zooplankton). Suspended inorganic matter mainly comes from surface soils through river discharge and coastal erosion. In coastal waters, particulate scattering and backscattering are complex and appear dependent on specific nature and relative composition of the particles that are present (Snyder et al. 2008).

5.2.2 Light Absorbance and Scattering of Estuarine Water

5.2.2.1 Inherent Optical Properties

Optical properties of the substances in water are quantified by both absorption (a) and scattering (b) coefficients, as well as by the volume scattering function (VSF). Both absorption and scattering coefficents measure a fraction of the incident radiation absorbed and scattered per unit of distance of light traveled in through an infinitesimally thin layer of water. Together, the sum of a and b defines the beam attenuation coefficient (c), which is the measurement of the loss of light due to absorption and scattering as the beam of light passes through the infinitesimally thin layer of water. The value c is closely and inversely related to the visual clarity of water (Davies-Colley et al. 2003). Accordingly, c is low for optically clear rivers and high for turbid rivers (Julian et al. 2008). The volume scattering function is defined as the ratio of the intensity of scattered light to the incident irradiance per unit of an infinitesimal volume of water at each angle from 0° to 180°. It describes the directional dependence of the scattering and is considered the sum of VSF of pure water, VSF of particulates, and VSF owing to turbulence (Agrawal 2005). The total scattering coefficient (b) is calculated by integrating VSF over all angles. The backscattering coefficient is the integration of the VSF over the angles from 90° to 180° (D'Sa and Miller 2005). Magnitudes of these coefficients described above depend only on the substance properties and wavelength of light but not on changes in the solar radiation field. Therefore, these optical properties are usually referred to as inherent optical properties (IOP). Physical properties of the substances that influence water optics include particle size, shape, composition, and chemical properties.

The general absorption and scattering spectra of each substance is characterized by its own features. Absorption of CDOM is strong at the ultraviolet and blue spectrum, and very low (if not absent) at the red end of the visible spectrum (Figure 5.1a) (Kirk 1994). Spatially, across the length of estuaries from the upper to the lower sector, CDOM absorption declines exponentially (Figure 5.1b) (Nieke et al. 1997). In Chesapeake Bay, the CDOM absorption coefficient has been reported to vary between 2.3 and 4.1 m^{-1} at 355 nm at the freshwater end member, and between 0.4 and 1.1 m^{-1} at 355 nm at the coastal end member (Rochelle-Newall and Fisher 2002). In contrast, the absorption of phytoplankton shows two major peaks with the primary maximum at approximately 435 nm and the second maximum at approximately 680 nm (Figure 5.1c) (Vahatalo et al. 2005). Absorption spectra by particles (strained through filters with nominal pore size 0.7 μm) typically decreases monotonically with increasing wavelength in the streams (Figure 5.1d) (Vahatalo et al. 2005). In terms of fraction of light absorption by the substances in estuaries, CDOM absorption accounts for a large fraction of total light absorption at the blue end of the spectrum (Figure 5.1e and f) (Bowers et al. 2004; Vahatalo et al. 2005). Nevertheless, at the red maximum (approximately 680 nm), water and mineral suspended solids (MSSs) contribute to a higher total fraction of absorption (Figure 5.1f) (Bowers et al. 2004). Overall, in the Conwy Estuary in North Wales, the averaged absorption across the wavelength range from 400 to 700 nm illustrates that suspended matter absorption is the highest (38%), followed by CDOM (37%), mineral suspended solids (22%), and chlorophyll (3%) (Bowers et al. 2004).

5.2.2.2 Apparent Optical Properties

In contrast to IOP, apparent optical properties (AOPs) of natural waters change with ambient light and inherent optical properties. AOPs include the most widely used diffuse attenuation coefficient (k_d), remote sensing reflectance (R_{rs}), and

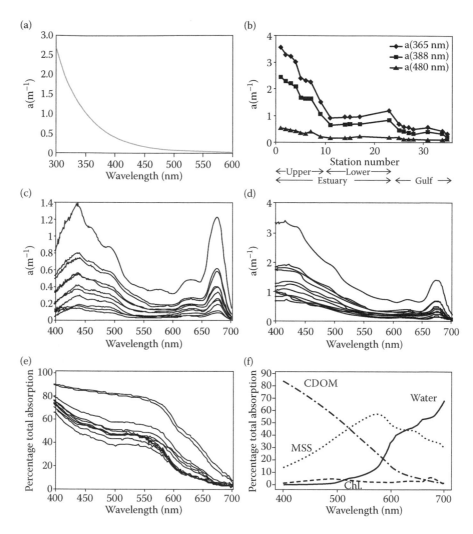

FIGURE 5.1 (a) Absorption spectrum of organic matter between 300 and 750 nm. (From http://www.cis.rit.edu/research/thesis/bs/2000/salmon/body.html. With permission.) (b) CDOM absorption coefficients across the Estuary and Gulf of St. Lawrence system in Canada. (From Nieke, B., Reuter, R., Heuermann, R., Wang, H., Babin, M., and Therriault, J. C., *Continental Shelf Research*, 17, 235–252, 1997. With permission.) (c) Absorption coefficient of phytoplankton. (d) Absorption coefficient of particles. (e) Percentage of CDOM absorption to the total absorption at multiple stations in the Neuse River Estuary. (From Vahatalo, A. V., Wetzel, R. G., and Paerl, H. W., *Freshwater Biology*, 50, 477–493, 2005. With permission.) (f) Fractions of absorption by substances in the Conwy estuary in North Wales. (From Bowers, D. G., Evans, D., Thomas, D. N., Ellis, K., and Williams, P. J. L., *Estuarine, Coastal and Shelf Science*, 59, 13–20, 2004. With permission.)

Secchi disk depth. k_d is essentially the attenuation experienced by sunlight as a beam of light propagates through water, its value depending on depth, sun angle, sky conditions, and shadowing by objects on the surface (Zheng et al. 2002). The general trend of k_d is high between 40 and 500 nm and between 700 and 800 nm, and is relatively flat between 500 and 700 nm as found in the Delaware Estuary (Figure 5.2a) (Wang et al. 1996). k_d can be estimated by the deployment depth of a Secchi disk, an instrument used to measure the transparency of a water body. Many studies have found a linear relationship between k_d and reverse Secchi disk depth as Liu et al. (2005) reported in their review. For example, at station 16 in the Delaware Estuary, the lowest Secchi disk depth shows the highest light attenuation (Figure 5.2a and b) (Wang et al. 1996). k_d is influenced by concentration of an optical substance in water and varies within and between estuaries. In the Indian River Lagoon on the East Coast of Florida, Christian and Sheng (2003) found that tripton (i.e., nonalgal particulate matter calculated from TSS and chlorophyll-a corrected for pheophytin) had the strongest effect on light attenuation, accounting for 59% to 78% of k_d (PAR). In the Danshuei River–Keelung River Estuarine

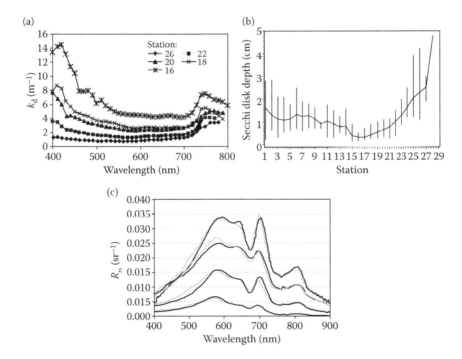

FIGURE 5.2 (a) Attenuation coefficients by stations. (b) Secchi disk depth in the Delaware Estuary. (From Wang, M. Z., Lyzenga, D. R., and Klemas, V. V., *Journal of Coastal Research*, 12, 211–228, 1996. With permission.) (c) Typical R_{rs} spectra measured (darker curves) and reproduced with the optical model (lighter curves) in the Tamar estuary during the 2004 summer period for TSS concentrations of 7, 20, 43, and 59 mg·L^{-1} with a(440) values in the range 0.5–0.9 m^{-1} and chlorophyll concentrations in the range 15–32 g·L^{-1}. (From Doxaran, D., Cherukuru, N., and Lavender, S. J., *Applied Optics*, 45, 2310–2324, 2006. With permission.)

system in Taiwan, Liu (2005) reported a seaward declining trend in the measured values of the light attenuation coefficient (k_d). Liu (2005) also found that the k_d value did not have a close relationship with TSS ($R^2 = 0.15$), but did with salinity ($R^2 = 0.61$). The spatial and temporal variability of light attenuation found by Phlips et al. (1995) within Florida Bay suggests that light attenuation in the water column caused by high levels of tripton and/or phytoplankton may have limited sea grass colonization in certain regions of the bay.

Remote sensing reflectance (R_{rs}) is an indicator of scattering in the water, associated with water brightness. It is defined as the ratio of water-leaving radiance ($L_w(\lambda)$) over downwelling irradiance ($E_d(\lambda)$) signal just above the water surface. It is described as follows (Doxaran et al. 2006; Ahn et al. 2008):

$$R_{rs}(0^+\lambda) = \frac{L_w(\lambda)}{E_d(0^+\lambda)} \tag{5.1}$$

In their study on three European estuaries, Doxaran et al. (2006) found that a typical R_{rs} signal increases with the increasing concentration of total suspended matter (Figure 5.2c). In the coastal waters near Long Island Sound, the peak reflectance was found in the wavelength range from 550 to 650 nm, and the reflectance increased with increasing salinity level (Szekielda et al. 2003). R_{rs} is often employed for predicting and mapping concentrations of suspended particulate matter and CDOM in estuaries from above-water optical measurements because the wavelength dependence of R_{rs} provides clues to the nature and amount of these materials (Snyder et al. 2008). For example, in the Tamar Estuary, United Kingdom, a close linear relationship ($R^2 = 0.96$) is obtained between total suspended matter concentration and ratio of the remote-sensing reflectance signal at 850 and 550 nm, as well as a power law relationship ($R^2 = 0.89$) between the CDOM concentration and the ratio at 400 and 600 nm (Doxaran et al. 2005).

5.3 REMOTE SENSING APPLICATIONS IN ESTUARINE SALINITY ASSESSMENT

5.3.1 A RETROSPECTIVE OVERVIEW

The use of satellite imagery to map salinity distribution in estuaries (specifically in the San Francisco Bay), was pioneered by Khorram (Khorram 1982). Subsequently, other studies emerged, including some which found divergent correlations between Landsat TM bands 2, 3, and 4 and *in situ* salinity measurements ($R^2 = 0.47$–0.86) (Lavery et al. 1993; Baban 1997; Dewidar and Khedr 2001). Correlations were found between surface salinity and SeaWiFS in both the Florida Bay (D'Sa et al. 2002), and in the East China Sea close to the Yangtze River Estuary (Ahn et al. 2008). In offshore oceanic waters, Pozdnyakov and Grassl (2003) discovered that inorganic salts at concentrations of 35 kg·m^{-3} (or 3.5%) account for about 20% to 30% of total spectral scattering. In contrast, in brackish estuaries with low salinity level, dissolved salts have little effect on spectrally averaged light absorption (Bowers et al.

2004). While salts have no strong color signal, remotely sensed salinity levels could be estimated through close relations of salinity levels with CDOM and/or TSS concentrations (Wang and Xu 2008). Both parameters are major colorant of estuarine waters and easily inferred from satellite color observations (Ahn et al. 2008; Bowers and Brett 2008).

Many empirical correlations between surface salinity level and light absorption of CDOM have also been established (Blough et al. 1993; Ferrari and Dowell 1998; Keith et al. 2002; Bowers et al. 2004; Hu et al. 2004). The relationship is especially evident in spring during periods of vertical mixing and large inputs from continental runoff (Ferrari and Dowell 1998). In Tampa Bay, Florida, it has been found that although the CDOM–salinity relationships vary at different parts of the bay, strong negative linear correlations have been identified ($R^2 = 0.97$–0.99) (Hu et al. 2004). In Chesapeake Bay, large amounts of CDOM absorption are observed in waters with lower salinity, linearly decreasing at the seaward end with an identifiable overall strong correlation ($R^2 > 0.9$) (Rochelle-Newall and Fisher 2002). In the Conwy Estuary in North Wales, CDOM absorption at 440 nm is strongly correlated with salinity ($R^2 = 0.98$) (Bowers et al. 2004).

In addition, a relationship between CDOM absorption and water-leaving radiance (reflectance) has also been identified. In their study conducted in Tampa Bay, Florida, Hu et al. (2004) found a good correlation ($R^2 = 0.8$) between the ratio of radiance at MODIS 465 over 555 nm and the CDOM absorption coefficient. In the East China Sea near the Yangtze River Estuary, Ahn et al. (2008) used a ratio of reflectance at 412 over 555 nm of SeaWiFS to reasonably predict ($R^2 = 0.67$) a CDOM (400) and a CDOM (412). In the Conwy Estuary in North Wales, Bowers et al. (2004) found a strong correlation ($R^2 = 0.94$) between CDOM absorption at 440 nm and the ratio of reflection coefficients in the red (670 nm) to blue-green (490 nm).

Collectively, the strong correlations among salinity, CDOM absorption, and remotely sensed water-leaving radiance (reflectance), as well as successful remote sensing of CDOM (Kahru and Mitchell 2001; Siegel et al. 2002; Kutser et al. 2005a, 2005b) denote that CDOM could actually be employed as the proxy in the remote sensing of salinity in coastal waters, as shown by D'Sa (2002), Binding and Bowers (2003), and Wang and Xu (2008).

5.3.2 LAKE PONTCHARTRAIN

This section concentrates on a salinity prediction study that uses satellite imagery to investigate saltwater intrusion by storm surge in Lake Pontchartrain, a large estuary north of New Orleans directly connected to the Northern Gulf of Mexico. The study (Wang et al. 2006) was conducted immediately after the landfall of Hurricane Katrina in August 2005. Cross-lake measurements on salinity and suspended solids were collected. Using the field data and additional Landsat images, we recalibrated and refined the previously developed salinity prediction model (Wang and Xu 2008) and applied it to estimate the lake salinity following four major hurricanes that occurred between 2002 and 2008.

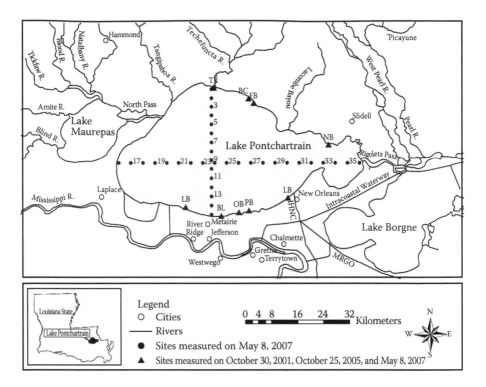

FIGURE 5.3 Lake Pontchartrain and salinity sampling locations. Water sample collection and field measurement of salinity started from site number 1 through 35. Filled triangles denote near-shore sampling sites by Lake Pontchartrain Basin Foundation and filled circles denote cross-sectional measurements conducted on May 8, 2007. (From Wang, F. G. and Xu, Y. J., *Journal of Hydrology*, 360, 184–194, 2008. With permission.)

Lake Pontchartrain, with a total surface water area of 1619 km², is the largest estuarine lake in the southeast United States (Figure 5.3). The city of New Orleans is on the lake's south shore; nearly 1.5 million people lived directly adjacent to Lake Pontchartrain before Hurricanes Katrina and Rita. The lake extends about 82 km from west to east at the longest point, and 39 km from north to south at the widest point, with an average water depth of 3.7 to 4.3 m. It connects to the Gulf of Mexico via the Rigolets Strait, Chef Menteur Pass, and the Mississippi River Gulf Outlet (MRGO), a 121 km (76 mile) long man-made navigation channel. Salinity in Lake Pontchartrain varies in response to estuarine circulation combined with tidal mixing, wind mixing, and the integration of varying volumes of freshwater that is discharged from three major rivers in the Upper Lake Pontchartrain Basin, including the Tangipahoa River, the Tickfaw River, and the Amite River (Wu and Xu 2007; Xu and Wu 2006). Diurnal tides of the Gulf of Mexico can push freshwater from the Pearl River, which typically discharges into Lake Borgne, into Lake Pontchartrain, also affecting its salinity dynamics. In addition to the saltwater intrusion regulated by daily tides, extreme weather events such as strong storm surges from Hurricane

TABLE 5.1
Characteristics of Four Major Hurricanes That Occurred in Southeast Louisiana during 2002–2008[a]

Name	Date	Local Time	Latitude	Longitude	Wind (km/h)	Category
Lili	October 3, 2002	6:00 a.m.	29.2	−92.1	148	H1
Katrina	August 29, 2005	6:00 a.m.	29.5	−89.6	204	H3
Rita	September 24, 2005	12:00 a.m.	29.4	−93.6	185	H3
Gustav	September 1, 2008	6:00 a.m.	28.8	−90.3	176	H2

[a] Characteristics at the locations represented by red dots in Figure 5.4.

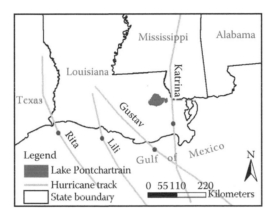

FIGURE 5.4 Tracking of four major hurricanes that struck on the northern Gulf of Mexico during the years 2002 to 2008. (From Wang, F. and D'Sa, E. J., *Remote Sensing*, 2, 1–18, 2010. With permission.)

Lili in 2002 can affect the salinity dynamics. Both Hurricanes Katrina and Rita in 2005 and Hurricane Gustav in 2008 also changed the normal salinity gradients and spatial distribution of the salinity in Lake Pontchartrain to a great extent within a short time period (Table 5.1 and Figure 5.4). Consequently, the immediate salinity oscillation could impact estuarine physical and chemical properties including salinity, temperature, turbidity, and concentrations of nutrients, sediment, and dissolved oxygen. Meanwhile, these potential salinity changes may exert long-term effects on the estuary ecosystem function and productivity.

5.3.3 MODEL DEVELOPMENT

The study was conducted in two steps using remote sensing, GIS, and statistical modeling techniques. First, a salinity prediction model was built by applying multivariate regression statistics to water-leaving reflectance derived from Landsat

TABLE 5.2
Model Parameter Estimates

Variable	Parameter Estimate	Standard Error	t Value	$p_r > |t|$	95% Confidence Limits	
Intercept	4.6127	0.4498	10.26	<.0001	3.7005	5.5248
TM1	106.44	42.765	2.49	0.0176	19.705	193.17
TM2	152.92	18.939	8.07	<.0001	114.51	191.33
TM3	−227.46	21.755	−10.46	<.0001	−271.6	−183.3
TM4	120.47	21.043	5.73	<.0001	77.794	163.15
TM5	−139.47	13.961	−9.99	<.0001	−167.8	−111.2

satellite images, and simultaneously, *in situ* salinity measurements. Then, the model was applied to multidate Landsat images to predict salinity levels over the entire lake. From the predictions, we compared multiple salinity levels and spatial patterns without hurricane disturbances on the lake, and assessed how the hurricanes' landfall perturbs the normal salinity in the lake.

Water salinity levels were measured at two transects across the entire lake by the U.S. Coast Guard Auxiliary on May 8, 2007, and along the shoreline by the Lake Pontchartrain Basin Foundation on October 30, 2001, October 25, 2005, and May 8, 2007 (Figure 5.3). Eighteen scenes of Landsat TM 5 imagery used in this study were obtained from United State Geological Survey (USGS) (http://glovis.usgs.gov/).

Twelve scenes, one image per month for each of 12 months January to December, were selected to evaluate normal seasonal salinity changes without hurricane storm surge impacts. The remaining six images were acquired following the landfall of Hurricanes Katrina, Rita, Gustav, and Lili. A salinity prediction model, as expressed below, was built with the backward ordinary least square (OLS) approach:

$$S = 4.6127 + 106.44 * TM1 + 152.92 * TM2 - 227.46 * TM3$$
$$+ 120.47 * TM4 - 139.47 * TM5 \quad\quad (5.2)$$

where S is salinity (ppt) and TM1 through TM5 are the reflectance values derived from Landsat TM bands 1 through 5 (Table 5.2). Details about image preprocessing and model development can be found in Wang and Xu (2008).

5.3.4 Salinity without Hurricane Disturbances

The model developed to simulate salinity levels in Lake Pontchartrain was tested using 20% of the field measurements as independent validation data. The results of this test suggest that the predictive power of the model is high ($R^2 = 0.89$ and root-of-mean-square error (RMSE) of validation = 0.49). The model was then used to

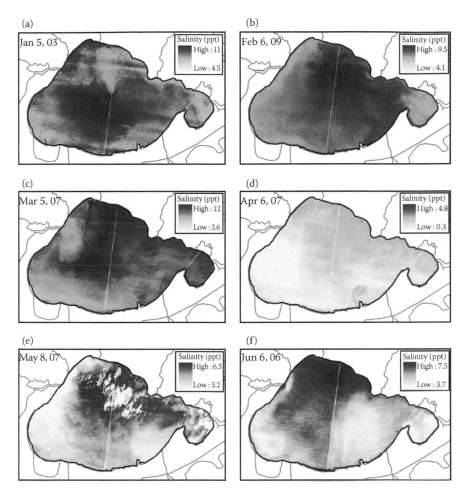

FIGURE 5.5 Normal salinity distributions (by month) in Lake Pontchartrain predicted by an empirical regression model developed with Landsat TM imagery and simultaneously *in situ* salinity measurements (a–l).

examine fluctuations of salinity levels within the lake under different scenarios. One of these indicated that there are noticeable intra-annual variations of salinity levels in Lake Pontchartrain under normal conditions that excluded hurricane impacts (Figure 5.5). Moreover, the spatial distribution of salinity levels could be broadly classified into three categories: (1) high salinity in the southern portion of the lake with low salinity along the northern and western shorelines (e.g., October 30, 2001), (2) low salinity in southern and western parts of the lake with high salinity along the northern shoreline (e.g., June 6, 2006), and (3) more evenly distributed salinity due to greater mixing of fresh and saline waters (e.g., December 1, 2001). Salinity levels in the lake did exhibit higher levels between November and March. It is likely that

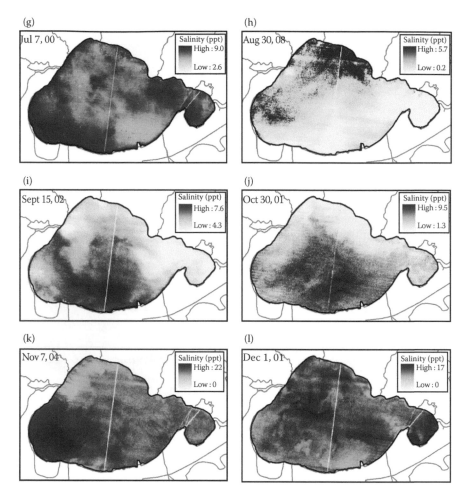

FIGURE 5.5   (Continued)

these patterns of salinity in the lake are primarily formed by freshwater discharges from the Upper Lake Pontchartrain Basin (Wang and Xu 2008), but other factors such as resuspension of bottom sediments caused by wind, mixing of freshwater and saltwater, and tidal dynamics could play significant roles in the spatial and temporal variability of salinity in the lake.

5.3.5   HURRICANE-INDUCED SALINITY CHANGES

The salinity maps following Hurricanes Katrina, Rita, Gustav, and Lili exhibited that the hurricane storm surges immediately altered salinity levels and patterns in the lake and that recovery to normal conditions were distinguishable (Figure 5.6). Freshwater discharge proved to be the main force in salinity recovery after Hurricanes Katrina and Rita. An apparent disturbance from Hurricane Gustav can

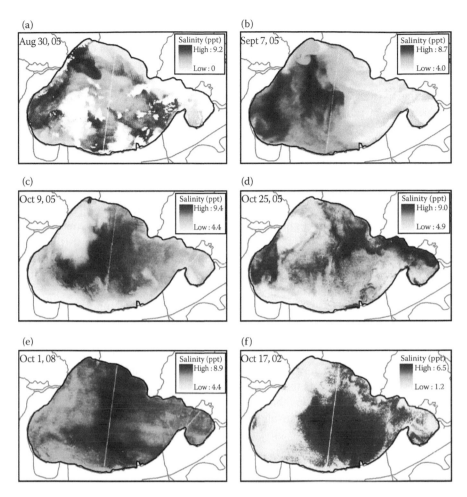

FIGURE 5.6 Predicted salinity levels in Lake Pontchartrain on different dates (on top left corner of each panel). The salinity was disturbed and affected by: (a) and (b) Hurricane Katrina (made its landfall on 30 Aug. 2005), (c) and (d) both Hurricane Katrina and Rita (Sept. 24, 2005), (e) Hurricane Gustav (Sept. 1, 2008), and (f) Hurricane Lili (Oct. 3, 2002).

be easily seen when comparing two salinity maps pre- and post-Gustav (maps on August 30, 2008, and October 1, 2008). In particular, the salinity levels increased from a range of 0.2 to 5.7 ppt pre-Gustav to 4.4 to 8.9 ppt post-Gustav. After Gustav's landfall, the large amount of freshwater discharge from the North Pass, Tangipahoa River, and Pearl River may have contributed to the low salinity levels found in the west and east, but salinity in central portions of the lake still remained high. Fourteen days following Hurricane Lili, salinity declined to 1.2 to 6.5 ppt (a much lower level) compared to 4.3 to 7.6 ppt on September 15, 2002. It can be concluded that the strong precipitation of the hurricane on the west side of the lake caused high freshwater discharge, consequently diluting the saltwater in the lake from the Gulf of Mexico.

5.4 CONCLUSIONS

Locally, estuarine water quality can be assessed with the analysis of water samples, but it would be unrealistic to collect enough field data to accurately assess water quality across an entire estuary. However, remote sensing offers the advantages in the course of overcoming spatial and temporal constraints while providing relatively inexpensive yet reliable information regarding water optics, which can be used to assess water quality (including salinity levels) across large estuarine water bodies. While dissolved salts have been found to have little effect on spectrally averaged light absorption in an estuary, the Lake Pontchartrain study has demonstrated that empirical models can be developed that relate the spectral reflectance of water to *in situ* salinity measurements. The strong relationship between salinity and water-leaving reflectance revealed by the model may be caused by close correlation of surface salinity level with light absorption of CDOM, as well as scattering of suspended solids. However, the optical properties can be affected by terrestrial properties within the estuarine drainage area including land use, land cover, soil texture, and soil type. These are all site-specific processes, requiring that an existing model be calibrated accurately for local use.

Once such a model has been calibrated and evaluated, spatiotemporal salinity maps can be produced to explore the affects that various scenarios have on salinity levels for a specific estuary. Overall, these maps can be especially helpful for evaluating potential habitat changes due to sea level rise and saltwater intrusion, assessing environmental stresses on the biodiversity of estuaries and their surrounding environments, and investigating the effects that land use changes to an estuary's catchment have on its waters. For example, maps that track changes following extreme weather events (such as hurricanes) would be invaluable for investigating the immediate impacts that saltwater intrusion (from storm surges) has on aquatic and coastal ecosystems.

REFERENCES

Adao, H., Alves, A. S., Patricio, J., Neto, J. M., Costa, M. J., and Marques, J. C. (2009). Spatial distribution of subtidal Nematoda communities along the salinity gradient in southern European estuaries. *Acta Oecologica—International Journal of Ecology*, 35, 287–300.

Agrawal, Y. C. (2005). The optical volume scattering function: Temporal and vertical variability in the water column off the New Jersey coast. *Limnology and Oceanography*, 50, 1787–1794.

Ahn, Y. H., Shanmugam, P., Moon, J. E., and Ryu, J. H. (2008). Satellite remote sensing of a low-salinity water plume in the East China Sea. *Annales Geophysicae*, 26, 2019–2035.

Arst, H. (2003). *Optical Properties and Remote Sensing of Multicomponental Water Bodies*. Springer, Parxis Publishing, Chichester, UK.

Baban, S. M. J. (1997). Environmental monitoring of estuaries estimating and mapping various environmental indicators in Breydon Water Estuary, UK, using Landsat TM imagery. *Estuarine, Coastal and Shelf Science*, 44, 589–598.

Binding, C. E. and Bowers, D. G. (2003). Measuring the salinity of the Clyde Sea from remotely sensed ocean colour. *Estuarine, Coastal and Shelf Science*, 57, 605–611.

Blough, N. V., Zafiriou, O. C., and Bonilla, J. (1993). Optical-absorption spectra of waters from the Orinoco River outflow—Terrestrial input of colored organic-matter to the Caribbean. *Journal of Geophysical Research—Oceans*, 98, 2271–2278.

Bowers, D. G. and Brett, H. L. (2008). The relationship between CDOM and salinity in estuaries: An analytical and graphical solution. *Journal of Marine Systems, 73*, 1–7.

Bowers, D. G., Evans, D., Thomas, D. N., Ellis, K., and Williams, P. J. L. (2004). Interpreting the colour of an estuary. *Estuarine, Coastal and Shelf Science*, 59, 13–20.

Bukata, R. P. (2005). *Satellite Monitoring of Inland and Coastal Water Quality: Retrospection, Introspection, Future Directions*. CRC Press, Taylor & Francis Group, Boca Raton, FL.

Calliari, D., Cervetto, G., Castiglioni, R., and Rodriguez, L. (2007). Salinity preferences and habitat partitioning between dominant mysids at the Rio de la Plata estuary (Uruguay). *Journal of the Marine Biological Association of the United Kingdom*, 87, 501–506.

Cardona, L. (2006). Habitat selection by grey mullets (Osteichthyes: Mugilidae) in Mediterranean estuaries: The role of salinity. *Scientia Marina*, 70, 443–455.

Christian, D. and Sheng, Y. P. (2003). Relative influence of various water quality parameters on light attenuation in Indian River Lagoon. *Estuarine, Coastal and Shelf Science*, 57, 961–971.

Clementson, L. A., Parslow, J. S., Turnbull, A. R., and Bonham, P. I. (2004). Properties of light absorption in a highly coloured estuarine system in south-east Australia which is prone to blooms of the toxic dinoflagellate Gymnodinium catenatum. *Estuarine, Coastal and Shelf Science*, 60, 101–112.

Colgan, C. S. (2009). The value of estuary regions in the U.S. economy. In *The Economic and Market Value of Coasts and Estuaries: What's at Stake?*, Pendleton, L. H. (ed.), Coastal Ocean Values Press, Washington, DC.

Cook, T. L., Sommerfield, C. K., and Wong, K. C. (2007). Observations of tidal and springtime sediment transport in the upper Delaware Estuary. *Estuarine, Coastal and Shelf Science*, 72, 235–246.

D'Sa, E. J., Hu, C., Muller-Karger, F. E., and Carder, K. L. (2002). Estimation of colored dissolved organic matter and salinity fields in case 2 waters using SeaWiFS: examples from Florida Bay and Florida Shelf. *Proceedings of the Indian Academy of Sciences—Earth and Planetary Sciences*, 111, 197–207.

D'Sa, E. J. and Miller, R. L. (2005). Bio-optical Properties of Coastal Waters. In *Remote Sensing of Coastal Aquatic Environments—Technologies, Techniques and Applications*, R. L. Miller, C. E. D. Castillo, and B. A. McKee (eds.), Springer, Dordrecht, the Netherlands, pp. 129–155.

Davies-Colley, R. J., Vant, W. N., and Smith, D. G. (2003). *Colour and Clarity of Natural Waters: Science and Management of Optical Water Quality*. Blackburn Press, Caldwell, NJ.

Dewidar, K. and Khedr, A. (2001). Water quality assessment with simultaneous Landsat-5 TM at Manzala Lagoon, Egypt. *Hydrobiologia*, 457, 49–58.

Diaz, R. J., and Rosenberg, R. (2008). Spreading dead zones and consequences for marine ecosystems. *Science*, 321, 926–929.

Doxaran, D., Cherukuru, N., and Lavender, S. J. (2006). Apparent and inherent optical properties of turbid estuarine waters: Measurements, empirical quantification relationships, and modeling. *Applied Optics*, 45, 2310–2324.

Doxaran, D., Cherukuru, R. C. N., and Lavender, S. J. (2005). Use of reflectance band ratios to estimate suspended and dissolved matter concentrations in estuarine waters. *International Journal of Remote Sensing*, 26, 1763–1769.

Dyer, K. R. (1997). *Estuaries—A Physical Introduciton*, 2nd Edition. John Wiley & Sons Ltd, New York.

Ferrari, G. M. and Dowell, M. D. (1998). CDOM absorption characteristics with relation to fluorescence and salinity in coastal areas of the southern Baltic Sea. *Estuarine, Coastal and Shelf Science*, 47, 91–105.

French, P. W. (2001). *Coastal Defences: Processes, Problems and Solutions*. Routledge Inc., London.

Fujii, T. and Raffaelli, D. (2008). Sea-level rise, expected environmental changes, and responses of intertidal benthic macrofauna in the Humber estuary, UK. *Marine Ecology—Progress Series*, 371, 23–35.

Garel, E., Pinto, L., Santos, A., and Ferreira, O. (2009). Tidal and river discharge forcing upon water and sediment circulation at a rock-bound estuary (Guadiana estuary, Portugal). *Estuarine, Coastal and Shelf Science*, 84, 269–281.

Godin, G. (1999). The propagation of tides up rivers with special considerations on the upper Saint Lawrence river. *Estuarine, Coastal and Shelf Science*, 48, 307–324.

Grabemann, H. J., Grabemann, I., Herbers, D., and Muller, A. (2001). Effects of a specific climate scenario on the hydrography and transport of conservative substances in the Weser estuary, Germany: A case study. *Climate Research*, 18, 77–87.

Greenwood, M. F. D., Matheson, R. E., McMichael, R. H., and MacDonald, T. C. (2007). Community structure of shoreline nekton in the estuarine portion of the Alafia River, Florida: Differences along a salinity gradient and inflow-related changes. *Estuarine, Coastal and Shelf Science*, 74, 223–238.

Hilton, T. W., Najjar, R. G., Zhong, L., and Li, M. (2008). Is there a signal of sea-level rise in Chesapeake Bay salinity? *Journal of Geophysical Research—Oceans*, 113, C09002, doi:10.1029/2007JC004247.

Hu, C. M., Chen, Z. Q., Clayton, T. D., Swarzenski, P., Brock, J. C., and Muller-Karger, F. E. (2004). Assessment of estuarine water-quality indicators using MODIS medium-resolution bands: Initial results from Tampa Bay, FL. *Remote Sensing of Environment*, 93, 423–441.

Julian, J. P., Doyle, M. W., Powers, S. M., Stanley, E. H., and Riggsbee, J. A. (2008). Optical water quality in rivers. *Water Resources Research*, 44, W10411, doi:10410.11029/12007WR006457.

Kahru, M. and Mitchell, B. G. (2001). Seasonal and nonseasonal variability of satellite-derived chlorophyll and colored dissolved organic matter concentration in the California Current. *Journal of Geophysical Research—Oceans*, 106, 2517–2529.

Keith, D. J., Yoder, J. A., and Freeman, S. A. (2002). Spatial and temporal distribution of coloured dissolved organic matter (CDOM) in Narragansett Bay, Rhode Island: Implications for phytoplankton in coastal waters. *Estuarine, Coastal and Shelf Science*, 55, 705–717.

Kemp, W. M., Boynton, W. R., Adolf, J. E., Boesch, D. F., Boicourt, W. C., Brush, G., Cornwell, J. C. et al. (2005). Eutrophication of Chesapeake Bay: Historical trends and ecological interactions. *Marine Ecology—Progress Series*, 303, 1–29.

Khorram, S. (1982). Remote-sensing of salinity in the San Francisco Bay delta. *Remote Sensing of Environment*, 12, 15–22.

Kirk, J. T. O. (1994). *Light and Photosynthesis in Aquatic Ecosystems*. Cambridge University Press, New York.

Kitheka, J. U., Obiero, M., and Nthenge, P. (2005). River discharge, sediment transport and exchange in the Tana Estuary, Kenya. *Estuarine, Coastal and Shelf Science*, 63, 455–468.

Kutser, T., Pierson, D., Tranvik, L., Reinart, A., Sobek, S., and Kallio, K. (2005a). Using satellite remote sensing to estimate the colored dissolved organic matter absorption coefficient in lakes. *Ecosystems*, 8, 709–720.

Kutser, T., Pierson, D. C., Kallio, K. Y., Reinart, A., and Sobek, S. (2005b). Mapping lake CDOM by satellite remote sensing. *Remote Sensing of Environment*, 94, 535–540.

Lane, R. R., Day, J. W., Marx, B. D., Reyes, E., Hyfield, E., and Day, J. N. (2007). The effects of riverine discharge on temperature, salinity, suspended sediment and chlorophyll a in a Mississippi delta estuary measured using a flow-through system. *Estuarine, Coastal and Shelf Science*, 74, 145–154.

Lavery, P., Pattiaratchi, C., Wyllie, A., and Hick, P. (1993). Water-quality monitoring in estuarine waters using the Landsat Thematic Mapper. *Remote Sensing of Environment*, 46, 268–280.

Lindeboom, H. (2002). The coastal zone: An ecosystem under pressure. In *Oceans 2020: Science, Trends, and the Challenge of Sustainability,* J. G. Field, G. Hempel, and C. P. Summerhayes (eds.), Island Press, Washington, DC, pp. 49–84.

Liu, W. C., Hsu, M. H., Chen, S. Y., Wu, C. R., and Kuo, A. Y. (2005). Water column light attenuation in Danshuei River estuary, Taiwan. *Journal of the American Water Resources Association,* 41, 425–435.

Lubben, A., Dellwig, O., Koch, S., Beck, M., Badewien, T., Fischer, S., and Reuter, R. (2009). Distributions and characteristics of dissolved organic matter in temperate coastal waters (Southern North Sea). *Ocean Dynamics,* 59, 263–275.

Lucas, L. V., Koseff, J. R., Cloern, J. E., Monismith, S. G., and Thompson, J. K. (1999a). Processes governing phytoplankton blooms in estuaries. I: The local production–loss balance. *Marine Ecology—Progress Series,* 187, 1–15.

Lucas, L. V., Koseff, J. R., Monismith, S. G., Cloern, J. E., and Thompson, J. K. (1999b). Processes governing phytoplankton blooms in estuaries. II: The role of horizontal transport. *Marine Ecology—Progress Series,* 187, 17–30.

Mallin, M. A. (1994). Phytoplankton ecology of North Carolina estuaries. *Estuaries,* 17, 561–574.

Mao, Q. W., Shi, P., Yin, K. D., Gan, J. P., and Qi, Y. Q. (2004). Tides and tidal currents in the pearl river estuary. *Continental Shelf Research,* 24, 1797–1808.

Martinez-Guijarro, R., Romero, I., Paches, M., del Rio, J. G., Marti, C. M., Gil, G., Ferrer-Riquelme, A., and Ferrer, J. (2009). Determination of phytoplankton composition using absorption spectra. *Talanta,* 78, 814–819.

McManus, J. (1998). Temporal and spatial variations in estuarine sedimentation. *Estuaries,* 21, 622–634.

Meehl, G. A., Stocker, T. F., Collins, W. D., Friedlingstein, P., Gaye, A. T., Gregory, J. M., Kitoh, A., et al. (2007). Global climate projections. In *Climate Change 2007: The physical science basis: Contribution of Working Group I to the Fourth Assessment Report of the Intergovernmental Panel on Climate Change,* S. Solomon, D. Qin, M. Manning, Z. Chen, M. Marquis, K. B. Averyt, M. Tignor, and H. L. Miller (eds.), Cambridge University Press, Cambridge, United Kingdom and New York, pp. 748–845.

Mitsch, W. J. and Gosselink, J. G. (1986). *Wetlands.* Van Nostrand Reinhold, New York.

Montagna, P. A., Estevez, E. D., Palmer, T. A., and Flannery, M. S. (2008). Meta-analysis of the relationship between salinity and molluscs in tidal river estuaries of southwest Florida, USA. *American Malacological Bulletin,* 24, 101–115.

Nieke, B., Reuter, R., Heuermann, R., Wang, H., Babin, M., and Therriault, J. C. (1997). Light absorption and fluorescence properties of chromophoric dissolved organic matter (CDOM), in the St. Lawrence Estuary (case 2 waters). *Continental Shelf Research,* 17, 235–252.

Nitsche, F. O., Ryan, W. B. F., Carbotte, S. M., Bell, R. E., Slagle, A., Bertinado, C., Flood, R., Kenna, T., and McHugh, C. (2007). Regional patterns and local variations of sediment distribution in the Hudson River Estuary. *Estuarine, Coastal and Shelf Science,* 71, 259–277.

Panfili, J., Mbow, A., Durand, J. D., Diop, K., Diouf, K., Thior, D., Ndiaye, P., and Lae, R. (2004). Influence of salinity on the life-history traits of the West African black-chinned tilapia (Sarotherodon melanotheron): Comparison between the Gambia and Saloum estuaries. *Aquatic Living Resources,* 17, 65–74.

Phlips, E. J., Lynch, T. C., and Badylak, S. (1995). Chlorophyll-a, tripton, color, and light availability in a shallow tropical inner-shelf lagoon, Florida Bay, USA. *Marine Ecology—Progress Series,* 127, 223–234.

Posey, M. H., Alphin, T. D., Harwell, H., and Allen, B. (2005). Importance of low salinity areas for juvenile blue crabs, Callinectes sapidus Rathbun, in river-dominated estuaries of southeastern United States. *Journal of Experimental Marine Biology and Ecology,* 319, 81–100.

Pozdnyakov, D. and Grassl, H. (2003). *Colour of Inland and Coastal Waters—A Methodology for Its Interpretation.* Praxis Publishing, Chichester, United Kingdom.

Pritchard, D. W. (1952). Salinity distribution adn circulation in the Chesapeake Bay Estuaries system. *Journal of Marine Research*, 11, 106–123.

Rabalais, N. N., Turner, R. E., and Wiseman, W. J. (2002). Gulf of Mexico hypoxia, A.K.A. "The dead zone." *Annual Review of Ecology and Systematics*, 33, 235–263.

Reis, J. L., Martinho, A. S., Pires-Silva, A. A., and Silva, A. J. (2009). Assessing the influence of the river discharge on the Minho Estuary tidal regime. *Journal of Coastal Research*, 2, 1405–1409.

Rochelle-Newall, E. J. and Fisher, T. R. (2002). Chromophoric dissolved organic matter and dissolved organic carbon in Chesapeake Bay. *Marine Chemistry*, 77, 23–41.

Siegel, D. A., Maritorena, S., Nelson, N. B., Hansell, D. A., and Lorenzi-Kayser, M. (2002). Global distribution and dynamics of colored dissolved and detrital organic materials. *Journal of Geophysical Research—Oceans*, 107, 3228, doi:3210.1029/2001JC000965.

Siegle, E., Schettini, C. A. F., Klein, A. H. F., and Toldo, E. E. (2009). Hydrodynamics and suspended sediment transport in the Camboriu Estuary—Brazil: Pre jetty conditions. *Brazilian Journal of Oceanography*, 57, 123–135.

Snyder, W. A., Arnone, R. A., Davis, C. O., Goode, W., Gould, R. W., Ladner, S., Lamela, G., et al. (2008). Optical scattering and backscattering by organic and inorganic particulates in US coastal waters. *Applied Optics*, 47, 666–677.

Szekielda, K.-H., Gobler, C., Gross, B., Moshary, F., Ahmed, S., and Wolff, J.-O. (2003). Spectral reflectance measurements of estuarine waters. *Ocean Dynamics*, 53, 98–102.

Vahatalo, A. V., Wetzel, R. G., and Paerl, H. W. (2005). Light absorption by phytoplankton and chromophoric dissolved organic matter in the drainage basin and estuary of the Neuse River, North Carolina (USA). *Freshwater Biology*, 50, 477–493.

Wang, F., and D'Sa, E. J. (2010). Potential of MODIS EVI in identifying hurricane disturbance to coastal vegetation in the Northern Gulf of Mexico. *Remote Sensing*, 2, 1–18.

Wang, F., Xu, Y. J., and Bourgeois-Calvin, A. (2006). Using Landsat imagery to model salinity change in Lake Pontchartrain due to Hurricanes Katrina and Rita. In *Coastal Environment and Water Quality,* Y. J. Xu and V. P. Singh (eds.), Water Resources Publications, Highlands Ranch, CO, pp. 467–480.

Wang, F. G. and Xu, Y. J. (2008). Development and application of a remote sensing-based salinity prediction model for a large estuarine lake in the US Gulf of Mexico coast. *Journal of Hydrology*, 360, 184–194.

Wang, M. Z., Lyzenga, D. R., and Klemas, V. V. (1996). Measurement of optical properties in the Delaware estuary. *Journal of Coastal Research*, 12, 211–228.

Watson, E. B. and Byrne, R. (2009). Abundance and diversity of tidal marsh plants along the salinity gradient of the San Francisco Estuary: Implications for global change ecology. *Plant Ecology*, 205, 113–128.

Wilson, M. A. and Farber, S. (2008). Accounting for ecosystem goods and services in coastal estuaries. In: *The Economic and Market Value of Coasts and Estuaries: What's at Stake?,* L. H. Pendleton (ed.), Coastal Ocean Values Press, Washington, DC.

Wolanski, E. (2007). *Estuarine Ecohydrology.* Elsevier Science, Amsterdam, the Netherlands.

Wu, K. and Xu, Y. J. (2007). Long-term freshwater inflow and sediment discharge into Lake Pontchartrain in Louisiana, USA. *Hydrological Sciences Journal*, 52, 166–180.

Xu, Y. J. (2006). Total nitrogen inflow and outflow from a large river swamp basin to the Gulf of Mexico. *Hydrological Sciences Journal*, 51, 531–542.

Xu, Y. J. and Wu, K. S. (2006). Seasonality and interannual variability of freshwater inflow to a large oligohaline estuary in the Northern Gulf of Mexico. *Estuarine, Coastal and Shelf Science*, 68, 619–626.

Yamamuro, M. (2000). Abundance and size distribution of sublittoral meiobenthos along estuarine salinity gradients. *Journal of Marine Systems*, 26, 135–143.

Yin, K. D., Harrison, P. J., Pond, S., and Beamish, R. J. (1995). Entrainment of nitrate in the Fraser River Estuary and its biological implications. 2. Effects of spring vs neap tides and river discharge. *Estuarine, Coastal and Shelf Science*, 40, 529–544.

Zheng, X. B., Dickey, T., and Chang, G. (2002). Variability of the downwelling diffuse attenuation coefficient with consideration of inelastic scattering. *Applied Optics*, 41, 6477–6488.

6 Multitemporal Remote Sensing of Coastal Sediment Dynamics

Paul Elsner, Tom Spencer,
Iris Möller, and Geoff Smith

CONTENTS

6.1 INTRODUCTION

The need for informed management of the coastal zone is increasingly pressing, considering climate change-related challenges and threats that are anticipated for coasts worldwide (Nicholls et al. 2007). In England and Wales alone, approximately 1.5 million people live in low-lying coastal areas that are protected by sea defenses (Jorissen et al. 2000). It has been estimated that 1.1 million properties in an area of 400,000 ha and a capital value of £137 billion are at risk of flooding or coastal erosion in England (DEFRA 2004). The expected acceleration of sea-level rise, in combination with the already observed loss of natural coastal defenses in the form of extensive areas of intertidal saltmarsh, is hence of significant concern for coastal managers (Turner 1995; Nicholls and de la Vega-Leinert 2008).

A central strategy for adapting to both salt marsh loss and an anticipated acceleration in the rate of sea-level rise is a shift from "hard engineering" approaches using rigid defense structures to "soft engineering" measures using beach nourishment and managed realignment to create natural coastal buffers (DEFRA 2002; Hanson et al. 2002). Managed realignment sites with elevations below the threshold for colonization by salt marsh plants will most probably initially convert to intertidal mudflats. Promotion of the rapid sedimentation and accretion of tidally imported mineral sediments is therefore a central objective in such management interventions, so that intertidal vegetation can subsequently colonize and stabilize such substrates (French 2006).

The sedimentation dynamics of mudflat environments are, compared to beaches and salt marshes, less well understood (Christie et al. 1999; Dyer et al. 2000). A central reason for this is their difficult accessibility, resulting in higher costs and overall research effort of field monitoring campaigns. Furthermore, most field-based methods represent point measurements, making it difficult to estimate the spatial variability of processes associated with a dynamic water body. Remote sensing approaches offer an alternative and supplemental technique in this respect because they collect synoptic and spatially coherent information. A number of studies have demonstrated the potential of such approaches to measure sediment concentration (Liedtke et al. 1995; Shimwell 1998; Froidefond et al. 2004). Most airborne coastal remote sensing applications rely on single imagery to capture sediment loading. Longer deployments of airborne platforms that target temporal sediment dynamics have rarely been implemented. A notable exception was a study of the Humber estuary, U.K. east coast. Here an airborne sensor was repeatedly flown along transects across the mouth of the estuary to derive sediment flux estimates (Robinson et al. 1999). Despite the promising results of this project, few follow-on projects have been implemented (Sterckx et al. 2006). This chapter presents the results of a coastal multitemporal monitoring campaign in which a series of overflights captured the inflow of tidal waters into a managed realignment site in South East England with a multispectral sensor.

6.2 FIELD SITE

The research presented here focused on sedimentation processes at the Tollesbury managed realignment project in Essex, South East England. The Tollesbury study area is located on the eastern shore of Tollesbury Fleet, a northern side arm of the Blackwater estuary (Figure 6.1). The Fleet is fronted by extensive estuarine fringing salt marsh (Allen 2000). The mean tidal range lies between 4.7 m at springs and 3.0 m on neap tides. The mean high water reaches 2.6 m above Ordnance Datum Newlyn (ODN, which approximates to mean sea level) at springs (MHWS) and 1.5 m at neaps (MHWN) (Reading et al. 2008).

The marshes are backed by seawalls and the marshes themselves are in places reinforced by brushwood groins to halt or slow erosion. Substantial marsh area has been lost over the past 100 years (van der Wal and Pye 2004). The earliest modern U.K. managed realignment site was established at Northey Island on the River Blackwater in 1991 (Leggett et al. 2004). Further interventions in this estuary have taken place at Abbots Hall (1992 and 2002), Tollesbury (1995), and Orplands (1995) (Figure 6.1).

The managed realignment site at Tollesbury was reclaimed more than 160 years ago. In August 1995, a 60 m wide gap was created in the outer seawall to reexpose an area of approximately 21 ha to the inflow and outflow of tidal waters. Prior to breaching, a new seawall was constructed along the 3-m (ODN) height contour at the east and southeast side of the site. In addition to this, a 100 m long and 2 m wide channel was cut to connect the old drainage network to the breach.

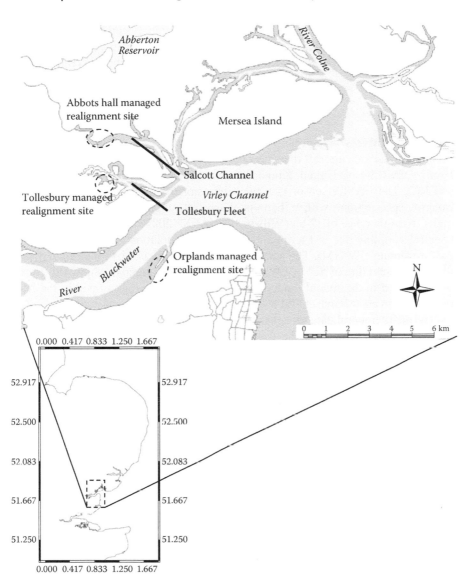

FIGURE 6.1 Overview of the Blackwater estuary and the location of managed realignment schemes. (Based on the Ordnance Survey, © Crown copyright. With permission.)

At the time of breaching, the site had a complex microtopography: its northeastern part was particularly low in the tidal frame with elevations of 1.0 m (ODN), compared to 1.20 m (ODN) toward the northwestern part of the site. There was a general rise in surface elevation in southwesterly direction, leading to heights of more than 2 m (ODN) in southern parts of the site. Most surface elevations within the site were therefore significantly lower compared to mature salt marshes at the seaward site of the seawalls, which typically have elevations in the range of 2.4 to

2.6 m (ODN) (Reading et al. 2008). This height differences can be attributed to the combined effect of dewatering and compaction of soils within the site since reclamation and continued vertical accretion of salt marshes outside the site (Hazeldon and Boorman 2001).

6.3 DATA

The data for this research project was acquired on July 29, 1999, by a CASI-2 sensor on board a research aircraft that was operated by the Airborne Research and Survey Facility (ARSF) of United Kingdom's Natural Environment Research Council (NERC). The CASI-2 sensor is a two-dimensional CCD array based pushbroom imaging spectrograph with a theoretical spectral range between 405 and 950 nm and a dynamic range of 12 bit (Wilson 1997; ARSF 2002). The potential theoretical spectral sampling rate is 1.8 nm and the spectral resolution is 2.2 nm full width at half maximum (FWHM). The across-track field of view is 54.4°, making it possible to record a scanline of 512 pixels. For the Tollesbury campaign, the CASI-2 sensor was operated in the spatial mode and 14 channels were programmed in the ARSF default ocean color bandset (Table 6.1).

The multitemporal data set consisted of a series of 18 overflights that captured the inflow during a spring tide at the Tollesbury managed realignment site. The first flight started at 10:05 GMT when the site was still in an unflooded state. Subsequently, the inflow of tidal waters was monitored with overflights every 8 to 10 minutes until near high tide at 12:30 GMT. Overflight 15 (12:04 GMT) did not capture the managed realignment site fully and therefore had to be omitted from further analysis. This omission created the biggest step in the time series, constituted by the 15 minutes interval between overflight 14 (11:56 GMT) and overflight 16 (12:11 GMT). The

TABLE 6.1
CASI Ocean Color Bandset

Band	Start nm	End nm
1	407.5	417.5
2	437.5	447.5
3	485	495
4	505	515
5	555	565
6	615	625
7	660	670
8	677.5	685
9	700	710
10	750	757.5
11	767.5	782.5
12	855	875
13	885	895
14	895	905

weather conditions during the entire flying campaign were clear. The flying height of 1400 m equated to a spatial resolution of 2.8 m. Concurrent ancillary data sets on the ground were not collected because the survey was scheduled at short notice as a target of opportunity. This constituted significant problems for the analysis of the remotely sensed data because established procedures using empirical models could not be applied.

6.4   DATA ANALYSIS

Due to the absence of concurrent ground reference data, it was necessary to establish a temporally robust hydro-optical model for the estuarine waters at Tollesbury that could then be employed for the analysis of the CASI images. This was done with the aid of the Water Colour Simulator (WASI), developed by Peter Gege from the Remote Sensing Technology Institute of the German Aerospace Center (DLR) (Gege 2001, 2003, 2005).

WASI is a physically based analytical model that uses nonlinear optimization procedures to analyze and simulate a wide range of hydro-optical parameters, including the concentration of suspended particulate matter. WASI links a series of models that describe the physical process of light traveling through: (a) the atmosphere, (b) the hydrosphere, and (c) the air–water interface. Although originally developed for freshwater applications, WASI offers the opportunity to parameterize the model for coastal Case-2 water environments such as those at Tollesbury.

To aid model parameterization and adaptation, extensive boat-based field spectroscopy was undertaken between August 2001 and May 2002, using a Geophysical and Environmental Research Corporation GER 1500 field spectrometer. Sampling was carried out in Eulerian mode from a moored position in the central part of the site. The sensor height above the water level was 1 m, resulting in a circular field of view of approximately 16 cm. The overall sampling procedure followed an approach similar to that of Doxaran et al. (2002) and included the extraction of water samples concurrent to spectral measurements. The water samples were subsequently analyzed for the concentration of its optically active ingredients (phytoplankton, yellow substance, and suspended particulate matter) using standard laboratory procedures (Edwards and Glysson 1999; Parsons et al. 1989).

The field measurements resulted in a set of 100 spectra and concurrent water samples that covered a wide range of tidal stages and suspended sediment concentrations (SSCs) at Tollesbury. This set was randomly split to provide independent data for model parameterization and validation, respectively.

In addition to boat-based spectroscopy, a number of terrestrial pseudoinvariant feature (PIF) spectra were collected that facilitated atmospheric correction and conversion of the CASI data to units of reflectance, using the empirical line method (Smith and Milton 1999).

The following WASI submodels were parameterized, based on the parameterization data set of *in situ* water spectra and concurrent water samples ($n = 50$): downwelling irradiance $E_d(\lambda)$, specular reflectance at the water surface $L_d^*(\lambda)$, and the subsurface irradiance reflectance spectrum $R(\lambda)$. $R(\lambda)$ was of particular interest for the eventual model inversion to determine suspended sediment concentration and was parameterized as a function of the absorption coefficient a and the backscattering coefficient b_b of the water body (Gordon et al. 1975; Sathyendranath and Platt 1997):

$$R(0) = f \frac{b_{bw} + Cb_{bc} + Sb_{bs}}{a_w + Ca_c + Ya_y + Sa_s + b_{bw} + Cb_{bc} + Sb_{bs}},$$ (6.1)

where f is a proportionality factor that is a function of the mean cosines for the downwelling and upwelling irradiances and the ratio of the upward-scattering coefficient and the backscattering coefficient, and the subscripts w, c, y, and s stand for water, phytoplankton, yellow substance, and suspended particulate matter, respectively. C, Y, and S stand for the corresponding concentrations. Both a and b_b exhibit some wavelength dependency but this property has been omitted here for display convenience.

The accuracy of the parameterized WASI model was determined with an absolute root-mean-square error (RMSE) of 22 mg·L⁻¹ and a normalized RMSE of 26% (Figure 6.2). Measured and modeled SSC had a coefficient of determination R^2 of 0.49 ($p < .01$). Further statistical analysis showed that neither mean difference nor relative error had a significant bias (Smith and Smith 2007). Additional validation was carried out with actual CASI airborne data collected during additional flights at Tollesbury in 2002. During these flights, concurrent ground reference data could be collected. The validation with this data had a much higher accuracy than results using field spectroscopy data. However, the size of the airborne CASI validation data set was very small ($n = 3$), which limited the statistical significance of the accuracy measures.

Model sensitivity was tested against parameter uncertainty for chlorophyll-a, yellow substance, suspended particulate matter, sunlight reflected at the water surface, anisotropy of the underwater light field, and water temperature. The analysis showed that the model was particularly sensitive to variations in reflection of direct sunlight at the water surface and changes in the underwater light field. Both parameters are a function of wave geometry that changes dynamically due to the local wave climate. Strong sensitivity existed also for negative variations of the backscattering coefficient for suspended particulate matter, b_{bs}. Little sensitivity was present to variations in the

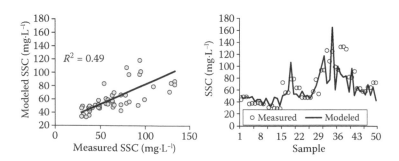

FIGURE 6.2 Accuracy of the calibrated hydro-optical model for the Tollesbury CASI data set.

concentration of chlorophyll-a, yellow substance, and suspended particulate matter (SPM) and to differences in water temperature. All vary substantially throughout the year. There are, therefore, no limitations to the use of the WASI model with spectral data from different seasons. More care, however, is needed when analyzing data collected under windy conditions and a correspondingly more dynamic wave climate. The calibrated model was eventually applied to invert the spectral water reflectance measured by the CASI sensor to estimates of SSC.

6.5   RESULTS

The inversion of the calibrated WASI model resulted in a series of maps that show estimates of suspended sediment concentration of the incoming tidal waters with a temporal resolution of between 7 and 15 minutes (Figure 6.3). The time series

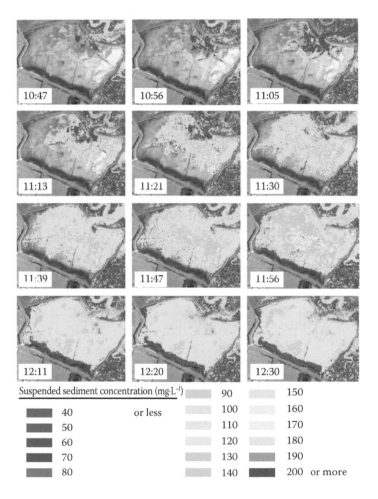

FIGURE 6.3   Times series of SSC maps at Tollesbury, depicting tidally driven sediment dynamics. Times (GMT) indicate timing of respective overflights.

illustrates the spatiotemporal sediment patterns within the Tollesbury managed realignment project during a summer spring tide. Three main stages can be identified: (1) tidal onset with increasing infilling of the external creek network and former drainage channels inside the managed realignment area until the bankfull stage was reached at approximately 11:00 GMT, with sediment loads under 100 mg·L^{-1}, (2) large-scale across-mudflat inflow into the site until 12:00 GMT with heterogeneous SSC patterns and loads of up to 200 mg·L and more, indicating considerable resuspension of sediment within the site, and (3) near-slack water phase until 12:31 GMT with more homogeneous SSC patterns in the range of 120 to 160 mg·L^{-1}.

The water height at respective overflights was determined by the waterline method (Lohani 1999; Foody et al. 2005) where a lidar digital elevation model (DEM), provided by the United Kingdom's Environment Agency, was combined with the measured near-infrared reflectance (CASI channels 12–14). This resulted in the estimation of the water levels at respective overflights with an RMSE of approximately 13 cm and made it possible to estimate tidal inflow and current velocity in the seawall breach for each interval. Figure 6.4 illustrates that the inflow in terms of water volume increased strongly until overflight 12 and remained at a comparatively high level thereafter. The tidal current velocity in the breach also peaked in the period between overflights 11 and 12, but decreased by more than 70% between overflights 12 and 18.

Assuming a fully mixed water column, the combination of these data sets with the modeled sediment concentration at the breach and within the site allowed imported sediment mass and total within-site suspended sediment to be inferred for all stages (Figure 6.5). This then served as the basis for interoverflight sediment balances that quantified the differences between total (cumulative) imported sediment and gain of overall suspended sediment within the site for each interval. If the increase in sediment mass in the site's water column is larger than the sediment mass imported through the breach, then this indicates a resuspension-dominated interval. The opposite case describes a deposition-dominated phase.

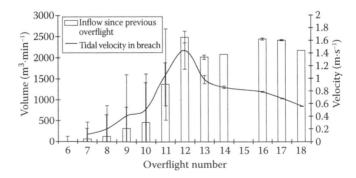

FIGURE 6.4 Tidal hydrodynamics estimated from water height measurements from successive images using the waterline method, normalized to m^3·min^{-1}. Current velocity in the breach calculated by assuming breach width of 60 m and bed elevation of 1 m. Error bars denote one RMSE.

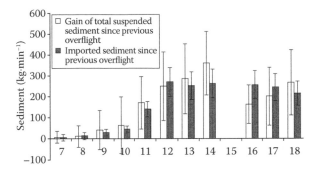

FIGURE 6.5 Relationship between gain of total suspended sediment in the water column and imported sediment between overflights. Error bars denote one RMSE.

The results illustrated in Figure 6.5 showed that nearly all intervals prior to over-flight 14 were resuspension dominated and that the phase between overflights 14 to 17 was deposition dominated. This correlates reasonably well with the estimated flow velocity in the breach and associated shear stress (Figure 6.4). There are two interesting exceptions to this trend. The interval between overflight 11 to 12 was deposition dominated. The last interval between overflights 17 and 18 was again resuspension dominated, despite experiencing a low tidal current velocity near slack water. A possible explanation for these exceptions might include the impact of wind-generated waves that are known to alter short-term tidal hydrodynamics and sediment processes (Christie et al. 1999) and have been reported to be significant across the Tollesbury site (Reading et al. 2008).

It should be noted, however, that the error bars indicate a considerable uncertainty. When modeling such derivatives, it is important to account for the uncertainty that is introduced by propagating errors of respective base data layers. In the framework of this project, these were the errors of the lidar DEM (RMSE 13 cm) and the relative error of the hydro-optical model (26% of SPM). The errors of the presented derivative data sets were estimated using standard error propagation theory for bivariate models (Burrough and McDonnell 1998). This included Monte Carlo analysis where, for respective overflights, 100 simulations of water height and 100 simulations of sediment load where constructed, resulting in 10,000 pairwise combinations.

In addition to analyzing total within-site bulk sediment budgets in the temporal domain, the modeled SSC maps were also analyzed spatially. An example parameter is the amount of sediment suspended in the water column of individual pixels. The choice for the specific locations was guided by the position of 20 sediment erosion bridges (SEBs) at which long-term accretion has been monitored since the breach in 1995 (Reading et al. 2008). We may display the tidal patterns for the three SEB locations together with the highest and lowest accretion, respectively (Figure 6.6). It is apparent that the water columns at high-accretion locations tended to contain more sediment at later stages of the tide, compared to low-accretion locations.

A subsequent linear regression between both parameters resulted in coefficients of determination (R^2) of 0.72 ($p < .05$) for the three highest and lowest locations.

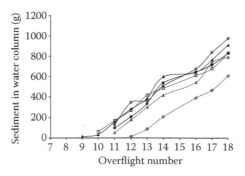

FIGURE 6.6   Time series of suspended sediment in water column at six example locations for which long-term accretion data was available. Solid lines represent locations with high accretion, and dashed lines indicate low-accretion locations.

When the same analysis was performed for all 20 transect locations where long-term accretion has been measured, the R^2 decreased to 0.33 but remained highly significant ($p < .01$). It was thus possible to establish a statistically significant relationship between single tide dynamics observed from the July 1999 CASI images and long-term accretion measurements on the ground between 1995 and 2000.

6.6   DISCUSSION

The accuracy of the WASI model demonstrates that it is possible to calibrate a temporally robust analytical hydro-optical model with good predictive power. It should be noted in this respect that estimations of SSC by traditional *in situ* methods are also not error free. For instance, Binding et al. (2005) found that laboratory-based gravimetric measurements of suspended sediment concentration had an average standard error of 12.4%. Measurements by optical beam transmissometer can also be sensitive to short-term variations of particle size and composition, resulting in errors of more than 20% (Jago and Bull 2000). It thus appears realistic to collect and analyze suspended sediment data from future airborne data of locations such as Tollesbury without the need for concurrent ground measurements.

The successful calibration of the WASI model also indicates that the Water Colour Simulator proved to be a versatile and powerful modeling and analysis tool that can be adjusted to aquatic conditions other than clear freshwater environments for which it originally was developed (Gege 2001, 2003). However, the sensitivity analysis demonstrated that the model is sensitive to changes in the geometric light field that is a function of surface waves. Thus the transferability of the calibrated model should be limited to data collected during relatively calm conditions.

The sensitivity analysis further showed that the established hydro-optical model lacks robustness against changes in the backscattering coefficient of inorganic particulate matter b_{bs}. This parameter is a function of particle size distribution, particle shape, and mineralogy (Bukata et al. 1995; Novo et al. 1989). Considering the tidal and seasonal dynamics of an environment such as the Blackwater estuary, where particle flocculation is significant and highly spatiotemporally variable, the assumption of unchanged sediment scattering properties appears unrealistic (Benson and French 2007; Doxaran et al. 2002; Manning and Bass 2006). Binding et al. (2005) found that scattering coefficients of inorganic sediments in the Irish Sea could vary to up to one order of magnitude. This was particularly significant as scattering of sediment particles dominates the spectral signal of such midturbid waters. Careful case-to-case parameterization of sediment scattering led to a substantial reduction of the errors of hydro-optical models (Binding et al. 2003, 2005). Similar observations have been made for the calibration of optical beam transmissometers when deployed for long-term monitoring projects (Jago and Bull 2000).

The model calibration identified for this research project appears to be temporally robust. This may be because all data sets were collected between May and September. It is not known how far the summer calibration of WASI as used in this project would perform for the analysis of data collected in the winter period. Apart from seasonal effects, tidal waters are also known for their short-term variability in suspended sediment concentrations. Observations for mudflats of The Wash, U.K. east coast have identified a strong quarter-diurnal signal, corresponding to changes in flow velocity and sediment resuspension and settlement processes (Jago and Bull 2000).

For the Tollesbury setting where data with a temporal resolution of <10 min is available, it would hence be interesting to know how far inherent optical properties (IOPs) might change during a tidal cycle at the same location and how strongly they vary spatially. Both factors directly affect the stability and accuracy of hydro-optical models. This issue links to the relationship between flocculation and sediment scattering. However, despite the awareness that flocculation leads to changes in spectral signature (Bukata et al. 1995), only limited information is available about the detailed optical behavior of flocs (Martin 2004; Mobley 1994).

The dynamics identified at Tollesbury are in good overall agreement with intertidal short-term sedimentation processes reported elsewhere (Christie et al. 1999; Jago and Bull 2000). It is possible to identify stages of resuspension, transportation/ relocation, and deposition/sediment settling. Of particular interest are the highly significant relationships between parameters from the period between overflights 14 and 16 and long-term measurements of accretion (Cahoon et al. 2000; Garbutt et al. 2006; Reading et al. 2008). This shows that in particular circumstances a close link may exist between processes observed over a single tide and long-term (>annual) accretion processes, if the spatial heterogeneity of such processes can be resolved. This study also confirms the added value that can be gained by utilizing and linking data sets from different research projects and approaches.

6.7 CONCLUSIONS

Multitemporal remote sensing offers significant potential to capture and quantify coastal processes such as intertidal suspended sediment dynamics that realistically cannot be obtained by traditional field-based sampling methods. Significant practical challenges have to be overcome, however, before such data collection campaigns can be realized. A central obstacle for implementing remote sensing as an operational research tool for monitoring dynamic coastal environments is logistical problems.

To make remote sensing a more integral part of the methodological suite employed in coastal research, it would therefore be necessary to have unrestricted access to the monitoring equipment when weather and tidal conditions open rare windows of opportunity. This appears to be unlikely in situations where sophisticated airborne platforms have to be shared with the wider remote sensing community. An alternative platform for many applications might be unmanned aerial vehicles (UAVs), which are equipped with relatively simple, low-cost digital multispectral sensors. This would provide a system that could be independently deployed in the field and thus makes remote sensing a more operational monitoring tool for environmental research.

ACKNOWLEDGMENTS

We wish to acknowledge the support of the Airborne Research and Survey Facility and the Field Spectroscopy Facility, U.K. Natural Environment Research Council. We thank the developer of the Water Colour Simulator, Dr. Peter Gege, German Aerospace Centre (DLR), for making this excellent software available to the scientific community. Adrian Hayes and his team from the Cambridge University Physical Geography Laboratories provided invaluable assistance by developing bespoke field equipment. Paul Elsner is grateful for financial support from the German Academic Exchange Service (DAAD) and for a Cambridge University Domestic Research Studentship.

REFERENCES

Allen, J. R. L. (2000). Morphodynamics of Holocene salt marshes: A review sketch from the Atlantic and Southern North Sea coasts of Europe. *Quaternary Science Reviews*, 19(12), 1155–1231.

ARSF (2002). *CASI-2 sensor. User Manual of the Airborne Research and Survey Facility of the National Environment Research Council*. Natural Environment Research Council, Oxford, United Kingdom.

Benson, T. and French, J. R. (2007). InSiPID: A new low cost instrument for in situ particle size measurements in estuaries. *Journal of Sea Research*, 58, 167–188.

Binding, C. E., Bowers, D. G., and Mitchelson-Jacob, E. G. (2003). An algorithm for the retrieval of suspended sediment concentrations in the Irish Sea from SeaWiFS ocean colour satellite imagery. *International Journal of Remote Sensing*, 24(19), 3791–3806.

Binding, C. E., Bowers, D. G., and Mitchelson-Jacob, E. G. (2005). Estimating suspended sediment concentrations from ocean colour measurements in moderately turbid waters; the impact of variable particle scattering properties. *Remote Sensing of Environment*, 94, 373–383.

Bukata, R. P., Jerome, J. H., Kondratyev, K. Y., and Pozdnyakov, D. V. (1995). *Optical Properties and Remote Sensing of Inland and Coastal Waters*. CRC Press, Boca Raton, FL.

Burrough, P. A. and McDonnell, R. A. (1998). *Principles of Geographic Information Systems*. Oxford University Press, Oxford, United Kingdom.

Cahoon, D. R., French, J. R., Spencer, T., Reed, D., and Moeller, I. (2000). Vertical accretion versus elevational adjustment in UK saltmarshes: An evaluation of alternative methodologies. In *Coastal and Estuarine Environments: Sedimentology, Geomorphology and Geoarchaeology*, K. Pye and J. R. L. Allen (eds.), Special Publication. The Geological Society, London, pp. 223–238.

Christie, M. C., Dyer, K. R., and Turner, P. (1999). Sediment flux and bed level measurements from a macro tidal mudflat. *Estuarine, Coastal and Shelf Science*, 49, 667–688.

DEFRA. (2002). *Managed Realignment Review: Project Report* (Policy Research Project FD 2008). Department of Environment, Food, and Rural Affairs, London.

DEFRA. (2004). *Making Space for Water. Developing a New Government Strategy for Flood and Coastal Erosion Risk Management in England. A Consultation Exercise.* Department of Environment, Food, and Rural Affairs, London.

Doxaran, D., Froidefond, J. M., Lavender, S., and Castaing, P. (2002). Spectral signature of highly turbid waters. Application with SPOT data to quantify suspended particulate matter concentrations. *Remote Sensing of Environment*, 81, 149–161.

Dyer, K. R., Christie, M. C., Feates, N., Fennessy, M. J., Pejrup, M., and van der Lee, W. (2000). An investigation into processes influencing the morphodynamics of an intertidal mudflat, the Dollart Estuary, the Netherlands: I. Hydrodynamics and suspended sediment. *Estuarine, Coastal and Shelf Science*, 50, 607–625.

Edwards, T. K. and Glysson, G. D. (1999). *Field Methods for Measurement of Fluvial Sediment.* U.S. Geological Survey, Reston, VA.

Foody, G., Muslim, A. M., and Atkinson, P. M. (2005). Super-resolution mapping of the waterline from remotely sensed data. *International Journal of Remote Sensing*, 26(24), 5381–5392.

French, P. (2006). Managed realignment: The developing story of a comparatively new approach to soft engineering. *Estuarine, Coastal and Shelf Science*, 67, 409–423.

Froidefond, J. M., Lahet, F., Hu, C., Doxaran, D., Guiral, D., Prost, M. T., and Ternon, J.-F. (2004). Mudflats and mud suspension observed from satellite data in French Guiana. *Remote Sensing of Environment*, 208(2–4), 153–168.

Garbutt, R. A., Reading, C. J., Wolters, M., Gray, A. J., and Rothery, P. (2006). Monitoring the development of intertidal habitats on former agricultural land after the managed realignment of coastal defences at Tollesbury, Essex, UK. *Marine Pollution Bulletin*, 53, 155–164.

Gege, P. (2001). The Water Colour Simulator WASI: A Software Tool for Forward and Inverse Modeling of Optical *in situ* Spectra, *IGARSS'01 Proceedings of IEEE Geoscience and Remote Sensing Symposium*, Sydney, Australia, pp. 2743–2745.

Gege, P. (2005). *The Water Colour Simulator WASI.* User manual for version 3, DLR-IB 564–1/2005, http://www.opairs.aero/publications_en.html.

Gege, P. (2003). WASI—A software tool for water spectra. *Backscatter*, 14(1), 22–24.

Gordon, H. R., Brown, O. B., and Jacobs, M. M. (1975). Computed relationships between the inherent and apparent optical properties of a flat homogeneous ocean. *Applied Optics*, 14, 417–427.

Hanson, H., Brampton, A., Capobianco, M., Dette, H. H., Laustrup, C., Lechuga, A., and Spanhoff, R. (2002). Beach nourishment projects, practices, and objectives—A European overview. *Coastal Engineering Journal*, 47(2), 81–111.

Hazeldon, J. and Boorman, L. A. (2001). Soils and "managed retreat" in South-East England. *Soil Use and Management*, 17, 150–154.

Jago, C. F. and Bull, C. F. J. (2000). Quantifications of errors in transmissometer-derived concentration of suspended matter in the coastal zone: implications for flux determinations. *Marine Geology*, 169, 273–286.

Jorissen, R., Litjens, J., and Menedez Lorenzo, A. (2000). Flooding Risk in Coastal Areas: Risks, Safety Levels and Probabilistic Techniques in Five Countries along the North Sea. Ministry of Transport, Public Works and Water Management, Koningskade 4, The Hague, the Netherlands.

Leggett, D. J., Cooper, N., and Harvey, R. (2004). *Coastal and Estuarine Managed Realignment—Design Issues.* DEFRA/Environment Agency Report FD2413/TR. CIRIA, London.

Liedtke, J., Roberts, A., and Luternauer, J. (1995). Practical remote sensing of suspended sediment concentration. *Photogrammetric Engineering & Remote Sensing*, 61, 167–175.

Lohani, B. (1999). Construction of a digital elevation model of the Holderness coast using the waterline method and Airborne Thematic Mapper data. *International Journal of Remote Sensing*, 20(3), 593–607.

Manning, A. J. and Bass, S. (2006). Variability in cohesive sediment settling fluxes: Observations under different estuarine tidal conditions. *Marine Geology*, 235, 177–192.

Martin, S. (2004). *Introduction to Ocean Remote Sensing*. Cambridge University Press, Cambridge, United Kingdom.

Mobley, C. D. (1994). *Light and Water. Radiative Transfer in Natural Waters*. Academic Press, San Diego, CA.

Nicholls, R. J., Wong, P. P., Burkett, V. R., Codignotto, J. O., Hay, J. E., McLean, R. F., Ragoonaden, S., and Woodroffe, C. D. (2007). Coastal systems and low-lying areas. In *Climate Change 2007: Impacts, Adaptation and Vulnerability. Contribution of Working Group II to the Fourth Assessment Report of the Intergovernmental Panel on Climate Change*, M. L. Parry, O. F. Canziani, J. P. Palutikof, P. J. van der Linden and C. E. Hanson (eds.), Cambridge University Press, Cambridge, United Kingdom, pp. 315–356.

Nicholls, R. J. and de la Vega-Leinert, A. (2008). Implication of sea-level rise for Europe's coasts: An introduction. *Journal of Coastal Research*, 24(2), 285–287.

Novo, E. M. M., Hansom, K. D., and Curran, P. J. (1989). The effect of sediment type on the relationship between reflectance and suspended sediment concentration. *International Journal of Remote Sensing*, 10(7), 1283–1289.

Parsons, P., Mita, Y., and Lalli, L. M. (1989). *A Manual of Chemical and Biological Methods for Seawater Analysis*. Pergamon Press, Oxford, United Kingdom.

Reading, C. J., Garbutt, A., Watts, C. W., Rothery, P., Turk, A., Yates, M., Boffey, C., Saunders, J., and Wolters, M. (2008). *Managed Realignment at Tollesbury*. CEH: Project Report Number C00356, R&D Technical Report FD 1922/TR. Department for Environment, Food, and Rural Affairs, London.

Robinson, M.-C., Morris, K. P., and Dyer, K. R. (1999). Deriving fluxes of suspended particulate matter in the Humber Estuary, UK, using airborne remote sensing. *Marine Pollution Bulletin*, 37(3–7), 155–163.

Sathyendranath, S. and Platt, T. (1997). Analytic model of ocean color. *Applied Optics*, 36(12), 2620–2629.

Shimwell, S. J. (1998). *Remote Sensing of Suspended Sediment off the Holderness Coast*. Proudman Oceanographic Laboratory, Plymouth, United Kingdom.

Smith, G. M. and Milton, E. J. (1999). The use of the empirical line method to calibrate remotely sensed data to reflectance. *International Journal of Remote Sensing*, 20, 2653–2662.

Smith, J. and Smith, P. (2007). *Environmental Modelling*. Oxford University Press, Oxford, United Kingdom.

Sterckx, S., Knaeps, E., Bollen, M., Trouw, K., and Houthuys, R. (2006). *Operational Remote Sensing Mapping of Estuarine Suspended Sediment Concentrations (ORMES), Hydro06—Evolutions in Hydrography*, Antwerp, Belgium.

Turner, K. M. (1995). Managed realignment as a coastal defense option. *Biological Journal of the Linnean Society*, 56, 217–219.

van der Wal, D. and Pye, K. (2004). Patterns, rates and possible causes of saltmarsh erosion in the Greater Thames area (UK). *Geomorphology*, 61, 373–391.

Wilson, A. K. (1997). An integrated system for airborne remote sensing. *International Journal of Remote Sensing*, 18, 1889–1901.

7 Estimating Total Phosphorus Impacts in a Coastal Bay with Remote Sensing Images and *in Situ* Measurements

Ni-Bin Chang and Kunal Nayee

CONTENTS

7.1   INTRODUCTION

7.1.1   GENERAL BACKGROUND

Eutrophication is a main concern in many coastal environments. About 65% of United States estuaries have moderate to high eutrophic conditions (Bricker et al. 1999; NOAA 2011); the most eutrophic estuaries are in the Gulf of Mexico (NOAA 2011). Nutrients are normally the key triggers of eutrophication, which can result in fish kills, loss of habitats, toxic algae blooms, commercially unproductive aquaculture, public health concerns, and loss of water resources (Figure 7.1). The primary and secondary symptoms are intimately tied with each other in response to

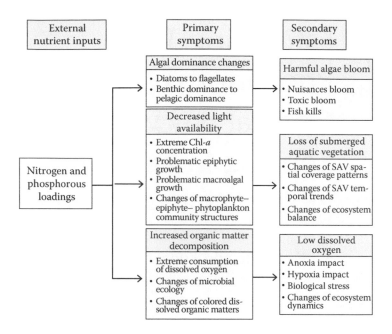

FIGURE 7.1 Summary of primary and secondary symptoms of eutrophication in coastal waters. (Based on Bricker, S. B., Clement, C. G., Pirhalle, D. E., Orlando, S. P., and Farrow, D. R. G., *National Estuarine Eutrophication Assessment, Effects of Nutrient Enrichment in the Nation's Estuaries*, The National Oceanic and Atmospheric Administration, p. 71, 1999.)

the changing nitrogen and phosphorus loadings in aquatic environments. They cohesively lead to the changes of submerged aquatic vegetation and macrophyte–epiphyte–phytoplankton community structures that eventually would cause the changes of ecosystem balance.

7.1.2 GEOGRAPHICAL ENVIRONMENT

Tampa Bay, the largest open-water estuary in Florida, extends approximately 56 km inland from the Gulf of Mexico and is 9 to 16 km wide along most of its length. Four segments make up the open-water section: Old Tampa Bay, Hillsborough Bay, Middle Tampa Bay, and Lower Tampa Bay (Figure 7.2). Spanning more than 900 km², with a drainage area nearly six times as large, Tampa Bay and its four major coastal watersheds stretch from the Hillsborough River to Alafia River, Little Manatee River, and Manatee River. The Greater Tampa Bay region included about 4.6 million permanent residents by 2010, with Tampa as the region's largest city in size and population. The Tampa Bay Basin also includes some of the state's most productive agricultural lands. Other significant economic components of this region include phosphate and other mining, power generation industries, and tourism and recreation.

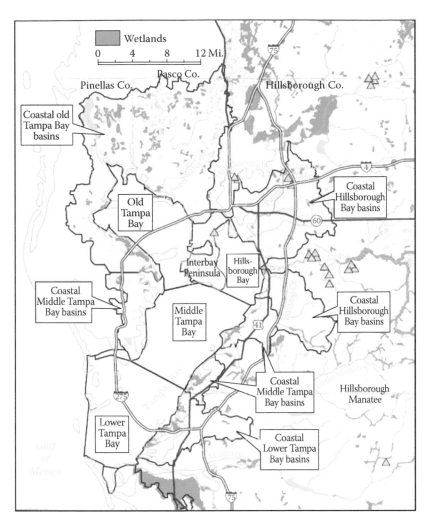

FIGURE 7.2 Tampa Bay Basin and bay segments. (From Florida Department of Environmental Protection [FDEP], http://tlhdwf2.dep.state.fl.us/basin411/tampa/status/Fig2-4.pdf.)

According to the Florida Department of Environmental Protection (FDEP 2011), Hillsborough Bay, the smallest of the four segments, receives runoff from a large portion of the city of Tampa. The Hillsborough and Alafia rivers drain into Hillsborough Bay, along with a number of smaller tributaries. Middle Tampa Bay receives runoff from the Little Manatee River and drainage from smaller tributaries along the Hillsborough and Pinellas county coastlines. Old Tampa Bay receives runoff from portions of Clearwater, St. Petersburg, and Tampa. Lower Tampa Bay, which has the largest volume of the four segments, connects the mouth of the bay to the Gulf of Mexico. The Manatee River, which receives runoff from the city of Bradenton, flows into the southern portion of this bay segment.

Overall, the Hillsborough, Alafia, Manatee, and Little Manatee rivers contribute significant flows to Tampa Bay (Figure 7.2). To the north, the Hillsborough River, with a drainage area of 1728 km^2, arises in the Green Swamp near the juncture of Hillsborough, Pasco, and Polk counties and flows 87 km through Pasco and Hillsborough counties to an outlet in the city of Tampa on Tampa Bay. In the middle, the Alafia River is 40 km long with a watershed of 870 km^2 in Hillsborough County flowing into Tampa Bay. The Little Manatee and Manatee rivers are located close to Middle Tampa Bay and Lower Tampa Bay, respectively. The Little Manatee River with an estimated drainage area of 570 km^2, flows almost 64 km from east of Fort Lonesome through southern Hillsborough County and northern Manatee County, Florida, into Tampa Bay. The Manatee River, a 97 km long river in Manatee County, Florida, has a watershed that spans 927 km^2. The river arises in the northeastern corner of Manatee County and flows into the Gulf of Mexico at the southern edge of Tampa Bay. Lake Manatee, an artificial reservoir, is located about midway in the river's course. The lower part of the river is an estuary, with Bradenton and other cities located along its banks.

The coastal area between the Little Manatee and Manatee rivers contains numerous bays, bayous, and tidal tributaries that drain into Tampa Bay (Figure 7.2). The most prominent of these surface waters are Cockroach Bay and Terra Ceia Bay, fed by major tributaries Cockroach Creek and Frog Creek, respectively. This small drainage basin includes a predominately urban area adjacent to the western side of the bay in Pinellas County and a predominately agricultural area on the eastern side of the bay in Hillsborough County. The basin also includes the smaller drainage basins north and south of the Little Manatee River and on the western side of the bay in the city of St. Petersburg.

7.1.3 WATER QUALITY CONDITIONS AND STUDY OBJECTIVES

Nutrients including nitrogen and phosphorus are essential for algal growth. Algal biomass is often expressed as chlorophyll-*a* (Chl-*a* hereafter) concentration. Increased nutrient supply, generally from anthropogenic sources such as wastewater treatment plants, leads to accelerated eutrophication (Figure 7.3). Elevated Chl-*a* concentrations are indicative of advanced trophic state. Algal biomass affects light attenuation, which in turn affects water clarity and dissolved oxygen concentrations. Monitoring water quality of the bay is critical to the success of the ecosystem restoration and protection. The Tampa Bay Estuary Program (TBEP) that provides essential field data for this study was established in 1991 to assist the community in developing a comprehensive plan to restore and protect Tampa Bay. The TBEP monitoring for Tampa Bay includes a water quality component designed to answer the following questions: (1) Are phytoplankton biomass levels (Chl-*a* hereafter) above, below, or consistent with established bay segment targets? (2) Are nutrient concentrations increasing, decreasing, or remaining stable? (3) Is water clarity increasing, decreasing, or remaining stable? (4) Is the areal extent of low dissolved oxygen concentrations (<2 mg·L^{-1}) increasing, decreasing, or remaining stable?

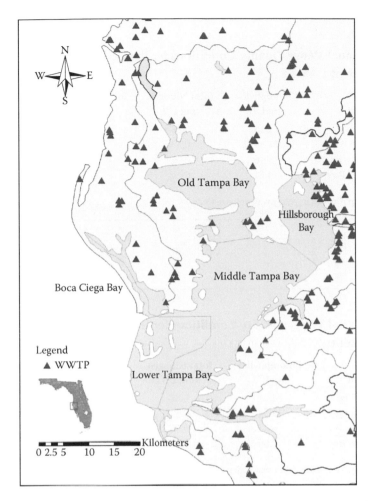

FIGURE 7.3 Permitted wastewater treatment facilities in Tampa Bay region. (Data from Florida Department of Environmental Protection [FDEP].)

The long-term trends of Chl-*a* and total phosphorus (TP) concentrations from 1974 to 1998 exhibit a decreasing trend across Hillsborough Bay, Old Tampa Bay, Middle Tampa Bay, and Lower Tampa Bay based on parametric trend tests (Janicki et al. 2001). Long-term improvements in trophic status of the bay have been observed (Janicki et al. 2001), yet Hillsborough Bay has reflected the poorest water quality conditions of the mainstem bay segments; the best water quality conditions are typically observed in Lower Tampa Bay (Janicki et al. 2001); and water quality along the western shore of Old Tampa Bay is poorer than that observed along the eastern shore of the segment (Janicki et al. 2001). Mean annual water quality conditions for periods 1996 to 1998, 2002 to 2003, and 2007 to 2008 associated with each bay segment (Tables 7.1 through 7.3) comparatively confirm a general improvement of water quality.

TABLE 7.1
Mean Annual Water Quality Conditions for the Period 1997–1998 for Each Bay Segment

Bay Segment	Chlorophyll-a ($\mu g \cdot L^{-1}$)	Total Nitrogen ($mg \cdot L^{-1}$)	Total Phosphorus ($mg \cdot L^{-1}$)	Secchi Disk Depth (m)
Old Tampa Bay	9.15	0.38	0.38	1.57
Hillsborough Bay	11.58	0.50	0.47	1.22
Middle Tampa Bay	7.46	0.36	0.36	1.80
Lower Tampa Bay	4.21	0.29	0.29	2.62

Source: http://www.tampabay.wateratlas.usf.edu/.

TABLE 7.2
Mean Annual Water Quality Conditions for the Period 2002–2003 for Each Bay Segment

Bay Segment	Chlorophyll-a ($\mu g \cdot L^{-1}$)	Total Nitrogen ($mg \cdot L^{-1}$)	Total Phosphorus ($mg \cdot L^{-1}$)	Secchi Disk Depth (m)
Old Tampa Bay	8.01	0.87	0.14	1.57
Hillsborough Bay	13.54	0.83	0.24	1.41
Middle Tampa Bay	6.45	0.76	0.15	2.20
Lower Tampa Bay	7.09	0.62	0.20	2.75

Source: http://www.tampabay.wateratlas.usf.edu/.

TABLE 7.3
Mean Annual Water Quality Conditions for the Period 2007–2008 for Each Bay Segment

Bay Segment	Chlorophyll-a ($\mu g \cdot L^{-1}$)	Total Nitrogen ($mg \cdot L^{-1}$)	Total Phosphorus ($mg \cdot L^{-1}$)	Secchi Disk Depth (m)
Old Tampa Bay	8.61	0.47	0.15	1.69
Hillsborough Bay	8.48	0.55	0.23	1.85
Middle Tampa Bay	5.87	0.40	0.16	2.01
Lower Tampa Bay	6.93	0.28	0.10	2.92

Source: http://www.tampabay.wateratlas.usf.edu/.

The spatiotemporal distributions of TP can be presented systematically on a seasonal basis in 2008 based on the online Web database (http://www.tampa bay.water atlas.usf.edu/) of Tampa Bay Water Atlas (Figures 7.4 through 7.7). Hillsborough Bay still reflects the poorest water quality conditions of the main-stem bay segments, while the best water quality conditions are typically observed in Lower Tampa Bay, and water quality along the western shore of Old Tampa Bay is still poorer than that observed along the eastern shore of the segment. However, daily flood and ebb currents in and out of the bay may compound the situation of nutrient loads. Locations of permitted wastewater treatment facilities in the Tampa Bay region indicate high levels of nutrient may be disposed of from these facilities

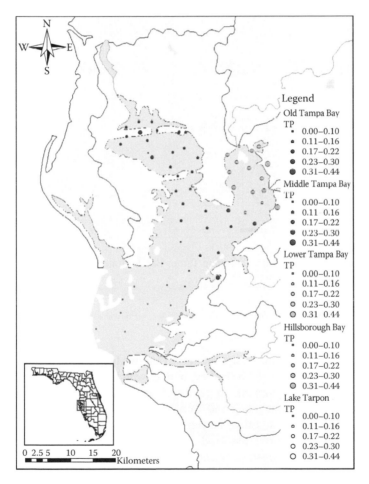

FIGURE 7.4   Average TP distribution in Tampa Bay, Spring 2008. (Data from http://www .tampabay.wateratlas.usf.edu/.)

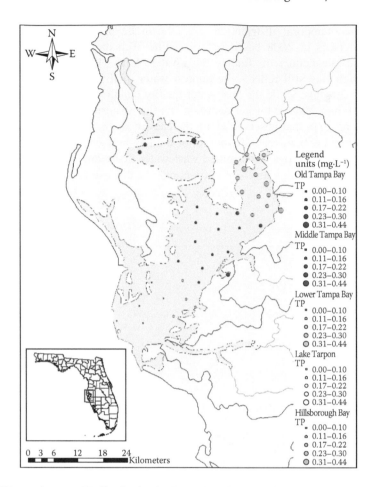

FIGURE 7.5 Average TP distribution in Tampa Bay, Summer 2008. (Data from http://www .tampabay.wateratlas.usf.edu/.)

and bottom stirring that may come from the Gulf of Mexico pouring into Tampa Bay may increase nutrient levels in the bay, both of which may be in concert with each other to trigger the phytoplankton growth. In contrast, ebb currents may flash out the accumulated nutrients along Old Tampa and Hillsborough bays and may alleviate the trophic state. In addition, nutrients from freshwater runoff into the bay may carry high levels of colored dissolved organic matter (CDOM) that could indirectly indicate the amount of freshwater entering the bay (Figure 7.8). Seasonal averages (Figures 7.4 through 7.7) cannot reflect the dynamics of daily changes of TP unless we can create detailed snapshots using remote sensing images to examine the spatiotemporal patterns of TP.

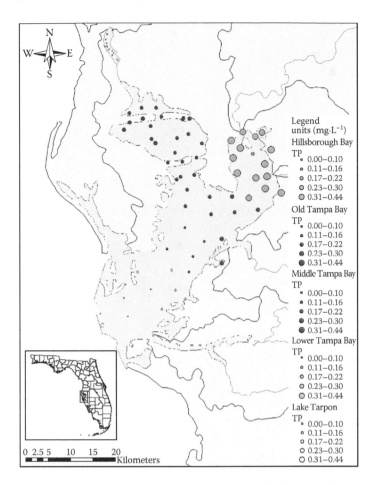

FIGURE 7.6 Average TP distribution in Tampa Bay, Fall 2008. (Data from http://www
.tampabay.wateratlas.usf.edu/.)

This chapter examines spatiotemporal patterns of TP concentrations in Tampa
Bay using the Moderate Resolution Imaging Spectroradiometer (MODIS) images
and inverse modeling, which is intimately tied to the changes in the areal extent of
Chl-*a* concentrations. Questions to be answered in this study include the following: (1) Can the TP maps in Tampa Bay be derived using remote sensing MODIS
images and data mining/machine learning algorithms? (2) What is the difference
between the seasonal patterns of TP concentrations and remote sensing–based
short-term TP distribution in Tampa Bay? These scientific questions can be better analyzed with the aid of remote sensing technology. The study therefore aims
to capture local variations of the spatiotemporal patterns of TP concentrations in
four prespecified time windows to fulfill a holistic comparison with seasonally

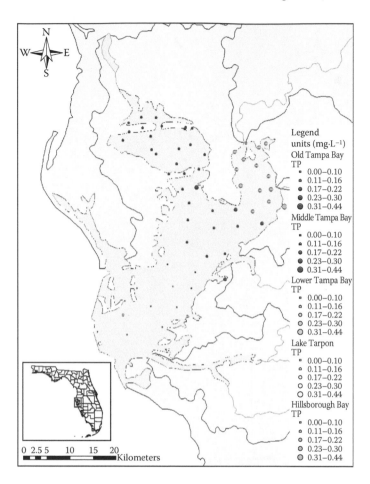

FIGURE 7.7   Average TP distribution in Tampa Bay, Winter 2008. (Data from http://www
.tampabay.wateratlas.usf.edu/.)

FIGURE 7.8   Ocean color satellite images. (Courtesy of Google Earth Map.)

averaged TP concentrations in regard to the variations of TP concentrations across the Tampa Bay in 2008.

7.2 MATERIALS AND METHODS

7.2.1 FIELD MEASUREMENTS, DATA COLLECTION, AND ANALYSIS

An understanding of phytoplankton populations and their distribution enables us to draw observations about an aquatic systems composition and health status. In spite of the fact that a detailed classification of algal species may be helpful, phytoplankton populations are typically estimated by measuring Chl-*a*, the primary photosynthetic pigment present in all forms of algae. Chl-*a* may be used as an ecosystem indicator correlated with the presence of TP that triggers occasional algal blooms when total nitrogen is also abundant. In this field study, we aimed to determine the water quality variations across two contiguous months (May and June) in summer season 2010 to explore spatiotemporal patterns of a various water quality constituents as well as differentiate the tidal effect.

We conducted the first field campaign on May 13, 2010, collecting 29 surface water samples from the bottom, middle, and top-left areas of Tampa Bay, and the second on June 29, 2010, collecting 39 surface water samples. All sampling sites were separated by at least 1 km in the Tampa Bay area. All samples were analyzed and measured for TP, *in vivo* Chl-*a*, CDOM, and turbidity. Supporting measurements included conductivity, pH, water temperature, dissolved oxygen, and Secchi disk

TABLE 7.4
Bands Related to Ocean Color/Phytoplankton/ Biogeochemistry

Band	Bandwidth[a]	Spectral Radiance[b]	Required SNR[c]
8	405 420	44.9	880
9	438–448	41.9	838
10	483–493	32.1	802
11	526–536	27.9	754
12	546–556	21	750
13	662–672	9.5	910
14	673–683	8.7	1087
15	743–753	10.2	586
16	862–877	6.2	516

Source: http://modis.gsfc.nasa.gov/about/specifications.php.

[a] Bands 1–19 are in nm, bands 20–36 are in μm.

[b] Spectral radiance values are (W·m^{-2}-μm-sr).

[c] SNR = Signal-to-noise ratio.

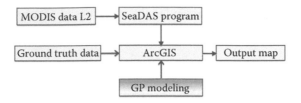

FIGURE 7.9   Overall flowchart of the inverse modeling analysis.

depth (SDD); CDOM, *in vivo* Chl-*a*, SDD, conductivity, temperature, and pH were measured on site. The CDOM was measured using an Aquafluor™ handheld field fluorometer. *In vivo* Chl-*a* fluorescence is a good indicator of algal levels because all algae contain Chl-*a*. For greatest accuracy of *in vivo* measurements, we took samples regularly for extraction to correlate with the *in vivo* readings. Fluorometric measurement of Chl-*a* was performed by the method of Holm-Hansen et al. (1965) using a Turner Laboratory Fluorometer as described by Parsons et al. (1984). Extracted Chl-*a* concentrations were measured by a certified water quality laboratory within 24 to 48 hours in Orlando, Florida, for comparison with *in vivo* Chl-*a* measurements. With this data, we performed a linear regression, modeling the *in vivo* Chl-*a* data and extracted Chl-*a* concentrations. Turbidity was measured at laboratories on the campus of the University of Central Florida (UCF) using turbidimeters. TP were also measured at laboratories on the campus of UCF using Hach Method 8190. The algal species samples collected in June 2010 were also analyzed by an external certified ecology laboratory to investigate which algal species dominate the coastal water environment in Tampa Bay.

TABLE 7.5
MODIS Bands Needed in Our Study

Band Name	Band Number	Key Use
nLw_412	8	Chlorophyll
nLw_443	9	Chlorophyll
nLw_469	3	Soil/vegetation differences
nLw_488	10	Chlorophyll
nLw_531	11	Chlorophyll
nLw_547		
nLw_555	4	Green vegetation
nLw_645	1	Absolute land cover transformation, vegetation chlorophyll
nLw_667	13	Atmosphere, sediments
nLw_678	14	Chlorophyll fluorescence
CDOM_index		
Chl-*a*		

7.2.2 Remote Sensing Images

The Ocean Color (MODIS/AQUA) data were downloaded from Ocean Color WEB (http://oceancolor.gsfc.nasa.gov/), NASA. The downloaded zip file contains the MODIS product of Chl-*a*. We also used TP data collected from our field campaigns to be concurrent with the MODIS data for inverse modeling. Our aim was to develop the relationships between *in situ* TP concentrations in the Tampa Bay area and the MODIS product of Chl-*a*, directly or indirectly related to ocean color, phytoplankton, and biogeochemistry (Table 7.4). Because existing products like CDOM and Chl-*a* were derived from similar bands of MODIS Aqua with different wavelengths which coexist and covary with CDOM and Chl-*a*, there is no need to include all relevant information of MODIS reflectance bands for inverse modeling. Thus, the relationships between TP concentration and Chl-*a* were examined by using a genetic

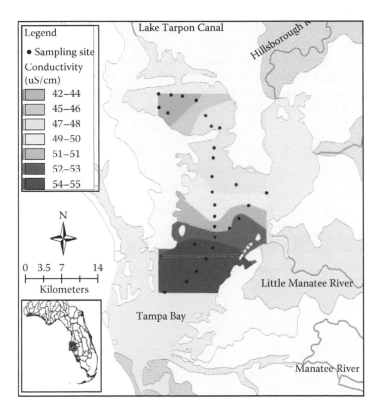

FIGURE 7.10 Salinity gradient in Tampa Bay based on the conductivity analysis on May 13, 2010. (From Chang, N. B., *Multi-Scale Water Infrastructure Characterization Study Using Remote Sensing*, Final report submitted to Water Supply and Water Resources Division, National Risk Management Research Laboratory, U.S. Environmental Protection Agency, Cincinnati, OH, 2011. With permission.)

programming (GP) model, which is considered as an inverse modeling tool in this study.

It is worthwhile to note that the principle of evolutionary computing (EC) is rooted in genetic algorithms (GAs), first developed by Holland (1975), and is similar to evolution strategies (ESs) (Rechenberg 1965; Back et al. 1997) and evolutionary programming (EP) developed by Fogel (1966). All three approaches were eventually combined into one entity called evolutionary computation (Gagne and Parizeau 2004). Under the EC framework, the well-known GP approach was invented by Koza (1992), which became the best advancement to create best-selective nonlinear regression models in terms of selected multiple independent variables. During model development in this study, model calibration and validation were conducted sequentially to increase model integrity. We used the Discipulus software package developed by Francone (1998) to perform GP modeling analysis for the estimation of Chl-*a* concentrations over the entire Tampa Bay.

A flowchart of inverse modeling analysis (Figure 7.9) shows the L2 data collected from MODIS sensor in HDF format have to be identified and examined at first. All raw MODIS/Aqua data were then processed by the Sea-viewing Wide Field-of-view Sensor (SeaWiFS) Data Analysis System (SeaDAS) software package. From the SeaDAS main menu, several products related to our study (Table 7.5) had to be imported to a geographical information system (GIS) (i.e., ArcGIS) as ASCII files (i.e., using ASCII import function in ArcGIS). These data being imported include the MODIS data with bandwidth between 405 and 683 nm and Chl-*a*. Then the TP data derived by the GP model can be used to compare the *in situ* data in GIS for further analysis.

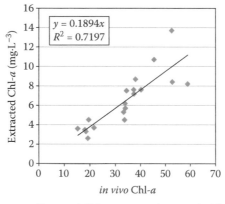

FIGURE 7.11 Calibration curve of *in vivo* Chl-*a* data and extracted Chl-*a* concentrations. (From Chang, N. B., *Multi-Scale Water Infrastructure Characterization Study Using Remote Sensing*, Final report submitted to Water Supply and Water Resources Division, National Risk Management Research Laboratory, U.S. Environmental Protection Agency, Cincinnati, OH, 2011.)

In summary, the workflow of MODIS/Aqua image processing and machine learning is as follows:

1. Sort and mine the *in situ* data obtained from the field campaigns
2. Arrange the locations and dates of the data points
3. Obtain remote sensing data that are temporally and spatially synchronous to the *in situ* data
4. Extract the remote sensing data (cloud cleaning, atmospheric correction, etc.)
5. Import transformed image pixel values and concurrent *in situ* data to GIS
6. Export the data into GP modeling platform
7. Perform nonlinear regression (inverse modeling) analyses

TABLE 7.6
Summary of Dominant Algal Species in Tampa Bay, FL

Sample	Unit	T31	T36	T56
Sampling date		June 10, 2010	June 10, 2010	June 10, 2010
Genus		*Dactyliosolen*	*Guinardia*	*Odontella*
Species		sp. 1	*Striata*	*Sinensis*
Algal group		Bacillariophyta	Bacillariophyta	Bacillariophyta
Counting unit		Cell	Chain	Cell
Cells/unit		1	3	1
Cell biovolume	(μm^3)	476.3	18,674.9	905,073.2
Species	Units·mL^{-1}	2086.7	18.1	18.1
Species	Cells·mL^{-1}	2086.7	54.4	18.1
Species biovolume	μm^3·mL^{-1}	993,933.6	1,016,619.7	16,423,385.8
Group total	Units·mL^{-1}	3446.5	1182.3	1020.4
Group total	Cells·mL^{-1}	3988.1	1421.0	1227.9
Group total biovolume	μm^3·mL^{-1}	5,152,696.1	3,448,752.4	18,339,435.2
%Total	Cells·mL^{-1}	54.0	38.4	8.7
%Total biovolume	μm^3·mL^{-1}	98.9	93.8	76.3
Sample total	Units·mL^{-1}	4776.4	2286.3	3200.5
Sample total	Cells·mL^{-1}	7384.4	3691.9	14,110.4
Sample total biovolume	μm^3·mL^{-1}	5,205,411.8	3,675,702.1	24,035,875.2

Source: Chang, N. B., *Multi-Scale Water Infrastructure Characterization Study Using Remote Sensing*, Final report submitted to Water Supply and Water Resources Division, National Risk Management Research Laboratory, U.S. Environmental Protection Agency, Cincinnati, OH, 2011. With permission.

Note: T31, T36, and T56 are three samples collected at the Old, Middle, and Lower Tampa bays, respectively.

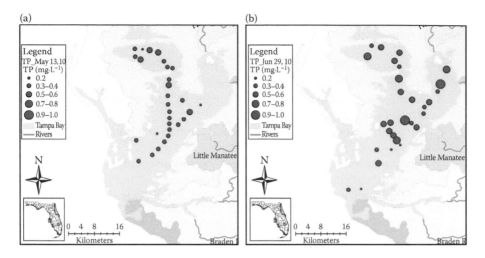

FIGURE 7.12 TP measurements in Tampa Bay in summer 2010. (a) May 13, 2010, (b) June 29, 2010. (From Chang, N. B., *Multi-Scale Water Infrastructure Characterization Study Using Remote Sensing*, Final report submitted to Water Supply and Water Resources Division, National Risk Management Research Laboratory, U.S. Environmental Protection Agency, Cincinnati, OH, 2011.)

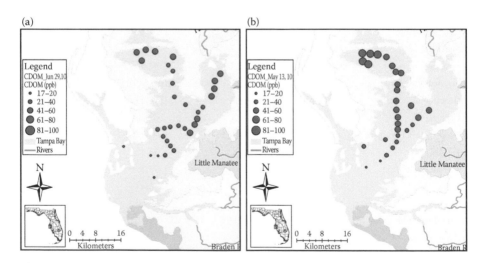

FIGURE 7.13 CDOM measurements in Tampa Bay in summer 2010. (a) May 13, 2010, (b) June 29, 2010. (From Chang, N. B., *Multi-Scale Water Infrastructure Characterization Study Using Remote Sensing*, Final report submitted to Water Supply and Water Resources Division, National Risk Management Research Laboratory, U.S. Environmental Protection Agency, Cincinnati, OH, 2011.)

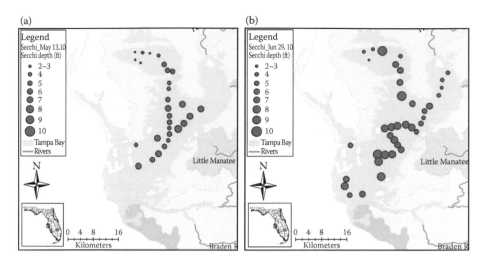

FIGURE 7.14 SDD measurements in Tampa Bay in summer 2010. (a) May 13, 2010, (b) June 29, 2010. (From Chang, N. B., *Multi-Scale Water Infrastructure Characterization Study Using Remote Sensing*, Final report submitted to Water Supply and Water Resources Division, National Risk Management Research Laboratory, U.S. Environmental Protection Agency, Cincinnati, OH, 2011.)

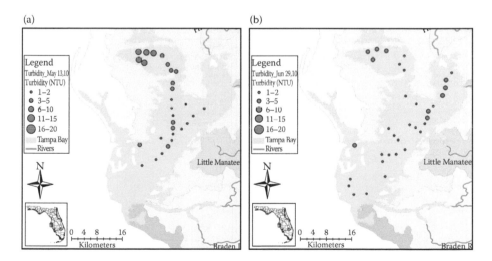

FIGURE 7.15 Turbidity measurements in Tampa Bay in summer 2010. (a) May 13, 2010, (b) June 29, 2010. (From Chang, N. B., *Multi-Scale Water Infrastructure Characterization Study Using Remote Sensing*, Final report submitted to Water Supply and Water Resources Division, National Risk Management Research Laboratory, U.S. Environmental Protection Agency, Cincinnati, OH, 2011.)

7.3   RESULTS AND DISCUSSION

7.3.1   Water Quality and Algal Biomass Analysis

The salinity gradient in Tampa Bay based on the conductivity analysis on May 13, 2010 generally indicates the effect of flood and ebb currents (Figure 7.10). Good agreement between the *in vivo* Chl-*a* data and extracted Chl-*a* concentrations can be confirmed by GP modeling (Figure 7.11). A summary of dominant algal species in Tampa Bay, including *Dactyliosolen* sp., *Guinardia striata*, and *Odontella sinensis* (Table 7.6), indicates that microzooplankton grazing on phytoplankton (as measured by Chl-*a*) inside and outside of Tampa Bay may affect the Chl-*a* concentrations along with other dynamic factors that may warrant further investigation in the future.

The surface water measurements showed that TP was higher in June than in May 2010 in Tampa Bay, probably due to higher storm activity (i.e., precipitation) in June 2010 (Figure 7.12). Relatively higher TP concentrations across Hillsborough Bay and Old Tampa Bay reveal that high values of CDOM were commonly detected in the upper bay from Old Tampa Bay to Hillsborough Bay (Figure 7.13). Higher SDD values mean less light attenuation (Figure 7.14); Tampa Bay water is clearer in June compared to that in May 2010 (Figure 7.15). Locations with higher turbidity (Figure 7.15) are generally consistent with locations having lower SDD (Figure 7.16), a pattern also highly consistent with the CDOM data. The upper sections of the bay normally have higher light attenuation than lower sections, a pattern of SDD

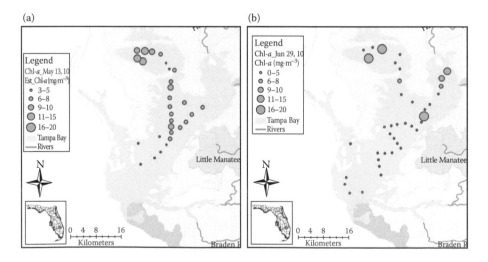

FIGURE 7.16   Chl-*a* measurements in Tampa Bay in summer 2010. (a) May 13, 2010, (b) June 29, 2010. (From Chang, N. B., *Multi-Scale Water Infrastructure Characterization Study Using Remote Sensing*, Final report submitted to Water Supply and Water Resources Division, National Risk Management Research Laboratory, U.S. Environmental Protection Agency, Cincinnati, OH, 2011.)

that also agrees with the CDOM pattern as shown previously in Figure 7.13. Because Chl-*a* measurements can be tied to several factors, including turbidity, light penetration, and TP concentrations, a unique spatial pattern is exhibited (Figure 7.16) with Chl-*a* measurements generally higher in upper bay regions from Old Tampa Bay to Hillsborough Bay, even extending into Middle Tampa Bay in May 2010.

7.3.2 REMOTE SENSING–BASED GP MODEL

Two datasets were used to create nonlinear regression models to estimate TP. This yields a single partition, generating two subgroups with equal amount of data or

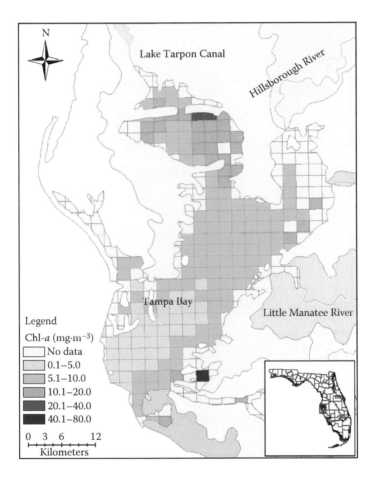

FIGURE 7.17 MODIS-based Chl-*a* maps in Tampa Bay on June 26, 2010. (From Chang, N. B., *Multi-Scale Water Infrastructure Characterization Study Using Remote Sensing*, Final report submitted to Water Supply and Water Resources Division, National Risk Management Research Laboratory, U.S. Environmental Protection Agency, Cincinnati, OH, 2011.)

with 70% and 30% of the whole data set, which was prepared for model calibration and validation. *R*-square values of both calibration and validation of the final model below are equal to 0.51 based on the subgroups with equal amount of data. The GP model (i.e., a nonlinear regression model) is designed to produce a TP estimation algorithm by expressing the *in situ* TP measurements as a nonlinear function of grid-based MODIS band images, average pixel values of Chl-*a* concentrations, and average pixel values of CDOM as described by the following convoluted equation (Chang 2011):

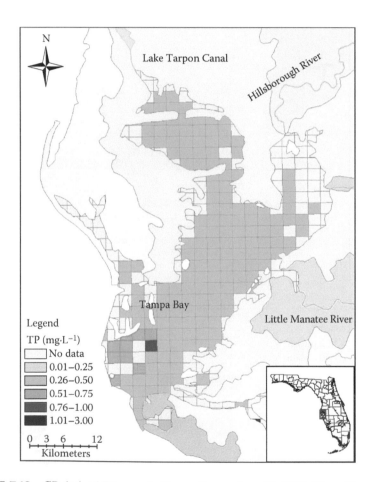

FIGURE 7.18 GP-derived TP maps in Tampa Bay on June 26, 2010. (From Chang, N. B., *Multi-Scale Water Infrastructure Characterization Study Using Remote Sensing*, Final report submitted to Water Supply and Water Resources Division, National Risk Management Research Laboratory, U.S. Environmental Protection Agency, Cincinnati, OH, 2011.)

$$Total\ Phosphorous\left(\frac{mg}{L}\right) = (0.2877938747406006) \cdot (\sqrt{Y1} - 0.002621650695800781)$$

$$Y1 = \frac{-1.259177207946777}{Y4+Y2} + 1.366016626358032$$

$$Y2 = (Y3)(2)^{Integer(Y6)}$$

$$Y3 = -[(Y5)(Y7 - 1.641227006912231)^2]$$

$$Y4 = Y5 + [(Y5)(Y7 - 1.641227006912231)^4]$$

$$Y5 = V0 - 2(Y7 - 1.641227006912231)^4$$

$$Y6 = (Y7 - 1.641227006912231)^2$$

$$Y7 = \frac{(-0.259177207946777)(V0)}{-0.8057427406311035}$$

$$V0 = \text{Extracted chlorophyll-}a\ (mg/m)^3$$

in which *Total Phosphorus* is estimated based on the GP nonlinear function, mainly in terms of Chl-*a*.

After calibration and validation using limited *in situ* TP measurements, the GP model was derived to explore the nonlinearity of the production of Chl-*a*. During the modeling process, MODIS bands were dropped by the GP modeling process automatically. The estimation of TP depends on only Chl-*a*. The combination of Chl-*a* and CDOM showed no significant difference, although both parameters were used as inputs via the GP algorithm. By using the GP model of MODIS-based Chl-*a* maps in Tampa Bay (Figure 7.17) that is similar to the field observations in Figure 7.16, the spatial distribution of TP can then be constructed (Figure 7.18).

7.4 CONCLUSIONS

This study explores the estimation of TP concentrations in Tampa Bay, Florida, using the MODIS-based GP model that takes into account the highly nonlinear structure between Chl-*a* and TP concentrations in coastal waters. It confirms that the MODIS images of Chl-*a* can be correlated with the estimated TP values through a highly convoluted nonlinear regression equation—the GP-derived TP model. Based on a direct comparison between the long-term historical trends and data collected from our field campaign, it is indicative that both findings if discovered through remote sensing analysis and filed investigation fall into the same spatiotemporal water quality patterns in which Old Tampa Bay and Hillsborough Bay are comparatively poor.

Overall, satellites can uniformly sample the entire Tampa Bay region with spatial coverage unobtainable with field campaigns alone, although remote sensing images often provide limited temporal coverage. In comparison, surface measurements can provide continuous surface water quality data at frequent intervals with limited spatial coverage. In any circumstance, the two sets of measurements combined complement each other to enrich a few prescribed scenarios in our study. The advanced comparative analysis between remote sensing images and *in situ* surface measurements implies that possible discrepancies of spatiotemporal patterns of some water quality constituents at specific locations might be salient. Bias may be introduced if we use only *in situ* point measurements for environmental management decision making. Therefore, in real-world applications, spatiotemporal analysis based on the generation of thematic information using remote sensing is highly recommended in concert with time series *in situ* surface measurements.

ACKNOWLEDGMENT

We wish to acknowledge the field and modeling support from Dr. Ammarin Daranpob in this study.

REFERENCES

Back, T., Hammel, U., and Schwefel, H. P. (1997). Evolutionary computation: Comments on the history and current state. *IEEE Transactions on Evolutionary Computation*, 1(1), 3–17.

Bricker, S. B., Clement, C. G., Pirhalle, D. E., Orlando, S. P., and Farrow, D. R. G. (1999). *National Estuarine Eutrophication Assessment, Effects of Nutrient Enrichment in the Nation's Estuaries*. The National Oceanic and Atmospheric Administration (NOAA), p. 71.

Chang, N. B. (2011). *Multi-Scale Water Infrastructure Characterization Study Using Remote Sensing*. Final report submitted to Water Supply and Water Resources Division, National Risk Management Research Laboratory, U.S. Environmental Protection Agency, Cincinnati, OH.

Florida Department of Environmental Protection (FDEP). (2011). *Florida's Water*, http://www.protectingourwater.org/watersheds/map/tampa_bay/ (accessed April 2011).

Fogel, L. J., Owens, A. J., and Walsh, M. J. (1966). *Artificial Intelligence through Simulated Evolution*. John Wiley & Sons, New York.

Francone, F. D. (1998). *Discipulus™ Software Owner's Manual, Version 3.0 DRAFT*, Machine Learning Technologies, Inc., CO.

Gagne, C. and Parizeau, M. (2004). *Genericity in Evolutionary Computation Software Tools: Principles and Case-Study*. Technical Report RT-LVSN-2004-01. Laboratoire de Vision et Systemes Numerique (LVSN), Universite Laval, Quebec, Canada.

Holm-Hansen, O., Lorenzen, J. C., Holmes, W. R., and Strickland H. J. D. (1965). Fluorometric determination of chlorophyll. *Journal of Marine Science*, 30(1), 3–15.

Holland, J. M. (1975). *Adaptation in Natural and Artificial Systems*. University of Michigan Press, Ann Arbor, MI.

Janicki, A., Pribble, R., Janicki, S., and Winowitch, M. (2001). *An Analysis of Long-term Trends in Tampa Bay Water Quality*. Technical Report, Tampa Bay Estuary Program, submitted by Janicki Environmental, Inc. http://www.tampabay.wateratlas.usf.edu/upload/documents/AnalysisLongTermTrendsTampWaterQuality.pdf.

Koza, J. R. (1992). *Genetic Programming: On the Programming of Computers by Means of Natural Selection*. MIT Press, Cambridge, MA.

National Oceanic and Atmospheric Administration (NOAA). (2011). *State of Coastal Environment.*http://oceanservice.noaa.gov/websites/retiredsites/sotc_pdf/EUT.PDF (accessed March 2011).

Parsons, T. R., Maita, Y., and Lalli, C. M. (1984). *A Manual of Chemical and Biological Methods for Seawater Analysis.* Pergamon Press, Oxford, United Kingdom, p. 173.

Rechenberg, I. (1965). *Cybernetic Solution Path of an Experimental Problem.* Royal Aircraft Establishment, Translation No. 1122, Ministry of Aviation, Farnborough Hants, United Kingdom.

8 Monitoring and Mapping of Flood Plumes in the Great Barrier Reef Based on *in Situ* and Remote Sensing Observations

Michelle Devlin, Thomas Schroeder,
Lachlan McKinna, Jon Brodie,
Vittorio Brando, and Arnold Dekker

CONTENTS

8.1   REVIEW OF *IN SITU* FLOOD PLUME OBSERVATIONS IN THE GREAT BARRIER REEF AND PLUME RELATED REMOTE SENSING TECHNIQUES

8.1.1   HISTORY OF RIVERINE PLUME SAMPLING IN THE GREAT BARRIER REEF

Riverine flood plumes are the major conduit between land use and the marine waters, providing an important transport mechanism of pollutants into the Great Barrier Reef (GBR). Measurement of the extent of plume waters coupled with information on water quality concentrations can be useful in quantifying the impact of higher concentrations of nutrient and sediments on marine ecosystems. Ocean color associated with these potential changes in water quality has been identified as a useful tool for eutrophication monitoring and assessment of coastal environments. Riverine plumes are visible as turbid, low salinity waters easily visible in the early stages by aerial photography or satellite imagery (Figure 8.1).

FIGURE 8.1   Aerial images of flood plumes in the Great Barrier Reef. (a) Edge of Fitzroy River plume (2008), (b) river plume wrapping around Great Keppel Island, (c) plume discharging from the Russell-Mulgrave catchment (2001).

The distribution of flood plume waters in the GBR has been studied oppor-
tunistically over the last 30 years in some detail. Observations of river water in
the GBR lagoon had been noted and documented at many times earlier in the
20th century (Table 8.1). For example low salinity water was recorded at Low
Isles (16 km offshore) during the 1928–1929 British Museum Great Barrier Reef
Expedition coinciding with flooding in the adjacent Barron and Daintree Rivers
(Orr 1933). The effects of low salinity water on the reefs of the Whitsundays, associated with major cyclones near Mackay in 1918 were reported by Hedley (1925)
and Rainford (1925). In the 1960s, low salinity water was noted in the wet season
well offshore in the Cairns area (Pearson and Garrett 1978). Davies and Hughes
(1983) noted terrigenous sedimentation in 1982 at Boulder Reef (15 km offshore)
associated with flooding in the Endeavour River. Wolanski and associates tracked
plumes using salinity measurements in both the 1979 (Wolanski and Jones 1981)
and 1981 Burdekin floods (Wolanski and van Senden 1983). Burdekin plume water
was shown to move north from the river mouth and was detectable up to 300 km
from the mouth (Wolanski and van Senden 1983). Plume water distribution was
governed by geostrophic forces—particularly the wind regime and Coriolis effect
(Wolanski 1994).

One of the biggest events in recent years was the 1991 Fitzroy flood associated
with Cyclone Joy, which was the third largest flood event on record for the Fitzroy
River in Southern GBR waters. Plume concentrations and the reef impacts were
monitored in one of the first collaborative monitoring efforts of food plume waters
in the GBR (Brodie and Mitchell 1992). The effects from the 1991 Fitzroy River
flood on reefs impacted by the river plume were dramatic. Low salinity, high suspended solids, and nutrient-rich water surrounded reefs of the Keppel Islands group
(20 km offshore) for a period of 3 weeks (Brodie and Mitchell 1992; O'Neill et al.
1992) and reached the northern reefs of the Capricorn-Bunker group (75 km offshore) for a few days (Devlin et al. 2001). Coral mortality in the Keppels was high
(van Woesik et al. 1995) with some mortality in the Capricorn-Bunkers (Devlin et
al. 2001).

Following the Fitzroy flood, a more formal investigation of flood plumes in the
GBR lagoon was instituted (Devlin and Lourey 1996; Steven et al. 1996; Devlin et
al. 1998, 2001) with the objectives of mapping the spatial limits of the influence of
riverine waters, quantifying the concentrations of key parameters in plume waters
at various time in the life of the plume, and determining the fate of materials discharged from the rivers.

In the 1990s, flood plume monitoring was the focus of water quality monitoring activities carried out by the Great Barrier Reef Marine Park Authority.
Dedicated surveys of plume events occurred between 1991 and 2001. Mapping
of the plume extents was carried out by aerial surveys and corresponding *in situ*
water quality sampling within the plume, typically focusing on surface sampling.
The results of such work were widely disseminated and published (Devlin et al.
2001; Devlin and Brodie 2005). From 2001, plume monitoring was more sporadic
as there was no one organization responsible for plume mapping and monitoring.
However, events were still sampled periodically between 2002 and 2007 (Brodie
et al. 2004; Fabricius et al. 2008). It was during this time that satellite remote

TABLE 8.1

Details of GBR Flood Plume Events Prior to 1991 and Cited Impacts

Year	Cyclone	Duration of Flooding	Rivers	Wind Speed/Direction	Observed Effects
1918	Mackay	January 22–29, 1918	Pioneer Don Burdekin	Lowest pressure recorded was 933 hPa	Storm surge: 4 m / Freshwater at sea (8 miles from land) / Loss of coral cover on Stoney Reef (Hedley 1925; Rainford 1925)
1932			Daintree		Low Isles (Orr 1933)
1934		March 12, 1934		SE–NE	Changes to geomorphology and death of coral (*Montipora, Porites*) and clams (*Hippopus*) / Changes in community structure post event (Moorhouse 1936)
1943					Observations of physical damage to various reefs (Gleghorn 1947)
1977	Otto and Keith		Wet tropics		Phytoplankton blooms (Pickard et al. 1977)
1977–1978	Peter	January 1–10, 1979	Wet tropics		Phytoplankton blooms (Revelante and Gilmartin 1982)
1979	Kerry	March 6–13, 1979	Burdekin	Wind gusts reached 76 knots	Measurements of low salinity (Wolanski and Jones 1981)
1981	Dominic		Burdekin		Measurements of low salinity (Wolanski and van Senden 1983)
1983					Sedimentation at Endeavour Reef (Davies and Hughes 1983)
1986	Winifred			175 km/hr Central pressure of 958 hPa SW-SE (1/2/86) NE-N (2/2/86)	Sediment resuspended and moved across shelf (Gagan et al. 1987, 1990) / Phytoplankton blooms (Furnas and Mitchell 1986)
1988	Charlie			Cape Bowling Green recorded 981 hPa wave heights up to 3.1 m	Brodie and Furnas 1996; Liston 1990
1989	Aivu		Pioneer and Proserpine Rivers	959 hPa recorded 20 km from coast	3 m storm surge in Upstart Bay

sensing of water quality parameters also became available for the GBR region, presenting a valuable new tool for the study of plume extent and exposure (e.g., Brando et al. 2006; Schroeder et al. 2008). From 2007, flood plume monitoring and the use of remotely sensed imagery became part of the marine monitoring program associated with assessing the performance of the Reef Water Quality Protection Plan in the GBR. The long-term goal of the Reef Plan is to ensure that the quality of water entering the GBR from adjacent catchments has no detrimental impact on its ecosystem health and resilience. This involves a monitoring and evaluation strategy that must be robust enough to evaluate the efficiency and effectiveness of implementation and progress toward this goal. The use of remote sensing data coupled with traditional *in situ* water quality monitoring is a key data source for this evaluation.

8.1.2 Riverine Plumes and Remote Sensing Products

Satellite remote sensing products provide an excellent tool for monitoring coastal water processes with periodic overpasses, allowing the routine and cost-effective collection of a variety of observations over large and often inaccessible expanses of coastal waters. The high frequency of information and large spatial coverage makes remote sensing imagery an extremely useful tool in the monitoring and mapping of flood plumes, which are logistically difficult to sample and costly to cover in their entirety. In plume events, riverine flow and cyclonic conditions stir the water column, bringing colored dissolved organic matter (CDOM) and nutrients to the surface and creating detectable ocean-color features (Shi and Wang 2007). Plumes of turbid water are a frequent phenomenon in coastal waters, in particular in shallow soft-bottom seas and in the mouth of rivers and estuaries. Turbidity is caused by a variety of particles in the water, some of mineral origin such as clay particles and others of organic origin. Collectively these particles are referred to as total suspended matter (TSM), which can be mapped through the use of remote sensing imagery. Ocean color, production, and patterns are also clearer and broadly visible through the use of detailed satellite ocean color studies. Catchment runoff events involve space scales ranging from hundreds of meters to kilometers and time scales from hours to weeks, and thus the use of remote sensing in monitoring marine indicators at appropriate time and space scales can be used as key indicators of cause and effect in these systems. Concentrations of TSM, CDOM, and chlorophyll can be used to track plume distribution, biological uptake, increases in production, and sedimentation of particulate matter.

8.1.3 Mapping of Riverine Plumes Using Remote Sensing Techniques

Recent studies on river and flood plumes using remote sensing techniques were carried out on the Meso American Barrier Reef system (Andrefouet et al. 2002), Florida (Hu et al. 2003, 2004), the North American west coast (Thomas and Weatherbee 2006; Lahet and Stramski 2010; Nezlin et al. 2008), Gulf of Mexico (Shi and Wang 2009), Japan (Lihan et al. 2008) and Moreton Bay, Australia (Roelfsema et al. 2006). River flood plume and influence studies are associated with the Meso American

Barrier Reef system (Sheng et al. 2007; Chérubin et al. 2008; Paris and Chérubin 2008). The work of Soto et al. (2009) is particularly instructive in such a type of assessment, using many years of remote sensing data collected from the Sea-viewing Wide Field-of-view Sensor (SeaWiFS) sensor, as they are able to show quantitatively the exposure of reefs in the region to river discharged materials. Salinity is an ideal parameter for mapping freshwater discharge and first direct ocean salinity measurements from space are now becoming available from the Soils Moisture and Ocean Salinity (SMOS) mission launched by the European Space Agency's (ESA). However, SMOS' spatial resolution of about 50 km per pixel will be too coarse for most coastal applications, which often require at least 1 km spatial resolution. Burrage et al. (2003) reported on a successful application of an airborne L-band passive microwave radiometer to map salinity in the GBR on a regional scale of a few hundred square kilometers.

The dispersion of river plumes in the GBR lagoon has been mapped using aerial photography and salinity sampling from surface vessels for over 25 years (Wolanski and van Senden 1983; Devlin et al. 2001; Brodie et al. 2004). Dispersion has also been modeled based on some of the early salinity measurements (King et al. 2001). Attempts have also been made to use satellite remote sensing to map the extent of plumes, and hence draw conclusions about the degree of exposure of GBR ecosystems to river waters and the pollutant load of these waters. However, these attempts have not been successful in the past (Devlin and Brodie 2005) due to a limited amount of available ocean color imagery and heavy cloud cover that is common at the time of river discharge. Today, there are far more ocean-color sensors available with spectral bands able to distinguish flood plume water from clear ocean water and these in all have an almost daily overpass frequency. The most common sensor systems include Sea-viewing Wide Filed-of-view Sensor (SeaWiFS), the Moderate Resolution Imaging Spectroradiometer (MODIS), and Medium Resolution Imaging Spectrometer (MERIS). Low temporal, higher spatial resolution sensors such as Landsat 7, Système Probatoire d'Observation de la Terre (SPOT) 5, Advanced Land Observation Satellite (ALOS), Ikonos, and QuickBird provide more spatial detail, but do not overpass as frequently (i.e., once every 5 to 60 days depending on sensor path design and geographic location) and have a much smaller swath coverage. Assessing the exposure of marine ecosystems to land-sourced pollution is thus now far more feasible using a combination of *in situ* water quality sampling and satellite remote sensing.

The use of remote sensing products in GBR waters is still relatively new; however, it is becoming a key tool in the Marine Monitoring Program. Recent work by Brando et al. (2010) demonstrated the use of remote sensing information to inform and guide water quality assessments through the comparison of seasonal and annual means against guideline values.

Brodie et al. (2010) reported the tracking of an extensive phytoplankton bloom in GBR waters in early 2005 that resulted from high nutrients discharging from rivers in the Mackay Whitsunday region in the central section of the GBR. Global standard satellite products were used to measure the dispersion of the plume and associated bloom. Through the use of satellite images, the plume was tracked over 9 days, and shown to impinge on a number of inner-shelf reefs of the central GBR. This type of long-term detail on the movement of plumes has not been available prior to the use of remote sensing products.

8.2 IMPACT OF FLOOD PLUMES ON COASTAL ECOSYSTEMS

8.2.1 IMPACTS OF AGRICULTURAL EXPANSION

At present, there exists uncertainty of the extent of transport and potential influence of catchment-derived contaminants in the GBR system. Previous studies of flood plumes and coastal sediment transport off the Burdekin River (Wolanski and van Senden 1983; McCulloch et al. 2003; Devlin et al. 2001; Wolanski and Jones 1981), the Wet tropic rivers (Orpin et al. 1999; Devlin et al. 2001; McCulloch et al. 2003b; Devlin and Brodie 2005), and the Fitzroy River (Brodie and Mitchell 1992; Devlin and Brodie 2005) have revealed exposure of inshore ecosystems including coral reefs and sea grass to a range of nutrients associated with dissolved and fine particulate fractions of the river load.

Agricultural activities and urban expansion on catchments adjacent to reef systems have resulted in increased erosion, the destruction of wetlands and stream bank vegetation, and runoff of sediment, fertilizer, and chemical residues, having a major impact on water quality in reef systems (Brodie et al. 2008a; Fabricius et al. 2005). Throughout the world the problems of leakage of nitrogen and phosphorus from agricultural, industrial, and urban systems to fresh, estuarine, and marine water bodies, and resulting eutrophication are increasing exponentially (Tilman 1999) with potentially damaging impacts on the diversity, composition, and functioning of the marine systems.

Increases of the nutrient supply to coastal marine waters can enhance the growth of phytoplankton, turf algae, and macroalgae, leading to a dominance of phytoplankton algal blooms and fast-growing filamentous macroalgae. Decreased light penetration and increased sedimentation of organic matter results in higher benthic oxygen consumption and alterations in benthic communities. Affected systems around the world are characterized by rapid increases in intensive agriculture and urbanization with significant losses of riparian and wetland areas over the catchment area (Furnas 2003; Tilman et al. 1999; Brodie et al. 2008a).

8.2.2 THE RIVERS AND CATCHMENTS OF THE GREAT BARRIER REEF

Rivers discharging into the GBR lagoon are one of the main mechanisms for inputting new sources of nutrients and sediments into the reef, though the actual distribution and movement of the individual constituents varies considerably between the wet and dry tropic rivers. Rainfall and runoff are highly seasonal with over two-thirds occurring during the summer (December–April) wet season in contrast to the dry season (May–November). The coastal region adjoining the Great Barrier Reef is divided into a number of wet and dry tropical catchments (Gilbert and Brodie 2001). Most catchments are small (<10,000 km^2), but the Burdekin (133,000 km^2) and Fitzroy River catchments (143,000 km^2) are among the largest in Australia. Wet tropic rivers have limited freshwater and saline mixing in the dry season with little input into the GBR and high freshwater flow in the wet season with rapid flushing times. The consequence of this is predominately freshwater flow to the mouth of the river, where the riverine waters discharged over and into the adjacent coastal seawater. Dry tropic rivers have negligible or no flow during the dry season and can act as a tidal bay with tidal intrusions from the seawater end. Flood plumes move in response to the prevailing weather conditions over the coastal shelf with the plume waters acting as an estuary itself with mixing processes from the freshwater end (mouth of the river) to the seawater end (i.e., end of plume).

8.2.3 HIGH FLOW EVENTS IN THE GREAT BARRIER REEF

In periods of high flow, typically associated with low-pressure systems or cyclonic events, most of the rivers of northeastern Queensland flow fresh to the mouth and estuarine processes take place on the continental shelf rather than in a traditional estuary (Devlin and Brodie 2005). Following discharge, plumes develop in the GBR lagoon and most commonly spread to the north and offshore under the influence of Coriolis force and the prevailing southeasterly wind regime (Devlin et al. 2001; Devlin and Brodie 2005). In times of high flow conditions, which can vary between catchments, the river waters pulse into the marine environment as a freshwater plume, influencing a wide range of chemical, geological, biological, and physical conditions in the adjacent coastal waters. These turbid riverine plumes are biologically rich, spatially complex water masses characterized by strong horizontal and vertical salinity gradients (McManus and Fuhrman 1990; McKee et al. 2004; Dagg et al. 2004) and can be visually recognized in surface waters through the sharp color change and accumulation of foam and flotsam at the interface of the riverine plume and oceanic waters (McManus and Fuhrman 1990; Brodie and Furnas 1996; Devlin and Brodie 2005).

Previous work and modeling studies (Wolanski and Jones 1981; Wolanksi and van Senden 1983; King et al. 1997, 2001) suggest that plumes are constrained close to the coast by oceanographic conditions imposed by Coriolis forces and the prevailing wind regime, producing a net northerly water movement (Burrage et al. 1997). It is thought that these overall forcing components generally drive the river plumes northward and toward the coast in GBR waters (Wolanski and Jones 1981; Gagan et al. 1987), primarily influencing the reefs and coastal systems that lie within this area of inshore coastal system. Despite this northerly inshore movement, plumes can still potentially move offshore (Gagan et al. 1987; Brodie and Mitchell 1992; Devlin et al. 2001; Devlin and Schaffelke 2009) and reach midshelf reefs that lie closer to the coast, such as those in the Cairns area (Furnas et al. 1995; Devlin and Schaffelke 2009). These plumes, characterized by lowered salinity (compared to seawater), turbidity from clay particles discharged from the river and phytoplankton blooms enhanced by the nutrients in the river discharge, may persist in the lagoon for periods of days to weeks. In low wind and tidal current conditions the plumes are strongly stratified, with a thin (0.5 to 2 m) plume floating on seawater, while when the wind and tidal currents are strong the plume water may be well mixed through the water column (Devlin et al. 2001) (Figure 8.1).

8.2.4 IMPACTS OF ALTERED RIVERINE (PLUME) WATER QUALITY

The ecology of corals and coral reefs is directly influenced by the "quality" of the water they live in. Waters washing over and around reefs deliver and remove dissolved and particulate nutrients, sediments, prey, and propagules, and generally protect reef organisms from extreme fluctuations in dissolved gases, temperature, and salinity. Under natural conditions, a wide range of GBR coral species live or once lived on nearshore and coastal reefs along the north Queensland coast (Veron 1996), in low nutrient environments, but possibly still fairly turbid conditions. Evidence of the increase in nutrients and sediment and pesticides entering the reef is well established, with load calculations (Brodie et al. 2009a), catchment monitoring (e.g., Hunter and Walton 1997, 2008), and catchment modeling (Webster et al. 2008) all showing increasing loads of contaminants reaching GBR rivers and plume waters. Brodie and Mitchell (2005) demonstrate correlation between percentage catchment or subcatchment development and dissolved inorganic nitrogen (nitrate + ammonia) concentrations in river flood flow for a number of rivers on the GBR catchment. In general, concentrations of dissolved inorganic nitrogen (DIN) increase by a factor of 3 to 50 times on rivers draining to the GBR from highly developed compared to undeveloped or lightly developed catchments.

The effects of variability in river influence on inner shelf ecosystems is still being investigated under a considerable research and monitoring efforts (Johnson et al. 2010). However, correlations between relative distance from the coast or relative distance across the shelf and diversity and/or abundance in taxa such as soft corals (Fabricius and De'ath 2001a) and crustose coralline algae (Fabricius and De'ath 2001b) are now a key factor in understanding the impacts of water quality. Such correlations are attributed to turbidity, sedimentation, and nutrient gradients with distance across the shelf (Fabricus and De'ath 2004; Fabricus et al. 2005). Long-term effects of eutrophication on some inner shelf coral reefs of the GBR are now evident. In the Whitsundays, a nutrient/suspended sediment gradient from the Proserpine River has been correlated with reduction in coral cover, species richness, and abundance combined with increased coral recruit mortality (van Woesik et al. 1999). Synergistic effects of nutrients and sediment (Fabricius and Wolanski 2000) in association with the acute effects of cyclones, bleaching, and crown of thorns starfish (Fabricius and De'ath 2004; Brodie et al. 2005) are the cause of the widespread reef degradation in inner shelf areas of the central GBR. At Green Island off Cairns, the large expansion in the area of seagrass meadows on reefal areas normally without seagrass has been shown to be a result of increased nutrient supply from mainland river discharge (Udy et al. 1999). Brodie et al. (2011) identified a number of factors related to large expansion in the area of seagrass meadows. They include persistently high nutrient concentrations, intense and extensive phytoplankton blooms following nutrient-enriched river discharge events with chlorophyll-a concentrations 4 to 100 times above the GBR water quality guidelines, and excessive macroalgal abundance on reefs with high nutrient inputs and replacement of corals in high exposure areas with coralline algae, filamentous algae, macroalgae, and/or a variety of filter feeders. All of these show that some parts of the GBR are potentially eutrophic.

8.3 METHODOLOGY FOR MAPPING AND MONITORING OF FLOOD EVENTS IN THE GREAT BARRIER REEF

8.3.1 SUMMARY

This section provides a chronologic overview of different methods and strategies that have been applied by the authors to characterize, map, and monitor flood events in the GBR over the last 20 years (Figure 8.2). It prepares the reader for a detailed case study, described in the next section, where different techniques are applied to convert remote sensing data into relevant plume information. These techniques and their resulting products evolved in complexity with time, from basic aerial photography in combination with *in situ* monitoring to the application of advanced regional parameterized ocean color algorithms.

8.3.2 *IN SITU* SAMPLING OF GBR WATER QUALITY

Flood plumes in the GBR were sampled during a series of dedicated field campaigns since the 1990s and on a regular basis since 2007 within the framework of a Reef Rescue Marine Monitoring Program (RRMMP). Discrete water samples were collected from multiple sites within the plume waters and locations of samples were dependent on which rivers were flooding and the areal extent of the plume. Generally, samples were collected in a series of transects away from the river mouth. Surface water samples are usually taken over a period of days to weeks, dependent on the event. Timing of sampling was also dependent on the type of event and the

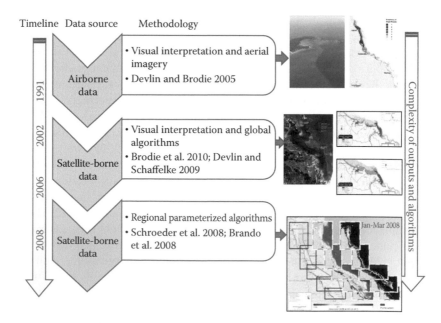

FIGURE 8.2 The evolution of remote sensed imagery in the mapping and monitoring of plume waters in the Great Barrier Reef.

logistics of vessel deployment. The majority of samples were collected inside the visible area of the plume, though some samples were taken outside the edge of the plume for comparison. The objective was to sample the development and extent of the plume waters and to identify concentration gradients of water quality parameters (i.e., salinity, temperature, dissolved inorganic, organic and particulate nutrients, suspended solids, CDOM, and chlorophyll-*a*). Salinity and temperature depth profiles were recorded at each site with a conductivity–temperature–depth (CTD) SeaBird profiler.

8.3.3 LABORATORY AND DATA ANALYSIS

Surface water samples were collected, filtered onboard, and stored for further analysis in the laboratory. Volumes filtered for all analyses were dependent on the turbidity of the water. Within the laboratory, phytoplankton chlorophyll pigments were analyzed by fluorescence following maceration of algal cells and pigment extraction in acetone (Parsons et al. 1984). The absorption of CDOM was measured, using a dual beam spectrophotometer. Nutrient samples were filtered into 10 mL sample tubes and stored on ice to be analyzed within 24 hours. Samples were analyzed for concentrations of dissolved inorganic nutrients (NH_4, NO_2, NO_3, $NO_2 + NO_3$, PO_4, and Si) by standard procedures (Ryle et al. 1981) implemented on a Skalar 20/40 autoanalyzer, with baselines run against artificial seawater. Analyses of total dissolved nutrients (e.g., total dissolved nitrogen (TDN) and total dissolved phosphorus (TDP)) were carried using persulfate digestion of water samples (Valderrama 1981), which are then analyzed for inorganic nutrients, as above. Dissolved organic nitrogen (DON) and dissolved organic phosphorus (DOP) were calculated by subtracting the separately measured inorganic nutrient concentrations (above) from the TDN and TDP values. Particulate nitrogen concentrations of the particulate matter collected on Whatman GF/F filters were determined by high-temperature combustion using an ANTEK Model 707 Nitrogen Analyzer. The filters were freeze-dried before analysis. Following primary (650°C) and secondary combustion (1050°C), the nitrogen oxides produced were quantified by chemiluminescence.

Plume water quality data collected over the range of salinity (0–35) data was used to investigate transport of the materials in the plume by mixing profiles, which relate concentrations of water quality constituents to salinity. These are commonly used to analyze processes in flood plumes (Boyle et al. 1974), such as estimating conservative or nonconservative mixing processes (Eyre 2000). Identification of the transport and uptake change rapidly in river/plume interface such as the rapid deposition of particulate matter in the lower salinity zones (Loder and Reichard 1981). The edge of river plume from sampling vessels were mapped using a combination of color (green/brown water in the plume, blue water outside the plume) and salinity (<35 in the plume, >35 outside plume).

8.3.4 THE USE OF AERIAL PHOTOGRAPHY FOR PLUME MAPPING

High rainfall and the resultant freshwater discharge extend the estuarine waters beyond the river mouth. Within these plumes, materials are transported, transformed, and deposited according to a complex myriad of internal interactions (Dagg et al. 2004). It is these processes that shape the visual properties of a plume area. It was recognized in the 1990s that the visual signal could be used to identify plume extent. Sampling of flood plumes in the GBR in the 1990s was a combination of sporadic *in situ* water quality sampling and aerial flyovers (Devlin et al. 2001, 2002). Aerial photography and the interpretation of resulting color images were used to identify and map flood plume waters. Airborne surveillance was usually carried out within 24 to 48 hours after a large flow event and aerial examination was used to define the extent and geographical limits of the plume, and in some instances movement of the plume, over a period of days. The visible edge of the plume was followed at an altitude of 1000 to 2000 m in a light aircraft and mapped using a Global Positioning System (GPS). Where individual rivers flooded simultaneously, as often happens in the wet tropics, adjacent plumes merge into a continuous area. In these cases, efforts were made to distinguish the edge of the individual river plumes through color differences. The vertical distribution of plume water and its depth stratification were studied in a limited number of cases by depth sampling (Taylor 1997). The results of each mapping exercise were transferred to a geographic information system (GIS) on which the subsequent spatial analysis was based.

Multiple images from aerial flyovers between 1991 and 2002 were overlaid using GIS spatial analysis techniques and merged into one map (Figure 8.3). The darker shades of this frequency map denote the areas that experience plumes on a high frequency basis (i.e., every 1 to 2 years). This map forms the basis of the first attempt at identifying the risk of exposure to inshore ecosystems from enhanced concentrations found in plume waters (Devlin and Brodie 2005).

The combination of analog aerial photography with GIS and *in situ* measurements was a useful approach for providing detailed information on plume composition at a limited number of locations and of plume extent at a regional scale of a few hundred kilometers at maximum. Plume extent can be measured from an aircraft even under complete cloud coverage, a clear advantage compared to satellite measurements in the visible spectral range. However, this approach is relatively costly due to aircraft operation and may provide only limited RGB (red–green–blue) spectral information of the surface targets. However, more sophisticated multi- and hyperspectral airborne spectrographic imagers such as CASI, AISA, or AVIRIS exist, which allow for surface parameter estimation based spectral data acquired in hundreds of bands.

8.3.5 The Use of Ocean-Color Observations for Plume Mapping

The large-scale spatial features of plumes are often difficult to observe during *in situ* sampling. Aerial imagery using RGB techniques can only distinguish the high sediment carrying plume waters. Limitations of aerial surveys are evident when the plume starts to move further offshore into a secondary phase and becomes dominated by chlorophyll and CDOM. The large-scale spatial features become difficult to observe by aerial imagery and more difficult to sample over the larger extent.

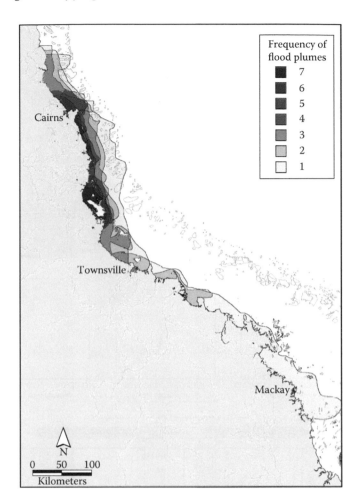

FIGURE 8.3 Frequency plume map created from all plume imagery mapped within the period of 1991 to 2002 (Airbourne data). Areas delineated with dark colors have highest frequency of plume intersections. This map was used as a preliminary basis to define the spatial risk of exposure from plume waters to inshore marine ecosystems.

However, the high suspended sediment, high chlorophyll, and high CDOM properties of the plume waters can be identified by appropriate ocean-color algorithms (Nezlin et al. 2005; Nezlin and DiGiacomo 2005).

Information from the satellite imagery can assist greatly in determining the extent and location of plume boundaries and how these change over time. In the GBR region, the use of satellite remote sensing imagery has allowed substantively more plume measurements to be included in the estimation of plume exposure. Furthermore, the spectral data that enables the retrieval of water quality parameters such as chlorophyll, TSM, and CDOM is infeasible to obtain by aerial photography.

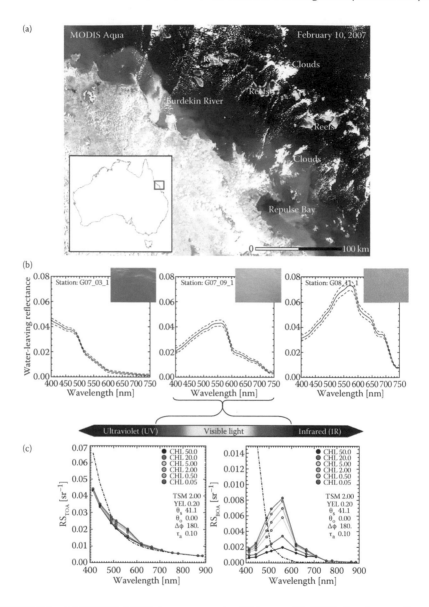

FIGURE 8.4 (a) MODIS AQUA imagery acquired February 10, 2007, showing a sediment-dominated flood plume of the Burdekin River and a dissolved organic matter–dominated plume in Repulse Bay. (b) *In situ* measured above water reflectance spectra in the visible spectral range for clear blue open ocean waters (left), chlorophyll-dominated coastal waters (center), and chlorophyll- and sediment-dominated waters (right) with associated photographs taken by a standard digital camera. (c) Simulated reflectance spectra for the MERIS instrument above coastal waters at the top of the atmosphere (left) and at the bottom of the atmosphere (right) accounting for six chlorophyll concentrations between 0.05 and 50 mg m–3. Atmospheric scattering corresponds to an aerosol optical thickness of 0.1 at 550 nm and a surface pressure of 1040 hPa (note different ordinate scaling).

The application of remote sensing data has changed the perception that plumes are nearly always constrained to the coast, with recognition that plume waters with elevated concentrations of chlorophyll and CDOM can be mapped at large distances offshore. Gradients of change within a plume is a dynamic movement, with TSM concentrations dropping out rapidly closer to the coast in lower salinity waters (Devlin and Brodie 2005; Brodie et al. 2010). Light is limited in these lower salinity waters, thus inhibiting production by primary producers. Reduction in turbidity occurs as the heavier particulate material deposits to the sea floor with a corresponding increase in dissolved nutrient availability. This leads to the appropriate conditions to support accelerated growth of phytoplankton. The later and extended stages of plume waters can still be visible by remote sensing algorithms with ongoing elevation of the CDOM concentrations. This variability on a spatial and temporal level is more easily monitored using spectral data acquired by ocean-color remote sensing sensors.

Ocean-color sensors such as MERIS, MODIS, or SeaWiFS measure the upward reflected sunlight off the earth's surface in up to 15 spectral bands in the visible and near-infrared spectral range (400–900 nm). These polar orbiting satellites overpass and scan the entire GBR within ten minutes from an approximate altitude of 700 to 800 km at a spatial resolution of up to 300 m. Daily overpasses and detailed spectral information make ocean-color remote sensing an excellent tool for large-scale environmental monitoring. The spectral information collected by these instruments can be translated into water quality information by appropriate ocean-color algorithms (IOCCG 2000).

The optical complexity and variability of GBR coastal waters is illustrated by a MODIS true color (RGB) composite acquired on February 10, 2007, covering the catchment of the Burdekin River and Repulse Bay of the Mackay-Whitsunday Region of Queensland, Australia (Figure 8.4a). Intense wet season rainfall caused rivers in this region to produce large discharges to the GBR lagoon. The image captures the full variation of color, or more precisely spectral reflectance, ranging from deep blue open ocean waters to more green and brownish coastal waters. This satellite image illustrates as well the influence of the land use on the composition of the floodwaters. In the north, the Burdekin River discharges high loads of inorganic sediments into the lagoon, while further south, Repulse Bay, with regional land use dominated by sugarcane cultivation and beef grazing, receives high loads of dissolved organic matter.

Examples of *in situ* measured reflectance spectra covering blue open ocean waters and more turbid chlorophyll- and sediment-dominated waters are shown in Figure 8.4b together with their associated color taken by a standard digital camera.

The color or spectral reflectance of the water is according to Gordon (1988) directly proportional to the backscattering and inversely proportional to the sum of backscattering and absorption. These inherent optical properties can be translated by an appropriate algorithm into concentrations of water constituents. The most common approach for the retrieval of water constituents from ocean-color observations is composed of two main processing or algorithm steps. First, an atmospheric correction procedure is applied to the satellite data to remove the disturbing effects of atmospheric absorption and scattering and to obtain the water-leaving radiance or reflectance. In the second step, the obtained reflectance spectra are used to retrieve the water quality parameters.

In the visible spectral range of the solar spectrum (Figure 8.4b), atmospheric scattering processes may contribute up to 90% to the total signal measured by a satellite sensor at the top of the atmosphere (TOA). Scattering obscures the spectral signal of the underlying water body that needs to be inverted to estimate water quality variables such as chlorophyll, total suspended solid (TSS), or CDOM. It is beneficial to illustrate the spectral variability of coastal waters with varying chlorophyll concentrations simulated for the MERIS sensor at the TOA and the bottom of the atmosphere (BOA) (see Figure 8.4c) (Schroeder 2005). The difference between TOA and BOA spectra is mainly caused by molecular (Rayleigh) and aerosol (Mie) scattering with wavelength dependencies of the resultant scattering of λ^{-4} and $\lambda^0-\lambda^{-4}$, respectively. Poor atmospheric correction is often one of the major limiting factors in the retrieval of water constituents. Without a carefully adapted atmospheric correction, reliable retrieval is not possible and multitemporal water quality studies based on satellite imagery would be biased by the influence of the highly variable atmosphere.

Some standard atmospheric correction algorithms designed for open ocean waters tried to decouple the oceanic and the atmospheric signals from the total measured spectral information at the top of the atmosphere (Lee et al. 2002). It is assumed the ocean color to be black (i.e., complete absorption of incident radiation) in the near-infrared spectral region (>700 nm) and therefore often fail to be detected above turbid coastal waters because of the influence of highly scattering water constituents or bottom reflections (Lee et al. 2002). Other, more suitable, attempts to correct for the atmospheric influence above coastal waters are based on iterative fitting approaches (Land and Haigh 1996) or on artificial neural network (ANN) techniques (Zhang et al. 2002; Zhang 2003; Jamet et al. 2004; Schroeder et al. 2007; Schroeder et al. 2008). Another approach developed by Wang and Shi (2007) for the MODIS sensor makes use of additional infrared spectral information at 1240 and 2130 nm to reestablish the "black pixel" assumption for turbid waters. At these shortwave infrared (SWIR) wavelengths, the absorption of water dominates the in-water scattering of particles and can be used to decouple the atmospheric from the water signal. A comprehensive overview of recent advances in ocean-color and atmospheric correction algorithm development can be found in the reports of the International Ocean-Colour Coordinating Group (IOCCG 2000; IOCCG 2010).

8.3.6 THE USE OF GLOBAL OCEAN COLOR ALGORITHMS IN PLUME MAPPING

In addition to the challenges of atmospheric correction above coastal waters, a large variability of in-water optical properties and concentration ranges, especially during flood events (Figure 8.4a), frequently cause empirical ocean-color algorithms to fail. These algorithms, like the default MODIS OC3 or SeaWiFS OC2 (O'Reilley et al. 1998), have been designed for open ocean waters, in which the optical properties are determined solely by phytoplankton, their degradation products, and the water itself. Simple reflectance ratios of two or more bands in the blue (443–490 nm) and green

(550–565 nm) spectral region are used by these algorithms to estimate the concentration of chlorophyll. Coastal waters however, are usually influenced in addition by riverine inputs of terrestrial originated CDOM and inorganic suspended material as well as tidal resuspension. The spectral absorption features of these substances partly overlap with the absorption features of phytoplankton and cause a frequent overestimation of chlorophyll from these ratio algorithms.

In GBR coastal waters, the global semianalytical ocean-color algorithms such as the GSM01 algorithm for chlorophyll (Garver and Siegel 1997; Maritorena et al. 2002) have been found more accurate than the empirical band ratio approach (Qin et al. 2007).

8.3.7 The Use of Regional Parameterized Ocean-Color Algorithms in Plume Mapping

In GBR coastal waters, especially during flood events in the dry tropics, we observe two distinct optically extreme cases of water types causing global algorithm failure. One is a highly scattering sediment-dominated water type (Burdekin plume, Figure 8.4a) and the other is a highly absorbing one dominated by colored dissolved organic material (Repulse Bay, Figure 8.4a). The standard algorithms that have traditionally been applied to GBR waters have difficulties in mapping due to this complexity of the inshore GBR waters, including bottom visibility, proximity to coral reefs, and sea grass beds, which can cause errors in the algorithm outputs.

To overcome these limitations associated with the use of global ocean-color algorithms in GBR optically complex coastal waters, a regional algorithm was developed. This new approach is based on an inversion scheme that couples an ANN atmospheric correction (Schroeder et al. 2008) with an in-water algorithm that is based on a variable parameterization of *in situ* measured inherent optical properties (Brando et al. 2008). This recently developed ANN algorithm does not need to uncouple atmosphere and ocean signals, but uses the full spectral information as measured at TOA (~400–900 nm) and can be adapted to other satellite sensors.

The application presented here was developed for MODIS imagery by inverse modeling of radiative transfer (RT) calculations within a coupled ocean–atmosphere system by utilizing ANN techniques. The algorithm was implemented similar to an approach developed by Schroeder et al. (2007) for MERIS, but with a different inverse model capable of generating more complex network architectures. Within this model-based approach, the ANNs were able to deal with the optical-complex coastal waters. The accuracy of the atmospheric correction was assessed through matchup analysis with *in situ* reflectance data. Comparisons between SeaDAS v5.1.1 derived from atmospherically corrected spectra and the ANN approach can be made possible (Schroeder et al. 2008). The ANN approach showed significant accuracy improvement especially in the blue part of the spectrum (412 nm) (Schroeder et al. 2008). Errors of 19% can be found with the proposed ANN algorithm, whereas SeaDAS resulted in errors up to 48% (Schroeder et al. 2008). A comparison between ANN and SeaDAS atmospheric corrected spectra is shown in Figure 8.5.

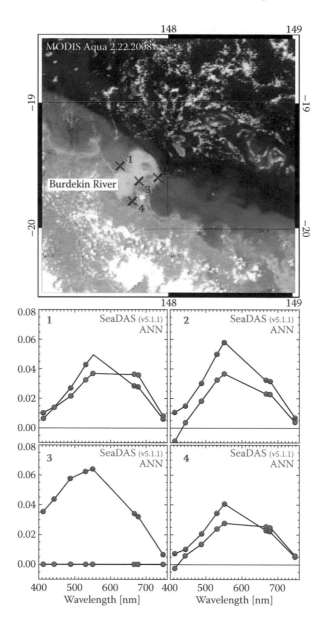

FIGURE 8.5 Comparison of NASA's standard near-infrared atmospheric correction (SeaDAS v5.1.1) and CSIRO's artificial neural network derived reflectance spectra for a MODIS Aqua scene acquired on February 22, 2008, covering the Burdekin River, Queensland, Australia.

Atmospheric corrected spectra are then used to derive the inherent optical properties and the concentrations of optically active constituents, namely, chlorophyll-*a*, nonalgal particulate matter (NAP), and the absorption of CDOM at 440 nm, by applying a linear matrix inversion (LMI) algorithm developed by Hoge and Lyon (1996). The implementation of the LMI method (Brando et al. 2008) uses the below-water remote sensing reflectance spectrum of the eight MODIS bands 8 to 15 (412–748 nm) as input to a semianalytical model developed by Gordon et al. (1988) to simultaneously derive the above three water constituents in an algebraic manner. The algorithm has been validated against *in situ* data and improves in accuracy compared to available standard (global) algorithms (Brando et al. 2010).

Use of the LMI algorithm is still in testing phases and undergoing further validation; however, these outputs have been used to deliver greater accuracy in the Reef Rescue Marine Monitoring Program, Australia. These advanced remote sensing products have been applied to derive data for the use of compliance monitoring and water quality management as described in Section 8.5.

8.4 MONITORING AND MAPPING OF FLOOD PLUMES IN THE TULLY-MURRAY MARINE REGION, GREAT BARRIER REEF: A CASE STUDY

8.4.1 INTRODUCTION

The monitoring of flood plumes in the GBR and the improved understanding of the internal biogeochemical processes is essential for understanding the effects of long-term exposure of marine ecosystems to excess nutrients and sediments and to other anthropogenic contaminants. Water quality is an important driver of coral reef health at local (reviewed in Fabricius 2005), regional (van Woesik et al. 1999; Fabricius et al. 2005), and GBR-wide scales (De'ath and Fabricius 2008). The effects of various water quality constituents are manifold, ranging from disturbance by sedimentation, light reduction by increased turbidity, reduced calcification rates by excess inorganic nutrients, inhibition of photosynthesis by herbicide exposure, and generally affect early life history stages more than adult corals (e.g., Fabricius 2005; Negri et al. 2005; Cantin et al. 2007). This section presents remotely sensed imagery from a single catchment and adjacent marine area that has been used to analyze the riverine plume extent and frequency over 11 years. The plume monitoring and biological data determines which physical characteristics of the floods (i.e., size of flood, direction of plume movement, shape of hydrograph) most influence the flood plume water quality and areal extent.

8.4.2 STUDY AREA

The Tully and Murray catchments are located within the wet tropics region of North Queensland and drain wet tropical rainforest in the upper reaches, beef grazing along the mid reaches, and a large coastal floodplain with a series of interconnected significant wetlands that have been extensively modified to support sugarcane and banana production as well as urban centers (Figure 8.6). The considerable floodplain

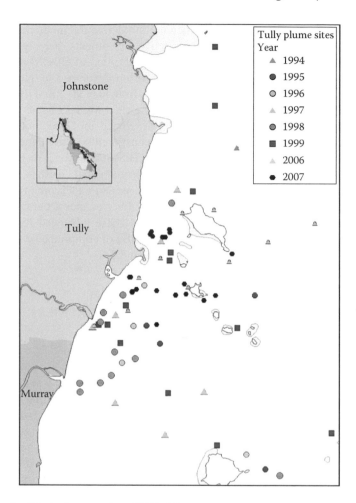

FIGURE 8.6   The inshore marine of Tully-Murray area under investigation, primarily influenced by the Tully and Murray rivers, located midpoint between Cairns and Townsville. Shapes indicate the location of water quality samples, differentiated by sampling year.

network transports sediments, nutrients, and herbicides into the GBR, either directly through these wetlands or via the larger Tully and Murray Rivers. The Tully River is among Australia's least variable rivers, representing the wet tropical climate of the region. It floods regularly one to four times per year, with riverine discharge extending into the adjacent marine waters. The coastal and inshore zone supports significant intertidal and subtidal sea grass beds. During the wet season, the coastal and inshore areas adjacent to the Tully catchment are regularly exposed to floodwaters from the Tully River, and to a lesser extent from the Herbert River via the Hinchinbrook Channel, carrying excess suspended solid and nutrients and herbicides into the marine environment.

Aerial images from 1994 to 1999 were combined with remote sensing images from 2002 to 2008 to describe the full extent of riverine plumes from the Tully River during 11 events over that period. River plumes monitored from 1994 to 1999 were mapped using aerial survey techniques. Over the monsoon season, weather reports were collected closely, and when plumes formed, aerial surveys were conducted once or twice during the event. Plumes were readily observable as brown turbid water masses contrasting with clearer seawater. The visible edge of the plume was followed at an altitude of 1000 to 2000 m in a light aircraft and mapped using GPS. Where individual rivers flooded simultaneously, adjacent plumes merge into a continuous area. In these cases, efforts were made to distinguish the edge of the individual river plumes through color differences. In all other years, the extents of the combined plumes were mapped using remote sensing techniques. The mapping outcomes were transferred to GIS software for subsequent spatial analysis.

Single images were selected based on their image quality and transposed from geo-referenced true color images and/or CDOM measurements into GIS shape files. The MODIS imagery was rereferenced to conform to Geocentric Datum of Australia 1994 (GDA-94), MGA projection (i.e., a metric rectangular grid system) by applying the imagery geographic coordinate values to the MGA-94 projected values (meters) to achieve a simple bilinear solution (i.e., Universal Transverse Mercator (UTM)). True color imagery of before, during, and after each plume have been identified where there was low cloud cover and reasonably good visualization of the plume area. Water characteristics were identified in each image, with the primary water type, characterized by high turbidity, high-sediment plume discharging relatively close to the river mouth. Plumes characterized by lower turbidity (i.e., low values of TSM) and higher production identified by elevated chlorophyll values are usually measured in the middle salinity ranges (e.g., 5–25 ppt). Turbidity (i.e., measured as TSM) may change through these secondary waters as a result of the offshore transport of the finer particulate material and desorption processes. Plume water types moving further offshore are typically characterized by elevated CDOM values, indicating some influence of freshwater, but TSM and chlorophyll reducing to baseline concentrations. This area can be mapped much further offshore and north of the river mouth than the visually evident primary and secondary waters.

The extent of the enrichment and increased production is hard to define by true color imagery only, and requires the application of a suite of algorithms, as well as the visual examination of the true color processing. Total absorption at 443 nm may be used as an indicator of organic material and CDOM absorption at 443 nm as an indicator of riverine freshwater extent. Application of appropriate chlorophyll algorithms can also be helpful in the offshore areas to identify the extent of the higher primary production over the whole plume area.

The derived CDOM and detritus absorption (CDOM+D) at 443 nm combined with careful examination of quasi-true color and chlorophyll images provided the information used to derive simple qualitative indices for separating the different stages of plume movement, or water "types," and extent. Flood plume categories were defined based on parameters that are readily derived from ocean-color remote sensing. Thus, the spatial extent of the different water types can be mapped.

Plume types were classified qualitatively using the following criteria:

1. Primary water type was defined as having a high TSS load, minimal chlorophyll, and high values of colored dissolved organic material plus detrital matter (CDOM+D).
2. Secondary plume type was defined as a region where CDOM+D are still high; however, the TSM has been reduced. In this region, the water is characterized by increased light and nutrient availability, which can prompt phytoplankton growth. Thus, secondary plume waters exhibit high chlorophyll, high CDOM+D, and low TSS.
3. Tertiary plume type is the region of the plume that exhibits no elevated TSS and reduced amounts of chlorophyll and CDOM+D when compared with that of the secondary plume, but still above ambient conditions. This region can be described as being the transition between a secondary plume and ambient conditions.

CDOM+D imagery was cross validated with the true-color images for a visual check of the extent of the primary, high sediment, plume. Combining true-color information and appropriate algorithm application identifies the three plume types with a suitable degree of confidence. In the areas where clouds had completely obscured the plume, estimations of the plume extents were achieved by assessing the plume patterns from consecutive imagery epochs in the following days.

8.4.3 OUTPUTS OF PLUME MAPPING IN THE TULLY-MURRAY REGION

The extent of a visible plume extends past the inshore reefs in the Tully-Murray marine region (Figure 8.7). The edge of the plume is delineated by eye using the "true color" of water and in addition the use of the total absorption at band 443 nm as a proxy for the presence of organic material (i.e., phytoplankton, CDOM, and detrital material). Much of the color is due to phytoplankton validated by high *in situ* chlorophyll concentrations, and hence indicative of the extent of the algal bloom. The colocation of the plume in lowered salinity water and phytoplankton bloom are not certain but there will be a fair degree of overlap (Devlin and Brodie 2005) as the dissolved nutrients move with the water and eventually stimulate the algal bloom.

For the final imagery classification and interpretation, two products were provided. The initial classification method as described above allowed us to map the three main plume densities (e.g., primary, secondary, and tertiary water types) based on CDOM+D absorption, and second, the true-color images allowed for a visual correlation of the classified values. In the areas where clouds had completely obscured the plume, an estimation of the plume extents were achieved by correlating the plume patterns from any other imagery epochs in the following days. Riverine plume extents from the Tully-Murray River over 12 plume events are shown in Figures 8.8a through 8.8c.

Extent of the plumes varied considerably through the years. Average floods occurred in 1994, 1996, 1997, 1998, and 2004 and 2008. The 1994 and 1997 plumes covered a very large area due to northerly winds. In contrast, the 1996 and 1998

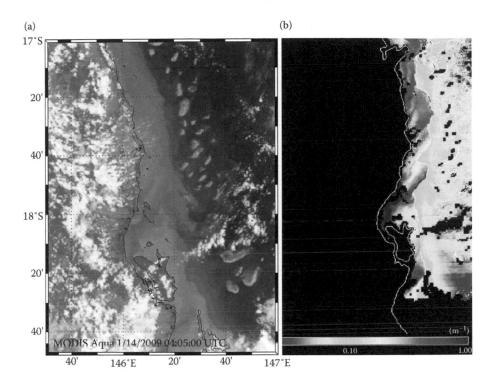

FIGURE 8.7 MODIS Aqua 250 m resolution images of the Herbert and Tully River flood plumes on January 14, 2009 following a high rainfall event. (a) The quasi-true color image of the flood event. Notice the high suspended sediments near the mouth of the Tully and Herbert Rivers and the extremely high chlorophyll biomass along the coast that appears highly green. (b) Absorption by colored dissolved and detrital matter (CDOM+D) using the QAA algorithm at 443 nm. A clear plume boundary is evident running parallel to the coast.

floods with similar flows covered a much smaller area due to prevailing SE winds. The W/SW winds in the 2004 flood period moved the primary plume further offshore. The 2008 flow event had a very large spatial extent with plume waters reaching into the Coral Sea. This was partly due to the prevailing SW winds, but also due to the very large flow event of a large river located south of the Tully (Burdekin River) in early 2008.

Large flow events, calculated as being above the 75th percentile of the long-term discharge record, occurred in 1999, 2000, and 2007. However, prevailing southeasterly winds during the flood periods in 1999 and 2000 constrained the plume extents to the coast. In contrast, the large flow volume of the 2007 flood event, coupled with S/SW winds, resulted in a large plume extent moving toward the midshelf reefs.

8.4.4 COMPOSITE PLUME MAPPING

A qualitative analysis was applied to each plume using the characteristics of the plume (discharge volume, certain wind conditions) to interpret the extent and direction of the

(a)

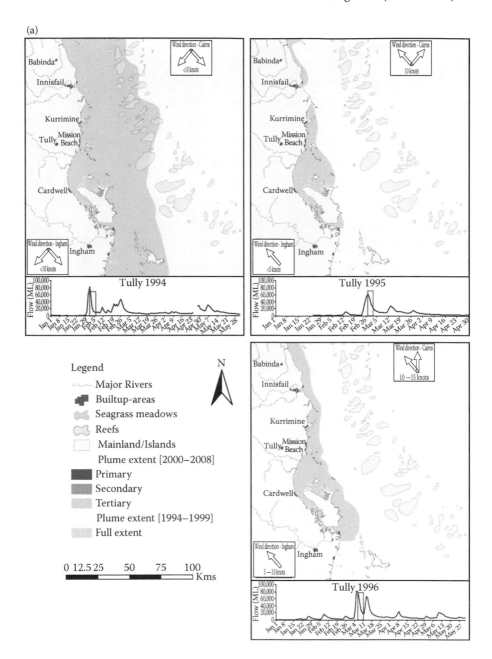

FIGURE 8.8 (a) Riverine plume extents estimated from aerial surveys for the period 1994 to 1996 in the Tully marine region. Hydrographs are shown for the period January to May, and the red box denotes the date of the aerial flyover.

(b)

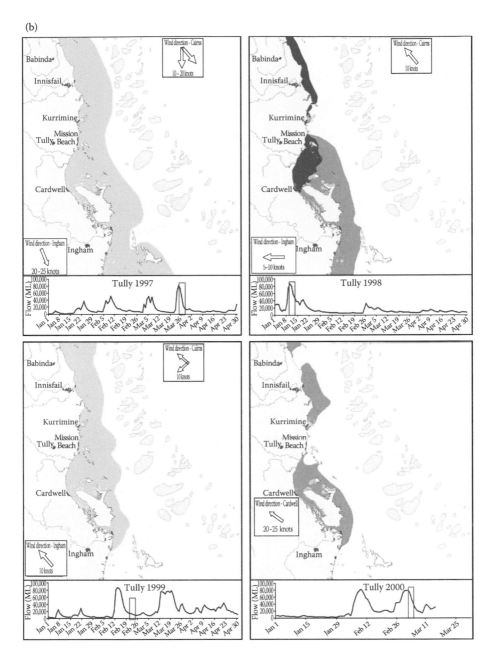

FIGURE 8.8 **(Continued)** (b) Riverine plume extents extracted from aerial flyovers for the period 1997 to 2000 in the Tully marine region. Hydrographs are shown for the period January to May, and the red box denotes the date of the aerial flyover.

(c)

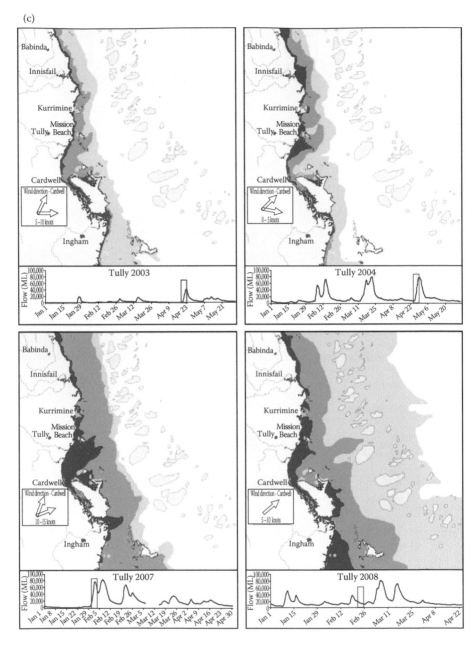

FIGURE 8.8 (Continued) (c) Riverine plume extents and plume types estimated using remote sensing images for the period 2003 to 2008 in the Tully marine region. Hydrographs are shown for the period January to May, and the red box denotes the date of the aerial flyover.

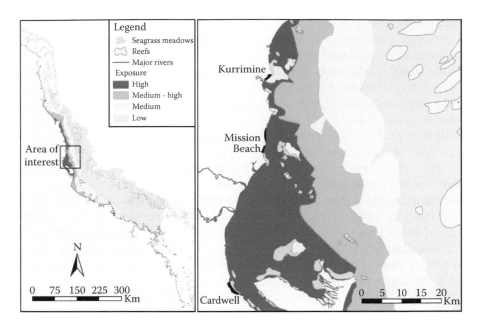

FIGURE 8.9   Plume exposure map for the Tully-Murray marine area. Exposure is calculated along a high to low gradient, where the high exposure relates to a higher frequency of plume intersections.

plume relative to the size of the flow event (small, average, or large based on the discharge volume). A plume exposure map was produced using a combination of plume extents and GIS software geoprocessing. Using the plume characteristics described above, each polygon was assigned a numeric value, with primary waters designated as 3, secondary type designated as 2, and tertiary water types as 1. A combined data set and map was then produced as a composite table of each plume index. The plume exposure values was overlaid on the Tully marine area to calculate the frequency of exposure of water type for the reefs and sea grass beds influenced or impacted by the Tully plume waters. The plume exposure map for Tully (Figure 8.9) was calculated from the intersection of the plume image and type from both the aerial surveys (1995–2000) and remote sensing images (2003–2009) for the marine area.

8.4.5 IMPACT OF PLUME WATERS ON THE TULLY MARINE AREA

Thirty-seven reefs and 14 seagrass beds in the Tully marine area were exposed to some degree to riverine plume waters during 11 flood events from the period 1994 to 2007. The number of reefs and seagrasses exposed to the plume waters varies from year to year, and is dependent on the type of plume. Over the 11 years, a minimum of 11 reefs (30%) and a maximum of 37 reefs (100%) were inundated by either a primary or secondary plume, indicating that it is likely that at least a third of the reefs are exposed to plume waters every year. In years where data has been collected to allow validation of the plume type and extent (1998, 2003–2008), we estimated that 6 to 15 reefs were inundated by primary plumes carrying high sediment loads, which is up to 41% of the inshore reefs in the Tully marine area, and 5 to 16 reefs (43%) were inundated by secondary plumes with elevated nutrient and chlorophyll concentrations. A smaller number of inshore reefs were inundated by a tertiary flood plume in three flood events. Note that tertiary plume extents and associated exposure of reefs may had been underestimated in the years where plume extent was estimated from aerial images only (1995–2000), based on a color change between the fresh and marine waters. Out of the 14 seagrass beds within the Tully marine area, at least 13 were inundated by either a primary or secondary plume in 10 of the events, with the exception of 2000 where only eleven seagrass beds were impacted.

This type of work illustrates how remote sensing products can be utilized to describe and define areas of risk in the GBR. The effects of excess nutrients and sediments in the marine environment are being increasingly understood (De'ath and Fabricius 2008). However, less well known are physical and biogeochemical processes transporting and transforming land-derived materials in the marine environment, as well as the hydrodynamics of the GBR inshore area that controls residence times. The missing links between catchment and marine processes hampers the implementation of management options for specific water quality constituents. A primary use of results from this type of study will be to set targets connecting end-of-river loads of particular materials to an intermediate end point target such as chlorophyll (Brodie et al. 2009b) or, in the future, to an ecological end point target such as a composite indicator for coral reef health (Fabricius et al. 2005).

8.5   TOWARD MANAGEMENT RELEVANT WATER QUALITY ASSESSMENTS

8.5.1   Mapping and Management Outputs

If environmental managers are to take full advantage of remote sensing capabilities, then products that translate remotely sensed scenes into useful information for managers are required. From primary remote sensing data, it is possible to produce a number of derived products suited to the specific needs of end users or to particular geographic regions in a number of outputs. Maps are the most common, and depending on user requirements, any number of primary variables or derived indices and attributes can be mapped over specified spatial aggregations or over timescales from days to years. A few good management products are those providing compliance information for environmental reporting.

8.5.2   Water Quality Guideline Values for GBR Marine Waters

A set of water quality guideline values and objectives was released in 2009 by federal and state government for the GBR waters to promote regionally and locally relevant water quality guideline values for Queensland waters. Regionally specific environmental values and objectives have been set and tested in some areas through the development of Water Quality Improvement Plans (WQIPs). The Great Barrier Reef Marine Park Authority (GBRMPA) has released water quality guidelines for the Marine Park identifying five water types: enclosed coastal, open coastal, inshore, offshore, and Coral Sea (De'ath and Fabricius 2008). The GBRMPA guidelines provide triggers for management action where exceedance occurs based on current condition and trend monitoring threshold levels for analysis. Besides, Queensland guidelines are to be adopted for all waters inshore of and within the enclosed coastal zone; and GBR guidelines will apply offshore from the enclosed coastal zone and within waters of the GBR Marine Park.

8.5.3 Water Quality Compliance Monitoring Using Remote Sensing

The regionally parameterized ocean-color algorithms discussed in Section 8.3.7 were applied to MODIS Aqua data (Brando et al. 2008, 2010; Schroeder et al. 2008). The outputs from this application show how remote sensing data can be used to identify exceedance of water quality guidelines for water quality variables data (Figures 8.10 and 8.11) for the region between Townsville and Cairns for a year, from November 2007 to October 2008.

Chlorophyll-*a* mean values for the year going from November 2007 to October 2008 were presented (Figure 8.10), accompanied by the map of the number of days with (error-free) data for that period. Chlorophyll-*a* exceedance maps (Figure 8.11), as defined by the guidelines when mean values for the year exceed the thresholds, were presented with the chlorophyll-*a* exceedance probability (EP). The EP provides the number of days where the concentration exceeded the threshold divided by number of days with (error-free) data for that period (Brando et al. 2010).

For most of the coastal waters area of the mean values of chlorophyll-*a* (Figure 8.11, left) are higher than the guideline trigger value of 0.45 μgL^{-1}, resulting in an almost complete exceedance according to the guidelines and chlorophyll-*a* EP close to or higher than 50%. For the inshore area, the mean values of chlorophyll-*a* ranged ~0.1 to 0.3 μgL^{-1} results in "no exceedance" against water quality guidelines and a chlorophyll-*a* EP ranging from 10% to 50%. For the offshore area, the mean values of chlorophyll-*a* ranged from ~0.3 to 0.4 μgL^{-1}, resulting in no exceedance, using a lower guideline trigger value of 0.4 μgL^{-1}, and with chlorophyll-*a* EP ranging from 0% to 10%.

The map of chlorophyll-*a* EP provides detailed information on the exceedances of chlorophyll relative to a water quality threshold and shows clearly the inshore to offshore gradient. Exceedances are more likely to occur within the inshore area during high flow events. Furthermore, the maps of chlorophyll-*a* EP can be used in a risk analysis framework similar to those developed for CDOM+D in the previous sections.

The metrics used in this study to evaluate compliance are meant to provide a demonstration of the use of remotely sensed data in the assessment of exceedance to the guidelines. These metrics are based on a high number of observations ranging from hundreds of thousands valid observations for open coastal in the wet season to millions for the offshore area in the dry season.

Further work in designing the exceedance and compliance metrics and how to combine the assessment over more variables is needed to provide a high degree of confidence in these results. This will enable these data sets to meet the requirements

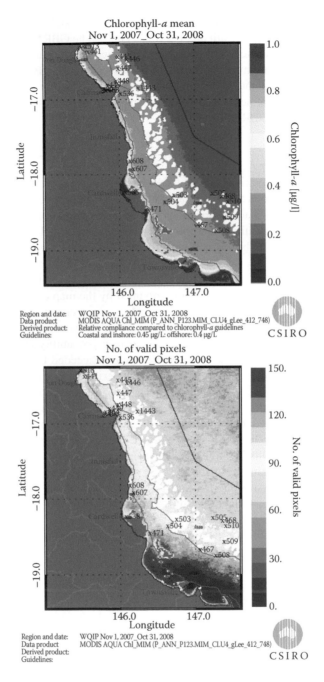

FIGURE 8.10 Maps of chlorophyll mean values for the year going from November 2007 to October 2008 (top), and the number of days with (error-free) data for that period (bottom).

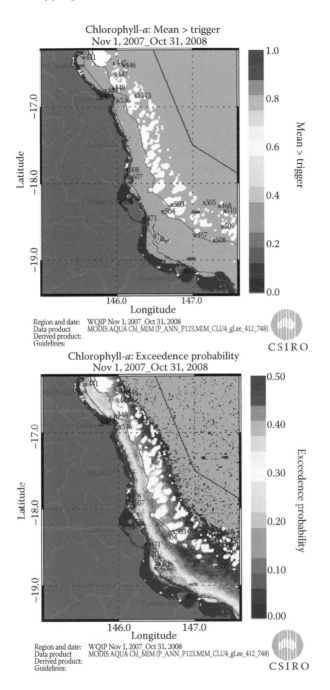

FIGURE 8.11 Maps of chlorophyll exceedance for the year going from November 2007 to October 2008 as defined by the GBRMPA guidelines (top), and of exceedance probability (bottom).

of the reasonable assurance statements and the monitoring and modeling strategies for the environmental planning of the different regional areas.

8.6   FUTURE DIRECTIONS

8.6.1   INCREASED AVAILABILITY OF REMOTE SENSING IMAGERY

Mapping and monitoring of river plumes from GBR allow the dispersion of river plumes in the GBR region to be followed over a period of weeks using satellite data. Opportunities to use historical imagery now exist where additional remotely sensed imagery is available from SeaWiFS and Landsat data. Processing of these historical images may allow us to hindcast the movement of plumes relative to prevailing weather and flow conditions over much longer time frames. There are now substantially more remote sensing products available as there are now many more satellites in operation with sensors able to "map" plume water, allowing a greater chance of the plumes to be free of cloud cover. More historical information and greater access to compliant imagery and algorithm outputs will give a greater confidence in the mapping of plume extents and concentrations.

8.6.2   USE OF QUANTITATIVE THRESHOLDS IN DEFINING WATER TYPES

Plume extents have been identified by true-color and CDOM+D absorbance at 443 nm using remote sensing techniques. Figure 8.12 (part 1) shows the true-color image of the Burdekin plume on the January 14, 2009. The two consecutive images are of the calculated CDOM+D and chlorophyll images taken at the same time. Figure 8.12 (part 2) illustrates the primary and secondary plume associated with the Burdekin flood waters. The very turbid inshore plume can be seen moving north and offshore from the Burdekin mouth, almost reaching the offshore reefs. There is also a secondary plume visible in the left-hand side of the picture, moving north.

The extremely high CDOM+D absorption during such an event is a consequence of elevated dissolved pigments introduced into marine waters from estuarine and terrestrial sources via flood plumes. At present, plume waters are classified into one of three discrete groups: (1) primary, (2) secondary, and (3) tertiary plumes. Preliminary work on this classification method indicated that a combination of TSM, chlorophyll, and CDOM+D imagery was extremely useful in identifying plume boundaries; however, work to date has focused on the qualitative determination of water type extent using true-color imagery and validated information from algorithm application.

Definition of quantitative thresholds for each water type (primary, secondary, and tertiary) will allow us to map the extent and frequency of water types with far greater confidence and identify the full range of spectral properties within each type. Further elucidation of the water types, using thresholds of the water quality concentrations (e.g., CDOM+D and TSM) can be made by defining the range of optical properties within the broad categories of water types. Monitoring information should include in situ optical data as a requirement. Water types can be further defined by the use of the optical properties inherent within the characteristics of the water.

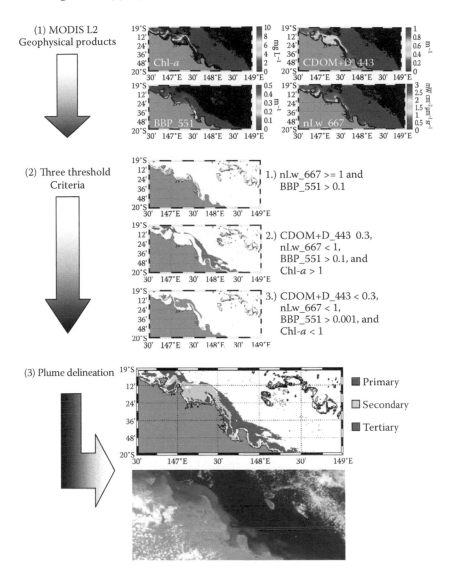

FIGURE 8.12 Use of thresholds to identify the plume water type (primary, secondary, and tertiary). Red denotes the primary, high-sediment carrying waters; green, the secondary, high-color, and elevated production waters; and blue, elevated CDOM+D waters indicative of the full extent of the freshwater plume.

Further elucidation of the movement of different water types (i.e., the high sediment waters as compared to the secondary clearer productive waters) allows better estimate of risk and exposure for biological ecosystems. Figure 8.12 (part 3) shows an example of mapping plume extents based on a series of spectral measurements that correlated to water quality thresholds. The use of spectral thresholds to delineate spatial areas can be correlated with the water type. More work is required on

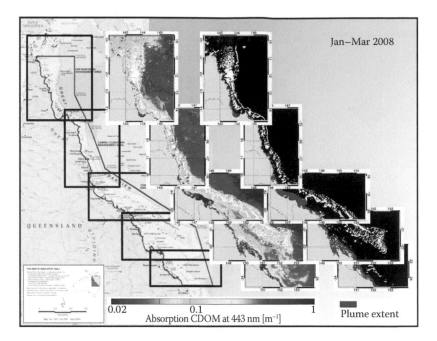

FIGURE 8.13 Great Barrier Reef World Heritage Area subdivided into five management regions (left). Maximum CDOM absorption from regional parameterized ocean-color algorithm mapped for the period January to March 2008 (center). Freshwater plume extent mapped by applying a CDOM threshold derived from linear regression of *in situ* CDOM and salinity measurements.

testing of the water quality thresholds related to biological ecosystems and the level of impact. Links with existing water quality guidelines must be considered.

8.6.3 Correlation of Full Plume Extent through CDOM Thresholds

Another new approach to map the extent of freshwater discharge to the GBR is currently under development by using only the regional parameterized CDOM product applied to MODIS data as a surrogate for low salinity waters. A maximum CDOM absorption map is generated from January to March of each year through aggregation of daily CDOM imagery. By applying a CDOM cutoff threshold previously defined from linear regression of *in situ* CDOM and salinity measurements, freshwater extent can be mapped as illustrated in Figure 8.13. Ongoing work on the relationship between CDOM and salinity will be useful in the further validation of this mapping method.

8.7 CONCLUSIONS

This chapter shows a method for the mapping of GBR riverine plumes and how visual interpretation of extent and duration can be overlaid to create maps of plume frequency and exposure. Spatial mapping of these coastal plumes can also identify the intersection with GBR inshore ecosystems. A combination of field and satellite image mapping was applied as an alternative, as flood plumes have been mapped successfully from remotely sensed data in a number of different coastal environments around the world. Water samples collected from within the plume are analyzed for the contaminants of concern and estimates of the length of exposure and concentrations experienced by biological systems can be assessed.

This work follows the evolution of plume mapping, from the use of aerial imagery to define the plume extent, mapping of plumes using true-color imagery, the use of global algorithms to define water quality parameters in plume waters, and finally the development and application of regionally specific algorithms for mapping of GBR plume waters. The use of true-color imagery and water quality algorithms has advanced our understanding of the full extent of plume waters with parameters such as ocean color, primary production, and suspended sediment being calculated or inferred from algorithm outputs.

Remote sensing is more cost-effective and more informative for a variety of detection, monitoring, and processes understanding tasks. Aerial imagery was utilized in the early 1990s to map riverine plumes in GBR waters, with the first spatial risk map estimated from plume extents identifying an area of concern for the central inshore GBR. There are limitations to this approach due to the relatively high costs involved with aircraft operation and the limited spectral information available from the visualization of plumes from aircraft, where only the visible plume can be mapped.

The development and application of ocean-color sensors allowed a greater aerial coverage of plume movement and extent through the mapping of true-color imagery and spectral data that enables the retrieval of water quality parameters. True-color composites of plume waters can capture the full variation of color or spectral information within a plume, with the characterization and contrast of the green and brown coastal plume waters with the deep blue ocean-color water. Spatial analysis of the true-color composites focused on one GBR catchment (Tully-Murray), providing a series of plume extents that identify water types within the plume area. The water types are based on true-color imagery and validated through the application of the CDOM+D algorithm highlighting the area of high organic material (phytoplankton, yellow substance, and detrital material). Aggregation of the spatial extents measured over time (1994–2008) resulted in a plume exposure map for Tully marine waters, identifying the coral reef and seagrass systems that were more likely to be impacted by high-sediment, high-nutrient plume waters.

Plume waters are characterized by optically complex coastal waters that can result in a large number of errors in the algorithm outputs. The development of regional algorithms based on a variable parameterization of *in situ* measured inherent properties gives a significant accuracy improvement over the SeaDAS algorithms. The use of these more appropriate regional algorithms is still in the testing phase, but has already been used successfully in water quality compliance monitoring and will benefit future risk and exposure mapping with a more accurate representation of water types based on spectral information.

The use of remotely sensed imagery allows us to identify areas of exposure in the Great Barrier Reef. Knowledge of high exposure areas will be useful in the links between catchment characteristics and reef health. This will link in with the current marine monitoring programs and provide an invaluable monitoring technique for the assessment of water quality and reef health in GBR waters.

ACKNOWLEDGMENTS

We acknowledge the MODIS mission scientists and associated NASA personnel for the production of the data used in this research effort. Further, we are grateful to the CSIRO National Research Flagship Wealth from Oceans for funding regional algorithm development. We acknowledge the Great Barrier Reef Marine Park Authority water quality program (1991–1998) and the Marine Monitoring Program (2007 to present) that has funded the *in situ* plume sampling programs.

REFERENCES

Andrefouet, S., Mumby, P. J., McField, M., Hu, C., and Muller-Karger, F. E. (2002). Revisiting coral reef connectivity. *Coral Reefs*, 21, 43–48.

Brando, V., Dekker, A., Marks, A., Qin, Y., and Oubelkheir, K. (2006). *Chlorophyll and Sediment Assessment in Microtidal Tropical Estuary Adjacent to the Great Barrier Reef: Spatial and Temporal Assessment Using Remote Sensing.* Technical Report 74, Cooperative Research Centre for Coastal Zone, Estuary & Waterway Management, Technical Report 74, CSIRO, Australia.

Brando, V. E., Dekker, A. D., Schroeder, Th., Park, Y. J., Clementson, L. A., Steven, A., and Blondeau-Patissier, D. (2008). Satellite retrieval of chlorophyll, CDOM and NAP in optically complex waters using a semi-analytical inversion based on specific inherent optical properties. A case study for the Great Barrier Reef coastal waters. In *Proceedings of the XIX Ocean Optics Conference*, Barga, Italy.

Brando, V. E., Steven, A., Schroeder, Th., Dekker, A. G., Park, Y.-J., Daniel, P., and Ford, P. (2010). *Remote Sensing of GBR Waters to Assist Performance Monitoring of Water Quality Improvement Plans in Far North Queensland*, Final Report to the Department of the Environment and Water Resources, Australia.

Brodie, J. E. and Furnas, M. (1996). Cyclones, river flood plumes and natural water extremes in the central Great Barrier Reef. In *Downstream Effects of Land Use*, H. M. Hunter, A. G. Eyles and G. E. Rayment (eds.), Queensland Department of Natural Resources, Brisbane, Australia, pp. 367–375.

Brodie, J. E. and Mitchell, A. W. (1992). Nutrient composition of the January (1991) Fitzroy River flood plume. In *Workshop on the Impacts of Flooding*, G. T. Byron (ed.), Workshop Series No. 17, Great Barrier Reef Marine Park Authority, Townsville, Australia, pp. 56–74.

Brodie, J. E. and Mitchell, A. W. (2005). Nutrients in Australian tropical rivers: Changes with agricultural development and implications for receiving environments. *Marine and Freshwater Research*, 56(3), 279–302.

Brodie, J., Binney, J., Fabricius, K., Gordon, I., Hoegh-Guldberg, O., Hunter, H., O'Reagain, P., Pearson, R., Quirk, M., Thorburn, P., Waterhouse, J., Webster, I., and Wilkinson, S. (2008a). *Synthesis of Evidence to Support the Scientific Consensus Statement on Water Quality in the Great Barrier Reef.* The State of Queensland (Department of Premier and Cabinet) Brisbane, Australia.

Brodie, J., Pearson, R., Lewis, S., Bainbridge, Z., Waterhouse, J., and Prange, J. (2008b). *MTSRF GBR Catchments and Lagoon WQ Synthesis Report. Year 1 Summary—2007.* Draft report to the Reef and Rainforest Research Centre, September 2007.

Brodie, J., Fabricius, K., De'ath, G., and Okaji, K. (2005). Are increased nutrient inputs responsible for more outbreaks of crown-of-thorns starfish? An appraisal of the evidence. *Marine Pollution Bulletin*, 51, 266–278.

Brodie, J., Faithful, J., and Cullen, K. (2004). *Community Water Quality Monitoring in the Burdekin River Catchment and Estuary: 2002–2004*. ACTFR Technical Report No. 03/16, Australian Centre for Tropical Freshwater Research, James Cook University, Townsville, Australia.

Brodie, J., Waterhouse, J., Lewis, S., Bainbridge, Z., and Johnson, J. (2009a). *Current Loads of Priority Pollutants Discharged from Great Barrier Reef Catchments to the Great Barrier Reef*, ACTFR Report No. 09/02, CSIRO, last edited May 6, 2010.

Brodie, J. E., Lewis, S. E., Bainbridge, Z. T., Mitchell, A., Waterhouse, J., and Kroon, F. (2009b). Target setting for pollutant discharge management of rivers in the Great Barrier Reef catchment area. *Marine and Freshwater Research*, 60, 1141–1149.

Brodie, J., Schroeder, Th., Rohde, K., Faithful, J., Masters, B., Dekker, A., Brando, V., and Maughan, M. (2010). Dispersal of suspended sediments and nutrients in the Great Barrier Reef lagoon during river discharge events: Conclusions from satellite remote sensing and concurrent flood plume sampling, *Marine and Freshwater Research*, 61(6), 651–664.

Brodie, J. E., Devlin, M. J., Haynes, D., and Waterhouse, J. (2011). Assessment of the eutrophication status of the Great Barrier Reef lagoon (Australia). *Biogeochemistry*, doi: 10.1007/s10533-010-9542-2.

Boyle, E., Collier, R., Dengler, A. T., Edmond, J. M., Ng, A. C., and Stallard, R. F. (1974). On the chemical mass-balance in estuaries. *Geochimica Cosmochimica Acta*, 38, 1719–1728.

Burrage, D., Steinberg, C., Bode, L., and Black, K. (1997). Long-term current observations in the Great Barrier Reef. In *Proceedings of a Technical Workshop of State of the Great Barrier Reef World Heritage Area Workshop*, D. Wachenfeld, J. Oliver, and K. Davis (eds.), Workshop No. 23, Great Barrier Reef Marine Park Authority, Townsville, Australia, pp. 21–45.

Burrage, D. M., Heron, M. L., Hacker J. M., Miller, J. L., Strieglitz, T. C., Steinberg, C. R., and Prytz, A. (2003). Structure and influence of tropical river plumes in the Great Barrier Reef: Application and performance of an airborne sea surface salinity mapping system. *Remote Sensing of Environment*, 85, 204–220.

Burrage, D. M., Heron, M. L., and Hacker J. M. (2002). Evolution and dynamics of tropical river plumes in the Great Barrier Reef: An integrated remote sensing and in situ study. *Journal of Geophysical Research—Oceans*, 107(C12), 1–22.

Cantin, N. E., Negri, A. P., and Willis, B. L. (2007). Photoinhibition from chronic herbicide exposure reduces reproductive output of reef-building corals. *Marine Ecology Progress Series*, 344, 81–93.

Chérubin, L. M., Kuchinke, C. P., and Paris, C. B. (2008). Ocean circulation and terrestrial runoff dynamics in the Mesoamerican region from SeaWiFS data and a high resolution simulation. *Coral Reefs*, 27(3), 503–519.

Dagg, M., Benner, R., Lohrenz, S., and Lawrence, D. (2004). Transformation of dissolved and particulate materials on continental shelves influenced by large rivers: Plume processes. *Continental Shelf Research*, 24, 833–858.

Davies, P. J. and Hughes, H. (1983). High-energy reef and terrigenous sedimentation, Boulder Reef, Great Barrier Reef, BMR. *Journal of Australian Geology and Geophysics*, 8, 201–209.

De'ath, G. and Fabricius, K. E. (2008). *Water Quality of the Great Barrier Reef: Distributions, Effects on Reef biota and Trigger Values for the Conservation of Ecosystem Health*. Research Publication No. 89. Great Barrier Marine Park Authority, Report to the GBRMPA and published by the GBRMPA, Townsville, Australia, p. 104.

Devlin, M. J. and Lourey, M. J. (1996). *Water Quality—Field and Analytical Procedures. Long-term Monitoring of the Great Barrier Reef.* Standard Operational Procedure No. 4, Australian Institute of Marine Science, Townsville, Australia.

Devlin, M. J., Taylor, J. P., and Brodie, J. (1998). Flood plumes, extent, concentration and composition. In *Reef Research*, 8(3–4), Great Barrier Reef Marine Park Authority, Townsville, Australia.

Devlin, M., Waterhouse, J., Taylor, J., and Brodie, J. (2001). *Flood Plumes in the Great Barrier Reef: Spatial and Temporal Patterns in Composition and Distribution.* GBRMPA Research Publication No 68, Great Barrier Reef Marine Park Authority, Townsville, Australia.

Devlin, M., Brodie, J., Waterhouse, J., Mitchell, A., Audas, D., and Haynes, D. (2002). Exposure of Great Barrier Reef Inner-shelf Reefs to River-borne Contaminants. *Proceedings of the 2nd National Conference on Aquatic Environments: Sustaining our Aquatic Environments—Implementing Solutions.* Queensland Department of Natural Resources and Mines, Brisbane, Australia.

Devlin, M. and Brodie, J. (2005). Terrestrial discharge into the Great Barrier Reef Lagoon: Nutrient behaviour in coastal waters. *Marine Pollution Bulletin*, 51(1–4), 9–22.

Devlin, M. J. and Schaffelke, B. (2009). Spatial extent of riverine flood plumes and exposure of marine ecosystems in the Tully coastal region, Great Barrier Reef. *Marine and Freshwater Research*, 60, 1109–1122.

Eyre, B. D. (2000). Regional evaluation of nutrient transformation and phytoplankton growth in nine river-dominated sub-tropical east Australian estuaries. *Marine Ecology Progress Series*, 205, 61–83.

Fabricius, K. E. (2005). Effects of terrestrial runoff on the ecology of corals and coral reefs: Review and synthesis. *Marine Pollution Bulletin*, 50, 125–146.

Fabricius, K. E. and De'ath, G. (2001a). Biodiversity on the Great Barrier Reef: Large-scale patterns and turbidity-related local loss of soft coral taxa. In *Oceanographic Processes of Coral Reefs: Physical and Biological Links in the Great Barrier Reef*, E. Wolanski (ed.), CRC Press, London, pp. 127–144.

Fabricius, K. E. and De'ath G. (2001b). Environmental factors associated with the spatial distribution of crustose coralline algae on the Great Barrier Reef. *Coral Reefs*, 19, 303–309.

Fabricius, K. E. and De'ath, G. (2004). Identifying ecological change and its causes: A case study on coral reefs. *Ecological Applications*, 14, 1448–1465.

Fabricius, K. E. and Wolanski, E. J. (2000). Rapid smothering of coral organisms by muddy marine snow. *Estuarine Coastal and Shelf Science*, 50, 115–120.

Fabricius, K. E., De'ath, G., McCook, L. J., Turak, E., and Williams, D. (2005). Coral, algae and fish communities on inshore reefs of the Great Barrier Reef: Ecological gradients along water quality gradients. *Marine Pollution Bulletin*, 51, 384–398.

Fabricius, K. E., De'ath, G., Puotinen, M. L., Done, T., Cooper, T. F., and Burgess, S. (2008). Disturbance gradients on inshore and offshore coral reefs caused by a severe tropical cyclone Larry. *Limnology and Oceanography*, 53(2), 690–704.

Fabricius, K. E. and De'ath, A. G. (2004). Identifying ecological change and its causes: A case study on coral reefs. *Ecological Applications*, 14, 1448–1465.

Furnas, M. J. and Mitchell, A. (1986). Oceanographic aspects of cyclone Winifred. In *The Offshore Effects of Cyclone Winifred*, I. M. Dutton (ed.), Workshop Series No. 7, Great Barrier Reef Marine Park Authority, Townsville, Australia, pp. 41–42.

Furnas, M. (2003). *Catchments and Corals: Terrestrial Runoff to the Great Barrier Reef.* Australian Institute of Marine Science, Townsville, Australia, p. 334.

Furnas, M. J., Mitchell, A. W., and Skuza, M. (1995). *Nitrogen and Phosphorus Budgets for the Central Great Barrier Reef Shelf*, Research Publication No. 36, Great Barrier Reef Marine Park Authority, Townsville, Australia.

Gagan, M. K., Chivas, A. R., and Herczeg, A. L. (1990). Shelf-wide erosion, deposition and suspended sediment transport during Cyclone Winifred, central Great Barrier Reef, Australia. *Journal of Sedimentary Petrology*, 60(3), 456–470.

Gagan, M. K., Sandstrom, M. W., and Chivas, A. R. (1987). Restricted terrestrial carbon input to the continental shelf during cyclone Winifred: Implications for terrestrial run-off to the Great Barrier Reef province. *Coral Reefs*, 6, 113–119.

Garver, S. A. and Siegel, D. A. (1997). Inherent optical property inversion of ocean colour spectra and its biogeochemical interpretation 1. Time Series from the Sargasso Sea, *Journal of Geophysical Research*, 102, 18607–18625.

Gilbert, M. and Brodie, J. E. (2001). *Population and Major Land Use in the Great Barrier Reef Catchment Area: Spatial and Temporal Trends*. 72pp. Great Barrier Reef Marine Park Authority, Townsville, Australia.

Gleghorn, R. J. (1947). Cyclone damage on the Great Barrier Reef. *Reports of the Great Barrier Reef Committee*, 6(1), 17–19.

Gordon, H. R., Brown, O. B., Evans, R., Brown, J., Smith, R. C., Baker, K. S., and Clark, D. C. (1988). A semianalytical model of ocean colour. *Journal of Geophysical Research*, 93(D9), 10909.

Hedley, C. (1925). *The Natural Destruction of a Coral Reef*. Transactions of the Royal Geographical Society (Queensland). *Reports of the Great Barrier Reef Committee*, 1, 35–40.

Hoge, F. E. and Lyon, P. E. (1996). Satellite retrieval of inherent optical properties by linear matrix inversion of oceanic radiance models: An analysis of model and radiance measurements errors. *Journal of Geophysical Research*, 101(C7), 16631–16648.

Hu, C., Hackett, K. E., Callahan, M. K., Andrefouet, S., Wheaton, J. L., Porter, J. W., and Muller-Karger, F. E. (2003). The 2002 ocean color anomaly in the Florida bight: A cause of local coral reef decline? *Geophysical Research Letters*, 30, 1151, doi:10.1029/2002GL016479.

Hu, C., Muller-Karger, F. E., Vargo, G. A., Neely, M. B., and Johns, E. (2004). Linkages between coastal runoff and the Florida Keys ecosystem: A study of a dark plume event. *Geophysical Research Letters*, 31, L15307, doi:10.1029/2004GL020382.

Hunter, H. M. and Walton, R. S. (1997). *From Land to River to Reef Lagoon: Land Use Impacts on Water Quality in the Johnstone River Catchment*. Queensland Department of Natural Resources, Brisbane, Australia.

Hunter, H. M. and Walton, R. S. (2008). Land-use effects on fluxes of suspended sediment, nitrogen and phosphorus from a river catchment of the Great Barrier Reef, Australia. *Journal of Hydrology*, 356, 131–146.

IOCCG. (2000). *Remote Sensing of Ocean Colour in Coastal, and Other Optically-Complex Waters*, S. Sathyendranath (ed.). Reports of the International Ocean Colour Coordinating Group, No. 3, IOCCG, Dartmouth, Canada.

IOCCG. (2010). *Atmospheric Correction for Remotely-Sensed Ocean-Colour Products*, M. Wang (ed.). Reports of the International Ocean-Colour Coordinating Group, No. 10, IOCCG, Dartmouth, Canada.

Jamet, C., Moulin, C., and Thiria, S. (2004). Monitoring aerosol optical properties over the Mediterranean from SeaWiFS images using a neural network inversion. *Geophysical Research Letters*, 31, L13107, doi:10.1029/2004GL019951.

Johnson, J., Waterhouse, J., Maynard, J., and Morris, S. (Writing Team). (2010). *Reef Rescue Marine Monitoring Program: 2008/2009 Synthesis Report*. Report prepared by the Reef and Rainforest Research Centre Consortium of monitoring providers for the Great Barrier Reef Marine Park Authority. Reef and Rainforest Research Centre Limited, Cairns, Australia, p. 158.

King, B., McAllister, F., Wolanski, E., Done, T., and Spagnol, S. (2001). *River Plume Dynamics in the Central Great Barrier Reef*. In Oceanographic Processes of Coral reefs: Physical and Biological Links in the Great Barrier Reef, E. Wolanski (ed.), CRC Press, Boca Raton, FL, pp. 145–160.

King, B., Spagnol, S., Wolanski, E., and Done, T. (1997). Modelling the Mighty Burdekin River in Flood, *Fifth International Conference on Coastal and Estuarine Modelling*, Washington, DC.

Lahet, F. and Stramski, D. (2010). MODIS imagery of turbid plumes in San Diego coastal waters during rainstorm events. *Remote Sensing of Environment*, 114, 332–344.

Land, P. E. and Haigh, J. D. (1996). Atmospheric correction over case 2 waters with an iterative fitting algorithm. *Applied Optics*, 35, 5443–5451.

Liston, P. W. (1990). *Spatial Variability and Covariability of Chlorophyll and Zooplankton on the Great Barrier Reef*. Ph.D. thesis, James Cook University of North Queensland, Australia.

Lee, Z. P., Carder, K. L., and Arnone, R. (2002). Deriving inherent optical properties from ocean color: A multi-band quasi-analytical algorithm for optically deep waters. *Applied Optics*, 41, 5755–5772.

Loder, T. C. and Reichard, R. P. (1981). The dynamics of conservative mixing in estuaries. *Estuaries*, 4, 64–69.

Maritorena, S., Siegel, D. A., and Peterson A. (2002). Optimization of a semi-analytical ocean colour model for global scale applications. *Applied Optics*, 41, 2705–2714.

McCulloch, M. T., Fallon, S., Wyndham, T., Hendy, E., Lough, J., and Barnes, D. (2003). Coral record of increased sediment flux to the inner Great Barrier Reef since European settlement. *Nature*, 421, 727–730.

McKee, B. A., Aller, R. C., Allison, M. A., Bianchi, T. S., and Kineke, G. C. (2004). Transport and transformation of dissolved and particulate materials on continental margins influenced by major rivers: Benthic boundary layer and seabed processes. *Continental Shelf Research*, 24, 859–870,

McManus, G. B. and Fuhrman, J. A. (1990). Mesoscale and seasonal variability of heterotrophic nanoflagellate abundance in an estuarine outflow plume. *Marine Ecology Progress Series*, 61(3), 207–213.

Moorhouse, F. (1936). The Cyclone of 1934 and Its Effect on Low Isles, with Special Observations on Porites. *Reports of the Great Barrier Reef Committee*, 4(2), 36–47.

Negri, A., Vollhardt, C., Humphrey, C., Heyward, A., Jones, R., Eaglesham, G., and Fabricius, K. (2005). Effects of the herbicide diuron on the early life history stages of coral. *Marine Pollution Bulletin*, 51, 370–383.

Nezlin, N. P., Digiacomo, P. M., Stein, E. D., and Ackerman, D. (2005). Stormwater runoff plumes observed by SeaWiFS radiometer in the Southern California Bight. *Remote Sensing of Environment*, 98, 494–510.

Nezlin, N. P. and DiGiacomo, P. M. (2005). Satellite ocean color observations of stormwater runoff plumes along the San Pedro Shelf (southern California) during 1997 to 2003. *Continental Shelf Research*, 25, 1692–1711.

Nezlin, N. P., DiGiacomo, P. M., Diehl, D. W., Jones, B. H., Johnson, S. C., Mengel, M. J., Reifel, K. M., Warrick, J. A., and Wang, M. (2008). Stormwater plume detection by MODIS imagery in the southern California coastal ocean, Estuarine. *Coastal and Shelf Science*, 80, 141–152.

O'Neill, J. P., Byron, G. T., and Wright, S. C. (1992). Some physical characteristics and movement of (1991) Fitzroy River flood plume. In *Workshop on the Impacts of Flooding, G. T. Byron (ed.)*, Workshop Series No. 17, Great Barrier Reef Marine Park Authority, Townsville, Australia, pp. 36–51.

O'Reilly, J. E., Maritorena, S., Mitchell, G., Siegel, D. A., Kendall, L. C., Garver, S. A., Karhu, M., and McClain, C. (1998). Ocean color chlorophyll algorithms for SeaWiFS. *Journal of Geophysical Research*, 103, C11, 24937–24953.

Orpin, A. R., Ridd, P. V., and Stewart, L. K. (1999). Assessment of the relative importance of major sediment-transport mechanisms in the central Great Barrier Reef lagoon. *Australian Journal of Earth Sciences*, 46, 883–896.

Orr, A. P. (1933). Physical and chemical conditions in the sea in the neighbourhood of the Great Barrier Reef. In *British Museum Great Barrier Reef Expedition 1928–1929 Science Report*, 2(3), pp. 37–86.

Paris, C. B. and Chérubin, L. M. (2008). River–reef connectivity in the Meso-American Region. *Coral Reefs*, 27, 773–781.

Parsons, T. R., Maita, Y., and Lalli, C. (1984). *A Manual of Chemical and Biological Methods for Seawater Analysis*, Pergamon, London.

Pearson, R. G. and Garret, R. N. (1978). *Acanthaster planci* on the Great Barrier Reef: Swains reefs and northern surveys in 1975. *Micronesia*, 14, 259–272.

Pickard, G. L., Donguy, J. R., Henin, C., and Rougerie, F. (1977). *A Review of the Physical Oceanography of the Great Barrier Reef and Western Coral Sea*, Australian Institute of Marine Science Monograph Series 2, Australian Institute of Marine Science, Townsville, Australia.

Qin, Y., Brando, V., Dekker, A., and Blondeau-Patissier D. (2007). Validity of SeaDAS water constituents retrieval algorithms in Australian tropical coastal waters. *Geophysical Research Letters*, 34, L21603, doi:10.1029/2007GL030599, 2007.

Rainford, E. H. (1925). *Destruction of the Whitsunday Group Fringing Reefs*. The Australian Museum.

Revelante, N. and Gilmartin, M. (1982). Dynamics of phytoplankton in the Great Barrier Reef lagoon. *Journal of Plankton Research*, 4, 47–76.

Roelfsema, C. M., Phinn, S. R., Dennison, W. C., Dekker, A. G., and Brando, V. E. (2006). Monitoring toxic cyanobacteria *Lyngbya majuscula* (Gomont) in Moreton Bay, Australia by integrating satellite image data and field mapping. *Harmful Algae*, 5, 45–56.

Ryle, V. D., Mueller, H. R., and Gentien, P. (1981). *Automated Analysis of Nutrients in Tropical Seawaters*. Science Data Report No. 3, Australian Institute of Marine Science, Townsville, Australia.

Schroeder, Th. (2005). Remote Sensing of Coastal Waters with MERIS Based on Explicit and Implicit Atmospheric Correction Algorithms. Doctoral thesis (German), Free University Berlin, www.diss.fu-berlin.de/ (last accessed 29 June 2010).

Schroeder, Th., Behnert, I., Schaale, M., Fischer, J., and Doerffer, R. (2007). Atmospheric correction algorithm for MERIS above case-2 waters. *International Journal of Remote Sensing*, 28(7), 1469–1486.

Schroeder, Th., Brando, V. E., Cherukuru, N., Clementson, L., Blondeau-Patisier, D., Dekker, A. G., and Fischer, J. (2008). Remote sensing of apparent and inherent optical properties of Tasmanian coastal waters: Application to MODIS Data. In *Proceedings of the XIX Ocean Optics Conference*, Barga, Italy.

Sheng, J., Wang, L., Andréfouët, S., Hu, C., Hatcher, B. G., Muller-Karger, F. F., Kjerfve, B., Heyman, W. D., and Yang, B. (2007). Upper ocean response of the Mesoamerican Barrier Reef System to Hurricane Mitch and coastal freshwater inputs: A study using Sea-viewing Wide Field-of-view Sensor (SeaWiFS) ocean color data and a nested-grid ocean circulation model. *Journal of Geophysics Research*, 112, C07016, doi:10.1029/2006JC003900.

Shi, W. and Wang, M. (2007). Observations of a Hurricane Katrina-induced phytoplankton bloom in the Gulf of Mexico. *Geophysical Research Letters*, 34(11), L11607, doi:10.1029/2007GL029724.

Shi, W. and Wang, M. (2009). Satellite observations of flood-driven Mississippi River plume in the spring of 2008. *Geophysical Research Letters*, 36, L07607, doi:10.1029/2009GL037210.

Soto, I., Andréfouët, S., Hu, C., Muller-Karger, F. E., Wall, C. C., Sheng, J., and Hatcher, B. G. (2009). Physical connectivity in the Mesoamerican Barrier Reef System inferred from 9 years of ocean color observations. *Coral Reefs*, 28, 415–425.

Steven, A. D. L., Devlin, M., Brodie, J., Baer, M., and Lourey, M. (1996). Spatial influence and composition of river plumes in the central Great Barrier Reef. In *Downstream Effects of Land Use,* H. M. Hunter, A. G. Eyles, and G. E. Rayment (eds.), Queensland Department of Natural Resources, Brisbane, Australia, pp. 85–92.

Taylor, J. (1997). *Nutrient Distribution in the Barron River and Offshore during Cyclone Sadie, in Cyclone Sadie Flood Plumes in the Great Barrier Reef Lagoon: Composition and Consequences,* A. D. L. Steven (ed.), Workshop Series No. 22, Great Barrier Reef Marine Park Authority, Townsville, Australia, pp. 17–26.

Tilman, D. (1999). Global environmental impacts of agricultural expansion: The need for sustainable and efficient practices. *Proceedings of the National Academy of Sciences of the United States of America,* 96, No. 11, 5995–6000.

Thomas, A. C. and Weatherbee, R. (2006). Satellite-measured temporal variability of the Columbia River plume. *Remote Sensing of Environment,* 100, 167–178.

Udy, J. W., Dennison, W. C., Lee Long, W. J., and McKenzie, L. J. (1999). Responses of seagrasses to nutrients in the Great Barrier Reef, Australia. *Marine Ecology Progress Series,* 185, 257–271.

Valderrama, J. C. (1981). The simultaneous analysis of total nitrogen and total phosphorus in natural waters. *Marine Chemistry,* 10, 109–122.

van Woesik, R., DeVantier, L. M., and Glazebrook, J. S. (1995). Effects of cyclone "Joy" on nearshore coral communities of the Great Barrier Reef. *Marine Ecology Progress Series,* 128, 261–270.

van Woesik, R., Tomascik, T., and Blake, S. (1999). Coral assemblages and physico-chemical characteristics of the Whitsunday Islands. *Marine and Freshwater Research,* 50, 427–440.

Veron, J. E. N. (1996). *Coral reefs—An Overview.* In *The State of the Marine Environment for Australia Technical Annex*: 1, L. P. Zann (ed.), Great Barrier Reef Marine Park Authority, Townsville, Australia.

Wang, M. and Shi, W. (2007). The NIR-SWIR combined atmospheric correction approach for MODIS ocean colour data processing. *Optics Express,* 15, 24, 15722–15733.

Webster, I., Brinkman, B., Parslow, J. S., Prange, J., Waterhouse, J., and Steven, A. D. L. (2008). *Review and Gap Analysis of Receiving-water Water Quality Modelling in the Great Barrier Reef.* Water for a Healthy Country National Research Flagship report, CSIRO, Canberra, Australia, p. 156.

Wolanski, E. (1994). *Physical Oceanographic Processes of the Great Barrier Reef.* CRC Press, Boca Raton, FL.

Wolanski, E. and Jones, M. (1981). Physical properties of Great Barrier Reef Lagoon waters near Townsville. I. Effects of Burdekin River floods. *Australian Journal of Marine and Freshwater Research,* 32, 305–319.

Wolanski, E. and van Senden, D. (1983). Mixing of Burdekin river flood waters in the Great Barrier Reef. *Australian Journal of Marine and Freshwater Research,* 34, 49–63.

Zhang, T., Fell, F., and Fischer, J. (2002). Modeling the backscattering ratio of marine particles in case-2 waters. *Proceedings of the Ocean Optics XVI,* Santa Fe, NM, published on CDROM.

Zhang, T. (2003). *Retrieval of Oceanic Constituents with Artificial Neural Network Based on Radiative Transfer Simulation Techniques.* Doctoral thesis, Free University Berlin, Germany.

Part II

Sensing and Monitoring
for Land Use Patterns,
Reclamation, and Degradation

9 Satellite Remote Sensing for Landslide Prediction

Yang Hong, Zonghu Liao,
Robert F. Adler, and Chun Liu

CONTENTS

9.1 INTRODUCTION

Landslides are caused primarily by prolonged heavy rainfall on saturated hill slopes or triggered by earthquakes (Baum et al. 2002). Rain-induced landslides rank among the most devastating natural disasters, causing billions of dollars in property damages and thousands of deaths in most years around the world. In the United States alone, landslides can occur in every state, causing an estimated $2 billion in damage and 25 to 50 deaths each year (USGS 2006). Annual average loss of life from landslide hazards in Japan is more than 100 (Sidle and Ochiai 2006). The situation is much worse in developing countries and remote mountainous regions due to lack of financial resources and inadequate disaster prevention capacity. For example, Hurricane Mitch caused catastrophic landslides throughout the Caribbean and Central America area in October 1998. It was reported that 6600 persons were killed and 8052 injured. Approximately 1.4 million people were left homeless. More than 92 bridges were destroyed and nearly 70% of crops were damaged. A landslide triggered by La Niña rains buried an entire village on the Philippines Island of Leyte on February 17, 2006, with at least 1800 reported deaths and only three houses left standing of the original 300. In a postanalysis, Lagmay et al. (2006) reported 500 mm of heavy rainfall fell on that area using National Aeronautics and Space Administration (NASA)'s Tropical Rainfall Measuring Mission (TRMM; http://trmm.gsfc.nasa.gov). Hereafter the term landslides are used to refer to these types of landmass movement.

Landslides triggered by rainfall can possibly be predicted by modeling the relationship between rainfall intensity–duration and landslide occurrence (Keefer and Wilson 1987). In particular, landslide warning systems can save lives and reduce damages if properly implemented in populated areas of landslide-prone nations (Sidle and Ochiai 2006). Society has dealt with landslide hazards primarily by trying to site development away from potential hazard zones based on experience or conventional field survey. Growing populations plus related environmental impacts such as deforestation have put a growing number of people at risk from landslides. At the same time, the required information infrastructure and data analysis capabilities to minimize injuries and deaths due to landslides are not yet practical in most developing countries.

The need to develop a more effective spatial coverage of landslide susceptibility (LS) and hazard early warning for vulnerable countries and remote areas remains apparent and urgent (Sidle and Ochiai 2006). However, analyzing landslide potential that might occur in a large area is very difficult and expensive in terms of time and money. This is especially true in developing countries where expensive ground observation networks are prohibitive and in mountainous areas where access is difficult. Currently no system exists at a global scale to identify rainfall conditions that may trigger landslides, largely due to lack of field-based observing networks in many parts of the world. In particular, developing countries usually lack expensive ground-based monitoring networks. Thus, for many countries around the world, remote sensing information may be the only possible source available for such studies. Currently available satellite data may provide useful and accurate information on Earth surface features and dynamic processes involved in landslide occurrence. The challenge facing our science community is to better understand the surface and meteorological processes leading to landslides and determine how new technology and techniques might be applied to reduce the risk of landslides to people across the globe.

Recent advances in satellite-based precipitation observation technology and increasing availability of high-resolution geospatial products at global scale are providing an unprecedented opportunity to develop a real-time prediction system for a global view of rainfall-triggered landslides. Drawing on the heritage of a space-based global precipitation observation system and remotely sensed land surface characteristic products, this chapter describes a framework for detecting rainfall-triggered landslides in a near real-time manner, with an eye toward developing a system to detect or forecast such events on a global basis. In the case study, we are primarily concerned with rainfall-triggered shallow landslides or debris flows.

9.2   A CONCEPTUAL FRAMEWORK

In this study, emphasis was placed on shallow landslides that involve poorly con-
solidated soils or colluviums on steep hill slopes. Shallow landslides, sometimes
referred to as debris flows, mudslides, mudflows, or debris avalanches, are rapidly
moving flows of mixes of rocks and mud, which have the potential to kill people
and destroy homes, roads, bridges, and other property. This study addresses those
landslides caused primarily by prolonged, heavy rainfall on saturated hill slopes
characterized by high permeability. Rainfall-triggered landslides may mobilize into
fast-moving mudflows, which generally present a greater hazard to human life than
slow-moving, deep-seated slides. Although most parts of the world had experienced
major socioeconomic losses related to landslide activity (Sidle and Ochiai 2006),
currently no system exists at either a regional or a global scale to identify rainfall
conditions that may trigger landslides.

Useful assessment of landslide hazards requires, at the minimum, an understanding of both where and when the landslides may occur (Hong et al. 2007a). Landslides
result from a combination of factors that can be broadly classified into two categories
according to (Dai and Lee 2002): (1) preparatory variables that make the land surface
susceptible to failure without triggering it, such as slope, soil properties, elevation,

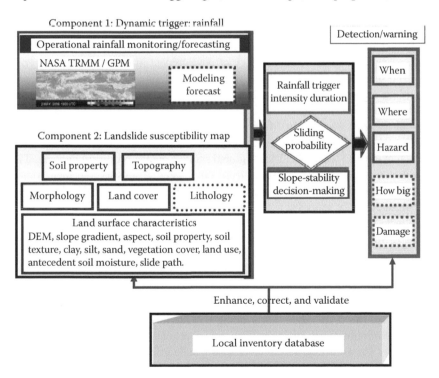

FIGURE 9.1   The conceptual framework of real-time identifying/warning system for
rainfall-triggered landslides at the global scale. Note that dashed-line boxes are important
components but not covered in this study. (From Hong, Y. and R. F. Adler, Special Issue of
International Journal of Sediment Research, 23, 3, 249–257, 2008. With permission.)

aspect, land cover, and lithology, and (2) the triggering events that induce mass movement, such as heavy rainfall. For rainfall-triggered landslides, at least two conditions must be met: (1) the areas must be susceptible to failure under certain saturated conditions, and (2) the rainfall intensity and duration must be sufficient to saturate the ground to a sufficient depth. Therefore, to diagnose the landslide occurrence, the proposed framework must link two major components: real-time precipitation analysis and LS information, as shown in Figure 9.1. The LS map empirically shows part of the "where" and the rain intensity–duration primarily determines the "when" information. In use, the "where" LS map is overlaid with a real-time satellite-based rainfall "when" layer to detect landslide hazards as a function of time and location. In this framework, the first-order control on the spatial distribution (the "where") of landslides is the topographic slope of the ground surface, elevation, soil types, soil texture, vegetation, and land cover classification while the first-order control on the temporal distribution (the "when") of shallow landslides is the space–time variation of rainfall, which changes the pore-pressure response in the soil or colluviums to infiltrating water (Iverson 2000).

9.3 THE DYNAMIC TRIGGER: SPACEBORNE REAL-TIME RAINFALL ESTIMATION

9.3.1 Rainfall and Landslides

The spatial distribution, duration, and intensity of precipitation play an important role in triggering landslides. A long history of development in the estimation of precipitation from space has culminated in sophisticated satellite instruments and techniques to combine information from multiple satellites to produce long-term products useful for climate monitoring (Adler et al. 2003). A fine time resolution analysis, such as the Tropical Rainfall Measuring Mission (TRMM) Multisatellite Precipitation Analysis (TMPA) (Huffman et al. 2007), is the key data set for the proposed landslide monitoring system in this study. The TMPA global rainfall map is produced by using TRMM to calibrate, or adjust, the estimates from other satellite sensors and then combining all the estimates into the TMPA final analysis. The coverage of the TMPA depends on input from different sets of sensors. First, precipitation-related passive microwave data are collected by a variety of low-Earth-orbit satellites, including the TRMM Microwave Imager (TMI) on TRMM, special sensor microwave/imager on Defense Meteorological Satellite Program (DMSP) satellites, advanced microwave scanning radiometer for the Earth Observing System (AMSR-E) on Aqua, and the Advanced Microwave Sounding Unit B (AMSU-B) on the National Oceanic and Atmospheric Administration (NOAA) satellite series. The second major data source for the TMPA is the window-channel ($-10.7 \mu m$) infrared (IR) data that is being collected by the international constellation of geosynchronous-Earth-orbit satellites, which provide excellent time–space coverage (half-hourly 4×4-km equivalent latitude/longitude grids) after merged by the Climate Prediction Center of NOAA (Janowiak et al. 2001). The IR brightness temperatures are corrected for zenith-angle viewing effects and intersatellite calibration differences.

The TMPA is a TRMM standard product at fine time and space scales and covers the latitude band 50°N-S for the period 1998 to the delayed present. A real-time version of the TMPA merged product was introduced in February 2002 and is available on the NASA TRMM Web site (http://trmm.gsfc.nasa.gov). Early validation results indicate reasonable performance at monthly scales, while at finer scales the TMPA is successful at approximately reproducing the surface-observation–based histogram of instantaneous precipitation over land, as well as reasonably detecting large daily events (Huffman et al. 2007). The availability of this type of rainfall information quasi-globally provides an opportunity to derive empirical rainfall intensity–duration thresholds related to landslides and to examine antecedent precipitation accumulation continuously in time and space.

In order to identify when landslide-prone areas receive heavy rainfall, the operational TMPA rainfall estimates can be accumulated at various space–time scales to estimate the storm magnitude and antecedent precipitation index. Then the influence of rainfall characteristics on the timing and occurrence of several landslides, which occurred during or just following periods of relatively substantial rainfalls, may be predicted (Figure 9.2). In Figure 9.2a, it shows the TMPA rainfall intensity (bar) and accumulation (line) of one devastating landslide (>2000 deaths) that occurred at Casita Volcano, Nicaragua, on October 30, 1998. Following 1 week of heavy rainfall (>700 mm), the landslide swept over the towns of El Provenira and Rolando Rodriguez on the day of peak rainfall as Hurricane Mitch moved across Central America. The fast-moving mud mixture then eroded additional sediment to transform into a debris flow containing more than 80% sediment by weight.

Not all slope failures, however, occur during peak or immediately after heavy rainfall. In Figure 9.2b, it indicates that the Philippines's Leyte landslide (Lagmay

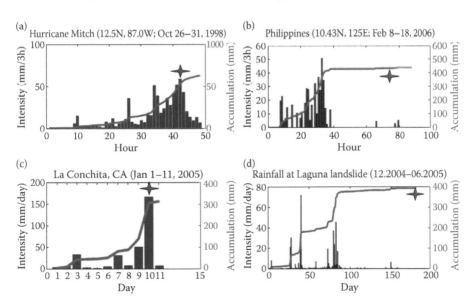

FIGURE 9.2 Influence of rainfall characteristics on the timing and occurrence of landslides. Bars show the rainfall intensity and the star denotes the timing of landslide occurrence at four different locations (a) Guatemala, (b) Philippines, and (c,d) La Conchita and Laguna, California. (From Hong, Y., Adler, R. F., and Huffman, G., *Geophysical Research Letter*, 33, L22402, doi:10.1029/2006GL028010, 2006. With permission.)

et al. 2006) occurred 3 days after heavy rainfall (>400 mm) battered the area. The excessive rains in this case were linked to La Niña. In December, cooler ocean waters began to emerge in the central equatorial Pacific, signaling the onset of a La Niña, which can significantly enhance rainfall across the West Pacific region. Warnings had been issued for this potential excessive rainfall scenario but no landslide warning system existed there to alert possible initiation of landslide. This landslide buried an entire village on the Philippines island of Leyte at about 2:45 a.m. UTC October 17, 2006, leaving at least 200 people dead and 1500 reported still missing.

TMPA precipitation data is available both in real-time and in after–real-time versions, which are useful for predicting landslides triggered by short-term intense storms or by antecedent seasonal rainfall. For example, in Figure 9.2c, it shows the antecedent 10-day rainfall intensity and accumulation, which caused widespread shallow landslides during the winter of 2005 in suburban La Conchita, a small unincorporated community in western Ventura County, California, on U.S. Route 101 just southeast of the Santa Barbara county line. The village of La Conchita is along a portion of the coast prone to mudslides and sits beneath a geologically unstable formation. In 1909, a devastating mudslide occurred approximately one-half mile north of the town, and in 2005, slides closed Highway 101 in both directions, trapping residents. Sandwiched between a steep, unstable hillside with the La Conchita Ranch Company situated on the plateau directly over the community and the Pacific Ocean, La Conchita was the site of a massive mudslide that buried four blocks of the town with over 9 m of Earth on January 10, 2005 at 12:30 a.m. In this event, 10 people were killed by the slide and 14 were injured. Of the 166 homes in the community, 50 homes were destroyed or tagged by the county as uninhabitable. La Conchita Ranch Company was sued by those affected by the 2005 landslide and a settlement was reached giving the plaintiffs the company's assets and 5 million dollars. The tragedy became a defining moment in its winter's wet season in populated Southern California delivered by the "Pineapple Express" weather system. Saturation by the heavy winter (from December 2004 to January 1–2, 2005) rainfall (>350 mm) persisted into late spring and triggered a larger, deeper but slower landslide that destroyed at least 11 homes in Laguna Beach (Figure 9.2d). By that time, the antecedent winter rainfall (391 mm), exceeding the minimum cumulative rainfall (250 mm) required to induce landslides, had been soaking the region.

9.3.2   SATELLITE-BASED RAINFALL INTENSITY–DURATION THRESHOLD

Landslide hazard assessment based on relationships with rainfall intensity–duration has been applied at both global (Caine 1980) and regional scales (Canuti et al. 1985; Larsen and Simon 1993; Godt 2004). Empirical rainfall intensity–duration thresholds have been developed for Seattle (Godt 2004), Puerto Rico (Larsen and Simon 1993), and worldwide (Caine 1980). Using the precipitation information from TMPA, Hong et al. (2006) derived the first satellite-based rainfall intensity–duration threshold curve from landslide cases in various climate and geological locations, in parallel to the previous rain-gauge–based studies. Note that the TMPA-based threshold falls below Caine's threshold, likely because the TMPA is an area-average value rather than a point accumulation (Figure 9.3).

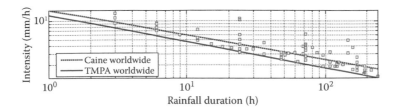

FIGURE 9.3   Satellite-based rainfall intensity–duration threshold curve for triggering land-slides (red; Intensity = 12.45 duration$^{-0.42}$) for landslides (squares) that occurred around the globe in the period 1998–2005 and rain-gauge–based threshold curve from Caine (1980), in blue. (After Hong, Y., Adler, R. F., and Huffman, G. J., Satellite remote sensing for landslide monitoring on a global basis, *American Geophysical Union EOS* Featured Cover Article, 88, 37, 357–358, 2007.)

9.4   CREATION OF GLOBAL LANDSLIDE SUSCEPTIBILITY MAP

9.4.1   THE GLOBAL LANDS SURFACE DATASETS

Remote sensing products can be utilized for deriving various parameters related to landslide controlling factors, as shown in Figure 9.1. The primary data sets used in this study are briefly described below.

9.4.1.1   Digital Elevation Model Data and Its Derivatives

The basic digital elevation model (DEM) data set used in this study includes the NASA Shuttle Radar Topography Mission (SRTM; http://www2.jpl.nasa.gov/srtm/) data set. The SRTM data is a major breakthrough in digital mapping of the world (i.e., with 30 m horizontal spatial resolution and vertical error less than 16 m) and provides a major advance in the availability of high-quality elevation data for large portions of the tropics and other areas of the developing world. SRTM data is distributed in two levels: (1) SRTM1 (for the United States and its territories and possessions) with data sampled at 1 arc-second intervals in latitude and longitude, and (2) SRTM3 (60°N-60°S) sampled at 3 arc-seconds. The horizontal resolution of SRTM1 has about a 30-m resolution and SRTM3 has a 90-m resolution in equatorial regions. A description of the SRTM mission can be found in Farr and Kobrick (2000). DEM data can be used to derive topographic factors other than simply elevation, including slopes, aspects, hill shading, slope curvature, slope roughness, slope area, and qualitative classification of landforms (Fernandez et al. 2003). DEM data can also be used to derive hydrological parameters (flow direction, flow path, and basin and river network basin).

9.4.1.2   Land Cover Data

MODIS is a key instrument aboard the Terra and Aqua satellites. MODIS views the entire Earth's surface every 1 to 2 days, acquiring data in 36 spectral bands or groups of wavelengths (http://modis.gsfc.nasa.gov/index.php). The MODIS land cover classification map is available at the highest resolution possible, 250 m. This land-cover product uses the classification scheme proposed by the International Geosphere

Biosphere Programme (IGBP). The MODIS land cover products describe the geographic distribution of the 17 IGBP land cover types based on an annual time series of observations (Fridel et al. 2002). For each spatial resolution there is a land-cover type classification layer (i.e., with numbers from 0 to 17), a classifier confidence assessment layer, and 17 associated layers that provide the percentage from 0 to 100 of each land-cover type per cell. The data set also provides the fraction of each of the 17 classes within the coarser resolution cells.

9.4.1.3 FAO Digital Soil Map and Soil Characteristics

Information on soil properties is obtained from Digital Soil of the World published in 2003 by Food and Agriculture Organization (FAO) of the United Nations (http://www.fao.org/nr/land/soils/harmonized-world-soil-database/en/). The soil parameters available include soil type classification, clay mineralogy, soil depth, soil moisture capacity, soil bulk density, soil compaction, and so forth. This product is not based on satellite information directly, but is based primarily on ground surveys and national databases.

9.4.1.4 International Satellite Land Surface Climatology Project

A second nonsatellite database is the International Satellite Land Surface Climatology Project (ISLSCP) Initiative II Data Collection (http://www.gewex.org/islscp.html), which provides gridded data of 18 selected soil parameters. These data sets are distributed by the Oak Ridge National Laboratory Distributed Active Archive Center (http://daac.ornl.gov/) at quarter degree resolution. One important parameter for this study is the soil texture. Following the U.S. Department of Agriculture soil texture classification, the 13 textural classes reflect the relative proportions of clay (granules size less than 0.002 mm), silt (0.002–0.05 mm), and sand (0.05–2 mm) in the soil. Three textural categories are recognized among the 13 original texture classes: (1) coarse: sands, loamy sands, and sandy loams with less than 18% clay and more than 65% sand; (2) medium: sandy loams, loams, sandy clay loams, silt loams, silt, silty clay loams, and clay loams with less than 35% clay and less than 65% sand (the sand fraction may be as high as 82% if a minimum of 18% clay is present); and (3) fine: clay, silty clays, sandy clays, and clay loams with more than 35% clay. Note that these soil texture classes are interpolated to the highest DEM spatial scale.

9.4.2 The Global Landslide Susceptibility Map

Based on the aforementioned geospatial datasets, a number of landslide-controlling parameters are derived including elevation, slope, aspect, curvature, concavity, percentage of soil types (including clay, foam, silt, and sand, etc.), soil texture, land use classification, and hydrological variables (i.e., drainage density, flow accumulation, and flow path). All parameters have been downscaled or interpolated to the SRTM elemental 30-m horizontal scale. Due to the lack of global landslide occurrence data, landslide factor selection and assignment of numerical values are based on the referenced studies and on information availability. For example, using an approach published by Larsen and Torres-Sanchez (1998), land cover can be discretized into several general categories: forested land, shrub land, grassland, pasture and/or cropland, and developed land and/or road corridors. These land use/land cover categories describe a continuum of increasing susceptibility to shallow landslides. In this study, following the same approach, the 17 MODIS land-cover types are classified into 11 categories that describe increasing landslide susceptibility to shallow landslides. The landslide susceptibility values from zero to one are assigned to each category, respectively. The effect of slope, soil type, and soil texture on landslides was widely documented by Dai et al. (2002) and Lee and Min (2001). In many regions, elevation according to Coe et al. (2004) is approximately a proxy for mean rainfall that increases with height due to orographic effects and high elevation areas are preferentially susceptible to landslides because they receive greater amounts of rainfall than areas at lower elevations. Drainage density provides an indirect measure of groundwater conditions and has an important role to play in landslide activity (Sarkar and Kanungo 2004). Sarkar and Kanungo (2004) also found an inverse relationship between landslides and drainage density that may be due to high infiltration in weathered gneisses causing more instability in the area. Based on these previous studies, assignment of landslide susceptibility values for other parameters is based on several empirical assumptions: (1) higher slope, higher susceptibility, (2) coarser and looser soil, higher susceptibility, (3) higher elevation, higher susceptibility, and (4) decreasing susceptibility for larger drainage density. Under assumption (1) for example, the slope map units are given a zero susceptibility value for a class of flat slopes and a susceptibility value of one for the class of steepest slopes.

A global landslide susceptibility map was then derived from the aforementioned geospatial data based on each factor's relative significance to the sliding processes using a weighted linear combination approach (Hong et al. 2007a). The slope and lithology are the primary factors, while land-cover type and soil properties are secondary in importance. This approach considers the integration of remote sensing and GIS techniques, given that most current models of the hazard prediction and landslide zoning are GIS-based or with the support of GIS (Metternicht et al. 2005). Corresponding thematic data layers were first generated and then stored in GIS; and finally the global susceptibility map was computed by performing a weighted linear combination function. The best model obtained was the one with weight determination (0.3, 0.2, 0.2, 0.1, 0.1, and 0.1) for the six parameters (slope, type of soil, soil texture, elevation, MODIS land-cover type, and drainage density), respectively. The consequent range in susceptibility values is normalized from zero to one. The larger the susceptibility value, the greater the potential to produce landslide. The consequent range in susceptibility values is normalized from zero to 100. The larger the susceptibility value, the greater the landslide potential at that location. The landslide susceptibility values are then classified into several landslide susceptibility categories (Sarkar and Kanungo 2004). A judicious way for such classification is to search for abrupt changes in values (Davis 1986). The category boundaries are drawn at significant changes in the histogram of the landslide susceptibility values. As a result, the global landslide susceptibility map is classified into several categories, ranging from negligible to hot spot (Figure 9.4). Excluding permanent snow/ice regions, it shows in Figure 9.4 that the low LSI areas cover about half of the land (52%). The very high and high susceptibility categories account for 2.8% and 18.6% of land areas, respectively. Figure 9.4 also demonstrates the hot spots of the high landslide potential regions: the Pacific Rim, the Alps, the Himalayas and South Asia, Rocky

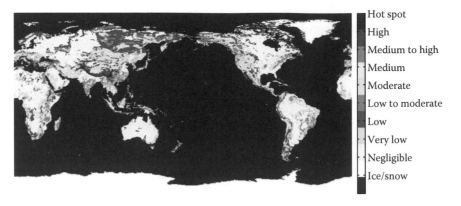

Hot spot
High
Medium to high
Medium
Moderate
Low to moderate
Low
Very low
Negligible
Ice/snow

FIGURE 9.4 Global landslide susceptibility map derived from surface multigeospatial data. (From Hong, Y., Adler, R. F., and Huffman, G. J., Satellite remote sensing for landslide monitoring on a global basis, *American Geophysical Union EOS* Featured Cover Article, 88, 37, 357–358, 2007.)

Mountains, Appalachian Mountains, and parts of the Middle East and Africa. India, China, Nepal, Japan, the United States, and Peru are shown to be landslide-prone countries. These results are compatible to those reported by Sidle and Ochiai (2006). For a more detailed description of global landslide susceptibility map, please refer to Hong et al. (2007a, b).

9.5   A PROTOTYPE LANDSLIDE EARLY WARNING SYSTEM

9.5.1   LINKING REAL-TIME RAINFALL TRIGGERS WITH LANDSLIDE SUSCEPTIBILITY INFORMATION

When coupled with real-time rainfall data, rainfall intensity–duration thresholds might provide the basis for early warning systems for shallow landslides (Liritano et al. 1998). Knowledge of landslide susceptibility as displayed in Figure 9.4 (the "where" of the problem) and the ability to detect heavy rain events that meet threshold conditions as shown in Figure 9.3 (the "when" of the problem) provide the basis for exploring the potential and limitations of such approaches for analyzing and studying the occurrences of landslides, and even possibly forecasting them (Liritano et al. 1998). The rainfall intensity–duration threshold developed in this study was integrated with the global LS with the frequently updated TMPA satellite-based real-time precipitation information to identify when areas with high landslide potential are receiving heavy rainfall. A prototype prediction system for real-time landslide hazard assessment has been developed and a trial version of this operational system is displayed on the NASA Web site (http://trmm.gsfc.nasa.gov/publications_dir/potential_landslide.html). Real-time precipitation from the TMPA is compared with the rainfall intensity–duration thresholds, and antecedent rainfall accumulation can be computed from the TMPA database. Therefore, the locations receiving rainfall exceeding the thresholds are marked as a landslide hazard zone if the underlying susceptibility category is high or very high at that location. The locations and timing of predicted landslides can then be checked against firsthand accounts from the field or validated by news reports.

9.5.2 EVALUATION OF RESULTS

The prototype system for rainfall-triggered landslides was initially evaluated by comparing with reported landslide occurrences within the 8-year TRMM operational time period. For example, one landslide case was predicted by this prototype system on April 13, 2006, in Colombia. The rainfall accumulation for the previous 24 hours was 103 mm over central Colombia and the landslide susceptibility map indicated high susceptibility category at this area, so that the landslide hazard is color-coded as red on the Web-based graphical interface. Later news reports indicated that at least 34 people were missing and four villages were destroyed in a landslide near the Pacific port city of Buenaventura in southwestern Colombia. A preliminary evaluation demonstrates the potential effectiveness of this approach, at least for the 25 large events examined (Hong et al. 2007c). The probability of detection (POD) is 0.76, reflecting 19 successful detections out of 25 occurrences in the assessment. However, the results also indicated that this first-generation system fails to identify landslides triggered by short-duration heavy rainfall events (<6 hours) or by rain falling on snow. Improvements are expected in the analysis as well as satellite data processing and information retrievals associated with both the characterization of land surface and rainfall monitoring. Kirschbaum et al. (2009) presented a more detailed retrospective evaluation using the global landslide inventory database developed by Kirschbaum et al. (2010).

9.6 DISCUSSION

Although the empirical rainfall intensity–duration thresholds have limitations, they have been successfully implemented in several regions, including San Francisco Bay (Keefer et al. 1987), Rio de Janeiro, Brazil (d'Orsi et al. 2004), and Hong Kong (Chan et al. 2003). While such systems have proved very useful and may save lives and protect property, the requisite level of data collection, transmission, and warning is not yet practical in the most vulnerable regions of developing countries (Sidle and Ochiai 2006). Only the availability of the recent multisatellite rainfall monitoring system such as NASA TRMM offers an opportunity to develop/test a prediction system for landslides over large areas. This preliminary landslide detection system is promising when comparing its predictions to recent landslide events that occurred during the TRMM operational period (Table 9.1).

A global evaluation of this system has been underway through comparison with various field databases, Web sites, and news reports of landslide disasters (Kirschbaum et al. 2009). While the global landslide hazard algorithm framework may eventually be useful for forecasting landslide hazard conditions at the global scale, the approach can only be effective for understanding the relationship between landslide-controlling variables at present (Kirschbaum et al. 2010). The need for retrospective validation and improvement of this prototype system requires continued

TABLE 9.1

Evaluation of the Preliminary System by Retrospectively Comparing with Reported Major Landslides within the Last 8-Year TRMM Operational Period

Time	Country (State/Province)	Identified (Yes/No)	Causes/Types	Major Losses and Damage
August 22, 2006	Ban Thahan Village in Phang Nga, Thailand	No	Heavy rainfall/flash flood	Roads blocked
August 20, 2006	Holiday village of Gulval in Cornwall, UK	Yes	Heavy rain showers	Unknown
August 20, 2006	Surat Thani, Thailand	Yes	Heavy rainfall and flash flood	600 residents evacuated
August 19, 2006	Song Bang town of northern mountainous Cao Bang province, Vietnam	Yes	Caused by prolonged heavy rains	10 killed
July 31, 2006	Roer Gulch east of Telluride, CO	No	Heavy rainfall/flash flood	Unknown
July 9, 2006	South Korea	Yes	Typhoon Ewiniar, >300 mm	Widespread mudslide
June 28, 2006	Albany, upstate NY	Yes	Heavy rainfall, 400 mm/5 days	2 killed
June 25, 2006	Villages of Chamba District, Shimla, India	No	Storm, flash flood	6 houses swept away
June 20, 2006	Sinjai in South Island of Sulawesi, Indonesian	Yes	Heavy rainfall >250 mm	>200 deaths
May 17, 2006	The Schweitzer Mountain Ski resort, Sandpoint, ID	No	Rain on snow and snowmelt, rocks, mudslide, and debris flows	Condo buildings damaged
April 13, 2006	Buenaventura, Colombia	Yes	Rain storm, 103.04 mm/day	>34 deaths
January 4, 2006	Jakarta, Indonesia	Yes	Monsoon rains, 250 mm/3 days	>200 buried

Date	Location		Trigger	Consequences
October 8, 2005	Solola, Guatemala	Yes	Hurricane Stan, 300 mm/3days	>1800 deaths
September 5, 2005	Yuexi County, Anhui, China	Yes	Rainstorm, 450 mm/6 days	210,000 people affected; 10,000 houses flattened
August 5, 2005	Guwahati, India	No	Monsoon rain, 310 mm/3 days	5 killed
January 10, 2005	La Conchita, CA	Yes	Heavy rain season, 390 mm/14 days	12 deaths
October 2004	Miyagawa area, Mie prefecture, Japan	Yes	Heavy and intense rainfall; numerous landslides and debris flow	17 deaths, 9 injuries; 87 homes damaged; extensive forest damage
July 20, 2003	Minamata and Hishikari, southern Kyushu, Japan	Yes	Heavy and intense rainfall; debris avalanches and debris flows	25 deaths; 7 homes destroyed; roads, power and hot spring lines damaged
May 2003	Ratnapura and Hambantota Districts, Sri Lanka	Yes	Continual heavy rains; many landslides and debris flows	>260 deaths; > 24,000 homes/schools destroyed; 180,000 families homeless
May 11, 2003	Southwest Guizhou Province, China	Yes	Heavy rainfall and road construction; road-related landslides	35 road workers killed and 2 buildings and road destroyed
April 20, 2003	Kara Taryk, Kyrgyzstan	No	Rain-on-snow; large landslides in Soviet-era uranium mining area	38 deaths; 13 homes destroyed; potential pollution of a river
December 14–16, 1999	North coast of Venezuela near Caracas	Yes	Nearly, 1000 mm/3 days; widespread shallow landslides and debris flows along a 40-km coastal strip	About 30,000 deaths; 8700 residences' infrastructures destroyed; extreme damage
Oct. 30, 1998	Casita Volcano, Nicaragua	Yes	Hurricane Mitch, 720 mm/6 days	>2000 deaths
August 26–31, 1998	Nishigo, Shirakawa, and Nasu, Japan	Yes	5 days of heavy rainfall; >1000 landslides	9 deaths; many homes/buildings destroyed
August 17, 1998	Malpa, Northern India	Yes	4 days of heavy rainfall; large rockfall/debris avalanche	207 deaths; 5.2 million rupees direct cost and 0.5 million rupees indirect cost

Note: mm = millimeters.

collection of global landslide inventory data and development of a physics-based landslide prediction model (Liao et al. 2010).

One of the main limiting factors in physics-based landslide hazard prediction at larger spatial scales is the dearth of high resolution of satellite images and globally consistent rainfall, soil moisture, and land surface products. Future satellite missions such as Global Precipitation Measurement (GPM; http://gpm.gsfc.nasa.gov) and Soil Moisture Active Passive measurement (SMAP; http://smap.jpl.nasa.gov) will provide globally consistent forcing data for the landslide model. High-resolution, multitemporal topographic mapping from synthetic aperture radar interferometry (InSAR), light detection and ranging (lidar), and DESDynI (http://desdyni.jpl.nasa.gov/) instruments can provide another valuable way to characterize the deformational patterns from previous landslide activity or potential sliding surfaces from slight displacements on hillslopes. With an adequate understanding of when it rains and how the slope responds to rainfall infiltration and soil moisture dynamics, the high resolution InSAR and lidar data can potentially provide important clues to the location of probable or existing mass movements, characterizing when these slopes will become unstable according to largely a function of the hydrological conditions of the slope.

Future versions of the algorithm, incorporating the suggested changes, could make this approach more valuable for discerning areas of potential landslide activity and allow the research community to consider issues of landslide hazard, risk, and vulnerability in a broader context. Several future activities are under consideration:

1. More information, such as geologic factors, could be incorporated into this global landslide susceptibility map when they become available globally.
2. Finer resolution DEM data such as lidar-based and InSAR data can also improve the landslide susceptibility mapping, even if only available over small areas.
3. It is anticipated that the TRMM rainfall product will be enhanced by the GPM. GPM is envisioned as improving the quality and frequency of observations from the constellation of operational and dedicated research satellites in order to provide improved global precipitation monitoring for hydrology and water resources management.
4. Soil moisture conditions observed from NASA Aqua satellite with the Advanced Microwave Scanning Radiometer—EOS (AMSR-E) instrument and future SMAP or an antecedent precipitation index accumulated from TRMM/GPM will be examined for usefulness in this preliminary landslide detection/warning system.
5. The empirical rainfall intensity–duration threshold triggering landslides may be regionalized using mean climatic variables (e.g., mean annual rainfall).
6. We may finally apply the real-time TMPA rainfall estimates to physics-based slope-stability models (e.g., Baum et al. 2002; Dhakal and Sidle 2004) over broad regions to detect rainfall conditions that may lead to landslides.

9.7   CONCLUSIONS

The primary criteria that influence shallow landslides are precipitation intensity, slope, soil type, elevation, vegetation, and land-cover type. Drawing on recent advances in remote sensing technology and the abundance of global geospatial products has motivated the development of a conceptual framework of real-time identifying/warning system of landslides. Such a framework (Figure 9.1) leads to the provision of dynamic forecasts of landslide occurrence based on areas with increased potential for slope instability conditions in near real-time basis by combining a calculation of landslide susceptibility with satellite derived rainfall estimates. This can be achieved by combining the NASA TMPA precipitation information (http://trmm.gsfc.nasa.gov) and land surface characteristics to assess landslides. First, a prototype of a global landslide susceptibility map (Figure 9.4) is produced using high-spatial scale DEM, slope, soil type information downscaled from the Digital Soil Map of the World (sand, loam, silt, clay, etc.), soil texture, and MODIS land-cover classification. Second, the map is overlaid with satellite-based observations of rainfall intensity–duration (Figure 9.3) to identify the location and time of landslide hazards when areas with significant landslide susceptibility are receiving heavy rainfall.

A major outcome of this work is the availability of a global perspective on rainfall-triggered landslide disasters, only possible because of the utilization of global satellite products. This type of real-time prediction system for disasters could provide policy planners with overview information to assess the spatial distribution of potential landslides. Given the fact that landslides usually occur after a period of heavy rainfall, a real-time landslide prediction system can be readily transformed into an early warning system by making use of the time lag between rainfall peak and slope failure. Therefore, success of this prototype system gives rise to promise as an early warning system for global landslide disaster preparedness and hazard management. Additionally, it is possible that the warning lead-time of global landslide forecasts can be extended by using rainfall forecasts (1–3 days) from operational numerical weather forecast models. This real-time prediction system bears the promise to extend current local landslide hazard analyses into a global decision support system for landslide disaster preparedness and mitigation activities across the world. However, it remains a matter of research to implement these concepts into a cost-effective method for capacity building in landslide risk management for developing countries. A final challenge to the existing algorithm framework at the global or regional level is to adequately account for landslides triggered by seismic activity or anthropogenic impact. It is a worthy topic to address in the future.

ACKNOWLEDGMENTS

This research was carried out with support from NASA's Applied Sciences program and the University of Oklahoma.

REFERENCES

Adler, R. F., Huffman, G. J., Chang, A., Ferraro, R., Xie, P., Janowiak, J., Rudolf, B., Schneider, U., Curtis, S., Bolvin, D., Gruber, A., Susskind, J., and Arkin, P. (2003). The version 2 Global Precipitation Climatology Project (GPCP) monthly precipitation analysis (1979–present). *Journal Hydrometeorology*, 4(6), 1147–1167.

Baum, R. L., Savage, W. Z., and Godt, J. W. (2002). TRIGRS—*A FORTRAN Program for Transient Rainfall Infiltration and Grid-based Regional Slope-stability Analysis*. U.S. Geol. Surv. Open File Rep. 02-0424, 64 pp. (available at http://pubs.usgs.gov/of/2002/ofr-02-424/).

Caine, N. (1980). The rainfall intensity–duration control of shallow landslides and debris flows: *Geografiska Annaler*, 62A, 23–27.

Canuti, P., Focardi, P., and Garzonio, C. A. (1985). Correlation between rainfall and landslide. *Bulletin of International Association of Engineering Geology*, 32, 49–54.

Chan, R. K. S., Pang, P. L. R., and Pun, W. K. (2003). Recent developments in the Landslide Warning System in Hong Kong. In *Geotechnical Engineering—Meeting Society's Needs. Proceedings of the 14th Southeast Asian Geotechnical Conference*, K. K. S. Ho and K. S. Li (eds.), Brookfield, VT, pp. 219–224.

Coe, J. A., Godt, J. W., Baum, R. L., Bucknam, R. C., and Michael, J. A. (2004). Landslide susceptibility from topography in Guatemala. In *Landslides: Evaluation and Stabilization*, W. A. Lacerda, M. Ehrlich, S. A. B. Fontoura, and A. S. F. Sayao (eds.), Taylor & Francis Group, London, pp. 69–78.

Dai, F. C. and Lee, C. F. (2002). Landslide characteristics and slope instability modeling using GIS. Lantau Island, Hong Kong. *Geomorphology*, 42, 213–238.

Davis, J. C. (1986). *Statistics and Data Analysis in Geology*. John Wiley & Sons, New York, p. 646.

Dhakal, A. S. and Sidle, R. C. (2004). Distributed simulations of landslides for different rainfall conditions. *Hydrological Processes*, 18, 757–776.

d'Orsi, R. N., Feijo, R. L., and Paes, N. M. (2004). 2500 operational days of Alerta Rio System: History and technical improvements of Rio de Janeiro Warning System for severe weather. In *Landslides: Evaluation and Stabilization*, W. A. Lacerda, M. Ehrlich, S. A. B. Fontoura, and A. S. F. Sayao (eds.), Taylor & Francis, London, pp. 831–842.

Fernandez, T., Irigaray, C., El Hamdouni, R., and Chacon, J. (2003). Methodology for landslide susceptibility mapping by means of a GIS, application to the Contraviesa Area (Granada, Spain). *Natural Hazards*, 30, 297–308.

Friedl, M. A., McIver, D. K., Hodges, J. C. F., Zhang, X. Y., Muchoney, D., Strahler, A. H., Woodcock, C. E., Gopal, S., Schneider, A., Cooper, A., Baccini, A., Gao, F., and Schaaf, C. (2002). Global land cover mapping from MODIS: algorithms and early results. *Remote Sensing of Environment*, 83(1–2), 287–302.

Godt, J. (2004). *Observed and Modeled Conditions for Shallow Landslide in the Seattle, Washington Area*. Ph.D. dissertation, University of Colorado, Boulder, CO.

Hong, Y., Adler, R. F., and Huffman, G. (2006). Evaluation of the Potential of NASA Multi-satellite Precipitation Analysis in Global Landslide Hazard Assessment. *Geophysical Research Letter*, 33, L22402, doi:10.1029/2006GL028010.

Hong, Y., Adler, R. F., and Huffman, G. J. (2007a). Satellite Remote Sensing for Landslide Monitoring on a Global Basis, American Geophysical Union EOS Featured Cover Article, 88(37), 357–358.

Hong, Y., Adler, R. F., Huffman, G., and Negri, A. (2007b). Use of satellite remote sensing data in mapping of global shallow landslides susceptibility. *Journal of Natural Hazards*, 43, 2, doi: 10.1007/s11069-006-9104-z.

Hong, Y., Adler, R. F., and Huffman, G. (2007c). An experimental global prediction system for rainfall-triggered landslides using satellite remote sensing and geospatial datasets. *IEEE Transactions of Geoscience and Remote Sensing*, 45(6), doi:10.1109/TGRS.2006.888436.

Hong, Y. and R. F. Adler (2008). Predicting landslide spatiotemporal distribution: Integrating landslide susceptibility zoning techniques and real-time satellite rainfall. Special Issue of *International Journal of Sediment Research*, 23(3), 249–257.

Huffman, G. J., Adler, R. F., Bolvin, D. T., Gu, G., Nelkin, E. J., Bowman, K. P., Hong, Y., Stocker, E. F., and Wolff, D. B. (2007). The TRMM Multi-Satellite Precipitation Analysis: Quasi-global, multi-year, combined-sensor precipitation estimates at fine scale. *Journal Hydrometeor*, 8(1), 38–55.

Iverson, R. M. (2000). Landslide triggering by rain infiltration. *Water Resources Research*, 36, 1897–1910.

Janowiak, J. E., Joyce, R. J., and Yarosh, Y. (2001). A real-time global half-hourly pixel-resolution IR dataset and its applications. *Bulletin of American Meteorology Society*, 82, 205–217.

Keefer, D. K. and Wilson, R. C. (1987). Real-time landslide warning during heavy rainfall. *Science*, 238, 13, 921–925.

Kirschbaum, D., Adler, R., Hong, Y., Hill, S., and Lerner-Lam, A. (2010). A global landslide catalog for hazard applications—Method, results and limitations. *Journal of Natural Hazards*, 52(3), 561–575.

Kirschbaum, D., Adler, R., Hong, Y., and Lerner-Lam, A. (2009). Evaluation of a preliminary satellite-based landslide hazard algorithm using global landslide inventories. *Natural Hazards and Earth System Science*, 9, 673–686.

Lagmay, A. M. A., Ong, J. B. T., Fernandez, D. F. D., Lapus, M. R., Rodolfo, R. S., Tengonciang, A. M. P., Soria, J. L. A., Baliatan, E. G., Quimba, Z. L., Uichanco, C. L., Paguican, E. M. R., Remedio, A. R. C., Lorenzo, G. R. H., Valdivia, W., and Avila, F. B. (2006). *Scientists Investigate Recent Philippine Landslide*, EOS Transaction. American Geophysical Union, 87(12), Washington, DC, doi:10.1029/2006EO120001.

Larsen, M. C. and Torres Sanchez, A. J. (1998). The frequency and distribution of recent landslides in three montane tropical regions of Puerto Rico. *Geomorphology*, 24, 309–331.

Larsen, M. C. and Simon, A. (1993). A rainfall intensity–duration threshold for landslides in a humid-tropical environment. Puerto Rico: *Geografiska Annaler*, 75A, 13–23.

Lee, S. and Min, K. (2001). Statistical analysis of landslide susceptibility at Yongin. Korea. *Environmental Geology*, 40, 1095–1113.

Liao, Z., Hong, Y., Wang, J., Fukoka, H., Sassa, K., Karnawati, D., and Fathani, F. (2010). Prototyping an experimental early warning system for rainfall-induced landslides in Indonesia using satellite remote sensing and geospatial datasets. *ICL Landslides Journal*, 7, 3, 317–324.

Liritano, G., Sirangelo, B., and Versace, P. (1998). Real-time estimation of hazard for landslides triggered by rainfall. *Environmental Geology*, 35/2–3, 175–183.

Metternicht, G., Lorenz, H., and Radu, G. (2005). Remote sensing of landslides: An analysis of the potential contribution to geo-spatial systems for hazard assessment in mountainous environments. *Remote Sensing of Environment*, 98, 284–303.

Sarkar, S. and Kanungo, D. P. (2004). An integrated approach for landslide susceptibility mapping using remote sensing and GIS. *Photogrammetric Engineering & Remote Sensing*, 70(5), 617–625.

Sidle, R. C. and Ochiai, H. (2006). *Landslide Processes, Prediction, and Land Use*, American Geophysical Union, Washington DC, pp. 1–312.

USGS (United States Geological Survey) report. 2006, http://landslides.usgs.gov/.

10 Analysis of Impervious Surface and Suburban Form Using High Spatial Resolution Satellite Imagery

D. Barry Hester, Stacy A. C. Nelson,
Siamak Khorram, Halil I. Cakir,
Heather M. Cheshire, and Ernst F. Hain

CONTENTS

10.1 INTRODUCTION

Urban areas are spatially and ecologically complex, and planning for their growth is data intensive. This is particularly true in suburban areas dominated by continuous urban sprawl. Galster (2001) noted that this term is not easily defined but generally refers to the patterns, processes, causes, and consequences of low-density and segregated land use development in populated places. The potential of sprawling suburban development to impact water quality is well documented (Bowen and Valiela 2001; Interlandi and Crockett 2003; Tu et al. 2007). Some analysts had estimated that substantial declines in watershed health begin to occur with as little as 10% impervious cover (Arnold and Gibbons 1996). Sprawl poses a unique threat to urban surface water quality in part because it incorporates an extensive array of impervious features, such as roads, rooftops, and parking lots. This threat can add to the fiscal strain placed on growing municipalities already financing stormwater management with special fees and utilities (Kasperson 2001). On the other hand, low-density suburban areas can accommodate an arrangement of open space that actually mitigates stormwater volume and pollution (Moglen and Kim 2007). Such a result is indicative of a proper spatial understanding of the relationship between suburban land use and surface water quality (Brabec et al. 2002).

The objective of this study was to evaluate the degree to which suburban imperviousness is derived from three different landscape features associated with urban sprawl: driveways in low-density residential lots, roads serving only low-density residential areas, and commercial retail parking lots. These impervious features reflect the broad orientation of sprawling land use around individual automobile travel. Nonetheless, their contribution to suburban imperviousness, and thus their relevance to suburban planning, has not been the subject of extensive analysis. Manville and Shoup (2005) even discussed this data gap with respect to the infamous urban landscape of Los Angeles, criticizing as unsubstantiated a widely repeated estimate that as much as two-thirds of that city is "consumed" by the automobile.

This study used state-of-the-art high spatial resolution satellite imagery to examine these important elements of the urban "microlandscape." The study area, a portion of suburban Raleigh, North Carolina, is experiencing rapid development and related water quality problems typical of many growing urban areas. In estimating the contribution of low-density residential streets, low-density residential driveways, and retail parking lots to suburban imperviousness, this study provides an empirical framework for discussing alternatives to suburban land use. The use of this framework is illustrated in this chapter by reference to one emerging alternative to sprawl: compact mixed-use development.

10.1.1 Low-Density Residential Streets and
Driveways and Impervious Surface

While there are gaps in the data explaining the relationship between the compact mixed-use development and current suburban imperviousness, urban impervious surface is undoubtedly linked to individual automobile usage. Planners make transportation and traffic primary considerations as they zone for and approve of new development (Moudon et al. 2005). Critics of contemporary suburban planning deride this dynamic as "the tail wagging the dog," claiming that design with individual automobile usage in mind encourages low-density urban development leading to the rapid suburban sprawl (Handy 2005). The connection is so pronounced that some researchers describe sprawl principally in terms of the human–automobile relationship (Galster et al. 2001).

Like the other suburban landscape features, conventional suburban road and driveway designs are such planning staples that their cumulative environmental footprint has not been thoroughly scrutinized. A growing number of researchers, however, motivated by the fact that roads and driveways lie unused much of the time, have evaluated ways to reduce the environmental impact of these impervious surfaces without comprising their social utility (Brattebo and Booth 2003; Stone 2004). Increasingly popular New Urban or traditional neighborhood development (TND) designs have incorporated some of the alternative forms, such as permeable pavement and narrower streets, and their proponents tout aesthetic and cultural benefits in addition to their environmental gains (Thompson-Fawcett and Bond 2003; Brander et al. 2004; Garde 2004). Relevant design elements include mixed use, smaller lots, reduced street setbacks, and private residential parking in shared alleyways and along streets (Berke et al. 2003).

High-resolution remote sensing was used for large-scale classification of the built versus natural environments to characterize traditional urban or impervious land use and land cover within a metropolitan area (Golden et al. 2009; Sonmez et al. 2009; Townsend et al. 2009). Other remote sensing studies have been relatively successful performing semiautomated extractions of roads as land cover features, keying in on their texture (Gamba et al. 2006) and reflectance properties (Zhu et al. 2005). The specific goal of this analysis was to quantify an impervious area associated with low-density residential roads and driveways in our study area. However, residential roads were manually extracted as land use features because of the narrow focus of this analysis and because manual delineation was feasible in light of the size of our study area.

The extraction of driveway surfaces in this study, by contrast, was semiautomated and represents a simple but novel methodological approach. When derived from remote sensing data, driveways are often delineated manually (Lee and Heaney 2003). Jennings et al. (2004) manually developed a "driveway impervious area percent" estimate for a calculation of watershed imperviousness using a random sample of driveways associated with relevant rooftops as seen in aerial photography. Stone (2004) estimated residential parcel impervious surface area from driveways by multiplying the city-required street setback by an average driveway width per garage stall (i.e., based on a 100-parcel visual analysis of aerial photography) and number of garage stalls. The present study introduces a semiautomated analysis of remote sensing data to estimate the contribution of low-density residential driveways to overall imperviousness in our study area. Additionally, this portion of the study compares imperviousness in a low-density residential neighborhood to that in part of the compact Bedford neighborhood at Falls River for the purpose of comparison.

10.1.2 RETAIL PARKING LOTS AND SUBURBAN IMPERVIOUS SURFACE

The commitment of suburban land to retail parking lots is even more poorly documented than its devotion to roads and driveways despite the fact that most U.S. municipalities include minimum parking space requirements for new commercial and retail development (Shoup 1999; Manville and Shoup 2005). This has important environmental implications, as parking lots are typically contiguous expanses of impervious surface. Parking supply can also have a powerful effect on transportation and commuting behavior (Mildner et al. 1997; de Cerreño 2004).

The present use of data derived from high-spatial resolution satellite imagery represents a unique approach to empirically analyzing this suburban land use issue. Previous studies either have failed to specifically map parking lots or have not relied heavily on automated geographical information system (GIS) and remote sensing methods in doing so. Satellite imagery has frequently been used to map urban commercial and industrial land uses separate from residential and other areas (McCauley and Goetz 2004; Lu and Weng 2005; Lu and Weng 2006), but these studies, although highly automated, have not typically attempted to map parking lots. Even related studies utilizing high spatial resolution imagery have distinguished rooftops from paved surfaces but have been short of mapping parking lots (Thomas et al. 2003). On the other hand, Albanese and Matlack (1999) studied urban stream degradation adjacent to parking lots measuring lot size onsite by foot pacing. The authors also concluded that those parking lots were chronically underutilized, even during peak holiday season usage, suggesting that the runoff problems they caused were extraneous. Although imperviousness in total study area was not determined, the mean parking area size for the 10 sampled commercial and retail land uses was calculated as 5.86 ha.

In a more efficient but less precise analysis, Akbari et al. (2003) used Monte Carlo statistical sampling to analyze the fabric of urban imperviousness in Sacramento, CA. Those researchers visually inspected 400 to 600 random point locations using 0.3-m color digital orthophotos and classified the land use at each point as one of 30 explicit types, such as "sidewalk," "tree covering grass," and "parking area." They determined that in two sampled commercial areas, an exposition and a shopping center, parking areas covered 62.8% and 38.0% of all paved surfaces, respectively. The present study uses remote sensing and GIS-based methods in an even more automated approach in mapping suburban Raleigh's retail parking land use.

10.2 STUDY AREA AND MATERIALS

The study area incorporates 71.5 km^2 of adjacent drainage catchments and is located in the northeast part of the city of Raleigh, North Carolina (Figure 10.1). This area,

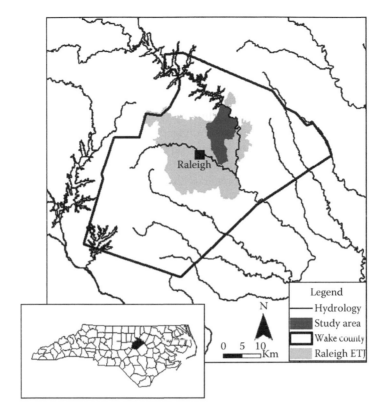

FIGURE 10.1   Study area in northeast Raleigh, Wake County, North Carolina, and Raleigh's extraterritorial jurisdiction (ETJ).

representing roughly 31% of Raleigh's incorporated land area, is composed of residential, commercial, and industrial land uses interspersed with large forested clusters and includes part of a major transportation corridor, US-1. The U.S. Census Bureau recently listed Wake County, which encompasses Raleigh, as one of the 100 fastest-growing counties in the United States (2003). Raleigh is located within the Neuse River watershed, the third largest river basin in the state of North Carolina. This basin drains all or part of 23 counties into the Albemarle-Pamlico Sound; its ecological integrity has been the subject of numerous studies in recent decades (Burkholder et al. 1992; McMahon and Lloyd 1995; Paerl et al. 1998). Both of the major streams within the study area, Marsh Creek and Perry Creek, have been included on North Carolina's federally mandated Clean Water Act, Section 303(d), Impaired Waters List each year as it has been compiled. Each time, the state has cited "urban runoff/ storm sewers" as the likely cause of impairment of water quality (NCDENR 2006; NCDENR 2007).

The land use/land cover (LU/LC) data used and generated in this study was based on QuickBird satellite imagery, a DigitalGlobe, Inc., product (www.digitalglobe

.com). Collecting panchromatic imagery at 0.61 m at nadir and multispectral imagery at 2.44 m at nadir, the QuickBird sensor currently produces one of the highest spatial resolution satellite data available on the commercial market. Multispectral and panchromatic study area imagery was collected in June 2002, and the data fusion algorithm developed by Cakir and Khorram (2008) was used to enhance the spatial resolution of a multispectral image to 0.61 m. This technique uses a multivariate correspondence analysis (CA) procedure to fuse QuickBird multispectral imagery with the complementary panchromatic imagery. In this method, the last component of a correspondence analysis image was replaced with the panchromatic imagery before the CA image was transformed back to the original image space. This procedure allowed the resultant fused image to retain most of the original image variance (94%–97%) while distortion was kept to a minimum. A hybrid per-pixel classification approach was used to complete the land cover classification (Hester et al. 2008). This approach employed a supervised classification step, which utilized a maximum likelihood algorithm and segmentation of visually problematic classes, and an unsupervised classification step within the segmented regions, which used iterative self-organizing data analysis technique (ISODATA) clustering, principal axis means computing, 25 maximum iterations, and a 95% convergence threshold. The unsupervised classification results were then recombined with the supervised classifications results and majority filters were applied to remove any remaining "salt-and-pepper" pixels within each land cover category (see Hester et al. 2008). The completed land cover classification produced six categories with an 89% overall accuracy and 90% user's accuracy for the impervious category. The five other mapped categories were water, bare/disturbed, deciduous, evergreen, and herbaceous; these are described in Table 10.1. In 10 selected commercial parcels, impervious surface area matched reference data with 96.9% accuracy. The resulting land cover map of our study area is

TABLE 10.1

Land Cover Categorization Scheme for QuickBird Image Classification

Class	Land Cover Type	Description
1	Impervious	Transportation infrastructure (roads, parking lots, etc.), rooftops, recreational areas (i.e., tennis courts), and all other man-made impervious surfaces that are paved or built
2	Water	Lakes, ponds, rivers, streams, pools, and all other natural and artificial surface waters
3	Bare/disturbed soil	Construction sites, landfills, gravel areas, or any other unpaved nonvegetated surface
4	Deciduous	Trees or shrubs that shed their leaves before the winter (mostly hardwood species)
5	Evergreen	Trees or shrubs that keep their leaves throughout the winter (mostly coniferous species)
6	Herbaceous	Urban grasses (yards, recreational fields, vegetated road medians, etc.) of varying degrees of maintenance

presented in Figure 10.2. The study area is primarily composed of vegetation (71%) but is substantially covered by impervious surface (24%).

Other data utilized in this study were GIS files for roads, property parcels, building footprints, and streams, all available from Wake County iMAPS server. The property parcel database includes a variety of individual lot data, including 2002 zoning classification, property value, and development build date. Raleigh utilizes Euclidean zoning and delineates residential lots by permissible density of dwelling units. It also sets minimum street setback requirements. Parcel types are subcategorized by use such as restaurant, retail-type buildings, and gas stations. The road database is a street segment file and does not characterize roads as residential or commercial. All GIS processing was performed in ArcGIS® 9.0 (Environmental Systems Research Institute), and remote sensing analyses were conducted by using ERDAS IMAGINE 8.7 (Leica Geosystems).

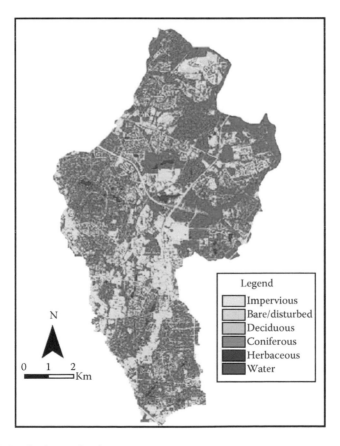

FIGURE 10.2   Study area land cover map.

10.3 METHODS

10.3.1 Estimating the Extent of Study Area Imperviousness from Low-Density Roads and Driveways

As an initial step in evaluating the contribution of suburban roads to study area imperviousness, residential road segment features in the study area were manually identified from the study area roads GIS layer using the QuickBird image as a backdrop. Road segments included in this selection were two-lane features weaving in and through single-family residential zones limited to six dwelling units per acre (14.8 dwelling units/ha) or less, including four- and two-unit-per-acre zones. This density threshold was utilized because it isolated the study area's most homogeneous and contiguous clusters of single-family residential development. The selected road segments exclusively served detached single-family dwellings, although segments constituting a unique through passage between major roads were not selected.

The total length of these road segments was multiplied by an average road width generated from a random sample of 30 road segments in the study area measured curb to curb. This calculation gave an estimate of the amount of impervious surface contributed by roads exclusively serving low-density residential housing. In order to reflect the road impervious surface obscured by tree canopy overhang, a 10% deduction was applied to the calculated estimate of road imperviousness. This deduction was intended to accord this estimate with the other calculations in this study, all of which were based on "above-the-canopy" satellite views, and it only incidentally implicates a potential effect of rainfall interception (Sanders 1986). This reduced figure was then used to calculate the percent contribution to overall impervious surface made by residential roads.

To estimate the percentage contribution of private, low-density residential driveways to the study area's total impervious surface area, a subset of residential parcels was selected. Selected parcels were those assigned an "R-4" zoning classification at the time of image capture, a designation that limits development to single-family detached housing not denser than four dwelling units per acre (9.9 dwelling units/ha). These parcels, exemplified by those in Figure 10.3, have the low density typically associated with suburban sprawl. Next, an unsupervised classification of the image data representing these parcels was performed, using 20 classes, the ISODATA algorithm, and a 95% convergence threshold in ERDAS IMAGINE 8.7. These 20 classes were aggregated into three unique categories: impervious building, impervious nonbuilding, and other. The relatively discrete reflectance properties of these three categories was a key part of this analysis, since dark rooftops and nonimpervious features, such as vegetation, have very different reflectance properties than impervious nonbuilding features. The accuracy of the three-category unsupervised classification was evaluated using a visual interpretation of 50 randomly selected R-4 parcels. In addition, the contribution of paved sidewalks and patios to average nonbuilding imperviousness was estimated using a random selection of 10 R-4 parcels.

FIGURE 10.3 Low-density residential development.

10.3.2 COMPARING IMPERVIOUSNESS IN LOW-DENSITY AND COMPACT RESIDENTIAL DEVELOPMENT

In order to compare the extent of imperviousness in low-density neighborhoods and alternative New Urban–type developments, two representative neighborhood sections were selected. This analysis was not meant to be an exhaustive statistical comparison of these two forms of development but rather a closer look at questions of form raised by the preceding analysis. It generally addresses the question of whether compact residential development in fact incorporates less impervious surface per household than conventional low-density residential development after accounting for associated transportation features. Analyzed sections included part of a low-density, curvilinear neighborhood of single-family detached dwellings and a part of the denser Bedford New Urban development discussed above. Housing construction was completed in both areas between 1997 and 2001, and each is a relatively stand-alone residential area having internal road segments. The low-density neighborhood section encompassed 21.9 ha (54.1 ac) and the New Urban section 15.8 ha (39 ac). These areas are displayed in Figure 10.4.

Within these areas, total imperviousness was calculated using the QuickBird-derived land cover map. The total number of households per area was extracted from the tax parcel database. As above, the relevant road segments were selected, and their widths were measured as appearing on the QuickBird image. This permitted the calculation of total imperviousness derived from roads in each neighborhood type.

FIGURE 10.4   Low-density (left) and compact neighborhoods.

10.3.3   STUDY AREA IMPERVIOUSNESS FROM RETAIL PARKING LOTS

A subset of commercial parcels was selected for the retail parking lot imperviousness analysis. Attributes in the parcel GIS layer were used to isolate retail parcels at which customer parking was provided. These 234 parcels were those occupied by restaurants, grocery stores, department stores, and similar retail uses. From these parcels, those for which the "year built" date was listed as 2001 or 2002 were excluded if construction appeared to be ongoing at the time of image capture. The final parcel selection included 195 parcels covering 1.4 km² of the study area. To summarize, development in these parcels was finished at the time of image capture, and they were being used for nonresidential activity and related customer parking. Representative retail parcels are displayed in Figure 10.5.

After updating the building footprint GIS layer using the QuickBird image, the amount of impervious surface in each parcel constituted by buildings was tabulated using a GIS zonal function and the QuickBird image-derived land cover map. In addition, an estimate of the amount of nonbuilding impervious surface devoted to loading, waste disposal, storage, or other uses necessary for commercial activity was generated through photointerpretation of the QuickBird image in a 30-parcel random sample. This sample included 10 parcels, each from among the selected parcels with the 66 smallest, 65 intermediate-sized, and 65 largest building footprints, and it was used to derive an average value of such nonbuilding, non–parking lot impervious surface within parcels of each building size class. Subtracting the building footprint and "loading" surfaces average developed for each building size class from the map estimate of nonbuilding impervious surface, an average value for the percent of commercial imperviousness derived from retail parking surfaces was generated.

FIGURE 10.5   Retail parcels and associated parking lots.

10.4   RESULTS

The results of the low-density residential imperviousness analysis are summarized in Figure 10.6. In all, imperviousness from roads and driveways together with retail parking lots accounted for 14.3% of study area imperviousness and 58.8% of the imperviousness associated with low-density residential land use. Residential roads and driveways, retail parking lots, and the neighborhood comparison findings are presented by the corresponding sections below.

10.4.1   IMPERVIOUS SURFACE FROM LOW-DENSITY RESIDENTIAL ROADS AND DRIVEWAYS

The results of the residential road imperviousness analysis are reported in Table 10.2. The total length of the selected residential road segments was 182.3 km, which

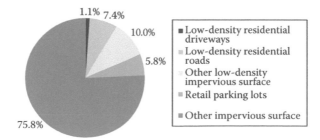

FIGURE 10.6   Summary of study area impervious surface (read clockwise from top).

TABLE 10.2

Results of Residential Road Segment Analysis

Residential Road Segment Length (km)	Percent of Study Area Road Length Made up of Residential Roads	Average Residential Road Width (Standard Deviation) (m)	Estimated IS Area from Roads (km²)	Estimated IS Area from Roads with Tree Canopy Deduction (km²)	Percent of Study Area Occupied by Residential Road IS	Percent of Study Area IS Made up of Residential Road IS
182.3	47%	7.7 (1.8)	1.40	1.26	1.8%	7.4%

Note: IS = impervious surface.

represents 47% of the total road length in the study area. Based on the 30-segment sample, the average residential road width in the study area was 7.7 m. These figures resulted in an estimated above-the-tree canopy impervious surface contribution from residential roads of 1.26 km². This is a contribution to total study area imperviousness (17.1 km²) of 7.4%.

Of those parcels completely bounded by the study area, 7704 parcels within the study area were zoned R-4 in 2002. A number of these parcels were excluded from analysis because they were either developed after the image date, contained more

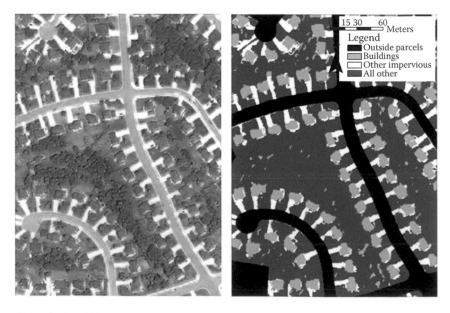

FIGURE 10.7   (a) Low-density residential parcel (left), and (b) land cover map of the same area.

TABLE 10.3
Accuracy Assessment of Residential Parcel Land Cover Map

	Reference	Map	Percent Error
Total IS	15359.5	17337.4	+12.8
Average parcel % IS	21.7	24.1	+2.4
Average parcel % building IS	14.4	17.1	+2.7
Average parcel % nonbuilding IS	7.3	7.0	−0.3
Average % of parcel IS from buildings	66.4	70.5	+4.1
Average % of parcel IS from nonbuildings	33.6	29.5	−4.1

Note: IS = impervious surface.

than 10% bare soil, were larger than 2.0 acres (0.81 ha) in size, or were being used for nonresidential purposes. These exclusions left 5436 parcels constituting 9.1 km^2 or 12.7% of the study area for this analysis. The impervious surface encompassed within these residential parcels totaled only 11.1% of total study area impervious surface, or 1.9 km^2.

A subset of the three-category land use map developed for this analysis is presented along with the corresponding QuickBird image in Figure 10.7. The accuracy assessment presented for this map is displayed in Table 10.3. Average parcel percent imperviousness was mapped with a positive difference from reference data of only 2.4%, while average parcel percent nonbuilding imperviousness was underestimated by 4.1. Nonbuilding, nondriveway paved surfaces made up an average of 16.1% of sampled residential parcel imperviousness. Subtracting this average from total imperviousness mapped as nonbuilding (26.8%), a final estimate from this analysis is that driveway surfaces comprised 10.7% of low-density residential development imperviousness and only 1.1% of study area imperviousness.

10.4.2 Low-Density and Compact Neighborhood Comparison

The results of neighborhood comparison are presented in Table 10.4 and highlighted by Figure 10.8. As expected, residential density was higher in the New Urban neighborhood (22.8 households/ha) than in the low-density neighborhood (19.6 households/ha). Total percent imperviousness was nearly double in the New Urban neighborhood (45.6%), and a greater percentage of impervious surface area was contributed by roads there (31.9%). However, factoring in all sources of imperviousness, including rooftop, driveway, sidewalk, and road surfaces, impervious surface per household was higher in the low-density neighborhood (511.2 compared to 436.5).

10.4.3 Retail Parking Lot Imperviousness Results

The selected commercial parcels constituted 2.4 km^2 or 3.4% of the study area but 1.4 km^2 or 8.1% of study area impervious surface. On average, these parcels were

TABLE 10.4
Low-Density and Compact Neighborhood Comparison

	Low-Density Neighborhood	Compact Neighborhood
Average parcel size (ha)	0.16	0.05
Total IS (ha)	5.7	7.2
Total area (ha)	21.9	15.8
Total area percent IS	26.0%	45.6%
Total roads (m)	1899.0	2246.7
Average road width (m)	6.9	10.3
Road IS (ha)	1.3	2.3
Percent of IS from roads	22.8%	31.9%
Households	112	164
Households per hectare	19.6	22.8
Total IS per household (m²/hh)	511.2	436.5

Note: IS = impervious surface.

covered by 68.3% impervious surface. In the 10 sampled commercial parcels from among those having the 66 smallest building footprints, an average of 7% of non-building imperviousness was devoted to loading, parking, and other nonbuilding, nonparking land uses. Those figures for the intermediate-sized and largest building footprint parcels were 3% and 8%, respectively. When those average values were subtracted from each parcel's mapped percent nonbuilding imperviousness, the resulting estimate of the average percent of commercial parcel imperviousness committed to retail parking lot surfaces was 68.0%. Thus, nearly 1 km² of the study area's 17.2 km² of impervious surface was comprised of retail parking lots. This represents 41.6% of the selected retail land usage itself. Figure 10.6 incorporates these results in summarizing the contribution of low-density residential imperviousness to overall study area imperviousness.

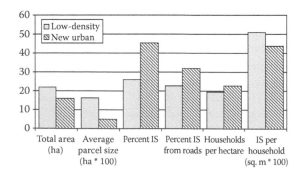

FIGURE 10.8 Summary of neighborhood comparison (scaled by 100 where indicated, IS = impervious surface).

10.5   DISCUSSION

10.5.1   SPRAWL AND TRANSPORTATION

The analysis of roads exclusively serving low-density residential development reveals that, in suburban Raleigh, such roads constitute nearly one-tenth of total study area imperviousness. More significantly, these features accounted for more almost one-third of the imperviousness associated with low-density residential land use. Their role in watershed health may be even greater than those figures reflect for several reasons. First, cul-de-sac features, an additional source of road-related imperviousness, did not figure into this estimate. Additionally, overhanging tree canopy obstructed as much as 10% of residential road surfaces from satellite image view, although this overhang can substantially mitigate increases in stormwater volumes and velocity that may result from underlying impervious surface (Sanders 1986). Finally, road surfaces are special conduits for pollutant transport given their collection of vehicle traffic contaminants and connectivity to streams via stormwater infrastructure (Van Bohemen and Van de Laak 2003; Walsh et al. 2004). Recent work has reinforced the related notion that the location of impervious surface may be determinative of watershed health than its extent (Snyder et al. 2005; Moglen and Kim 2007).

Low-density residential driveways contributed a trivial amount to total study area imperviousness. This result suggests that reduced street setback, a tenet of the New Urban design, may not appreciably reduce overall impervious surface in the study area. Accordingly, Stone (2004) recommended this policy change as a way to reduce watershed imperviousness but emphasized that reductions in lot size and street frontage would more significantly advance that goal. Nonetheless, like roads, driveways are also often uniquely connected to ditches, storm drains, and other concentrated means of runoff transport. This connectivity gives these land uses a potential role in determining watershed health disproportionate to their contribution to overall imperviousness. In that sense, strategically reducing small amounts of impervious surface through a modest policy change such as reduced street setback may indeed be desirable.

Retail parking lots constituted an equally important source of study area imperviousness and pose an additional threat to water quality because they concentrate vast tracts of imperviousness. The average parking lot within the 195 parcels studied was 0.51 ha (1.3 ac) in size. Such large tracts of contiguous impervious surface can result in high volumes and velocities of stormwater runoff. In addition, parking lot runoff is highly concentrated in vehicle emitted pollutants such as oil, grease, and heavy metals (Kim et al. 2007).

10.5.2   LOW-DENSITY RESIDENTIAL IMPERVIOUSNESS GENERALLY

In total, the low-density residential development analyzed here represented 17.8% of the study area and 24.3% of its impervious surface when accounting for roads, driveways, and retail parking lots, sources that constituted nearly two-thirds of this imperviousness. These sources are distinct from the other one-third in that they are closely associated with the use of private automobiles. This is a distinction that should be considered when alternative residential development designs are evaluated as in the portion of this study comparing a low-density neighborhood with its compact, New Urban counterpart.

10.5.3 Low-Density and Compact Neighborhood Comparison

The results of this study provide some quantitative support for the potential environmental benefits of compact residential development as an alternative to the development patterns associated with sprawl. The New Urban subdivision's wider roads result in a higher percentage of imperviousness from those surfaces. Nonetheless, the compact neighborhood studied here has a lower level of impervious surface per household. This is primarily because its shorter street setbacks minimize nonroad sources of impervious surface including driveways and sidewalks. On-street parking, in fact, replaces driveways in many New Urban designs. The denser development also required less street frontage per lot, a strategy identified by Stone (2004) as an effective way to reduce overall watershed imperviousness. The improved runoff management capacity of this type of development as compared to low-density designs has already been documented by Brander et al. (2004) and Berke et al. (2003).

Additionally, this neighborhood design presents a variety of opportunities to indirectly reduce the environmental impact of residential land use through reduction in resident automobile dependence. Possible related environmental benefits include open space preservation and enhanced suitability of residential areas to alternative transportation, including mass transit, pedestrianism, and bicycling. Compact mixed-use development can replace commercial parking lots outside residential areas with green spaces or parks because residents can walk to shopping destinations. A small grocery store, for example, built in the undeveloped portion of the New Urban neighborhood shown in Figure 10.4, may be able to substitute for both the store and the parking lot it would have required if built elsewhere, an open space savings measured in hectares. Such mixed use may even minimize the extent of new road construction. Residents may, however, prefer to have shared open space within their neighborhood, as it is in Figure 10.4.

Despite its potential environmental benefits, social critiques of New Urbanism cast it as being racially and socioeconomically exclusive (Al-Hindi 2001; Day 2003; Clarke 2005). The touted environmental benefits have been questioned. In particular, analysts have tested the hypothesis that pedestrianism is indeed heightened in such communities (Patterson and Chapman 2004). Greenwald (2003) found that New Urban development strategies did increase the substitution of walking for personal automobile usage in one city. Vojnovic (2006), however, observed that neighborhood self-selection may obscure the effect of such development on nonmotorized travel behavior, since those inclined to make active travel choices might specifically seek out development friendly to it. Nonetheless, mixed-use development, of which some analysts have identified New Urbanism as a remarketed form, is an economic and perhaps social necessity in areas where open space is precious, such as Hong Kong (Lau et al. 2005). This may be instructive to municipalities experiencing rapid growth and considering such development strategies but not yet approaching build-out.

10.5.4 Urban Planning and Environmental Management Implications

This study quantified the substantial contribution of the low-density residential development associated with sprawl to one suburban area's total imperviousness. This development was itself heavily vegetated but accounted for 18.5% of total study area imperviousness when both paved surfaces inside residential parcels and roads serving those parcels were considered. The low-density impervious landscape features, such as parking lots, houses, driveways, and roads are standard features of the suburban residential landscape, and they primarily serve the needs of private dwellings, automobile travel, recreation, commerce and commercial transport, and emergency response. Developers of commercial or retail space have economic incentives to avoid parking shortages, and, because they can externalize all of the costs of parking oversupply, they are often built to accommodate a peak volume that is rarely if ever seen. Advocates of mass transit, bicycling, pedestrianism, and other modes of alternative transportation see such oversupply an impediment to those causes (Deakin et al. 2004). Some critics of suboptimal parking supply argue that it generates negative environmental and economic externalities such that it should be specially taxed in order to mitigate its cost to communities and municipalities (Feitelson and Rotem 2004).

Within the municipality that this study was conducted, the City of Raleigh, North Carolina began collection of an annual stormwater utility fee in 2004 (City of Raleigh 2008). This fee was based on estimates of impervious surfaces for industrial, businesses, and other commercial properties that were calculated directly from aerial photographs. Impervious surface estimates for residential homes were based on a median impervious surface area of 210 square meters, which was determined from a statistical sampling of representative properties within the Raleigh city limits. Impervious surfaces used in the fee calculations of both residential and commercial properties included features such as roofs, driveways, parking lots, or any other hard surface located on a property. The fee structure was designed to reflect each property's stormwater runoff contribution into neighborhood culverts and streams. Funds generated from this fee are used to support increased demands on drainage infrastructure, public education and outreach, pollution and illicit discharge detection and elimination, and construction and postconstruction runoff control, as well as other stormwater management programs required by the Environmental Protection Agency's Clean Water Act.

Annual stormwater utility fees have become a common assessment tool in large metropolitan areas since the early 1990s. These fees are aimed at proportionately designating charges to properties that have greater runoff production. An analysis, such as the one offered by this study, has the potential to aid in development, permitting requests and urban/suburban planning decisions that take into consideration current impervious surface densities as well as best management practices. Some of these best management practices may include permeable pavement and curb systems, retention ponds and wetlands, reduced parking lot demand, narrower roads, and less-extensive impervious surfaces such as smaller driveways through decreased street setbacks (Brattebo and Booth 2003; Stone 2004).

10.6   CONCLUSIONS

The results of this study indicate that, while the compact development strategies incorporated by New Urbanism may only slightly reduce residential impervious surface area, related decreases in required retail parking space may significantly reduce overall watershed imperviousness. Moreover, these findings suggest that carefully planned compact mixed-use development can arrange impervious surface in a way that reduces the overall environmental footprint of residential development and enhances long-term suburban environmental quality. Municipalities reluctant to approve vast tracts of New Urban development but interested in water quality–friendly design may wish to merely decrease simple street setback requirements for residential development and minimum parking space requirements for retail development. Coupled with mixed use or increases in density, these changes might minimize the collateral role of parking lots as concentrated sources of impervious surface and vehicle pollutants. Tangentially, the results of this study support the continued use of population density as a measure of sprawl inasmuch as it inversely relates to the extent of transportation infrastructure and segregated land use.

Although the empirical findings of this study most reliably characterize the suburban Raleigh area, they are of broader interest for at least two important reasons. First, the study area is fairly representative of suburban areas throughout the United States. However, it should be noted that these specific results may not precisely characterize areas outside the temperate deciduous forest region in which this study was set. The second value of this study is in its presentation of transferable methods for using novel high-spatial resolution satellite imagery such as QuickBird in urban land use analysis. In this study, methods were developed for studying the urban "micro-landscape." Such a focus might be used in future studies to examine the relationship between discrete landscape features such as driveways or parking lots, and highly localized environmental parameters such as water quality. This may uniquely inform the degree to which stormwater drainage infrastructure functions as a source of point-source pollution.

This new way to study urban land use is made possible by the resolution and consistency of high-spatial resolution satellite imagery, and the methods described here permit a high level of analytical uniformity relative to traditional photointerpretation. Moreover, for smaller geographic coverage, the cost of tasking new imagery has become comparable to that of contracting for new aerial photography. High-profile usage of such imagery by Internet mapping giants such as Google® will only increase its availability, especially for urban study areas, and a number of new high-resolution satellites are scheduled to launch during the next few years. For these reasons, analysts and institutions currently buying and working with this imagery are developing databases and expertise that will exponentially increase in value. This is a promising sign of the continued synergy between remote sensing and landscape planning.

ACKNOWLEDGMENTS

This study was enabled in part through funding provided by the North Carolina Water Resources Research Institute (NC WRRI) for the project "Integration of High

Resolution Imagery in Cost-Effective Assessment of Land Use Practices Influencing Erosion and Sediment Yield (2004–2006)." Ancillary data was acquired from the Wake County and City of Raleigh GIS iMaps server.

DISCLAIMER

This work is not a product of the United States Government or the United States Environmental Protection Agency, and the author has not performed this work in any governmental capacity. The views expressed are those of the author only and do not necessarily represent those of the United States or the US EPA.

REFERENCES

Akbari, H., Rose, L. S., and Taha, H. (2003). Analyzing the land cover of an urban environment using high-resolution orthophotos. *Landscape and Urban Planning*, 63, 1–14.

Al-Hindi, K. F. (2001). The New Urbanism: Where and for whom? Investigation of an emergent paradigm. *Urban Geography*, 22, 202–219.

Arnold, C. and Gibbons, C. (1996). Impervious surface coverage—The emergence of a key environmental indicator. *Journal of the American Planning Association*, 62, 243–258.

Berke, P. R., MacDonald, J., White, N., Holmes, M., Line, D., Oury, K., and Ryznar, R. (2003). Greening development to protect watersheds—Does New Urbanism make a difference? *Journal of the American Planning Association*, 69, 391–413.

Bowen, J. L. and Valiela, I. (2001). The ecological effects of urbanization of coastal watersheds: Historical increases in nitrogen loads and eutrophication of Waquoit Bay estuaries. *Canadian Journal of Fisheries and Aquatic Sciences*, 58, 1489–1500.

Brabec, E., Schulte, S., and Richards, P. L. (2002). Impervious surfaces and water quality: A review of current literature and its implications for watershed planning. *Journal of Planning Literature*, 16, 499–514.

Brander, K. E., Owen, K. E., and Potter, K. W. (2004). Modeled impacts of development type on runoff volume and infiltration performance. *Journal of the American Water Resources Association*, 40, 961–969.

Brattebo, B. O. and Booth, D. B. (2003). Long-term stormwater quantity and quality performance of permeable pavement systems. *Water Research*, 37, 4369–4376.

Burkholder, J. M., Noga, E. J., Hobbs, C. H., and Glasgow, H. B. (1992). New phantom dinoflagellate is the causative agent of major estuarine fish kills. *Nature,* 358, 407–410.

Cakir, H. I. and Khorram, S. (2008). Correspondence analysis fusion. *Photogrammetric Engineering & Remote Sensing*, 74, 183–192.

City of Raleigh. (2008). *Stormwater Utility Fee Credit and Adjustment Manual*. Municipal Public Works—Stormwater Utility Division, Raleigh, North Carolina, 1–73. http://www.raleighnc.gov/services/content/PWksStormwater/Articles/StormwaterUtilityMainPage.html (accessed March 2, 2011).

Clarke, P. W. (2005). The ideal of community and its counterfeit construction. *Journal of Architectural Education*, 58, 43–52.

Day, K. (2003). New Urbanism and the challenges of designing for diversity. *Journal of Planning Education and Research*, 23, 83–95.

de Cerreño, A. L. C. (2004). Dynamics of on-street parking in large central cities. *Transportation Research Record*, 1898, 130–137.

Deakin, E., Bechtel, A., Crabbe, A., Archer, M., Cairns, S., Kluter, A., Leung, K., and Ni, J. (2004). Parking management and downtown land development in Berkeley, California. *Journal of Planning Education and Research*, 1898, 124–129.

Feitelson, E. and Rotem, O. (2004). The case for taxing surface parking. *Transportation Research Part D-Transport and Environment*, 9, 319–333.

Galster, G., Hanson, R., Ratcliffe, M. R., Wolman, H., Coleman, S., and Freihage, J. (2001). Wrestling sprawl to the ground: Defining and measuring an elusive concept. *House Policy Debate*, 12, 681–717.

Gamba, P., Dell'Acqua, F., and Lisini, G. (2006). Improving urban road extraction in high-resolution images exploiting directional filtering, perceptual grouping, and simple topological concepts. *IEEE International Geoscience & Remote Sensing*, 3, 387–391.

Garde, A. M. (2004). New Urbanism as sustainable growth? A supply side story and its implications for public policy. *Journal of Planning Education and Research*, 24, 154–170.

Golden, J., Chuang, W. C., and Stefanov, W. L. (2009). Enhanced classifications of engineered paved surfaces for urban system modeling. *Earth Interactions*, 13, 1–18.

Greenwald, M. J. (2003). The road less traveled—New Urbanist inducements to travel mode substitution for nonwork trips. *Journal of Planning Education and Research*, 23, 39–57.

Handy, S. (2005). Smart growth and the transportation—Land use connection: What does the research tell us? *International Regional Science*, 28, 146–167.

Hester, D. B., Khorram, S., Nelson, S. A. C., and Cakir, H. I. (2008). Per-pixel classification of high spatial resolution satellite imagery for urban land cover mapping. *Photogrammetric Engineering and Remote Sensing*, 74, 463–471.

Interlandi, S. J., and Crockett, C. S. (2003). Recent water quality trends in the Schuylkill River, Pennsylvania, USA: A preliminary assessment of the relative influences of climate, river discharge and suburban development. *Water Research*, 37, 1737–1748.

Jennings, D., Jarnagin, S., and Ebert, D. (2004). A modeling approach for estimating watershed impervious surface area from national land cover data 92. *Photogrammetric Engineering and Remote Sensing*, 70, 1295–1307.

Kasperson, J. (2001). The stormwater utility: Will it work in your community? *Stormwater*, 1, 213–254.

Kim, L. H., Ko, S. O., Jeong, S., and Yoon, J. (2007). Characteristics of washed-off pollutants and dynamic EMCs in parking lots and bridges during a storm. *Science of the Total Environment*, 376, 178–184.

Lau, S. S. Y., Giridharan, R., and Ganesan, S. (2005). Multiple and intensive land use: case studies in Hong Kong. *Habitat International*, 29, 527–546.

Lee, J. G. and Heaney, J. P. (2003). Estimation of urban imperviousness and its impacts on storm water systems. *Journal of Water Resources Planning and Management*, 129, 419–426.

Lu, D. S. and Weng, Q. H. (2005). Urban classification using full spectral information of Landsat ETM+ imagery in Marion County, Indiana. *Photogrammetric Engineering and Remote Sensing*, 71, 1275–1284.

Lu, D. S. and Weng, Q. H. (2006). Use of impervious surface in urban land-use classification. *Remote Sensing of Environment*, 102, 146–160.

Manville, M. and Shoup, D. (2005). Parking, people, and cities. *Journal of Urban Planning and Development*, 131, 233–245.

McCauley, S., and Goetz, S. (2004). Mapping residential density patterns using multi-temporal Landsat data and a decision-tree classifier. *International Journal of Remote Sensing*, 25, 1077–1094.

McMahon, G. and Lloyd, O. (1995). *Water Quality Assessment of the Albemarle-Pamlico Drainage Basin*, North Carolina and Virginia: Environmental Setting and Water-Quality Issues. U. S. Geological Survey Open File Report 95–136. U. S. Geological Survey, Raleigh, North Carolina, p. 72.

Mildner, G. C. S., Strathman, J. G., and Bianco, M. J. (1997). Parking policies and commuting behavior. *Transportation Quarterly*, 51, 111–125.

Moglen, G. E. and Kim, S. (2007). Limiting imperviousness—Are threshold-based policies a good idea? *Journal of the American Planning Association*, 73, 161–171.

Moudon, A. V., Kavage, S. E., Mabry, J. E., and Sohn, D. W. (2005). A transportation-efficient land use mapping index. *Transportation Research Record*, 1902, 134–144.

North Carolina Department of Environment and Natural Resources (NCDENR). (2006). North Carolina Water Quality Assessment and Impaired Waters List (2002 Integrated 305(b) and 303(d) Report). NC Department of Environment and Natural Resources. Division of Water Quality Planning Section. 1617 Mail Service Center. Raleigh, NC, p. 184.

North Carolina Department of Environment and Natural Resources (NCDENR) (2007). North Carolina Water Quality Assessment and Impaired Waters List (2006 Integrated 305(b) and 303(d) Report). NC Department of Environment and Natural Resources. Division of Water Quality Planning Section. 1617 Mail Service Center. Raleigh, NC, p. 94.

Paerl, H. W., Pinckney, J. L., Fear, J. M., and Peierls, B. L. (1998). Ecosystem responses to internal and watershed organic matter loading: Consequences for hypoxia in the eutrophying Neuse river estuary, North Carolina, USA. *Marine Ecology Progress Series*, 166, 17–25.

Patterson, P. K., and Chapman, N. J. (2004). Urban form and older residents' service use, walking, driving, quality of life, and neighborhood satisfaction. *American Journal of Health Promotion*, 19, 45–52.

Sanders, R. A. (1986). Urban vegetation impacts on the hydrology of Dayton, Ohio. *Urban Ecology*, 9, 361–376.

Shoup, D. C. (1999). The trouble with minimum parking requirements. *Transportation Research Part A—Policy and Practice*, 33, 549–574.

Snyder, M. N., Goetz, S. J., and Wright, R. K. (2005). Stream health rankings predicted by satellite derived land cover metrics. *Journal of American Water Resources Association*, 41, 659–677.

Sonmez, N. K., Onur, I., Sari, M., and Maktav, D. (2009). Monitoring changes in land cover/use by CORINE methodology using aerial photographs and IKONOS satellite images: A case study for Kemer, Antalya, Turkey. *International Journal of Remote Sensing*, 30, 1771–1778.

Stone, B. (2004). Paving over paradise: How land use regulations promote residential imperviousness. *Landscape and Urban Planning*, 69, 101–113.

Thomas, N., Hendrix, C., and Congalton, R. (2003). A comparison of urban mapping methods using high-resolution digital imagery. *Photogrammetric Engineering and Remote Sensing*, 69, 963–972.

Thompson-Fawcett, M. and Bond, S. (2003). Urbanist intentions for the built landscape: examples of concept and practice in England, Canada and New Zealand. *Program Planning and Management*, 60, 147–234.

Townsend, P. A., Lookingbill, T. R., Kingdon, C. C., and Gardner, R. H. (2009). Spatial pattern analysis for monitoring protected areas. *Remote Sensing of Environment*, 113, 1410–1420.

Tu, J., Xia, Z. G., Clarke, K. C., and Frei, A. (2007). Impact of urban sprawl on water quality in eastern Massachusetts, USA. *Environmental Management*, 40, 183–200.

U.S. Census Bureau (2003). Table HU-EST2002-07—County Housing Unit Estimates. Accessed online May 28, 2009: http://www.census.gov/popest/archives/2000s/vintage_2002/HU-EST2002-07.html

Van Bohemen, H. D. and Van de Laak, W. H. J. (2003). The influence of road infrastructure and traffic on soil, water, and air quality. *Environmental Management*, 31, 50–68.

Vojnovic, I. (2006). Building communities to promote physical activity: A multi-scale geographical analysis. Geografiska Annaler: Series B, *Human Geography*, 88, 67–90.

Walsh, C. J., Papas, P. J., Crowther, D., and Yoo, J. (2004). Stormwater drainage pipes as a threat to a stream-dwelling amphipod of conservation significance, *Austrogammarus australis*, in southeastern Australia. *Biodiversity and Conservation*, 13, 781–793.

Zhu, C., Shi, W., Pesaresi, M., Liu, L., Chen, X., and King, B. (2005). The recognition of road network from high-resolution satellite remotely sensed data using image morphological characteristics. *International Journal of Remote Sensing*, 26, 5493–5508.

11 Use of InSAR for Monitoring Land Subsidence in Mashhad Subbasin, Iran

Maryam Dehghani,
Mohammad Javad Valadan Zoej,
Mohammad Sharifikia, Iman Entezam,
and Sassan Saatchi

CONTENTS

11.1 INTRODUCTION

Interferometric synthetic aperture radar (InSAR) is an efficient technique to precisely measure deformation over large areas at high spatial resolution (Amelung et al. 1999; Crosetto et al. 2002; Fruneau and Sarti 2000; Strozzi et al. 2001; Tesauro et al. 2000). Highly correlated radar images are integrated together in order to generate interferograms containing precise deformation measurements. The InSAR technique is able to map the extent and pattern of the ground surface deformation. Moreover, time series analysis of a significant number of interferograms can be used in order to study the temporal evolution of the detected phenomena. InSAR time series analysis has been widely used to study the short-term as well as the long-term behavior of a continuous deformation phenomenon (Lanari et al. 2004; Berardino et al. 2002; Dehghani et al. 2009).

In this study, InSAR techniques were applied to monitor the temporal evolution of land subsidence caused by overexploitation of groundwater. It is known that groundwater extraction results in decreasing pore pressures and thus increases the effective stress, causing the fine-grained interbeds to become compressed (Hoffmann et al. 2003). A typical aquifer system that consists of a series of poorly permeable and highly compressible fine-grained interbeds was selected for this study in Iran.

A large number of areas in Iran dedicated to agricultural activities suffer from land subsidence due to the overexploitation of groundwater. Compaction of aquifer systems caused by the extraction of groundwater is a significant problem resulting in considerable consequences including earth fissures, surface runoff, and significant damage to engineered structures such as buildings, roadways, railways, and pipelines (e.g., Tolman and Poland 1940; Poland and Davis 1969; Galloway et al. 1998; Galloway et al. 1999). One of the areas subject to land subsidence caused by groundwater withdrawal is located in the northeast of the Mashhad subbasin. Mashhad, northeast of Iran, is the second largest city with 3.5 million residents. The main part of the subsidence area is covered by cultivated fields where excessive groundwater extraction from pumping wells to provide water for the agricultural activities has significantly lowered aquifer hydraulic heads. According to the unit hydrograph of the area, the groundwater table in the Mashhad subbasin will decline by 12 m within 16 years (Alizadeh 2004), which is the most probable reason for the land subsidence occurrence.

The subsidence rate was first measured by geodetic observations of precise leveling surveys between 1995 and 2005 across the area. In order to monitor the temporal behavior of the Mashhad subbasin subsidence, one permanent Global Positioning System (GPS) station was installed by the National Cartographic Center of Iran (NCC) in the subsidence area in 2005. However, it was only possible to measure the deformation locally at the GPS station with no spatial coverage. InSAR time series analysis using 12 ENVISAT Advanced Synthetic Aperture Radar (ASAR) images was then applied to study the short-term as well as the long-term behavior of the Mashhad subbasin subsidence with a large spatial coverage.

Simple time series analysis without taking into account the error sources was already applied on the radar images of Mashhad subbasin (Dehghani et al. 2009). However, some of the interferograms were strongly affected by the topographic artifacts due to large spatial baselines. The goal of this study was to measure land subsidence rate of the Mashhad subbasin using InSAR time series in which interferograms with small temporal baselines were processed to decrease the temporal decorrelation effect caused by the agricultural fields. Since we tried to generate as many interferograms as possible to prevent the solution from the rank deficiency though, the spatial baselines of the processed interferograms ended up not being as small as those produced by the conventional small baseline subset (SBAS) method. The spatial decorrelation caused by the large spatial baselines becomes insignificant compared to the temporal decorrelation in the interferograms processed from the available data sets of the study area. This fact was recognized by the calculation of the coherence matrix of the interferograms. The generated interferograms therefore can be correctly unwrapped due to the high degree of coherence despite the large spatial baselines. It should be mentioned that the interferograms were refined before they were used in the time series analysis in order to reduce the topographic artifacts produced by the large spatial baselines. Furthermore, we tried to reduce other phase components including the atmospheric signals, noise, and orbital errors through the time series analysis. The results obtained from the proposed method were then evaluated using the measurements from precise leveling survey techniques. Such time series analysis results were finally integrated with the groundwater level data sets by which the piezometric wells could be applied to investigate the stress–strain relationship and compressibility of the aquifer system.

11.2   STUDY AREA

Mashhad is located in the northeast region of Iran between the Binalood and Hezarmasjed mountains with northwest and southeast trailing. The Mashhad subbasin is a part of the Ghareghom basin (see Figure 11.1) with an area of 44,165 (km²). Various studies on groundwater resources have been started since 1963. These studies focused mostly on the measurements of groundwater level fluctuations by using observation wells. The main part of the subsidence area is covered by cultivated fields as shown in the Landsat ETM+ color composite image depicted in Figure 11.1.

The average temperature of the Mashhad subbasin is 12.5°C. According to the data acquired during a 30-year period, the average amount of annual rainfall is 315 and 258 mm in mountainous and plain areas, respectively. The main rainfall occurs in spring and winter seasons; however, the aquifer does not receive sufficient recharge, resulting in a continuous decrease of groundwater level. The unit hydrograph of the Mashhad subbasin shown in Figure 11.2 indicates that the decrease in groundwater table reaches 6 m within a 10-year horizon (Alizadeh 2004). As a result, land subsidence has occurred. This makes monitoring the aquifer stability in the study area a top priority.

11.3   RADAR INTERFEROMETRY TIME SERIES ANALYSIS

The deformation time series analysis was carried out by using InSAR techniques in this study. First, the time series analysis approach that had been previously used to measure groundwater-induced land subsidence in the Tehran basin (Dehghani et al. 2010) is introduced. Such an approach is aimed at decreasing the errors existing in the interferograms. In the next step, the coherence analysis along with interferogram generation is explained. The selected time series analysis method is then applied to deal with the land subsidence issue at the Mashhad subbasin. The results of selected time series analysis and associated evaluation are finally presented in Sections 11.3.3 and 11.3.4, respectively.

11.3.1   SBAS PROCESSING

Time series analysis using the interferometric data is one of the most effective methods to monitor the temporal evolution of surface deformation. Least squares

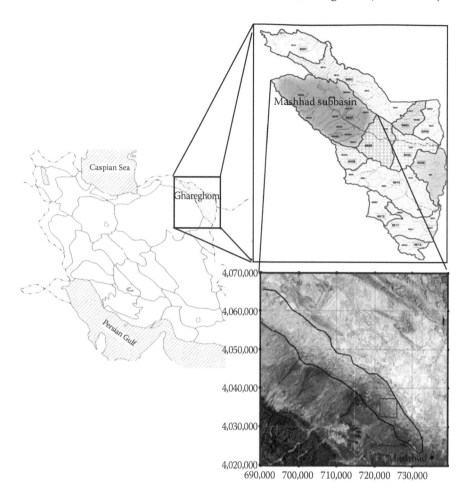

FIGURE 11.1 Location of the study area (Mashhad subbasin) in Ghareghom basin in the northeast of Iran. Bottom: Landsat ETM+ color composite image (R:7, G:4, B:2) of Mashhad subbasin. The black polygon demonstrates the approximate location of the subsidence area that consists of the cultivated fields. Blue rectangle indicates the area in which the averaged coherence matrix is calculated.

inversion is employed to obtain the time series deformation provided that there are at least as many linearly independent interferograms as acquisition dates and that the chain of interferograms is not broken at any point. In time series analysis, the deformation associated with each acquisition date can be estimated given that the deformation corresponding to the start date is known.

Assume that the interferometric phase, $\delta\phi_k$ (rad), computed from two synthetic aperture radar (SAR) images at times t_i and t_j (day) is given by

$$\delta\phi_k = \phi(t_i) - \phi(t_j) \quad k = 1,...,M; \quad i, j = 1,...,N \quad (11.1)$$

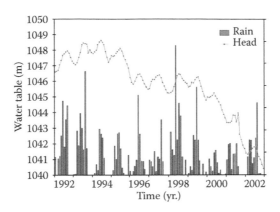

FIGURE 11.2 Unit hydrograph of Mashhad subbasin extracted from hydraulic head information from 1992 to 2002.

where M and N are the total number of processed interferograms and SAR images, respectively. $\phi(t_i)$ and $\phi(t_j)$ (rad.) associated with the deformation are the phase information of SAR images that are considered as unknown values.

In the case that there are M interferograms generated from N radar images, a system of M equations is

$$A\phi = \delta\phi, \tag{11.2}$$

where ϕ (rad) is the vector of N unknown phase values, $\delta\phi$, (rad.) is the vector of M known interferometric phases, and A is the design matrix in the least squares solutions.

The interferometric phase is affected by various phase components. It can be decomposed into more components, as shown below:

$$\delta\phi_k = \phi(t_i) - \phi(t_j)$$
$$\approx \Delta\phi_{def.k} + \Delta\phi_{topo.art.k} + \Delta\phi_{atm.k} + \Delta\phi_{n.k}. \tag{11.3}$$

The first component (i.e., $\Delta\phi_{def.k}$ [rad]) is surface deformation caused by land subsidence. It can be computed as

$$\Delta\phi_{def.k} = \frac{4\pi}{\lambda}\left[d(t_i) - d(t_j)\right], \tag{11.4}$$

where $d(t_i)$ and $d(t_j)$ are the cumulative deformation at times t_i and t_j with respect to the start time (m). In addition, t_0 is a reference time (day), and λ is the radar signal wavelength (m). The correction factor $\frac{4\pi}{\lambda}$ is used in order to convert the displacement into the phase value.

The second term of Equation 11.3 (i.e., $\Delta\phi_{topo.art.k}$ [rad.]) is the phase artifacts caused by an error Δz (m) in the knowledge of the topography. The effect of Δz on the interferometric phase depends on slant range r (m), incidence angle θ (rad.), and perpendicular baseline of the interferogram k, $B_{\perp k}$ (m):

$$\Delta\phi_{topo.art.k} = \frac{4\pi}{\lambda}\frac{B_{\perp k}\Delta z}{r\sin\theta}. \tag{11.5}$$

The third component of Equation 11.3 is referred to as the atmospheric effect; it is different for two SAR acquisitions at times t_i and t_j. Eventually, the last term $\Delta\phi_{n.k}$ (rad) is noise caused by various types of phenomena including baseline and temporal decorrelation.

In this study, we followed a processing algorithm illustrated in the block diagram of Figure 11.3 to generate the deformation time series. The algorithm was originally presented by Berardino et al. (2002). The innovation of their proposed method is to use a large number of SAR acquisitions distributed in small baseline subsets via the singular value decomposition (SVD) method (Berardino et al. 2002). Despite this fact, our study for the chain of interferograms is not broken at any point and only one data set exists. We can therefore deduce that the SVD algorithm is not required for the inversion solution.

In the conventional SBAS method, only the interferograms with small spatial baselines are processed. Since the temporal decorrelation is much more significant than the spatial decorrelation in our study area though, we had to use these interferograms that can only be characterized by the small temporal baseline. Topographic artifacts are then produced due to the large spatial baselines. The interferograms should hence be refined before they can be used in the time series analysis based on

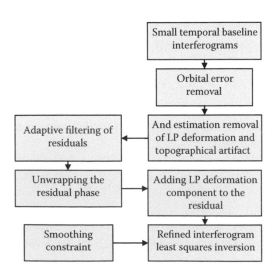

FIGURE 11.3 Block diagram of the implemented algorithm.

the method developed by Berardino et al. (2002). The refinement of the interferograms is included in the algorithm shown in Figure 11.3.

Orbital errors due to the tilts and offsets remaining in the interferograms can be removed by subtracting a plane fitted to the data in the far field, away from the deformation signal (e.g., Funning et al. 2005). The method developed by Berardino et al. (2002) was then used to remove the topographic artifacts caused by the large spatial baselines. It is assumed that the phase is a linear function of temporal low-pass (LP) components of the deformation (i.e., mean velocity, $\bar{v}$ [m·day^{-1}], mean acceleration, $\bar{a}$ [m·day^{-2}], and mean acceleration variation, $\Delta\bar{a}$):

$$\phi(t_i) = \bar{v} \cdot (t_i - t_0) + \frac{1}{2}\bar{a} \cdot (t_i - t_0)^2 + \frac{1}{6}\Delta\bar{a} \cdot (t_i - t_0)^3. \qquad (11.6)$$

The LP components of the subsidence were then jointly estimated with the topographic artifacts (Equation 11.5) using the least squares solution (see Berardino et al. 2002 for further details). The estimated LP phase patterns and topographic artifacts then subtracted modulo-2π from each input interferograms as shown in the block diagram of Figure 11.3. This results in the fringe rate reduction. An adaptive filter was applied on the residuals in order to reduce the phase noise leading to a high-performance unwrapping of the residuals. A refined unwrapped phase pattern is finally obtained by adding back the subtracted LP component of the deformation. The refined interferograms are inverted via a least squares solution of the equation system of Equation 11.2 to obtain the cumulative phase. A smoothing constraint is incorporated into the inversion problem to mitigate the atmospheric artifacts, noise, and unwrapping errors (e.g., Lundgren et al. 2001; Schmidt and Burgmann 2003). The smoothing constraint used here is based on the finite difference approximation for the second order differential of the time series applying the minimum curvature concept (i.e., constant velocity). The finite difference approximation for the second order differential of the time series for three sequent acquisition times, t_1, t_2, and t_3 (day) is written as

$$\gamma^2 \cdot \partial^2\phi/\partial t^2 \equiv \gamma^2 (t_3 - t_2) \phi_1 + \gamma^2 (t_1 - t_3) \phi_2 + \gamma^2 (t_2 - t_1) \phi_3 = 0, \qquad (11.7)$$

where ϕ_1, ϕ_2, and ϕ_3 (rad) are the phases of SAR images corresponding to t_1, t_2, and t_3 (day), respectively. γ is the smoothing factor to be determined optimally. By adding Equation 11.7 as an additional equation to Equation 11.2, we have

$$\begin{pmatrix} A \\ \gamma^2 \cdot \partial^2 / \partial t^2 \end{pmatrix} \phi = \begin{pmatrix} \delta\phi \\ 0 \end{pmatrix}. \qquad (11.8)$$

Its small values lead to a rough deformation time series while a large one will damp any nonlinear deformation. A trade-off methodology should hence be applied to select the most appropriate smoothing factor regarding reduction of the errors as well as preservation of the nonlinear seasonal deformation signal. One approach to estimate the most appropriate smoothing factor is to plot the overall least squares misfits (root-mean-square error, RSME) against various corresponding smoothing factors. The resulted plot is a curve with an "elbow." Although the choice of the most optimum smoothing factor is arbitrary depending on the plot scale, the middle point on the curve would be an estimate for the smoothing factor, acting as a good compromise between noisy fluctuations and their removal when preserving the non-linear seasonal deformation. The phase values corresponding to each acquisition time may be finally obtained using the least squares solution of the equation system in Equation 11.8 (Dehghani et al. 2010).

11.3.2 GENERATION OF INTERFEROGRAMS

With the selected method above (Dehghani et al. 2010), we should be able to process the interferograms as coherently as possible in conventional interferometry. Small baseline interferograms can be easily formed and individually phase-unwrapped for areas with high coherence. In this study, 18 coherent interferograms were calculated from 12 Level 0 ENVISAT ASAR images acquired from track 392 by applying GAMMA software (Wegmuller et al. 1995). The information of the constructed interferograms can then be summarized (Table 11.1).

As observed in Table 11.1, the spatial baselines of the processed interferograms are not as small as in the conventional SBAS method. Still, all the spatial baselines

TABLE 11.1
Constructed Interferograms with the Spatial and Temporal Baselines

Interferograms	Perpendicular Baseline (m)	Temporal Baseline (day)
2003.06.30–2003.09.08	610	70
2003.06.30–2003.12.22	180	175
2003.09.08–2003.12.22	444	105
2003.12.22–2004.06.14	246	175
2004.06.14–2004.07.19	20	35
2004.07.19–2004.11.01	110	105
2004.06.14–2004.12.06	55	175
2004.07.19–2004.12.06	62	140
2004.11.01–2004.12.06	172	35
2004.12.06–2005.09.12	39	280
2004.12.06–2005.05.30	267	175
2005.05.30–2005.09.12	274	105
2005.09.12–2005.11.21	158	70
2005.09.12–2005.10.17	118	35
2005.05.30–2005.10.17	184	140
2005.10.17–2005.11.21	261	35
2005.05.30–2005.08.08	232	70
2004.06.14–2004.11.01	111	140

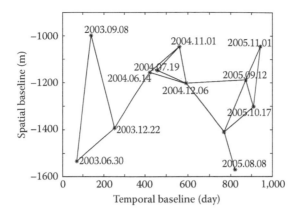

FIGURE 11.4   Acquisition geometry of the available radar data in the study area. The connections represent the processed interferograms mostly characterized by the small temporal and large spatial baselines. The nodes correspond to the radar acquisitions.

are smaller than the critical value (~1500 m). Temporal decorrelation is a typical phenomenon observed in the interferograms due to the agricultural fields (Zebker and Villasenor 1992). It can therefore be deduced that the interferograms with short time intervals should be processed to reduce the temporal decorrelation effects. The acquisition geometry of the available radar data is illustrated in Figure 11.4.

According to Figure 11.4, there is only one chain of the interferograms. In this figure, the connections show the processed interferograms while the nodes correspond to the radar acquisitions. The processed interferograms were characterized by small temporal baselines and relatively large spatial baselines. The decorrelation effect in all processed interferograms from the available dataset is insignificant. Rocca (2007) proposed an approach to study the coherence of a series of interferograms that is the calculation of coherence matrix. Each element in the matrix that represents the coherence of a pixel in a pair of images is calculated by sampled estimate of the coherence using an estimation window (Hanssen 2001). The principal diagonal is always unitary because every image is perfectly coherent with itself. It should be pointed out that the elements of the coherence matrix are ordered by the acquisition date.

In this study, the averaged coherence matrix for one specific area within the subsidence is calculated using the 5 × 5 estimation window (see Figure 11.5). The area that is highly affected by temporal decorrelation is located in the cultivated lands. Since only the coherent interferograms used in the time series analysis have to be applied to calculate the coherence matrices, some of the elements in the coherence matrices are zero. Other values of the elements are higher than 0.6. The coherence values of all calculated interferograms are high enough that the interferograms can be correctly unwrapped.

After processing the coherent interferograms, a digital elevation model (DEM) derived from the Shuttle Radar Topography Mission (SRTM) with the spatial resolution of 90 m was used in order to remove the topographic phase component. The time series analysis approach presented in Section 11.3.1 was then implemented on the

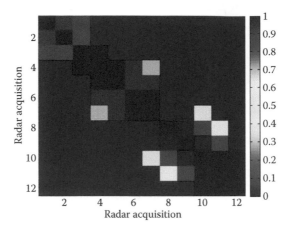

FIGURE 11.5 Averaged coherence matrices of a specific area located in the cultivated lands. The area is indicated in Figure 11.1.

interferometric data of the Mashhad subbasin subsidence. The results are presented in the next section.

11.3.3 TIME SERIES ANALYSIS RESULTS

The approach presented in Section 11.3.1 is applied to generate the deformation time series in the Mashhad subbasin. The time series results can be viewed in different ways. Line of sight (LOS) means a displacement velocity map indicating the major features of the subsidence and long-term behavior of the deformation is computed using the time series analysis results (Figure 11.6). Maximum deformation rate

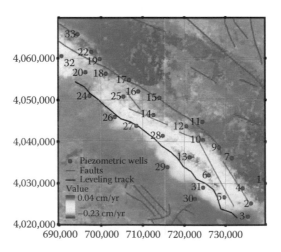

FIGURE 11.6 Mean displacement velocity map. The black line indicates the precise leveling track. Piezometric wells along the subsidence area are illustrated by the circles.

estimated from the mean displacement velocity map is 23 cm·yr⁻¹ in the subsidence area.

The subsidence trend is northwest toward southeast and resembles the topographic trend in the study area. The subsidence pattern is probably controlled by linear structures such as faults (see Figure 11.6). The faults and structural lineaments shown in Figure 11.6 are extracted from a geological map with the scale of 1:250,000 and satellite images. These faults and structures can act as barriers in the aquifer system that controls the groundwater flow. Furthermore, they may be considered as the boundaries of material zones with different controlling hydrogeological parameters that affect the subsidence, such as specific storage and hydraulic conductivity. Another method to represent the time series analysis results is to plot the deformation sequences that are explained in Section 11.4.

11.3.4 Results Validation

The results obtained from the time series analysis are compared to the precise leveling data. Precise leveling surveys were performed by NCC in 1995 and 2005. The leveling track shown in Figure 11.6 crosses the margin of the subsidence bowl. It therefore can be deduced that the leveling data does not contain the maximum amount of the deformation rate. The InSAR-derived subsidence rate along the leveling track is compared to the deformation rate extracted from leveling measurements during a 10-year time frame (Figure 11.7).

The RMSE between time series analysis results and precise leveling was estimated at 0.9 cm·yr⁻¹, which is an indication of high performance of the algorithm applied in this study. Although the time interval between InSAR data and leveling surveys is different, the deformation rates obtained from both methods are quite comparable. This fact shows that the long-term behavior of the subsidence has remained unchanged for at least 10 years. A mean displacement velocity map allows for the identification of long-term behavior of the aquifer compaction. In order to study the

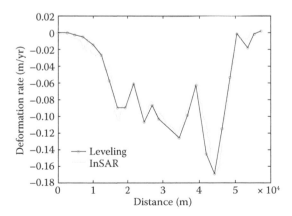

FIGURE 11.7 Comparison between time series analysis results and precise leveling measurements.

short-term behavior of the subsidence, the InSAR-derived deformation sequences were generated. The results are presented in the next section.

11.4 COMPARISON OF INSAR-DERIVED DEFORMATION WITH WATER LEVEL INFORMATION OF THE PIEZOMETRIC WELLS

The main strength of the InSAR time series analysis is its potential for detecting nonsteady deformation signals such as seasonal effects (e.g., Lanari et al. 2004). The short-term behavior of the aquifer compaction is studied by the generation of the chronological sequence of the deformation. Deformation time series computed at piezometric wells are shown in Figure 11.8a, b, and c. Moreover, water level information from the piezometric wells is analyzed in order to study the effective stress changes in the Mashhad subbasin (i.e., water level measurements have been collected on a monthly basis by the Water Management Organization since 1995). Figure 11.8d, e, and f present groundwater level fluctuations of the wells whose InSAR-derived deformation time series are plotted in Figures 11.8a, b, and c. The y-axis in these figures represents the water table. Any decrease in water table corresponds to the water level decline caused by groundwater withdrawal.

The locations of the piezometric wells shown in Figure 11.6 demonstrate that they are distributed in different parts of the study area. The deformation sequences of the piezometric wells located outside of the subsidence area (i.e., wells 11 and 15) are depicted in Figure 11.8a. As expected, no deformation signal due to the subsidence

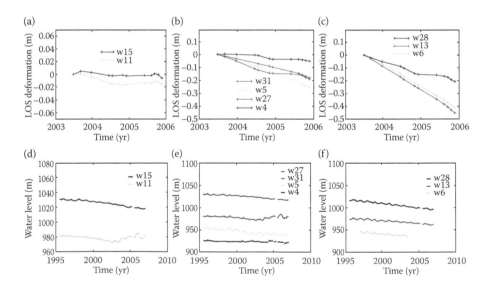

FIGURE 11.8 InSAR-derived deformation at piezometric wells located (a) outside the subsidence area, (b) along the margin of the subsidence area, and (c) in the middle of the subsidence area. Water level depth changes are shown at piezometric wells located (d) outside the subsidence area, (e) along the margin of the subsidence area, and (f) in the middle of the subsidence area.

phenomenon can be collected at these points. Moreover, no seasonal fluctuations were observed in the deformation time series. On the other hand, the water table decline is insignificant at well 11 while the water level of well 15 was lowered by approximately 10 m during a 10-year time frame (see Figure 11.8d).

InSAR time series plots of the second category of piezometric wells located in the margin of the subsidence bowl (wells 4, 5, 27, and 31) were presented in Figure 11.8b. Wells 4 and 27 show similar temporal behavior, since seasonal oscillation can be clearly observed in their deformation sequences. In the recovery season, which is from December 2003 to June 2004 and from December 2004 to June 2005, the deformation time series showed the decelerated subsidence signal. Furthermore, the ground surface is subsided at a higher rate in the discharge season spanning from June 2003 to December 2004 and from June 2004 to December 2005. On the other hand, the subsidence behavior of wells 31 and 5 does not contain any seasonal effect as the ground surface is lowering at a constant rate. According to Figure 11.8b and e, the water level behavior of the wells located in the margin of the subsidence area does not have significant correlation with the deformation rate. The sinusoidal seasonal effect can be clearly observed in water table information of well 5. Still, such fluctuations are not reflected in its corresponding deformation time series. In contrast, the water level has declined continuously in wells 4 and 27 without any seasonal fluctuations, although short-term seasonal behavior exists in their deformation sequences.

The above-mentioned low correlation between water level variations and short-term behavior of the deformation time series was repeated at wells 6, 11, and 28 located in the middle of the subsidence area (Figure 11.8c and f). The decreasing trend observed in the InSAR time series associated with wells 6 and 11 clearly indicates that the aquifer was compacted at a constant rate during the study period. However, the subsidence rates associated with the recharge and discharge seasons were different at well 28.

The compaction mechanism and the stress–strain relationship of the aquifer system can be derived from InSAR displacement time series and contemporaneous measurements of water level. An appropriate method to investigate the relationship between groundwater level fluctuations and surface displacements is to map both parameters in a unique plot. The water level variations representing the stress are plotted on the y-axis, whereas the InSAR deformation time series showing the aquifer compaction are plotted on the x-axis. According to Hoffmann et al. (2001), a rough estimate of the skeletal storage coefficient of the aquifer system can be obtained by the inverse slope of the best fitting line to the plotted points. The storage coefficient of an aquifer system presenting the responses of the aquifer and fine-grained interbeds to variations in hydraulic head can be also determined by pumping tests and from core samples. Still, the extracted storage coefficients indicate only the most permeable fraction of the aquifer system. Another method to estimate the storage coefficients can be applied in the laboratory from core samples. This is generally costly and the measurements are not representative of *in situ* conditions.

The storage coefficients of the aquifer system extracted by the relationship between the InSAR-derived displacements and groundwater levels yield spatially varying estimates of storage coefficients at the well locations. Furthermore, the estimated values of the slope of the best-fitting line enable us to predict the amount of subsidence caused by water level decline.

The water level variations plotted against the ground displacements for the piezometric wells are illustrated in Figure 11.9. Ground displacements were linearly interpolated corresponding to the water level observation dates since the water level observations were made on a monthly basis. The red lines in these plots represent the best-fitting line to the plotted points. The estimated storage coefficient of each piezometric well can be indicated in its corresponding plot in Figure 11.9. The lowest amounts of storage coefficients appeared at wells 11 and 15 as expected. This means the declining water level would not cause significant subsidence at these wells. A higher storage coefficient implies a higher response of the ground surface to water level variations as can be observed at wells 5 and 6. The storage coefficient of well 27 has not been estimated correctly probably due to the unreliable water level measurements.

Note that there is a time delay between the water table decline and compaction of the aquifer system due to the low vertical hydraulic conductivity of the compressible sediments in the aquifer system. The estimated values of the storage coefficients do not exactly reflect the compressibility of the aquifer system due to the time delay involved in the equilibration of the fine-grained interbeds. Still, the estimated storage

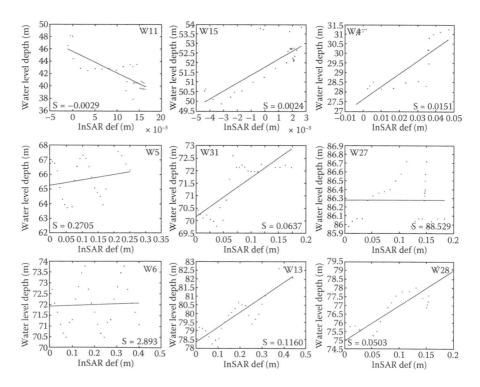

FIGURE 11.9 Storage coefficient calculated from stress–strain relationship. S represents the estimated storage coefficient corresponding to each well.

coefficients can be used to predict the surface subsidence as a function of water level decline in the near future, though a more reliable prediction can be performed by fitting a sinusoidal function to the plotted points. In order to estimate more reliable values of the storage coefficients, other information, including interbed thicknesses and their distributions in the aquifer system, is required. Furthermore, the InSAR-derived displacements can be applied to produce a map of storage coefficients using the hydraulic heads calculated from a groundwater flow model (Hoffmann et al. 2003).

11.5 CONCLUSIONS

In this study, the temporal behavior of the Mashhad subbasin was investigated using a multistep time series analysis approach based on InSAR measurements. The generated interferograms were characterized based on the small temporal baselines in order to decrease the temporal decorrelation. The spatial baselines of the interferograms however, are not as small as in the conventional SBAS method. Coherence analysis of the processed interferograms confirms that the spatial decorrelation in the study area is insignificant in comparison with the temporal decorrelation. As the interferograms with large spatial baselines were influenced by the topographic artifacts, they were refined before using in the time series analysis. Moreover, atmospheric-error–free deformation corresponding to every acquisition time was retrieved by applying the smoothing constraint into the least squares solution.

The obtained time series analysis results were used to extract the mean displacement velocity map in order to highlight the major deformation features of the subsidence. The maximum amount of subsidence rate is estimated at 23 cm·yr^{-1} in the study area. The chronological sequences of the estimated deformation for a selection of piezometric wells located in different parts of the subsidence area were then plotted. Some piezometric wells were subject to the seasonal fluctuations while others were subsiding at constant rates without significant seasonal effects. The time series analysis results were then compared to the precise leveling measurements for verification. The comparison demonstrates the high performance of the algorithm applied in this study. InSAR-derived deformation time series also indicated that seasonal fluctuations can be observed in some parts of the subsidence area; however, the short-term behavior of the subsidence shows only a questionable correlation with the variations of groundwater level. This observation indicates that land subsidence is a function of not only groundwater level but also other factors such as the types of sediments that constitute the aquifer. The compressibility of the aquifer system is finally investigated by quantitative integration of the InSAR-derived displacement measurements with observations of the hydraulic head fluctuations.

The InSAR-derived deformation and water level information were finally used to investigate the stress–strain relationship at the piezometric wells. The compressibility property of the aquifer system was obtained by the integration of deformation time series and water level measurements. It is indicative that the strain generally increases with the decrease in water level. Still, the nonlinear relation between the InSAR time series and the water level variations in the piezometric wells show that other important factors may control the subsidence rate in the area. Various hydrological and soil properties including hydraulic conductivity, permeability, depths and distributions of the fine-grained interbeds, and specific storage are required in order to study the compressibility of the aquifer system in the future.

ACKNOWLEDGMENTS

This study was supported by the Department of Earth Science, University of Oxford. The authors wish to thank the Remote Sensing group of the Geological Survey of Iran (GSI) and the European Space Agency (ESA) for providing us with the GAMMA software and radar data, respectively. We also convey our sincere gratitude to the Engineering Geology group of GSI for their helpful comments.

REFERENCES

Alizadeh, A. (2004). *Mathematical Models of Mashhad Study Area*. Preliminary Reports of the Applied Researches Committee, Regional Water Company of Khorasan, Ministry of Power, Iran.

Amelung, F., Galloway, D. L., Bell, J. W., Zebker, H. A., and Laczniak, R. J. (1999). Sensing the ups and downs of Las Vegas—InSAR reveals structural control of land subsidence and aquifer-system deformation. *Geology*, 27, 483–486.

Berardino, P., Fornaro, G., Lanari, R., and Sansosti, E. (2002). A new algorithm for surface deformation monitoring based on small baseline differential SAR interferograms. *IEEE Transactions on Geoscience and Remote Sensing*, 40, 2375–2383.

Crosetto, M., Crippa, C. C., and Castillo, M. (2002). Subsidence monitoring using SAR interferometry: Reduction of the atmospheric effects using stochastic filtering. *Geophysical Research Letters*, 29, 26–29.

Dehghani, M., Valadan Zouj, M. J., Biggs, J., Mansourian, A., Parsons, B., and Wright, T. (2009). RADAR interferometry time series analysis of Mashhad subsidence. *Journal of International Society of Remote Sensing (ISRS)*, 37, 191–200.

Dehghani, M., Valadan Zouj, M. J., Entezam, I., Saatchi, S., and Shemshaki, A. (2010). Interferometric measurements of ground surface subsidence induced by overexploitation of groundwater. *Journal of Applied Remote Sensing*, 4, 041864, doi:10.1117/1.3527999.

Fruneau, B. and Sarti, F. (2000). Detection of ground subsidence in the city of Paris using radar interferometry: Isolation from atmospheric artifacts using correlation. *Geophysical Research Letters*, 27, 3981–3984.

Funning, G. J., Parsons, B., Wright, T. J., and Jackson, J. A. (2005). Surface displacements and source parameters of the 2003 Bam (Iran) earthquake from ENVISAT advanced synthetic aperture radar imagery. *Journal of Geophysical Research*, 110, B09406, doi:10.1029/2004JB003338.

Galloway, D. L., Hudnut, K. W., Ingebritsen, S. E., Phillips, S. P., Peltzer, G., Rogez, F., and Rosen, P. A. (1998). Detection of aquifer system compaction and land subsidence using interferometric synthetic aperture radar, Antelope Valley, Mojave Desert, California. *Water Resources Research*, 34, 2573–2585.

Galloway, D. L., Jones, D. R., and Ingebritsen, S. E. (1999). *Land subsidence in the United States*. US Geological Survey Circular, 1182, 175.

Hanssen, R. F. (2001). *Radar Interferometry: Data Interpretation and Error Analysis*. Kluwer Academic Publishers, Dordrecht, the Netherlands.

Hoffmann, J., Galloway, D. L., Zebker, H. A., and Amelung, F. (2001). Seasonal subsidence and rebound in Las Vegas Valley, Nevada, observed by synthetic aperture radar interferometry. *Water Resources Research*, 37, 1551–1566.

Hoffmann, J., Leake, S. A., Galloway, D. L., and Wilson, A. M. (2003). *Ground-Water Model—User Guide to the Subsidence and Aquifer-System Compaction (SUB) Package*. U.S. Geological Survey, Tucson, Arizona, Open-File Report 03-233.

Lanari, R., Lundgren, P., Manzo, M., and Casu, F. (2004). Satellite radar interferometry time series analysis of surface deformation for Los Angeles, California. *Geophysical Research Letters*, 31, L23611, doi:10.1029/2004GL021294.

Lundgren, P., Usai, S., Sansosti E., Lanari, R., Tesauro, M., Fornaro, G., and Berardino, P. (2001). Modeling surface deformation observed with SAR interferometry at Campi Flegrei Caldera. *Journal of Geophysical Research*, 106, 19355–19367.

Poland, J. F. and Davis, G. H. (1969). Land subsidence due to withdrawal of fluids. *Reviews in Engineering Geology*, 2, 187–269.

Rocca, F. (2007). Modeling Interferogram Stacks. *IEEE Transactions on Geoscience and Remote Sensing*, 45(10), 3289–3299.

Schmidt, D. A. and Burgmann, R. (2003). Time-dependent land uplift and subsidence in the Santa Clara valley, California, from a large interferometric synthetic aperture radar data set. *Journal of Geophysical Research*, 108(B9), 2416, doi:10.1029/2002JB002267.

Strozzi, T., Wegmüller, U., Tosi, L., Bitelli, G., and Spreckels, V. (2001). Land subsidence monitoring with differential SAR interferometry. *Photogrammetric Engineering & Remote Sensing*, 67, 1261–1270.

Tesauro, M., Beradino, P., Lanari, R., Sansoti, E., Fornaro, G., and Franceschetti, G. (2000). Urban subsidence inside the City of Napoli (Italy) observed with synthetic aperture radar interferometry at Campi Flegrei caldera. *Journal of Geophysical Research*, 106, 19,355–19,366.

Tolman, C. F. and Poland, J. F. (1940). *Ground-Water, Salt-Water Infiltration, and Ground-Surface Recession in Santa Clara Valley, Santa Clara County, California*. Transactions—American Geophysical Union, 23–24.

Wegmuller, U., Werner Ch., Strozzi, T., Wiesmann, A., and Santoro, M. (1995). GAMMA software, GAMMA Remote Sensing Research and Consulting AG. Worbstrasse 225, CH-3073 Gumligen, Switzerland.

Zebker, H. A. and Villasenor, J. (1992). Decorrelation in interferometric radar echoes. *IEEE Transactions on Geoscience and Remote Sensing*, 30, 950–959.

12 Remote Sensing Assessment of Coastal Land Reclamation Impact in Dalian, China, Using High-Resolution SPOT Images and Support Vector Machine

Ni-Bin Chang, Min Han, Wei Yao, and Liang-Chien Chen

CONTENTS

12.1 INTRODUCTION

Urbanization, the physical growth of urban areas, and associated land use and land cover change (LUCC) is normally driven by a combination of socioeconomic development and political considerations. In turn, urbanization and LUCC may have numerous direct and indirect impacts on micro- and macro-level climatology, hydrology, and environmental ecology, some of which can be quite destructive, such as pollution and deforestation. The ultimate impact of urbanization will affect urban dwellers through climate change and constraints of the varying hydrological cycle. Understanding the compelling and encompassing causes and consequences of urbanization-related LUCC policies requires using new remote sensing satellite technologies.

Early studies of remote sensing image classification relied on statistical methods such as the maximum likelihood (ML) classifier (Lombardo and Oliver 2001), the K-nearest neighbor (Knn) algorithm (Blanzieri and Melgani 2008), and the K-means clustering approach (Jian et al. 2008). In recent years, methods based on machine learning techniques have been popular in this field. Various segmentation and classification approaches such as neural computing, fuzzy logic, evolutionary algorithms, and expert systems have been widely applied (Tso and Mather 1999; Giacinto and Roli 2001; Stefanov et al. 2001; Foody and Mathur 2004a,b; Canty 2009; Chen et al. 2009; Kavzoglu 2009). Neural computing methods are date-driven methods that have high fault tolerance in general (Giacinto and Roli 2001; Canty 2009; Kavzoglu 2009). Many successful applications of remote sensing image classification use classifiers based on various artificial neural networks (ANNs) such as backward propagation (BP), radial basis function (RBF), and self-organized mapping (SOM) (Heermann and Khazenie 1992; Hoi-Ming and Ersoy 2005; Suresh et al. 2008), and global optimization techniques such as support vector machine (SVM) (Foody and Mathur 2004a,b). In ANNs, an input space is mapped into a feature space through the hidden layer, resulting in a nonlinear classifier that outperforms most traditional statistical methods. However, these ANNs are all black box models, whose classification mechanisms are difficult to interpret. Problems such as overfitting, local minimum, and slow convergence speed are quite common for neural computing methods. SVM differs radically from ANNs because SVM training processes always give rise to a global minimum, and their simple geometric interpretation provides opportunities for advanced optimization. While ANNs are limited by multiple local minima, the solution to an SVM is global and unique; on the other hand, ANNs use empirical risk minimization whereas SVMs choose structural risk minimization.

Classifiers based on fuzzy logic are much easier to interpret because the classification is usually implemented according to rules summarized from the training data set. Most fuzzy logic methods are hybrid methods; for example, the fuzzy C-means (FCM) algorithm (Fan et al. 2009) is a hybrid between fuzzy logic and the statistical algorithm (c-means). Classifiers based on fuzzy neural networks (FNNs) (Chen et al.

2009) and Fuzzy ARTMAP (FA) (Han et al. 2004), which are combinations of fuzzy logic and neural networks, were also reported. Evolutionary algorithms are another category of machine-learning techniques that have been widely used in remote sensing image classification. Genetic algorithm (GA), evolutionary programming (EP), and genetic programming (GP) are some classical evolutionary algorithms with many successful applications (Agnelli et al. 2002; Ross et al. 2005; Makkeasorn et al. 2006; Awad et al. 2007; Makkeasorn and Chang 2009; Chang et al. 2009). Classifiers based on artificial immune system (AIS) (Yanfei et al. 2007) and swarm intelligence (Daamouche and Melgani 2009) can also be included in this category. In addition, classifiers such as those based on expert system theory (Stefanov et al. 2001) and decision tree techniques (Friedl and Brodley 1997) are also representative and important classification methods. The current progress in the literature can be summarized based on the above findings (Table 12.1).

The city of Dalian, China, is a fast-growing region both economically and socially, as evidenced by the rapid expansion of the Dalian Development Area (DDA) in the past decade (Figure 12.1). Using remote sensing satellite images to monitor the process-oriented movement of the urban sprawl and predict human influences on landscapes and ecosystems has become essential. The spatial and temporal patterns of LULC that shape and transform cities, suburbs, and rural areas can be identified to illuminate the interaction among urbanization, economic development, land use policy, urban planning, and natural resources in this region. With the aid of modern remote sensing technologies, essential information about urbanization processes that simultaneously converge and diverge can determine how urbanization alters the characteristics of natural disturbances, how disturbance regimes are altered through human influences along urban–rural gradients, and how the spatiotemporal aspects and management models of LUCC may help identify subsequent risks to human and natural environments during fast economic development. Such an endeavor links human dimensions aspects of LUCC with environmental and ecological aspects of LUCC along urban–rural interfaces.

TABLE 12.1
Summary of Classification Methods in Image Processing

	Statistical Methods	ML, Knn, K-means
Artificial intelligence	Neural networks	BP, RBF, SOM
	Global optimization	SVM
	Fuzzy logic	FCM, FA, FNN
	Evolutionary algorithms	GA, EP, GP, AIS
Other methods		Expert system, decision tree

Note: BP = backward propagation; RBF = radial basis function; SOM = self-organized mapping; SVM = support vector machine; fuzzy C-means = FCM; FA = fuzzy ARTMAP; FNN = fuzzy neural network; GA = genetic algorithm; EP = evolutionary programming; GP = genetic programming; AIS = artificial immune system.

FIGURE 12.1 The street view of the Dalian Development Area (DDA).

In our study, two scenes of Système Probatoire d'Observation de la Terre (SPOT-5) images with 2.5 m resolution were acquired and classified to capture the spatiotemporal LULC patterns from 2003 to 2007 to inform policy makers, natural resource professionals, and private citizens of possible changes to ecosystem goods and services along urban–rural gradients. To tackle the spatial and temporal heterogeneity of LULC patterns, additional features including the vegetation index and texture features were extracted for the classification process. An SVM classifier is constructed based on ground truth data collected beforehand. Experimental results show that following our multifeature classification strategy, the SVM classifier can provide remarkably high accuracy. Although the processing speed of the classifier is not as fast as simpler statistical methods, it is acceptable for processing high-resolution remote sensing images that usually contain large numbers of data points.

12.2   DATA PREPARATION

12.2.1   HIGH-RESOLUTION SPOT-5 IMAGE PRODUCTS

Due to technical limitations, high spatial resolution and large swath width of remote sensing images are difficult to achieve simultaneously. For rural areas, high spatial resolution is not necessary, and the large swath width is much more convenient. In the remote sensing of urban areas, however, higher spatial resolution is preferable because the objects of interest are usually on a smaller scale and are difficult to recognize in low-resolution images. Furthermore, the swath widths of high-resolution images are large enough for most urban areas.

Two scenes of SPOT-5/HRG images were collected for the research, originally acquired during the overpasses of the SPOT-5 satellite on May 5, 2003, and April 13, 2007. The SPOT-5 images used in our study cover an area larger than 39 × 40 km, while the region of interest is 20 × 20 km; therefore, we cut subsets from the whole images in advance. The resolution of the original SPOT-5/HRG images in the green, red, and near infrared channels is 10 m, and the 2.5-m images are resolution-enhanced image fusion products. The original SPOT-5/HRG images are in false color composite, and the true color composite images (Figure 12.2) are also

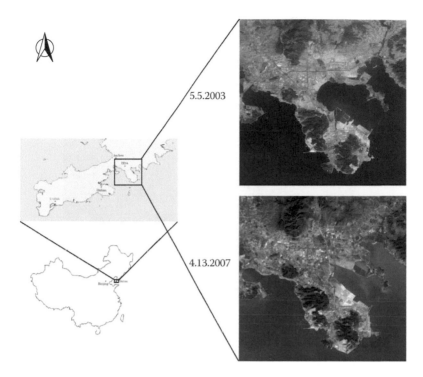

FIGURE 12.2   Location of the study area and the corresponding SPOT-5 images in 2003 and 2007.

image fusion products in which blue channel images are synthesized from the original SPOT-5/HRG images in green, red, and near infrared channels.

12.2.2   FIELD CAMPAIGNS FOR GROUND TRUTHING

The study area is a state-run complex (the DDA) administered by the city of Dalian. The DDA is adjacent to the seashore (about 121°E, 39°N) with a coastline of about 130 km (Figure 12.1). To the north of the urban area is Big Black Mountain. A comparison of two remote sensing images of the area (Figure 12.2) shows some obvious changes in LULC. Between 2003 and 2007, large-scale land reclamations along the shoreline occurred in this area, which became the major concern for this study. Quantitative information about these changes can be retrieved with the aid of the proposed classification method.

Six aggregated classes of land cover categorized for this study include buildings, roads, bare fields, grasslands, forests, and water bodies. Ground truth data collected during the two field campaigns (Figure 12.3) consisted of more than 20,000 pixels constructed in advance to generate both the training and the testing data sets for the classifiers. These ground truth samples covered all classes of LULC of interest. The differences within and between some classes of LULC were carefully considered during the sampling processes of the ground truth data; for example, a variety of buildings were classified as the same type of LULC.

FIGURE 12.3   Distribution of ground truth data points in the study area.

12.3   THE SVM CLASSIFIER

In recent years, kernel methods, such as SVM in particular, have become more popular for remote sensing image classification (Melgani and Bruzzone 2004; Filippi and Archibald 2009; Bovolo et al. 2010; Giacco et al. 2010; Mūnoz-Marí et al. 2010; Ratle et al. 2010; Tarabalka et al. 2010). SVM is an advanced and promising classification as well as a regression approach proposed by Vapnik and his group at AT&T Bell Laboratories in 1990s (Vapnik 2000).

As a representative kernel method, SVM relies on the definition of a distance measure between data points in a proper Hilbert data space (Courant and Hilbert 1954). Based on the idea of structural risk minimization, SVM not only minimizes the cost function but also controls the complexity of the classification function. This property improves SVMs robustness to filter noises in training processes and therefore enhances its classification accuracy over other widely used methods such as maximum likelihood, decision tree, and neural network–based approaches (Huang et al. 2002; Bazi and Melgani 2006). An SVM classifier is binary by nature, however, while multiple class results are usually desired in remote sensing image classification applications; therefore, how to build a multiclass SVM classifier is an issue that we have to consider particularly.

12.3.1   CLASSIFICATION BASED ON SVM

A simple linearly separable case can illustrate how to use SVM as a binary classifier. Training data points are located in a two-dimensional data space (Figure 12.4). The

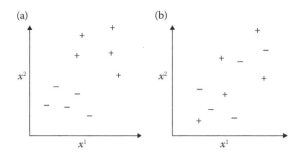

FIGURE 12.4 Two-dimension linearly separable (a) and linearly inseparable data spaces (b).

training data set can be expressed as $\{x_i, y_i\}$, $i = 1,2,\ldots, N$, where $x_i = \left(x_i^1, x_i^2\right)$, and $y_i \in \{-1,1\}$ are the two features and the corresponding class label of a data point in the data set. N is the size of the training data set. In the linearly separable case, these training data points can be separated by a line, which is formally named as a separating hyperplane because the dimensions of the data space could be much larger than two. A separating hyperplane can be defined as follows:

$$w \cdot x + b = 0, \tag{12.1}$$

in which the vector w represents the direction of the plane and the constant scalar b is the bias. When the data points are linearly separable, there are many options to define the separating hyperplane.

An SVM classifier seeks an optimal separating hyperplane (OSH) by which the training data points belonging to different classes can be separated by the largest margin; therefore, the training of an SVM classifier is a constraint optimization problem that can be expressed as

$$\min \frac{\|w\|^2}{2} \tag{12.2}$$

$$st\, y_i \, (w \cdot x_i + b) \geq 1, \, \forall i = 1,2,\ldots,N. \tag{12.3}$$

The minimization of $\|w\|$ is deduced from the maximization of the separating margin and represents the conception of structural risk minimization, which is why SVM yields a better generalization performance than other classic classifiers.

To solve this constraint optimization problem, we introduce the Lagrange function:

$$\min J(w,b,a) = \frac{1}{2} w^T w - \sum_{i=1}^{N} a_i \left[y_i(w \cdot x_i + b) - 1 \right], \tag{12.4}$$

where $a = (a_1, a_2, \ldots, a_N)$ and the nonnegative constant a_i is named the Lagrange multiplier. This optimization problem can be solved from its dual problem (Boyd and Vandenberghe 2004). If we denote the optimal solution as w_0 and b_0, then the prediction process for a new data point x' using the well trained SVM classifier can be expressed as

$$y = \text{sgn } (w_0 \cdot x + b_0),\qquad\qquad(12.5)$$

where $y \in \{-1,1\}$ is the class label of the data point x.

Commonly, the training data points are not linearly separable. In this case, we need to project these data points into a higher dimension data space in which they will become separable. This mapping can be done by introducing so-called kernel functions, which can be defined as

$$k(x, x_i) = \phi(x) \, \phi(x_i)\qquad\qquad(12.6)$$

in which $\phi(x)$ represents the mapping from the original data space to the linearly separable data space.

Using this mapping, the optimization problem behind the training process of the SVM classifier can be expressed as

$$\min\left[\frac{\| w \| 2}{2} + C\sum_{i=1}^{N} \xi_i \right]\qquad\qquad(12.7)$$

$$sty_i \, (w \cdot \phi(x_i) + b) > 1 - \xi_i, \forall i = 1,2,\ldots,N.\qquad\qquad(12.8)$$

In Equation 12.8, ξ_i is the relaxation term indicating that some data points lying on the wrong side of the OSH are acceptable. However, this situation will be controlled by the penalty term $C\sum_{i=1}^{N} \xi_i$ in Equation 12.7. The optimization problem given by these two equations can be solved following a similar procedure as the linearly separable case.

Mercer's theorem (Vapnik 2000) guarantees the existence of such a kernel function $k(x, x)$, and hence the existence of mapping $\phi(x)$. There are many kinds of kernels (Table 12.2). In our research, a radial basis function (also called the Gaussian kernel) is selected to be the kernel of the classifier.

TABLE 12.2

Some Types of Kernel Functions Commonly Used in SVM

Linear kernel	$k(x, x_i) = x \cdot x_i$
Polynomial kernel	$k(x, x_i) = (x \cdot x_i + 1)^d, d \in N$
Radial basis function	$k(x, x_i) = e^{-r\|x, xi\|2}, \gamma > 0$
Sigmoid function	$k(x, x_i) = \tanh (\gamma (x \cdot x_i) + c)$

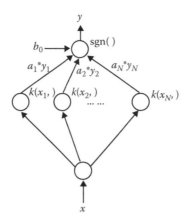

FIGURE 12.5   The network describing the SVM classification process.

SVM is developed from the statistic theory, so by nature it should be categorized as an advanced statistic learning method rather than a neural network; however, we can still describe the classification process of an SVM classifier using a three-layer network (Figure 12.5). The first and the third layer are the input and output layer, respectively; the middle layer consists of N nodes, where N is the size of the training data set. The output of an SVM classifier (Figure 12.5) can be expressed as

$$y = \text{sgn}\left(\sum_{i=1}^{N} a_i^* y_i k(x_i, x) + b_0 \right) \tag{12.9}$$

in which a_i^* is the optimal solution of the Lagrange multiplier.

12.3.2   MULTICLASS PROBLEM

The basic theory of SVM can only explain how to use SVM as a binary classifier; however, in more realistic applications, including remote sensing image classification, we often have to address problems involving $K > 2$ classes. To solve these multiclass problems, various methods have been proposed for combining multiple basic SVMs to build a multiclass classifier (Mathur and Foody 2008). The two most important approaches for combining multiple SVMs are the "one against all" approach (Bottou et al. 1994) and the "one against one" approach (Knerr et al. 1990).

   In the "one against all" approach, K binary SVM classifiers are trained to separate each class from the other K-1 classes. After the training process, K sets of optimized parameters $\{w_i, b_i\}$ $(i = 1,2,\ldots,K)$ will be obtained. For an unlabeled data point x, K predictions will be made by the K different binary classifiers following

$$Ny_i \, w_i \cdot x + b_i, \, i = 1,2,\ldots,K. \tag{12.10}$$

In Equation 12.10, $y_i > 0$ suggests that the data point x may belong to the ith class, and the larger the y_i, the higher the confidence. Finally, the data point will be allocated to the class i for which the largest decision value y_i was determined. This can be expressed as

$$y_i = \arg_i \max\{w_i x + b_i, i = 1,2,\ldots,K\}, \tag{12.11}$$

where y stands for the class label of the data point x given by the multiclass SVM classifier.

In the "one-against-one" approach, a set of classifiers are trained to separate each pair of classes; therefore, this approach requires the training of $K(K - 1)/2$ binary classifiers using data from every pair of classes. For an unlabeled data point x, predictions will be made by the $K(K - 1)/2$ well-trained classifier. Each binary classifier will give a "vote" to the class it chose from a pair. Finally, the class with the maximum number of votes will be assigned to the data point as its final label.

Other approaches can be used to set up a multiclass SVM classifier, such as DAGSVM (Platt et al. 2000), M-ary SVM (Sebald and Bucklew 2001), and ECC-SVM (Dietterich and Bakiri 1995). In our research, the "one against all" approach is applied because it is simple and effective enough to meet our demand on both classification accuracy and processing speed.

12.3.3 PARAMETER OPTIMIZATION

Previous studies show that SVM generally outperforms traditional statistical methods and neural networks in a wide range of classification problems. However, the good performances of SVM are heavily based on the optimization of its parameters (Bazi and Melgani 2006). An SVM classifier without optimized parameters can produce rather poor predictions. The most important parameters in SVM are γ and C (Foody and Mathur 2004a,b); C is the regularization parameter in Equation 12.7 that controls the penalty assigned to the classification errors, and γ is a parameter inversely proportional to the width of the Gaussian kernel. The accuracy of the classification results produced by an SVM is dependent on the magnitudes of the parameter γ and C, so the major focus will always be on these two parameters during the parameter optimization procedure of an SVM classifier.

The parameter optimization is implemented by applying cross-validation. The training data set is divided into several parts to train the classifier and validate it at the same time, and the optimization is aimed at achieving the best accuracy during the validation procedure. So iterative training and validating are involved to search for the optimal parameters, which makes the optimization a time-consuming process. Many advanced optimization approaches, such as GA and particle swarm optimization (PSO), can be used to implement the parameter optimization. In our research, we adopted a basic grid search approach; the optimization was proved effective by the classification accuracy, and the time cost of the process was acceptable. The SVM classifier we used was implemented based on the LibSVM toolbox (Chang and Lin 2001).

12.4 COMPARATIVE CLASSIFIERS

In addition to SVM, Naïve Bayes and decision tree methods were also employed in our research as comparative classifiers. While very popular in solving remote sensing image classification problems, the theorems of Naïve Bayes and decision tree are quite different from SVM.

12.4.1 THE NAÏVE BAYES CLASSIFIER

The Naïve Bayes classifiers have been widely used in remote sensing image classification problems. As proposed by Wu et al. (2008), the Naïve Bayes classifier is popular for several reasons. It is easy to construct because it does not require complicated iterative parameter estimation schemes and can be readily applied to huge data sets. It is easy to interpret, so users unskilled in classifier technology can understand its classification choices. Finally, it is surprisingly accurate. While it may not be the best possible classifier in any particular application, it can usually be relied on to be robust and to perform well.

The classification of a data point x by the Naïve Bayes classifier is achieved by calculating the probability of the data point x belonging to a certain class j ($j = 1,2,...,K$), namely, $P(j|x)$. A fundamental assumption of Naïve Bayes is that the feature values of a data point are conditionally independent given the class label of the data point.

The Naïve Bayes classifier assigns the unclassified data point with the class label j, which produces the largest posterior probability $P(j|x)$. To calculate the posterior probability of the data point x belonging to the class j, Bayes formula is used:

$$P(j \mid x) \propto P(j, x) = P(x \mid j)P(j), \tag{12.12}$$

where $P(j)$ is the *a priori* probability of data point belonging to class j, and $P(x|j)$ is the conditional probability of a data point within the class j taking the feature values of $x = (x_1, x_2, ..., x_p)$.

The priori probability $P(j)$ and the class-conditional probability $P(x|j)$ can be estimated from the training data set. The priori probability is simply equaled to the portion of samples within the training data set taking the class label j; therefore,

$$P(j) = \frac{N_j}{N}, \quad j = 1, 2, ... K, \tag{12.13}$$

where N is the number of the training dataset and N_j is the number of training samples with class label j. The estimation of the conditional probability $P(x|j)$ is based on the simplifying assumption that the feature values are conditionally independent given the class label. This conditional independence can be illustrated using a probabilistic graph model (Figure 12.6). According to this conditional independence, the multivariable probability can be decomposed as

$$P(x|j) = \prod_{i=1}^{p} P(x_i|j), \quad j = 1, 2, ..., K. \tag{12.14}$$

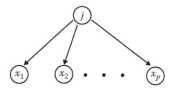

FIGURE 12.6 Probabilistic graph model describing the conditionally independence of features $x_1, x_2, \ldots, x_p$, given the class j.

The Naïve Bayes classifier can handle both discrete and continuous variables. If x_i is a discrete variable taking a finite number of possible values, then the estimation of $P(x_i|j)$ can easily be achieved by simply counting the proportion of training samples belonging to class j and taking the value of x_i for the ith feature. The estimation can be expressed as

$$P\left(x_i|j\right) = \frac{N_j(x_i)}{N_j}, \quad i = 1, 2, \ldots, p; \quad j = 1, 2, \ldots, K, \quad (12.15)$$

where $N_j(x_i)$ is the number of samples within the training dataset that take the value of x_i for the ith feature.

If x_i is a continuous variable, then the most direct strategy is discretization, and the more sophisticated approach is to assume a normal distribution for the decomposed conditional probabilities:

$$P\left(x_i|j\right) = N\left(x_i, \mu_j, \sigma_j^2\right), \quad i = 1, 2, \ldots, p; \quad j = 1, 2, \ldots, K, \quad (12.16)$$

where $N\left(x_i, \mu j, \sigma_j^2\right)$ is the conditional probability of the ith feature taking the form of a normal distribution. The parameters μ_j and σ_j can be estimated previously from the training data set; therefore, the training process is separated from the predicting process, while using the discretization approach the training process is implicitly carried on during the classification of every unclassified data point.

12.4.2 THE DECISION TREE CLASSIFIER

Rather than a specific kind of classification method, decision tree (DT) refers to a group of statistic techniques that recursively partition a data set into smaller subdivisions following a tree structure. Decision tree classifiers are popular in a wide range of pattern recognition problems including radar signal classification, character recognition, medical diagnosis, speech recognition, and remote sensing image classification (Safavian and Landgrebe 1991). An attractive feature of decision tree methods is the capability to break down a complex decision-making process into a collection of simpler decisions, thus providing a solution that is often easier to interpret.

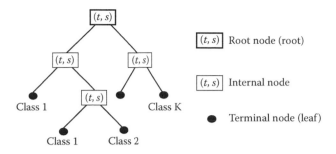

FIGURE 12.7   The structure of a decision tree (a binary tree).

Decision tree classifiers are very effective when dealing with complex classification problems because of their flexibility, intuitive simplicity, and computational efficiency. As argued by Friedl and Brodley (1997), decision tree classifiers have several advantages over traditional supervised classification methods such as maximum likelihood classification. In particular, decision tree classifiers are strictly nonparametric and do not require assumptions regarding the distributions of the input data. In addition, they handle nonlinear relations between features and classes, allow for missing values, and are capable of handling both numeric and categorical inputs in a natural fashion. Finally, decision trees have significant intuitive appeal because the classification is interpretable.

A decision tree is composed of a root node, a set of internal nodes, and a set of terminal nodes, which are also called leaves (Figure 12.7). Each node in a decision tree except the root has one parent node, and each node except the leaves has two or more descendant nodes. The root node of a decision tree corresponds to the entire data space. The leaves correspond to subspaces, which are not further partitioned and will be assigned with certain class labels. The class label of each leaf is that of the majority class in the leaf.

The splits defined at each internal node and the root node of a decision tree are estimated from the training data set by using a specific kind of learning algorithm. There are different kinds of learning algorithms for decision tree methods; we used the classification and regression tree (CART) algorithm, a binary recursive partitioning procedure (Waheed et al. 2006), in our research. In a binary tree, each node has two descendants, and the splitting rule of a node can be expressed as follows: "If a condition is true, the data point goes to the left descendant; otherwise, the data point goes to the right descendant."

In the CART algorithm, the splitting rule of a node is defined to achieve the best homogeneity of the training data in the resulting descendant nodes. Therefore, the goodness of split is measured by an impurity function defined for each node, and in most cases, the Gini diversity index will be a good choice for the impurity function. If a node is denoted as t, then the impurity measurement of the node t can be calculated as

$$i(t) = 1 - \sum_{j=1}^{K} p_j(t)^2 \qquad (12.17)$$

where $p_j(t)$ is the probability of class j within node t. This probability can be estimated from the number of training samples that go to node t and the number of training samples that go to nodes t with class label j;

$$p_j(t) = N_j(t) / N(t), \quad j = 1,2,\ldots,K. \tag{12.18}$$

Goodness of the split s for the node t is then defined by

$$\Delta i(s,t) = i(t) - p_R i(t_R) - p_L i(t_L), \tag{12.19}$$

where p_R and p_L are the proportions of the samples in node t that go to the right descendant t_R and the left descendant t_L, respectively. Goodness of the split s should be maximized to achieve the largest decrease of impurity during every step of split and eventually achieve greatest purity in the leaves of a tree.

The split process goes on until a certain stopping criterion is met. In the CART algorithm, the stopping criterion is defined by

$$\max_s \Delta i(s,t) < \beta, \tag{12.20}$$

where β is a chosen threshold. When the split process stops, the growth of the tree stops, and the training process of the decision tree classifier is complete.

Following the splitting rules of the decision tree, unclassified data points start from the root, go through some internal nodes, and finally arrive at a leaf where a class label is assigned to the data point according to the leaf into which the data point falls.

12.5 CLASSIFICATION STRATEGIES

12.5.1 CLASSIFICATION BASED ON MULTIPLE FEATURES

The false color SPOT-5 images contain only three channels, and the synthesized blue channel in the true color images is merely a linear combination of the original green, red, and near-infrared channels. Theoretically speaking, data points corresponding to these image pixels are in a three-dimensional feature space. Large numbers of data points in low-level feature spaces tend to cluster, making them more difficult to distinguish and correctly classify.

To add more features to the pixels, we introduced the normalized difference vegetation index (NDVI) and four commonly used texture features including angular second moment, contrast, correlation, and homogeneity (Sklansky 1978). Therefore, the dimension of the feature space for all the data points and pixels is extended to eight.

The NDVI can be calculated from the red and near-infrared channels of the SPOT-5 images as follows:

$$NDVI = \frac{NIR - RED}{NIR + RED}, \tag{12.21}$$

where *NIR* and *RED* denote the digital numbers (DNs) of the near-infrared channel and red channel images, respectively. The exact definition for NDVI is based on the spectral reflectance rather than the DNs; hence, Equation 12.12 is a simplification for the calculation of NDVI. This simplification is not compatible for quantitative assessments but will not result in negative consequences in a classification application.

During the classification procedures, different LULC often exhibit similar spectral patterns, especially for those images with low spectral resolutions (large spectral span and few spectral bands). However, these spectrally similar objects can be distinguished by their different texture patterns; therefore, texture features, usually as a complement to the spectral features, are functional in dealing with image classification in many cases (Jobanputra and Clausi 2006). In our study, the four kinds of texture features are calculated from a 9×9 gray level co-occurrence matrix (GLCM) with respect to every pixel.

12.5.2 CLASSIFICATION BASED ON SUBCLASSES

High-resolution remote sensing images of urban regions tend to show complex and diverse spectral characteristics, and this complexity and diversity can be further enlarged by including vegetation indexes and texture features into the image data sets. Based on their naturally diverse character, the data points (pixels) in an image data set can be divided into numerous categories that do not necessarily have clear meanings. In contrast, categories in a thematic map are defined according to human understanding.

The most similar data points should not always be divided into the same human-defined category, and the inner category diversity can be so large that it can "confuse" the classifier. For example, in terms of distances in the feature space, two data points belonging to different LULC types may show more similarity than two data points of the same LULC type.

To cope with this problem, a subclass strategy is applied. Data points in the image data set are first categorized as subclasses, a process based on the natural character of the data points. The subclasses are then summarized into fewer human-defined classes.

12.6 FURTHER EXPERIMENT

12.6.1 CLASSIFICATION STEPS

The proposed classification approach follows four major steps (Figure 12.8). First, additional features including NDVI and texture features are extracted from the original remote sensing images to construct multifeature image sets. Second, training data sets and testing data sets are defined and prepared based on the ground truth data. Third, the classifier is trained using the training data sets and evaluated using the testing data sets. Fourth, the full image sets are imported into the well-trained classifier to produce the final classification images of the study region.

The extraction of additional features previously described in detail supports both the training and testing data sets. Given that the type of LULC was divided into six

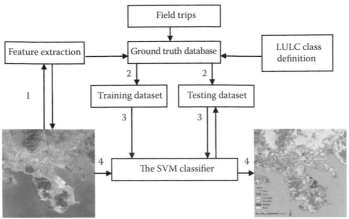

Major experimental steps:

1. Extract multiple features from the original remote sensing images;
2. Construct the training dataset and testing dataset based on the ground truth database;
3. Train the classifier with the training dataset, then test its performance with the testing dataset;
4. Classify the full scale image of the study area using the SVM classifier.

FIGURE 12.8 Flowchart of the proposed classification approach.

major categories (water bodies, forests, grasslands, bare fields, buildings, and roads), multitemporal variations of grasslands due to urban sprawl, water coverage due to coastal land reclamation, and buildings due to population increase and migration would reflect the main direction and connection of the urbanization process, which are the major concerns of our study.

However, the better the classifier is, the lower the odds of confusing similar objects within the same category. A major challenge is that remarkable similarity and variety exist within some of these six LULC classes, and these variances may impact the final performance of the classifier. To deal with this issue, during the classification process, the six major classes of LULC were further divided into 14 subclasses of LULC (Table 12.3).

To improve the prediction accuracy, 500 data points from the ground truth data set associated with each subclass were sampled to build the training data set, and another 500 associated with each subclass were sampled to form the testing data set. The SVM classifier can then be trained using the training data set and the performance of the proposed classification approach may be evaluated further by using the testing data set. Comparisons with other classifiers could clarify how good the SVM classifier could be, comparatively speaking. Once the credibility of the SVM classifier is confirmed after the intercomparisons among all applicable classifiers, the well-trained classifier can be applied to classify the full-scale remote sensing images of the study area. Once the task is completed, direct outputs of the classifier (expressed as 14 subclasses) can be reorganized into the six major classes of LULC for the final assessment of classification accuracy and final presentation of all thematic maps accordingly.

TABLE 12.3

Basic LULC Classes and Subclasses

Basic LULC Type	Subtype	Explanation
Water body	Water 1	Clean water
	Water 2	Turbid water
Forest	Forest 1	Dark green trees
	Forest 2	Yellow green trees
	Forest 3	Sparse forest
Grassland	Grass 1	Light green grass
	Grass 2	Dark green grass
Bare field	Bare field	No subclasses
Building	Building 1	Buildings in residential areas
	Building 2	Factory buildings with blue roofs
	Building 3	Factory buildings with orange roofs
	Building 4	Factory buildings with gray roofs
	Building 5	Factory buildings with white roofs
Road	Road	No subclasses

Using the confusion matrix, the accuracy of classification was first assessed with regard to the overall accuracy (OA), user's accuracy (UA), and producer's accuracy (PA) (Congalton 1991). The ultimate accuracy of classification can be carried out using Kappa statistics (Monserud and Leemans 1992) based on the confusion matrix. The Kappa coefficient is defined based on the confusion matrix as follows:

$$Kappa = \frac{N \sum_{k=1}^{r} x_{kk} - \sum_{k=1}^{r} \left(x_{k+} \times x_{+k} \right)}{N^2 - \sum_{k-1}^{r} \left(x_{k+} \times x_{+k} \right)}, \quad (12.22)$$

where r is the number of rows in the confusion matrix, x_{kk} is the number of observations in row i and column j, x_{k+} and x_{+k} are the marginal totals for row i and column j, respectively, and N is the total number of observations.

12.6.2 CLASSIFICATION OUTCOME

Predictions are generated by the SVM classifier, which is designed to sort out 14 subclasses of LULC. Aggregation of land use subclasses into major categories can then be done by following a hierarchical order (Table 12.3). The classification accuracy of the SVM classifier was evaluated based on the testing dataset of 2007 for the purpose of demonstration. According to the results (Table 12.4), the proposed classification approach is very effective, although the classification results are not equally good for all the LULC classes. The good performance of the proposed approach should be ascribed to the superiority of SVM itself, while the classification strategies that we developed in this study also contribute extensively.

TABLE 12.4
Evaluation of the SVM Classifier

		Predictions						PA
		Water Body	**Forest**	**Grassland**	**Bare Field**	**Building**	**Road**	**(%)**
Actual	Water body	995	2	0	0	3	0	99.5
conditions	Forest	0	1445	46	0	9	0	96.3
	Grassland	0	63	936	0	1	0	93.6
	Bare field	0	0.0	1	456	35	8	91.2
	Building	1	16	3	26	2398	56	95.9
	Road	0	0.0	0	3	44	453	90.6
	UA (%)	99.9	94.7	94.9	94.0	96.3	87.6	
OA (%)					95.6			
Kappa coefficient					0.9416			

To validate the effectiveness of the two classification strategies, we repeated the classification experiments with the same SVM classifier following different classification strategies. During the first test, the training and testing process were implemented based only on the basic spectral features of the remote sensing image data. During the second test, NDVI was then added, but the textural features were still left out of consideration. By comparing the results of these tests (Tables 12.5 and 12.6), we can confirm that the additional features improve the classification results (Tables 12.4 through 12.6). In particular, the involvement of NDVI raises the OA of the classification results by 6.6%.

During the third test, the subclass strategy was given up. The training of the classifier was based on the six LULC classes, and the well-trained classifier made

TABLE 12.5
Evaluation of the Classification Based Only on Spectral Features

		Predictions						PA
		Water Body	**Forest**	**Grassland**	**Bare Field**	**Building**	**Road**	**(%)**
Actual	Water body	996	3	0	0	1	0	99.6
conditions	Forest	12	1320	133	0	27	8	88.0
	Grassland	0	132	859	0	1	8	85.9
	Bare field	0	0	17	393	52	38	78.6
	Building	21	164	34	49	2064	168	82.6
	Road	0	8	24	11	105	352	70.4
	UA (%)	96.8	81.1	80.5	86.8	91.7	61.3	
OA (%)					85.5			
Kappa coefficient					0.8144			

TABLE 12.6
Evaluation of the Classification Based on NDVI and Spectral Features

		Predictions						
		Water Body	Forest	Grassland	Bare Field	Building	Road	PA (%)
Actual conditions	Water body	996	2	0	0	2	0	99.6
	Forest	1	1407	69	0	23	0	93.8
	Grassland	0	91	907	0	2	0	90.7
	Bare field	0	0	1	411	61	27	82.2
	Building	3	32	11	52	2310	92	92.4
	Road	0	0	5	11	69	415	83.0
	UA (%)	99.6	91.8	91.3	86.7	93.6	77.7	
OA (%)					92.1			
Kappa coefficient					0.8980			

predictions labeled as the six basic LULC classes directly. The OA and Kappa coefficients (Table 12.7) are lower than those reported for classification accuracies with respect to different classifiers (Table 12.4), so we concluded that the subclass strategy was also effective.

As discussed in Section 12.3.3, parameter optimization is critical to SVM. There is no such general guide for the selection of the most proper parameters in the SVM classifier, and the parameters should be optimized to fit the distribution characteristics of the specific data set. Statistics of OA, PA, and Kappa coefficient reported in Table 12.8 are based on the classification of an unoptimized SVM classifier. Obvious deterioration from the results reported in Table 12.4 can be observed, and these results can be considered as evidence for the necessity of the parameter

TABLE 12.7
Evaluation of the Classification Based Directly on the Six Basic LULC Classes

		Predictions						
		Water Body	Forest	Grassland	Bare Field	Building	Road	PA (%)
Actual conditions	Water body	997	2	0	0	1	0	99.7
	Forest	0	1433	54	0	12	1	95.5
	Grassland	0	60	937	0	3	0	93.7
	Bare field	0	0	1	442	48	9	88.4
	Building	1	17	3	35	2388	56	95.5
	Road	0	0	0	3	66	431	86.2
	UA (%)	99.9	94.8	94.2	92.1	94.8	86.7	
OA (%)					94.7			
Kappa coefficient					0.9314			

TABLE 12.8

Evaluation of the Unoptimized SVM Classifier

		Predictions						
		Water Body	Forest	Grassland	Bare Field	Building	Road	PA (%)
Actual conditions	Water body	993	2	0	0	5	0	99.3
	Forest	0	1342	114	2	42	0	89.5
	Grassland	0	128	861	0	11	0	86.1
	Bare field	0	0	4	396	81	19	79.2
	Building	5	22	15	59	2227	172	89.1
	Road	0	0	0	12	115	373	74.6
	UA (%)	99.5	89.8	86.6	84.4	89.8	66.1	
OA (%)					88.5			
Kappa coefficient					0.8513			

optimization procedure. As we developed in this study, a threefold cross-validation procedure was implemented to select the optimal parameters for the SVM classifier. In the classifier with no optimization, default values are assigned to these parameters. The defaults for the parameter γ and C are 1/8 and 1, respectively, while the optimized values are 2 and 32. Thus, it can be concluded that to achieve higher classification accuracy, a large penalty term and a narrow Gaussian kernel were preferred in our research.

The performance of the SVM classifier can be grossly compared with other representative supervised classifiers, such as radial basis function neural network (RBF), Naïve Bayes and decision tree, using the same training and testing data sets. Within these classifiers selected, RBF is another very important kernel method in addition to SVM, which is also a classic neural computing technique. The RBF classifier we used was implemented based on the Netlab toolbox (Nabney 2002). A comparison between SVM and RBF shows that SVM is a better kernel method for our classification task. NB is a representative probabilistic method based on the Bayesian theorem, which can often outperform other more sophisticated classification methods despite its simplicity (Domingos and Pazzani 1997). The decision tree classifier is established using the CART algorithm. Both the Naïve Bayes classifier and the decision tree classifier are implemented based on the statistic toolbox with Matlab (2011).

The classification accuracies (PA) for each major type of LULC were then compared with one another leading to the generation of overall accuracies and the Kappa coefficients. The SVM classifier achieved the best overall accuracy and the Kappa coefficient based on the testing data set (Table 12.9). When investigating the details, the SVM classifier obviously outperformed the other classifiers in almost every LULC type, while the biggest advantage of the SVM classifier was that it can distinguish road from the other LULC types much more effectively.

TABLE 12.9

Comparisons of Classification Accuracies with Respect to Different Classifiers

	SVM	RBF	NB	DT
Water body	99.5	99.4	98.6	99.2
Forest	96.3	89.9	85.3	91.5
Grassland	93.6	76.1	87.6	87.0
Bare field	91.2	75.6	70.4	81.2
Building	95.9	82.3	83.9	89.8
Road	90.6	61.0	67.8	74.2
OA (%)	95.6	83.5	84.7	89.4
Kappa	0.9416	0.7877	0.8045	0.8632

Note: RBF = radial basis function neural network model; NB = Naïve Bayes; DT = decision tree.

Comparisons of processing speeds between the SVM classifier and other classifiers were also made possible in terms of training and testing time periods. The SVM, RBF, Naïve Bayes, and decision tree classifiers can be all implemented in Matlab (2011). Experiments for these classifiers were carried out in Windows XP with Pentium E2180 2.00GHz Dual CPU and 2.00GHz DDR800 RAM; therefore, all these methods are comparable in these speed campaigns.

Using the training and testing data sets with 500 data points for each subclass of LULC, the training and testing times required for all classifiers were recorded and reported (Table 12.10). When compared with one another, the RBF classifier worked by an inefficient way during both the training and the testing processes. The training speed of the SVM was faster than the training speed of RBF, but the testing speed of the SVM could not outperform the testing speed of RBF. The processing speed of the NB and the DT classifiers is extremely fast compared to the SVM classifier and the RBF classifier, so generally speaking, SVM has no superiority in processing speed over the other classifiers. The slower processing speed of SVM can be considered

TABLE 12.10

Comparisons of Processing Speeds with Different Classifiers

	SVM	RBF	NB	DT
Training time (s)	1.236010	5.507882	0.030319	0.517039
Testing time (s)	2.912904	1.014806	0.044070	0.008494
Total time (s)	4.148914	6.511688	0.074389	0.525543

Note: RBF = radial basis function neural network model; NB = Naïve Bayes; DT = decision tree.

a sacrifice for higher classification accuracy, and the speed of the SVM classifier is acceptable for our image classification tasks.

The image of a subregion of the study area was segmented out from the 2007 SPOT-5 image for the purpose of illustration (Figure 12.9). This subregion image corresponds to the neighborhood of a small campus and was classified using the RBF, NB, DT, and SVM classifiers. The results can visually reveal a comparative analysis (Figure 12.9). The numbered circles on these LULC maps highlight the location where SVM classification shows relatively better quality. For example, the roads in circle 1 were misidentified as bare field using the RBF classifier, whereas they were correctly classified by the SVM classification approach.

The classification results illustrating the LULC conditions of DDA in 2003 and 2007, respectively, can be shown in Figure 12.10 using the SVM classifier. Table 12.11 summarizes the LUCC over the 5 years. These statistics are derived from the classification results as illustrated in Figure 12.10 with respect to the LUCC across the whole study area. The land reclamation and urban expansion processes can be verified by these statistics, while such LUCC can be easily captured based on visual comparisons. From 2003 to 2007, the area of water body had decreased by 12.63 km^2 as a result of land reclamations. In comparison, the building coverage in the region had increased by 8.85 km^2, which is strong evidence for the rapid urban expansion process. The LUCC from 2003 to 2007 can then be summarized in Table 12.11. The

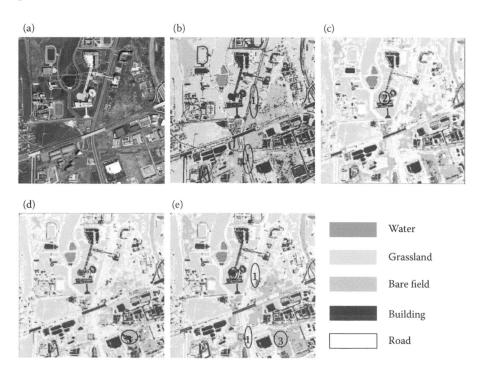

FIGURE 12.9 (a) The original image of a subregion and classification results produced using (b) the RBF classifier, (c) the NB classifier, (d) the DT classifier, (e) the SVM classifier, respectively.

(a)   (b)

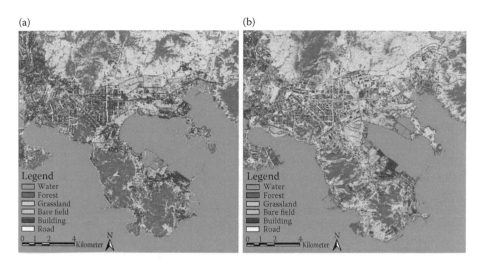

FIGURE 12.10   Final classification results produced from (a) the 2003 image and (b) the 2007 image using the SVM classifier.

statistics in Table 12.11 also imply that severe deforestation occurred during this time period.

The rapid increases of brown-colored areas (buildings) indicate a rapid urbanization process due to population growth and migration during a 4-year time period. The exorbitant decreases of water coverage due to the coastal land reclamation are a symbolic pattern of the urbanization process in DDA driven by the land management policy. In all circumstances, the drastic reduction of grassland during such a short time period implies a loss of ecosystem service in this urban region. Impermeable pavement due to urbanization resulted in a distorted hydrological cycle. At the same time, the high rate of the coastal land reclamation that improves the connectivity of the coastal industrial complex is also a cause of concern of seawater quality and soil salinity. However, the positive aspects of urbanization leading to regional economic development should not be overshadowed by deterioration in the physical environment and quality of life in the region as long as there is no widening gap between supply and demand for public infrastructures.

TABLE 12.11
Statistics of the LUCC in the Area from 2003 to 2007

Category	2003 (km²)	2007 (km²)	Changes (km²)
Water body	180.6278	167.9903	−12.6375
Forest	72.0060	61.6042	−10.4018
Grassland	44.5262	32.0620	−12.4642
Bare field	20.5095	44.5907	24.0812
Building	66.5971	75.4500	8.8529
Road	15.7331	18.3028	2.5697

12.7 FINAL REMARKS

In our current research, additional features including NDVI and textural features were utilized. In future work, data from other sources such as a digital elevation model (DEM) could also be integrated into the data set to improve the LULC classification and further the LULC change detection originally based only on multitemporal high-resolution remote sensing images. The prospects are good for including high-quality DEM data such as TanDEM-X (Krieger et al. 2007) in an urban LULC classification application. The DEM data produced by TanDEM-X can provide a vertical accuracy of 2 m (relative) and 10 m (absolute). These levels of elevation data will help distinguish between roads and buildings, which tend to show similar spectral and even textural features in the urban images.

A rapid change detection platform can therefore be established by integrating multitemporal high-resolution remote sensing images with DEM data and using these compound data in a fast intelligent classifier like the PL-ELM classifier. Such a platform for detecting rapid changes in urban regions cannot only be applied in urban dynamic monitoring, but it also shows a great potential for disaster monitoring, damage assessment, and emergency response. A powerful application of remote sensing technology is to conduct the rapid change detection immediately following a natural disaster, such as an earthquake (Ehrlich et al. 2009; Cao et al. 2010), tsunami (Tanathong et al. 2008), hurricane (Wang and Xu 2010), or landslide (Voigt et al. 2007; Xu et al. 2010). The damage caused by a disaster can then be assessed based on the multitemporal remote sensing images (Rau et al. 2007; Yamazaki and Matsuoka 2007), and emergency responses can then be made efficiently and accordingly (Chuan 2005).

12.8 CONCLUSIONS

Remote sensing plays an important role in urban environmental and hydrologic studies such as dynamic monitoring of urban sprawl and LUCC analyses. In our study, a novel LULC classification approach was developed based on a newly developed machine-learning algorithm, the SVM. The proposed approach was applied to two SPOT-5 image sets acquired in 2003 and 2007 to produce LULC maps based on six-category with 2.5 m resolution. To improve the prediction accuracy, additional features were extracted from the original remote sensing images and taken into account during the classification procedures. The classification accuracy of the proposed approach was assessed against the testing data set, which consisted of 500 ground truth data points, and the performance of the SVM classifier was fully verified in these computational experiments. Rapid urbanization processes, especially the coastal land reclamation process, can be recognized from the distinct changes in the two LULC maps.

REFERENCES

Agnelli, D., Bollini, A., and Lombardi, L. (2002). Image classification: An evolutionary approach. *Pattern Recognition Letters*, 23(1–3), 303–309.
Awad, M., Chehdi, K., and Nasri, A. (2007). Multicomponent image segmentation using a genetic algorithm and artificial neural network. *Geoscience and Remote Sensing Letters, IEEE*, 4(4), 571–575.

Bazi, Y. and Melgani, F. (2006). Toward an optimal SVM classification system for hyperspectral remote sensing images. *IEEE Transactions on Geoscience and Remote Sensing*, 44(11), 3374–3385.

Blanzieri, E. and Melgani, F. (2008). Nearest neighbor classification of remote sensing images with the maximal margin principle. *IEEE Transactions on Geoscience and Remote Sensing*, 46(6), 1804–1811.

Bottou, L., Cortes, C., Denker, J. S., Drucker, H., Guyon, I., Jackel, L. D., LeCun, Y., et al. (1994). Comparison of Classifier Methods: A Case Study in Handwritten Digit Recognition. *Pattern Recognition*, 1994. Vol. 2—*Proceedings of the 12th IAPR International Conference on Computer Vision & Image Processing*, Jerusalem, Israel, 77–82.

Bovolo, F., Bruzzone, L., and Carlin, L. (2010). A novel technique for subpixel image classification based on support vector machine. *IEEE Transactions on Image Processing*, 19(11), 2983–2999.

Boyd, S. P. and Vandenberghe, L. (2004). *Convex Optimization*, Cambridge University Press, Cambridge, United Kingdom.

Canty, M. J. (2009). Boosting a fast neural network for supervised land cover classification. *Computers & Geosciences*, 35(6), 1280–1295.

Cao, C., Chang, C., Xu, M., Zhao, J., Gao, M., Zhang, H., Guo, J. P. et al. (2010). Epidemic risk analysis after the Wenchuan Earthquake using remote sensing. *International Journal of Remote Sensing*, 31(13), 3631–3642.

Chang, N. B., Daranpob, A., Yang, J. and Jin, K. R. (2009). A comparative data mining analysis for information retrieval of MODIS images: Monitoring lake turbidity changes at Lake Okeechobee, Florida (2009). *Journal of Applied Remote Sensing*, 3, 033549, doi:10.1117/1.3244644.

Chang, C. and Lin, C. (2001). LIBSVM: A Library for Support Vector Machines. Technical Report, Department of Computer Science, National Taiwan University, Taipei, Taiwan, http://www.csie.ntu.edu.tw/~cjlin/papers/libsvm.pdf.

Chen, H. W., Chang, N. B., Yu, R. F., and Huang, Y. W. (2009). Urban land use and land cover classification using the neural-fuzzy inference approach with Formosat-2 data. *Journal of Applied Remote Sensing*, 3, 033558-033518, doi:10.1117/1.3265995.

Chuan, T. (2005). Approaches on emergency response system of abrupt geological hazard associated with urban areas. *The Chinese Journal of Geological Hazard and Control*, 3.

Congalton, R. G. (1991). A review of assessing the accuracy of classifications of remotely sensed data. *Remote Sensing of Environment*, 37(1), 35–46.

Courant, R. and Hilbert, D. (1954). Methods of mathematical physics: Vol. I. *Physics Today*, 7, 17.

Daamouche, A. and Melgani, F. (2009). Swarm intelligence approach to wavelet design for hyperspectral image classification. *IEEE Geoscience and Remote Sensing Letters*, 6(4), 825–829.

Dietterich, T. and Bakiri, G. (1995). Solving multiclass learning problems via error-correcting output codes. *Journal of Artificial Intelligence Research*, 2(1), 263–286.

Domingos, P. and Pazzani, M. (1997). On the optimality of the simple Bayesian classifier under zero–one loss. *Machine Learning*, 29(2), 103–130.

Ehrlich, D., Guo, H. D., Molch, K., Ma, J. W., and Pesaresi, M. (2009). Identifying damage caused by the 2008 Wenchuan earthquake from VHR remote sensing data. *International Journal of Digital Earth*, 2(4), 309–326.

Fan, J., Han, M., and Wang, J. (2009). Single point iterative weighted fuzzy C-means clustering algorithm for remote sensing image segmentation. *Pattern Recognition*, 42(11), 2527–2540.

Filippi, A. M. and Archibald, R. (2009). Support vector machine-based endmember extraction. *IEEE Transactions on Geoscience and Remote Sensing*, 47(3), 771–791.

Foody, G. M. and Mathur, A. (2004a). A relative evaluation of multiclass image classification by support vector machines. *IEEE Transactions on Geoscience and Remote Sensing*, 42(6), 1335–1343.

Foody, G. M. and Mathur, A. (2004b). Toward intelligent training of supervised image classifications: Directing training data acquisition for SVM classification. *Remote Sensing of Environment*, 93(1–2), 107–117.

Friedl, M. A. and Brodley, C. E. (1997). Decision tree classification of land cover from remotely sensed data. *Remote Sensing of Environment*, 61(3), 399–409.

Giacco, F., Thiel, C., Pugliese, L., Scarpetta, S., and Marinaro, M. (2010). Uncertainty analysis for the classification of multispectral satellite images using SVMs and SOMs. *IEEE Transactions on Geoscience and Remote Sensing*, 48(10), 3769–3779.

Giacinto, G. and Roli, F. (2001). Design of effective neural network ensembles for image classification purposes. *Image and Vision Computing*, 19(9–10), 699–707.

Han, M., Tang, X., and Cheng, L. (2004). An improved fuzzy ARTMAP network and its application in wetland classification. *IEEE International Geoscience and Remote Sensing Symposium Proceedings, 2004, IGARSS '04*, Anchorage, AK.

Heermann, P. D. and Khazenie, N. (1992). Classification of multispectral remote sensing data using a back-propagation neural network. *IEEE Transactions on Geoscience and Remote Sensing*, 30(1), 81–88.

Hoi-Ming, C. and Ersoy, O. K. (2005). A statistical self-organizing learning system for remote sensing classification. *IEEE Transactions on Geoscience and Remote Sensing*, 43(8), 1890–1900.

Huang, C., Davis, L. S., and Townshend, J. R. G. (2002). An assessment of support vector machines for land cover classification. *International Journal of Remote Sensing*, 23(4), 725–749.

Jian, Z., Zhanzhong, C., Liu, A., and Jia, Y. (2008). A K-Means Remote Sensing Image Classification Method Based on AdaBoost. *The Fourth International Conference on Natural Computation, 2008. ICNC '08*, Jinan, China.

Jobanputra, R. and Clausi, D. A. (2006). Preserving boundaries for image texture segmentation using grey level co-occurring probabilities. *Pattern Recognition*, 39(2), 234–245.

Kavzoglu, T. (2009). Increasing the accuracy of neural network classification using refined training data. *Environmental Modelling & Software*, 24(7), 850–858.

Knerr, S., Personnaz, L., and Dreyfus, G. (1990). Single-layer learning revisited: A stepwise procedure for building and training a neural network. *Optimization Methods and Software*, 1, 23–34.

Krieger, G., Fiedler, H., Zink, M., Hajnsek, I., Younis, M., Huber, S., Bachmann, M., Hueso-Gonzalez, J., Werner, M., and Moreira, A. (2007). TanDEM-X: A Satellite Formation for High-Resolution SAR Interferometry. *IEEE Transactions on Geoscience and Remote Sensing*, 45(11), 3317–3341.

Lombardo, P. and Oliver, C. J. (2001). Maximum likelihood approach to the detection of changes between multitemporal SAR images. *IEEE Proceedings of Radar, Sonar and Navigation*, 148(4), 200–210.

Makkeasorn, A., Chang, N. B., Beaman, M., Wyatt, C., and Slater, C. (2006). Soil moisture prediction in a semi-arid reservoir watershed using RADARSAT satellite images and genetic programming, *Water Resources Research*, 42, 1–15.

Makkeasorn, A. and Chang, N. B. (2009). Seasonal change detection of riparian zones with remote sensing images and genetic programming in a semi-arid watershed, *Journal of Environmental Management*, 90, 1069–1080.

Mathur, A. and Foody, G. M. (2008). Multiclass and binary SVM classification: Implications for training and classification users. *IEEE Geoscience and Remote Sensing Letters*, 5(2), 241–245.

Matlab (2011). http://www.mathworks.com/ (accessed June 2011).

Melgani, F. and Bruzzone, L. (2004). Classification of hyperspectral remote sensing images with support vector machines. *IEEE Transactions on Geoscience and Remote Sensing*, 42(8), 1778–1790.

Monserud, R. A. and Leemans, R. (1992). Comparing global vegetation maps with the Kappa statistic. *Ecological Modelling*, 62(4), 275–293.

Mūnoz-Marí, J., Bovolo, F., Gómez-Chova, L., Bruzzone, L., and Camp-Valls, G. (2010). Semisupervised one-class support vector machines for classification of remote sensing data. *IEEE Transactions on Geoscience and Remote Sensing*, 48(8), 3188–3197.

Nabney, I. (2002). *NETLAB: Algorithms for Pattern Recognition*, Springer Verlag, Germany.

Platt, J., Cristianini, N., and Shawe-Taylor, J. (2000). *Large Margin DAGS for Multiclass Classification*, MIT Press, Cambridge, MA, S. A. Solla, T. K. Leen and K.-R. Müller (eds.), 547–553.

Ratle, F., Camps-Valls, G., and Weston, J. (2010). Semisupervised neural networks for efficient hyperspectral image classification. *IEEE Transactions on Geoscience and Remote Sensing*, 48(5), 2271–2282.

Rau, J. Y. R., Chen, L. C., Liu, J. K., and Wu, T. H. (2007). Dynamics monitoring and disaster assessment for watershed management using time-series satellite images. *IEEE Transactions on Geoscience and Remote Sensing*, 45(6), 1641–1649.

Ross, B. J., Gualtieri, A. G., and Budkewitsch, P. (2005). Hyperspectral image analysis using genetic programming. *Applied Soft Computing*, 5(2), 147–156.

Safavian, S. R. and Landgrebe, D. (1991). A survey of decision tree classifier methodology. *IEEE Transactions on Systems, Man and Cybernetics*, 21(3), 660–674.

Sebald, D. J. and Bucklew, J. A. (2001). Support vector machines and the multiple hypothesis test problem. *IEEE Transactions on Signal Processing*, 49(11), 2865–2872.

Sklansky, J. (1978). Image segmentation and feature extraction. *IEEE Transactions on Systems, Man and Cybernetics*, 8(4), 237–247.

Stefanov, W. L., Ramsey, M. S., and Christensen, P. R. (2001). Monitoring urban land cover change: An expert system approach to land cover classification of semiarid to arid urban centers. *Remote Sensing of Environment*, 77(2), 173–185.

Suresh, S., Sundararajan, N., and Saratchandran, P. (2008). A sequential multi-category classifier using radial basis function networks. *Neurocomputing*, 71(7–9), 1345–1358.

Tanathong, S., Rudahl, K. T., and Goldin, S. E. (2008). Object Oriented Change Detection of Buildings after the Indian Ocean Tsunami Disaster. *The 5th International Conference on Electrical Engineering/Electronics, Computer, Telecommunications and Information Technology, ECTI-CON 2008*, Krabi, India.

Tarabalka, Y., Fauvel, M., Chanussot, J., and Benediktsson, J. A. (2010). SVM- and MRF-based method for accurate classification of hyperspectral images. *IEEE Geoscience and Remote Sensing Letters*, 7(4), 736–740.

Tso, B. C. K. and Mather, P. M. (1999). Classification of multisource remote sensing imagery using a genetic algorithm and Markov random fields. *IEEE Transactions on Geoscience and Remote Sensing*, 37(3), 1255–1260.

Vapnik, V. (2000). *The Nature of Statistical Learning Theory*, Springer Verlag, Germany.

Voigt, S., Kemper, T., Riedlinger, T., Kiefl, R., Scholte, K., and Mehl, H. (2007). Satellite image analysis for disaster and crisis-management support. *IEEE Transactions on Geoscience and Remote Sensing*, 45(6), 1520–1528.

Waheed, T., Bonnell, R. B., Prasher, S. O., and Paulet, E. (2006). Measuring performance in precision agriculture: CART-A decision tree approach. *Agricultural Water Management*, 84(1–2), 173–185.

Wang, F. and Xu, Y. (2010). Comparison of remote sensing change detection techniques for assessing hurricane damage to forests. *Environmental Monitoring and Assessment*, 162(1), 311–326.

Wu, X., Kumar, V., Quinlan, J. R., Ghosh, J., Yang, Q., Motoda, H., McLachlan, G. J. et al. (2008). Top 10 algorithms in data mining. *Knowledge and Information Systems*, 14(1), 1–37.

Xu, M., Cao, C., Zhang, H., Guo, J., Nakane, K., He, Q., Guo, J. et al. (2010). Change detection of an earthquake-induced barrier lake based on remote sensing image classification. *International Journal of Remote Sensing*, 31(13), 3521–3534.

Yamazaki, F. and Matsuoka, M. (2007). Remote sensing technologies in post-disaster damage assessment. *Journal of Earthquake and Tsunami*, 1(3), 193–210.

Yanfei, Z., Liangpei, Z., Gong, J., and Li, P. (2007). A supervised artificial immune classifier for remote-sensing imagery. *IEEE Transactions on Geoscience and Remote Sensing*, 45(12), 3957–3966.

13 Mapping Impervious Surface Distribution with the Integration of Landsat TM and QuickBird Images in a Complex Urban–Rural Frontier in Brazil

Dengsheng Lu, Emilio Moran,
Scott Hetrick, and Guiying Li

CONTENTS

13.1   INTRODUCTION

Impervious surfaces are generally defined as any man-made materials that water cannot infiltrate. They are primarily associated with human activities and habitation through the construction of transportation infrastructure and buildings (Slonecker et al. 2001; Bauer et al. 2008). Impervious surface area (ISA) has long been recognized as an important factor in many urban or environment related studies, including the improvement of urban land use and land cover (LULC) classification (Madhavan et al. 2001; Phinn et al. 2002; Lu and Weng 2006a), residential population estimation (Wu and Murray 2005; Lu et al. 2006), urban land use planning (Harbor 1994; Brabec et al. 2002), and urban environmental assessment, especially water quality (Schueler 1994; Arnold and Gibbons 1996; Zug et al. 1999; Brabec et al. 2002) and rainfall runoff (Lohani et al. 2002). Therefore, the timely and accurately mapping ISA distribution is of importance. The unique characteristics of remote sensing data in repetitive data acquisition, its synoptic view, and digital format suitable for computer processing make it the primary data source for ISA mapping. The research for extracting ISA from remotely sensed data has attracted great attention since the 1970s, especially in the most recent decade (Phinn et al. 2002; Gillies et al. 2003; Wu and Murray 2003; Yang et al. 2003a; Lu and Weng 2006a; Bauer et al. 2008; Mohapatra and Wu 2008; Wang et al. 2008; Xian et al. 2008; Hu and Weng 2009; Wu 2009; Weng et al., 2009; Yang et al. 2010). Since many new techniques for ISA mapping have been developed in recent decades, it is necessary to overview recent progress in order to provide guidelines for selecting a suitable technique for a specific study. This chapter briefly summarizes the major ISA mapping techniques that have appeared in recent literature and then provides a case study of ISA mapping with the integrated use of Landsat Thematic Mapper (TM) and QuickBird images in a complex urban–rural frontier in Brazil.

13.2   BACKGROUND

Urban landscapes can be regarded as a complex combination of buildings, roads, grass, trees, soil, water, and so on. In high spatial resolution images such as QuickBird and IKONOS, individual objects such as buildings and roads can be clearly identified; however, these features are less distinct in Landsat TM color composites because of Landsat's relatively coarse spatial resolution (i.e., 30 m) (see Figure 13.1). Different construction materials often result in high spectral variation, that is, ISA appears different colors in QuickBird image, making it difficult to automatically map ISA distribution based on spectral signatures. In coarse and medium spatial resolution images such as Landsat TM, mixed pixels can be a serious problem, diminishing the effective use of remotely sensed data in urban LULC classification and change detection. If per-pixel based methods are used for mapping urban LULC distribution based on medium or coarse spatial resolution images, large uncertainty may be introduced into the result, that is, urban ISA extent could be significantly overestimated, and ISA in rural areas could be significantly underestimated (Lu and Weng 2004). This situation worsens if multitemporal remote sensing data are used for urban LULC change detection, especially in the urban–rural frontiers.

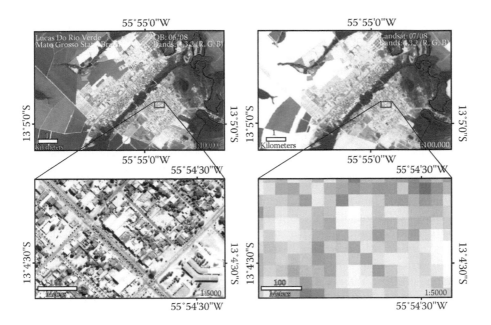

FIGURE 13.1 Comparison of color composites between QuickBird (left) and TM images (right), illustrating the mixed pixel problem in relatively coarse spatial resolution image.

In urban environments, land covers are assumed to be a linear combination of three components: vegetation, ISA, and soil (V-I-S) (Ridd 1995). The V-I-S model provides a guideline for decomposing urban landscapes and a link for these components to remote sensing spectral characteristics. Several studies have adopted this model as a basis for understanding the urban environment (Ward et al. 2000; Madhavan et al. 2001; Rashed et al. 2001; Phinn et al. 2002). From the view of remote sensing data, shade is also an important part affecting the spectral signature. Therefore, shade, green vegetation, soil, and ISA can almost explain all land covers in urban landscape (Lu and Weng 2004). In practice, ISA has high spectral variation; for example, bright building roofs may have very high spectral signatures that are confused with soils; dark roads and building roofs have very low spectral signatures that are often confused with shadow, water, and wetland (see Figure 13.1). Thus, direct extraction of ISA from remotely sensed data becomes very difficult because of the spectral confusion between ISA and other land covers. We can assume that urban landscape is composed of four components: high-albedo land covers, low-albedo land covers, vegetation, and soil. All land covers are a composition of these four components with linear or nonlinear relationship. Previous research has indicated that these four fraction images can be developed with spectral mixture analysis (Wu and Murray 2003; Lu and Weng 2006a,b). High-albedo fraction image highlights the land covers with high spectral reflectance, such as bright ISA and dry bare soils; low-albedo fraction image highlights the land cover with low spectral reflectance, such as dark ISA, shadow, water, and wetland. Soil fraction image highlights soil information, mainly located in agricultural lands; vegetation fraction image highlights the forest and plantation information. ISA is mainly concentrated on the high- and low-albedo fraction images; thus, ISA can be extracted from the addition of high- and low-albedo fraction images (Wu and Murray 2003; Lu and Weng 2006a,b).

Since the 1970s, many approaches for ISA mapping have been developed. Based on the achievements in the 1970s and 1980s, Slonecker et al. (2001) reviewed many of the approaches for ISA extraction from remotely sensed data and grouped them into three basic categories: interpretive applications, spectral applications, and modeling applications. Brabec et al. (2002) summarized four ways for ISA mapping: (1) using a planimeter to measure ISA on aerial photography, (2) counting the number of intersections on the overlain grid on an aerial photography, (3) conducting image classification, and (4) estimating ISA through the percentage of urbanization in a region. Recently, more advanced algorithms have been developed for quantitative extraction of ISA from satellite imagery. The major methods include per-pixel image classification (Lu and Weng 2009), subpixel classification (Ji and Jensen 1999; Phinn et al. 2002; Rashed et al. 2003), neural network (Mohapatra and Wu 2008; Wang et al. 2008; Hu and Weng 2009; Wu 2009), regression tree model (Yang et al. 2003a,b; Xian and Crane 2005; Xian 2008; Xian et al. 2008; Yang et al. 2009), the combination of high-albedo and low-albedo fraction images (Wu and Murray 2003; Wu 2004; Lu and Weng 2006a,b; Weng et al. 2008, 2009), and through the established relationship between ISA and vegetation cover (Gillies et al. 2003; Bauer et al. 2008). Major ISA mapping approaches can be summarized based on satellite images, which have appeared in recent literature (Table 13.1).

Image classification-based methods for mapping ISA are common (Slonecker et al. 2001; Brabec et al. 2002), but overestimation often occurs in urban extent and underestimation occurs in rural landscapes due to the limitation of spatial resolution in remotely sensed data and the heterogeneity in the urban environment (Lu and Weng 2004). The high spectral variation in ISA and similar spectral signatures between ISA and other nonvegetation land covers also make it difficult to select suitable training samples for the ISA class, resulting in misclassification. An alternative is to use ERDAS IMAGINE's subpixel classifier (Ji and Jensen 1999; Civco et al. 2002). However, the complexity of ISA materials often makes the subpixel classifier difficult to employ and leads to underestimation in ISA extraction. Since a high inverse correlation exists between vegetation cover and ISA in urban landscapes, the ISA can be estimated based on the established regression models with the vegetation indices, such as from tasseled cap greenness (Bauer et al. 2008) and fractional vegetation cover from the normalized difference vegetation index (NDVI) (Gillies et al. 2003). This approach has a drawback in that vegetation greenness varies with different seasons, which may result in large uncertainties of ISA estimation. Also this method cannot be directly transferred to other study areas or other dates of data sets due to local specificities regarding phenologies, climate conditions, and different composition of land covers in the urban landscape. Another common method for ISA estimation is based on the regression tree model (Yang et al. 2003a,b; Xian and Crane 2005; Xian 2008; Xian et al. 2008; Yang et al. 2009), which has been used for ISA mapping for continental United States based on Landsat TM images.

In recent years, spectral mixture analysis (SMA) has emerged as an important approach for ISA extraction from multispectral data. For example, ISA as one of the endmembers may be directly extracted from remotely sensed data (Rashed et al.

TABLE 13.1
Summary of Major Approaches for Impervious Surface Mapping

Category	Approach	Materials	References
Per-pixel based methods	Decision tree classifier, maximum likelihood	IKONOS, QuickBird	Goetz et al. 2003; Lu and Weng 2009; Lu et al. 2011
Subpixel based methods	Subpixel classifier	TM	Ji and Jensen 1999
	Artificial neural networks	ASTER	Hu and Weng 2009
	Subpixel proportional land cover information transformation (SPLIT)	TM, videography	Wang and Zhang 2004
	Standard SMA	IRS-1C; TM	Rashed et al. 2001; Phinn et al. 2002
	Addition of low- and high-albedo fractions derived from standard or normalized SMA	ETM+	Wu and Murray 2003; Wu 2004
	Modified approach based on low-albedo, high-albedo, and land surface temperature	ETM+	Lu and Weng 2006a,b
	Multiple endmember SMA	TM	Rashed et al. 2003
	Combination of support vector machines and geospatial analysis	ETM+, ATKIS vector data	Esch et al. 2009
	Regression tree model	ETM+	Xian and Crane 2005; Yang et al. 2003a,b
	Multiple regression analysis	DMSP-OLS, ancillary data	Elvidge et al. 2007; Sutton et al. 2009
Vegetation-based methods	A regression model based on impervious surface and tasseled cap greenness	TM	Bauer et al. 2008
	Based on fraction vegetation cover (Fr), i.e., 1-Fr for the developed areas	MSS and TM	Gillies et al. 2003
Other methods	A combination of image processing methods based on PCA and spatial morphological operators	IKONOS	Cablk and Minor 2003
	Combination of knowledge-based classification and spectral unmixing	MSS/TM/ ETM+	Powell et al. 2008

2001; Phinn et al. 2002). ISA estimation may be improved by the addition of high- and low-albedo fraction images, which both are used as endmembers in SMA (Wu and Murray 2003; Lu and Weng 2006a,b). In addition, multiple endmember SMA (MESMA) (Rashed et al. 2003) has also been applied in which several ISA endmembers can be extracted. However, ISA is often overestimated or underestimated when medium spatial resolution images are used, depending on the relative proportion of

ISA in a pixel (Wu and Murray 2003; Lu and Weng 2006a,b; Greenfield et al. 2009). Since high spatial resolution satellite images (e.g., IKONOS and QuickBird) have become available, more research has shifted to the use of these data (Mohapatra and Wu 2008; Lu and Weng 2009; Wu 2009; Lu et al. 2011). However, the use of high spatial resolution images also presents a challenge for automatically mapping ISA distribution (Lu and Weng 2009). The main problems include spectral confusion between ISA and other land covers due to limited spectral resolution [i.e., usually only visible and near-infrared (NIR) wavelengths are available in very high spatial resolution images], high spectral variation within the same land cover, and shadows caused by tall objects and the confusion between dark ISA and water/wetland (Dare 2005; Li et al. 2005; Chen et al. 2007; Zhou et al. 2009).

13.3 RESEARCH PROBLEMS AND OBJECTIVES

In a complex urban–rural landscape, traditional classification methods cannot effectively distinguish ISA from other land covers due to the spectral confusion between dark ISA and wetland/water. A similar issue happens among bright ISA, soils, and cropped fields. Sometimes, traditional classification methods cannot effectively distinguish ISA from other land covers due to the difficulty in selecting training sample plots. Although subpixel-based methods can improve area estimation accuracy, it is critical and often difficult to distinguish the pixels with ISA from other land covers. Rarely has research focused on the improvement of ISA estimation in a complex urban–rural environment with multiscale remote sensing data. Therefore, this case study attempts to develop a new method suitable for a complex urban–rural landscape for mapping ISA distribution with the combination of Landsat TM and QuickBird images.

13.4 DESCRIPTION OF THE STUDY AREA

The city of Lucas do Rio Verde (hereafter called Lucas) in Mato Grosso State, Brazil, has a relatively short history and small urban extent (see Figure 13.2). It was established in 1982 and has experienced rapid urbanization. Highway BR 163 connects Lucas in the north to Santarém, a river port city on the Amazon River, and in the south to the heart of the soybean growing area at Cuiabá, capital city of Mato Grosso state. Highway BR-163 is a major route for the transport of export bound commodities grown in the region. The economic base of Lucas is large-scale agriculture, including the production of soy, cotton, rice, and corn as well as poultry and swine, to take advantage of the grain feed to add value to production. Major poultry and meat producing industries are developing as are other industrial complexes to add value to the agricultural output, which is now substantial. The county is at the epicenter of soybean production in Brazil, and it is expected to grow in population threefold in the next 10 years (personal communication with secretariat for planning at Lucas). Because this relatively small town has complex urban–rural spatial patterns derived from its highly capitalized agricultural base, large silos and warehouses, and planned urban growth, Lucas is an ideal site for exploring the methods for rapidly mapping urban extent with remotely sensed data.

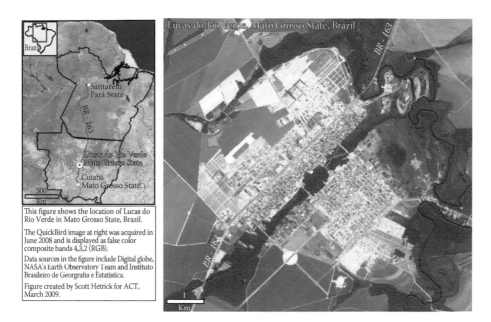

FIGURE 13.2 Study area, Lucas do Rio Municipio, Mato Grosso State, Brazil.

13.5 METHODOLOGY

Mapping ISA with Landsat images in a complex urban–rural frontier is challenging due to the mixed pixel problem and the confusion of ISA with other land covers. This research designed a comprehensive method based on the combined use of QuickBird and Landsat images, as illustrated in Figure 13.3. The major steps include: (1) mapping ISA with a hybrid method based on QuickBird imagery, (2) extracting per-pixel ISA images from Landsat images based on the thresholding of maximum and minimum filtering images and unsupervised classification, (3) mapping fractional images of high-albedo, low-albedo, vegetation, and soil endmembers with linear spectral

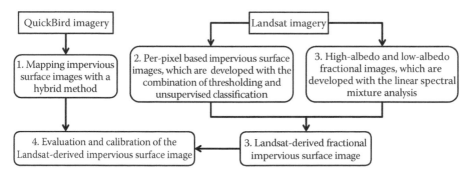

FIGURE 13.3 Strategy of integrating Landsat TM and QuickBird images for mapping impervious surface distribution.

mixture analysis (LSMA), and producing a fractional ISA image by adding high-albedo and low-albedo fraction images while removing non-ISA pixels by combining the per-pixel ISA image from step 2, and (4) establishing a regression model to calibrate the Landsat-derived ISA by using the QuickBird-derived ISA imagery.

13.5.1 Image Preprocessing

A Landsat TM image acquired on May 22, 2008 and a QuickBird image acquired on June 20, 2008 were used in this research. The TM image was geometrically registered into a Universal Transverse Mercator (UTM) projection with geometric error of less than 0.5 pixel. The nearest neighbor resampling method was used to resample the Landsat image into a pixel size of 30 m by 30 m during image-to-image registration. Radiometric and atmospheric correction was conducted on the TM image by utilizing the dark object subtraction (DOS) approach. The DOS model is an image-based procedure that standardizes imagery for the effects caused by solar zenith angle, solar radiance, and atmospheric scattering (Lu et al. 2002; Chander et al. 2009). Here are the equations used for Landsat image calibration:

$$R_\lambda = PI \cdot D \cdot (L_\lambda - L_{\lambda haze})/(Esun_\lambda \cdot COS\,(\theta)) \tag{13.1}$$

$$L_\lambda = gain_\lambda \cdot DN_\lambda + bias_\lambda, \tag{13.2}$$

where L_λ is the apparent at-satellite radiance for spectral band λ, DN_λ is the digital number of spectral band λ, R_λ is the calibrated reflectance, $L_{\lambda.haze}$ is path radiance, $Esun_\lambda$ is exoatmospheric solar irradiance, D is the distance between the Earth and Sun, and θ is the Sun zenith angle. The path radiance for each band is identified based on the analysis of water bodies and shades in the images. The $gain_\lambda$ and $bias_\lambda$ are radiometric gain and bias corresponding to spectral band λ, respectively, and they are often provided in an image head file or metadata file, or calculated from maximal and minimal spectral radiance values (Lu et al. 2002).

13.5.2 Mapping ISA with QuickBird Imagery

QuickBird imagery is used to develop ISA data that is used as a reference for establishing a calibration model for refining the Landsat-derived ISA image. QuickBird imagery has four multispectral bands with 2.4-m spatial resolution and one panchromatic band with 0.6-m spatial resolution. In order to make full use of both the multispectral and panchromatic images, the wavelet merging technique (Lu et al. 2008) was used to merge the QuickBird multispectral bands and panchromatic band into a new multispectral image with 0.6-m spatial resolution. A hybrid method that consisted of thresholding, unsupervised classification, and manual editing was used to produce the ISA image from the fused QuickBird imagery (Lu et al. 2011) (see Figure 13.4). The fact that vegetation has significantly different spectral features comparing with ISA in the NDVI image and clear and deep water bodies have much lower spectral values than ISA in the NIR wavelength image, the vegetation and

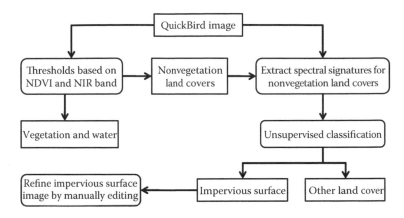

FIGURE 13.4 Framework of impervious surface mapping with the hybrid method.

water pixels can be masked out with selected thresholds on NDVI and NIR images. The major steps for the hybrid approach include (1) producing the NDVI image from QuickBird red and near-infrared images and then masking vegetation out with the selected threshold on the NDVI image, and masking water out with the selected thresholds on the NIR image, (2) extracting spectral signatures of the nonvegetation pixels, using an unsupervised classification algorithm to classify the extracted spectral signatures into 50 clusters and merging the clusters into ISA and other classes, and (3) manually editing the extracted ISA image (Lu et al. 2011). Although unsupervised classification can separate most ISA from bare soils and wetlands, some confusion still remains between bare soil and bright ISA, and among dark ISA, shadowed ISA, wetlands, or shadows from tree crowns. Therefore, visually interpreting the extracted ISA image is necessary to further refine the ISA image quality by eliminating the confused pixels (e.g., bare soils, non-ISA shadows, and wetlands).

Accuracy assessment is conducted with the error matrix approach (Congalton and Mead 1983; Congalton 1991; Janssen and van der Wel 1994; Foody 2002; Congalton and Green 2008). A total of 450 test samples were selected with the random sampling method. The analyst then examined each test sample plot to decide whether it was correctly classified as ISA or not based on visual interpretation on the QuickBird color composite. When the accuracy is satisfied, the ISA image with spatial resolution of 0.6 m is finally aggregated to 30 m to generate fractional ISA image for use as reference data.

13.5.3 Developing Per-Pixel Based ISA from Landsat Images

Per-pixel ISA mapping is often based on the image classification of spectral signatures (Shaban and Dikshit 2001; Lu et al. 2011), but per-pixel based classification methods based on medium or coarse spatial resolution images cannot effectively extract the ISA because of the spectral confusion between ISA and other land covers and the mixed pixel problems, especially in a complex urban–rural frontier (Wu and Murray 2003; Lu and Weng 2006a). This research for mapping per-pixel ISA was based on the combination of filtering images and unsupervised classification of Landsat spectral signatures. The fact that the red-band image in Landsat TM has high spectral values for ISA but has low spectral values for vegetations and water/wetlands provides a potential way to rapidly map ISA. The minimum and maximum filter with a window size of 3 × 3 pixels was separately applied to the Landsat red-band image. The image differencing between maximum and minimum filtering images was used to highlight linear features (mainly roads) and other ISAs. Examination of the difference image indicated that a threshold value of 13 can extract the ISA image. The spectral signature of the initial ISA image was then extracted and was further classified into 60 clusters using an unsupervised classification method to refine the ISA image by removing the non-ISA pixels. Finally, manual editing of the ISA image was conducted to make sure that all ISAs, especially in urban regions, were extracted. The final per-pixel based ISA image was overlain on the TM color composite to visually examine the ISA mapping quality in order to make sure all urban area and major roads were properly extracted.

13.5.4 MAPPING FRACTIONAL ISA DISTRIBUTION

As per-pixel methods based on medium or coarse spatial resolution often overestimate or underestimate ISA, it is important to estimate fractional ISA images in order to improve area estimation. Of the many methods for mapping ISA (Slonecker et al. 2001; Brabec et al. 2002), the LSMA-based method has proven valuable for extracting fractional ISA from Landsat images (Wu and Murray 2003; Lu and Weng 2006a,b). LSMA is regarded as a physically based image-processing tool. It supports repeatable and accurate extraction of quantitative subpixel information (Smith et al. 1990). The LSMA approach assumes that the spectrum measured by a sensor is a linear combination of the spectra of all components (endmembers) within the pixel and that the spectral proportions of the endmembers represent proportions of the area covered by distinct features on the ground (Adams et al. 1995; Mustard and Sunshine 1999). The mathematical model can be expressed as

$$R_i = \sum_{k=1}^{n} f_k R_{ik} + \varepsilon_i, \tag{13.3}$$

where i is the number of spectral bands used; $k = 1, ..., n$ (number of endmembers), R_i is the spectral reflectance of band i of a pixel that contains one or more endmembers, f_k is the proportion of endmember k within the pixel, R_{ik} is known as the spectral reflectance of endmember k within the pixel on band i, and ε_i is the error for band i. For a constrained least squares solution, f_k is subject to the following restrictions:

$$\sum_{k=1}^{n} f_k = 1 \text{ and } 0 \le f_k \le 1 \tag{13.4}$$

$$\text{RMSE} = \sqrt{\left(\sum_{i=1}^{m} \varepsilon_i^2 / m \right)}. \tag{13.5}$$

The root-mean-square error (RMSE) is used to assess the fit of the model. It is calculated for all image pixels. The larger the RMSE is, the worse the fit of the model is. So the error image can be used to assess whether the endmembers are properly selected and whether the number of selected endmembers is sufficient.

In the LSMA approach, endmember selection is a key step, and many approaches have been developed (Lu et al. 2003; Theseira et al. 2003). In practice, image-based endmember selection methods are frequently used because image endmembers can be easily obtained and they represent the spectra measured at the same scale as the image data. Image endmembers can be derived from the extremes of the image feature space, assuming that they represent the purest pixels in the images (Mustard and Sunshine 1999). In order to effectively identify image endmembers and to achieve high-quality endmembers, different image transform approaches such as principal component analysis (PCA) and minimum noise fraction (MNF) may be used to transform the multispectral images into a new data set (Green et al. 1988; Boardman and Kruse 1994). Endmembers are then selected from the feature spaces of the transformed images (Garcia-Haro et al. 1996; Cochrane and Souza 1998; van der Meer and de Jong 2000; Small 2001, 2002, 2004). In this research, four endmembers (i.e., high-albedo objects, low-albedo objects, vegetation, and soil) were selected from the feature spaces formed by the MNF components. A constrained least squares solution was then used to unmix the Landsat TM multispectral image into four fractional images and one error image.

Previous research indicates that the high-albedo fraction image contains the bright ISA objects such as building roofs with high spectral values, and low-albedo fraction image contains the dark ISA objects such as dark roads and streets with low spectral values (Lu and Weng 2006a,b). Therefore, the overall ISA image can be extracted from the addition of high-albedo and low-albedo fraction images. One critical step in mapping ISA is to remove the non-ISA pixels. The strategy of mapping fractional ISA can be illustrated based on the LSMA-based methods and per-pixel based method (Figure 13.5). By combining per-pixel ISA and high- and low-albedo fraction images, a fractional ISA image was then generated with the following rules: if the pixel is ISA in the per-pixel based ISA image, the pixel value is then extracted from the sum of high-albedo and low-albedo fraction images; otherwise, assign zero to the pixel.

13.5.5 Refining ISA by Integrated Use of Landsat- and QuickBird-Derived ISA Images

Previous research also indicates that the developed ISA data set from Landsat TM images is often overestimated or underestimated, depending on the proportion of ISA in a pixel (Wu and Murray 2003; Lu and Weng 2006a). In the complex urban–rural landscape, ISA estimation with LSMA-based method often overestimates its area

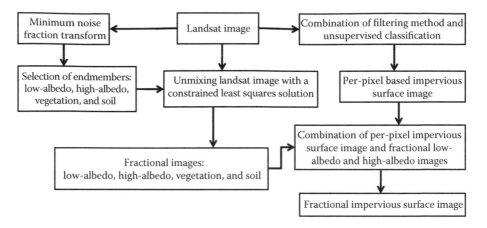

FIGURE 13.5 Strategy of developing fractional impervious surface images based on the combination of linear spectral mixture analysis and per-pixel based method.

statistics. It is necessary to calibrate this bias. One method is to develop a regression model to calibrate the TM-derived ISA images. In this research, the overlap area between the 2008 QuickBird and the corresponding Landsat-derived ISA images was used for sample collection based on this rule: select one pixel for every five intervals on the overlapped images. Because many pixels were non-ISA, they had zero values. After removal of all samples with zero values, 1512 samples were used to develop the calibration model. A scatterplot of these samples was used to examine the relationship between the Landsat-derived and QuickBird-derived ISA images. A regression model was developed to conduct the calibration.

13.6 RESULTS AND DISCUSSION

13.6.1 ISA MAPPING WITH QUICKBIRD IMAGERY

A high-quality ISA data set from QuickBird imagery is required because it is used as reference data for calibrating Landsat-derived ISA data set. The accuracy assessment based on randomly selected 450 sample plots indicated that 98% overall accuracy was achieved, according to visual interpretation on the QuickBird color composite. The spatial patterns of ISA distribution can be shown in Figure 13.6. The QuickBird-derived ISA data were then aggregated from 0.6 to 30 m cell sizes to generate fractional ISA data for use in linking it to Landsat TM image for calibration.

13.6.2 EVALUATION OF PER-PIXEL ISA IMAGE FROM LANDSAT TM IMAGERY

Evaluation of the per-pixel ISA image based on overlaying it on the TM color composite indicates that a combination of filtering images and unsupervised classification method developed in this research can effectively extract the pixel-based ISA data set in the complex urban–rural frontier. In per-pixel based results, each

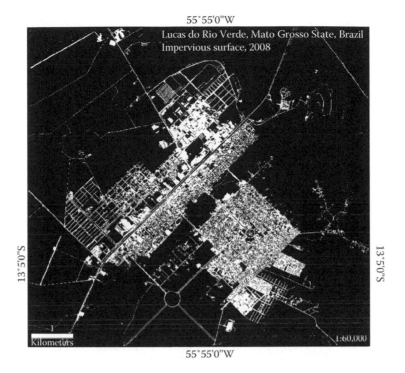

FIGURE 13.6   Impervious surface image developed from QuickBird imagery.

extracted ISA pixel is assumed to be 100% ISA. This data set is useful for visually interpreting the ISA distribution, but not suitable for area statistics. Therefore, it is necessary to produce a subpixel ISA data set for area statistical purpose in order to remove the impact of the mixed pixel problem.

13.6.3   Development of Fractional ISA Image from Landsat TM Imagery

The four fraction images developed from Landsat TM imagery with the LSMA approach have physical meanings as described in Section 13.2. ISA is mainly concentrated on high-albedo and low-albedo fraction images; thus, the fractional ISA data can be directly extracted through the addition of the high-albedo and low-albedo fraction images. However, some roads appear in the soil fraction image due to the confusion of their spectral signatures (see Figure 13.7), and some non-ISA pixels, such as water, are also included in the initial fractional ISA image. Therefore, the per-pixel based ISA image is used to mask out the non-ISA pixels from the fractional ISA image. A comparison of per-pixel ISA and fractional ISA images (Figure 13.8) indicated that the area amount may be significantly overestimated by the per-pixel ISA image. Because of the limitation of the LSMA method in extracting ISA information from other land covers, underestimation or overestimation of ISA is common, as shown in previous research (Lu and Weng 2006a,b);

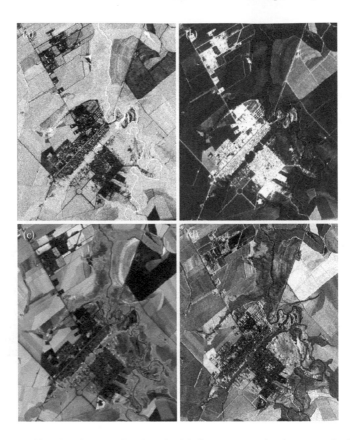

FIGURE 13.7 Fraction images developed with linear spectral mixture analysis based on Landsat TM multispectral data: (a) low-albedo, (b) high-albedo, (c) vegetation, and (d) soil.

thus, it is necessary to further calibrate the fractional ISA image to improve the area estimation accuracy.

13.6.4 CALIBRATION OF THE LANDSAT-DERIVED FRACTIONAL ISA WITH THE QUICKBIRD-DERIVED ISA IMAGE

In theory, if the ISA data are accurately estimated from both Landsat TM and QuickBird images, the scatterplot between both fractional ISA data sets should show a very good linear relationship. As shown in Figure 13.9, the ISA image developed in this research demonstrates a reasonably good result, although overestimation occurred when the ISA accounted for a relatively small proportion in a pixel and underestimation occurred when the ISA accounted for a large proportion in a pixel. This trend is similar in other previous research (Wu and Murray 2003; Lu and Weng 2006; Greenfield et al. 2009). Overall, a good linear relationship exists between the fractional ISA images developed independently from 2008 Landsat TM

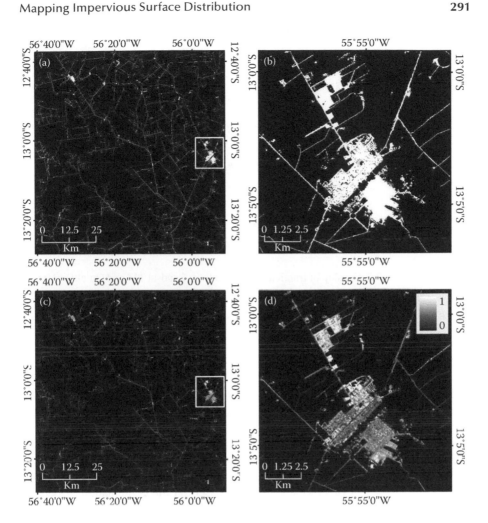

FIGURE 13.8 Comparison of impervious surface results with pixel-based (a and b) and subpixel-based (c and d) methods.

and QuickBird images. Based on the samples from QuickBird- and Landsat-derived ISA images, a linear regression model is established as follows:

$$y = 1.0674x - 0.0119; R^2 = 0.45, \tag{13.6}$$

where x is the fractional ISA values from 2008 Landsat TM image, and y is the calibrated fraction ISA values. For the non-ISA pixels (i.e., x equals zero), assign a zero value to that pixel; otherwise, this equation was used to calibrate the entire image of 2008 Landsat TM-derived fractional ISA image. When the ISA was calculated from the per-pixel ISA image, the ISA for Lucas município (see Figure 13.2) was 99.87 km^2, accounting for 1.291% of the study area. However, when the ISA was calculated from the calibrated fractional ISA data set, the ISA became only 56.86 km^2,

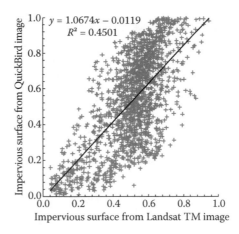

FIGURE 13.9   Relationship of fractional ISA data from Landsat TM and QuickBird images.

accounting for only 0.735%. This implies that development of fractional ISA data set is required for ISA measurement, especially in an urban–rural frontier.

13.7   CONCLUSIONS

The complexity of ISA in the urban landscape and the mixed pixel problem in medium and coarse spatial resolution images make the mapping of ISA a challenging task. Traditional per-pixel based image classification methods cannot effectively handle the mixed pixel problem, and subpixel-based methods cannot effectively separate the pixels of ISA from other land covers; thus, underestimation or overestimation of the ISA are common, depending on the proportion of ISA in a pixel. The method described in this chapter, which is based on the combination of per-pixel based ISA mapping with filtering and unsupervised classification, and subpixel-based method with the LSMA approach, can effectively map ISA distribution with Landsat images. The calibration with QuickBird-derived ISA data set can further reduce the bias caused by mixed pixel problems and thus improve ISA mapping performance. The major advantage of the method described in this chapter is that the ISA estimation can be considerably improved comparing with per-pixel based results in a complex urban–rural frontier. More research is needed to apply this method to other study areas and other dates of Landsat images to examine its transferability and robustness.

ACKNOWLEDGMENT

The authors wish to thank the National Institute of Child Health and Human Development at NIH (grant # R01 HD035811) for the support of this research, addressing population-and-environment reciprocal interactions, in several regions of the Brazilian Amazon. Any errors are solely the responsibility of the authors and not of the funding agencies.

REFERENCES

Adams, J. B., Sabol, D. E., Kapos, V., Filho, R. A., Roberts, D. A., Smith, M. O., and Gillespie, A. R. (1995). Classification of multispectral images based on fractions of endmembers: Application to land cover change in the Brazilian Amazon. *Remote Sensing of Environment*, 52, 137–154.

Arnold, C. L. and Gibbons, C. J. (1996). Imperious surface coverage: the emergence of a key environmental indicator. *Journal of the American Planning Association*, 62(2), 243–258.

Bauer, M. E., Loffelholz, B. C., and Wilson, B. (2008). Estimating and mapping impervious surface area by regression analysis of Landsat imagery. In *Remote Sensing of Impervious Surfaces*, Q. Weng (ed.), Taylor & Francis Group, LLC, Boca Raton, FL, pp. 3–19.

Boardman, J. W. and Kruse, F. A. (1994). Automated spectral analysis: A geological example using AVIRIS data, North Grapevine Mountains, Nevada. In *Proceedings, ERIM Tenth Thematic Conference on Geologic Remote Sensing*, Ann Arbor, MI, pp. 407–418.

Brabec, E., Schulte, S., and Richards, P. L. (2002). Impervious surface and water quality: A review of current literature and its implications for watershed planning. *Journal of Planning Literature*, 16, 499–514.

Cablk, M. E. and Minor, T. B. (2003). Detecting and discriminating impervious cover with high-resolution IKONOS data using principal component analysis and morphological operators. *International Journal of Remote Sensing*, 24, 4627–4645.

Chander, G., Markham, B. L., and Helder, D. L. (2009). Summary of current radiometric calibration coefficients for Landsat MSS, TM, ETM+, and EO-1 ALI sensors. *Remote Sensing of Environment*, 113, 893–903.

Chen, Y., Wen, D., Jing, L., and Shi, P. (2007). Shadow information recovery in urban areas from very high resolution satellite imagery. *International Journal of Remote Sensing*, 28, 3249–3254.

Civco, D. L., Hurd, J. D., Wilson, E. H., Arnold, C. L., and Prisloe, Jr., M. P. (2002). Quantifying and describing urbanizing landscapes in the northeast United States. *Photogrammetric Engineering and Remote Sensing*, 68, 1083–1090.

Cochrane, M. A. and Souza, Jr. C. M. (1998). Linear mixture model classification of burned forests in the eastern Amazon. *International Journal of Remote Sensing*, 19, 3433–3440.

Congalton, R. G. (1991). A review of assessing the accuracy of classification of remotely sensed data. *Remote Sensing of Environment*, 37, 35–46.

Congalton, R. G. and Mead, R. A. (1983). A quantitative method to test for consistency and correctness in photo interpretation. *Photogrammetric Engineering and Remote Sensing*, 49, 69–74.

Congalton, R. G. and Green, K. (2008). *Assessing the Accuracy of Remotely Sensed Data: Principles and Practices*, Second Edition, CRC Press, Taylor & Francis Group, Boca Raton, FL, p. 183.

Dare, P. M. (2005). Shadow analysis in high-resolution satellite imagery of urban areas. *Photogrammetric Engineering and Remote Sensing*, 71, 169–177.

Elvidge, C. D., Tuttle, B. T., Sutton, P. C., Baugh, K. E., Howard, A. T., Milesi, C., Bhaduri, B., and Nemani R. (2007). Global distribution and density of constructed impervious surfaces. *Sensors*, 7, 1962–1979.

Esch, T., Himmler, V., Schorcht, G., Thiel, M., Wehrmann, T., Bachofer, F., Conrad, C., Schmidt, M., and Dech, S. (2009). Large-area assessment of impervious surface based on integrated analysis of single-date Landsat-7 images and geospatial vector data. *Remote Sensing of Environment*, 113, 1678–1690.

Foody, G. M. (2002). Status of land cover classification accuracy assessment. *Remote Sensing of Environment*, 80, 185–201.

Garcia-Haro, F. J., Gilabert, M. A., and Melia, J. (1996). Linear spectral mixture modeling to estimate vegetation amount from optical spectral data. *International Journal of Remote Sensing*, 17, 3373–3400.

Goetz, S. J., Wright, R., Smith, A. J., Zinecker, E., and Schaub, E. (2003). IKONOS imagery for resource management: Tree cover, impervious surface, and riparian buffer analyses in the mid-Atlantic region. *Remote Sensing of Environment*, 88, 195–208.

Gillies, R. R., Box, J. B., Symanzik, J., and Rodemaker, E. J. (2003). Effects of urbanization on the aquatic fauna of the Line Creek watershed, Atlanta—A satellite perspective. *Remote Sensing of Environment*, 86, 411–422.

Green, A. A., Berman, M., Switzer, P., and Craig, M. D. (1988). A transformation for ordering multispectral data in terms of image quality with implications for noise removal. *IEEE Transactions on Geoscience and Remote Sensing*, 26, 65–74.

Greenfield, E. J., Nowak, D. J., and Walton, J. T. (2009). Assessment of 2001 NLCD percent tree and impervious cover estimates. *Photogrammetric Engineering and Remote Sensing*, 75, 1279–1286.

Harbor, J. M. (1994). A practical method for estimating the impact of land use change on surface runoff, groundwater recharge and wetland hydrology. *American Planning Association*, 60, 95–108.

Hu, X. and Weng, Q. (2009). Estimating impervious surfaces from medium spatial resolution imagery using the self-organizing map and multi-layer perceptron neural networks. *Remote Sensing of Environment*, 113, 2089–2102.

Janssen, L. F. J. and van der Wel, F. J. M. (1994). Accuracy assessment of satellite derived land-cover data: a review. *Photogrammetric Engineering and Remote Sensing*, 60, 419–426.

Ji, M. and Jensen, J. R. (1999). Effectiveness of subpixel analysis in detecting and quantifying urban imperviousness from Landsat Thematic Mapper. *Geocarto International*, 14, 31–39.

Li, Y., Gong, P., and Sasagawa, T. (2005). Integrated shadow removal based on photogrammetry and image analysis. *International Journal of Remote Sensing*, 26, 3911–3929.

Lohani, V., Kibler, D. F., and Chanat, J. (2002). Constructing a problem solving environment tool for hydrologic assessment of land use change. *Journal of the American Water Resources Association*, 38, 439–452.

Lu, D., Mausel, P., Brondízio, E., and Moran, E. (2002). Assessment of atmospheric correction methods for Landsat TM data applicable to Amazon basin LBA research. *International Journal of Remote Sensing*, 23, 2651–2671.

Lu, D., Moran, E., and Batistella, M. (2003). Linear mixture model applied to Amazonian vegetation classification. *Remote Sensing of Environment*, 87, 456–469.

Lu, D. and Weng, Q. (2004). Spectral mixture analysis of the urban landscapes in Indianapolis with Landsat ETM+ imagery. *Photogrammetric Engineering and Remote Sensing*, 70, 1053–1062.

Lu, D. and Weng, Q. (2006a). Use of impervious surface in urban land use classification. *Remote Sensing of Environment*, 102, 146–160.

Lu, D. and Weng, Q. (2006b). Spectral mixture analysis of ASTER images for examining the relationship between urban thermal features and biophysical descriptors in Indianapolis, United States. *Remote Sensing of Environment*, 104(2), 157–167.

Lu, D., Weng, Q., and Li, G. (2006). Residential population estimation using a remote sensing derived impervious surface approach. *International Journal of Remote Sensing*, 27(16), 3553–3570.

Lu, D., Batistella, M., Moran, E., and de Miranda, E. E. (2008). A comparative study of Landsat TM and SPOT HRG images for vegetation classification in the Brazilian Amazon. *Photogrammetric Engineering and Remote Sensing*, 70, 311–321.

Lu, D. and Weng, Q. (2009). Extraction of urban impervious surface from an IKONOS image. *International Journal of Remote Sensing*, 30(5), 1297–1311.

Lu, D., Hetrick, S., and Moran, E. (2011). Impervious surface mapping with QuickBird imagery. *International Journal of Remote Sensing*, 32(9), 2519–2533.

Madhavan, B. B., Kubo, S., Kurisaki, N., and Sivakumar, T. V. L. N. (2001). Appraising the anatomy and spatial growth of the Bangkok Metropolitan area using a vegetation-impervious-soil model through remote sensing. *International Journal of Remote Sensing*, 22, 789–806.

Mohapatra, R. P. and Wu, C. (2008). Subpixel imperviousness estimation with IKONOS imagery: An artificial neural network approach. In *Remote Sensing of Impervious Surfaces*, Q. Weng (ed.), Taylor & Francis Group, LLC, Boca Raton, FL, pp. 21–37.

Mustard, J. F. and Sunshine, J. M. (1999). Spectral analysis for earth science: Investigations using remote sensing data. In *Remote Sensing for the Earth Sciences: Manual of Remote Sensing*, Third Edition, Volume 3, A. N. Rencz (ed.), John Wiley & Sons Inc., New York, pp. 251–307.

Phinn, S., Stanford, M., Scarth, P., Murray, A. T., and Shyy, P. T. (2002). Monitoring the composition of urban environments based on the vegetation-impervious surface–soil (VIS) model by subpixel analysis techniques. *International Journal of Remote Sensing*, 23, 4131–4153.

Powell, S. L., Cohen, W. B., Yang, Z., Pierce, J. D., and Alberti, M. (2008). Quantification of impervious surface in the Snohomish Water Resources Inventory Area of Western Washington from 1972–2006. *Remote Sensing of Environment*, 112, 1895–1908.

Rashed, T., Weeks, J. R., Gadalla, M. S., and Hill, A. G. (2001). Revealing the anatomy of cities through spectral mixture analysis of multispectral satellite imagery: A case study of the Greater Cairo region, Egypt. *Geocarto International*, 16, 5–15.

Rashed, T., Weeks, J. R., Roberts, D., Rogan, J., and Powell, R. (2003). Measuring the physical composition of urban morphology using multiple endmember spectral mixture models. *Photogrammetric Engineering and Remote Sensing*, 69, 1011–1020.

Ridd, M. K. (1995). Exploring a V-I-S (vegetation-impervious surface-soil) model for urban ecosystem analysis through remote sensing: Comparative anatomy for cities. *International Journal of Remote Sensing*, 16, 2165–2185.

Schueler, T. R. (1994). The importance of imperviousness. *Watershed Protection Techniques*, 1, 100–111.

Shaban, M. A. and Dikshit, O. (2001). Improvement of classification in urban areas by the use of textural features: The case study of Lucknow city, Uttar Pradesh. *International Journal of Remote Sensing*, 22, 565–593.

Slonecker, E. T., Jennings, D., and Garofalo, D. (2001). Remote sensing of impervious surface: A review. *Remote Sensing Reviews*, 20, 227–255.

Small, C. (2001). Estimation of urban vegetation abundance by spectral mixture analysis. *International Journal of Remote Sensing*, 22, 1305–1334.

Small, C. (2002). Multitemporal analysis of urban reflectance. *Remote Sensing of Environment*, 81, 427–442.

Small, C. (2004). The Landsat ETM+ spectral mixing space. *Remote Sensing of Environment*, 93, 1–17.

Smith, M. O., Ustin, S. L., Adams, J. B., and Gillespie, A. R. (1990). Vegetation in Deserts: I. A regional measure of abundance from multispectral images. *Remote Sensing of Environment*, 31, 1–26.

Sutton, P. C., Anderson, S. A., Elvidge, C. D., Tuttle, B. T., and Ghosh, T. (2009). Paving the planet: Impervious surface as proxy measure of the human ecological footprint. *Progress in Physical Geography*, 33, 510–527.

Theseira, M. A., Thomas, G., Taylor, J. C., Gemmell, F., and Varjo, J. (2003). Sensitivity of mixture modeling to endmember selection. *International Journal of Remote Sensing*, 24, 1559–1575.

Van der Meer, F. and de Jong, S. M. (2000). Improving the results of spectral unmixing of Landsat Thematic Mapper imagery by enhancing the orthogonality of end-members. *International Journal of Remote Sensing*, 21, 2781–2797.

Wang, Y. and Zhang, X. (2004). A SPLIT model for extraction of subpixel impervious surface information. *Photogrammetric Engineering and Remote Sensing*, 70, 821–828.

Wang, Y., Zhou, Y., and Zhang, X. (2008). The SPLIT and MASC models for extraction of impervious surface areas from multiple remote sensing data. In *Remote Sensing of Impervious Surfaces,* Q. Weng (ed.), Taylor & Francis Group, LLC, Boca Raton, FL, pp. 77–92.

Ward, D., Phinn, S. R., and Murray, A. L. (2000). Monitoring growth in rapidly urbanizing areas using remotely sensed data. *Professional Geographer*, 53, 371–386.

Weng, Q., Hu, X., and Lu, D. (2008). Extracting impervious surface from medium spatial resolution multispectral and hyperspectral imagery: A comparison. *International Journal of Remote Sensing*, 29, 3209–3232.

Weng, Q., Hu, X., and Liu, H. (2009). Estimating impervious surfaces using linear spectral mixture analysis with multitemporal ASTER images. *International Journal of Remote Sensing*, 30(18), 4807–4830.

Wu, C. and Murray, A. T. (2003). Estimating impervious surface distribution by spectral mixture analysis. *Remote Sensing of Environment*, 84, 493–505.

Wu, C. (2004). Normalized spectral mixture analysis for monitoring urban composition using ETM+ imagery. *Remote Sensing of Environment*, 93, 480–492.

Wu, C. and Murray, A. T. (2005). A cokriging method for estimating population density in urban areas. *Computers, Environment and Urban Systems*, 29, 558–579.

Wu, C. (2009). Quantifying high-resolution impervious surfaces using spectral mixture analysis. *International Journal of Remote Sensing*, 30(11), 2915–2932.

Xian, G. and Crane, M. (2005). Assessments of urban growth in the Tampa Bay watershed using remote sensing data. *Remote Sensing of Environment*, 97, 203–215.

Xian, G. (2008). Mapping impervious surfaces using classification and regression tree algorithm. In *Remote Sensing of Impervious Surfaces*, Q. Weng (ed.), Taylor & Francis Group, LLC, Boca Raton, FL, pp. 39–58.

Xian, G., Crane, M. P., and McMahon, C. (2008). Quantifying multitemporal urban development characteristics in Las Vegas from Landsat and Aster data. *Photogrammetric Engineering and Remote Sensing*, 74, 473–481.

Yang, L., Xian, G., Klaver, J. M., and Deal, B. (2003a). Urban land-cover change detection through sub-pixel imperviousness mapping using remotely sensed data. *Photogrammetric Engineering and Remote Sensing*, 69, 1003–1010.

Yang, L., Huang, C., Homer, C., Wylie, B., and Coan, M. (2003b). An approach for mapping large-area impervious surface: synergistic use of Landsat 7 ETM+ and high spatial resolution imagery. *Canadian Journal of Remote Sensing*, 29, 230–240.

Yang, L., Jiang, L., Lin, H., and Liao, M. (2009). Quantifying sub-pixel urban impervious surface through fusion of optical and InSAR imagery. *GIScience and Remote Sensing*, 46(2), 161–171.

Yang, F., Matsushita, B., and Fukushima, T. (2010). A pre-screened and normalized multiple endmember spectral mixture analysis for mapping impervious surface area in Lake Kasumigaura Basin, Japan. *ISPRS Journal of Photogrammetry and Remote Sensing*, 65, 479–490.

Zhou, W., Huang, G., Troy, A., and Cadenasso, M. L. (2009). Object-based land cover classification of shaded areas in high spatial resolution imagery of urban areas: A comparison study. *Remote Sensing of Environment*, 113, 1769–1777.

Zug, M., Phan, L., Bellefleur, D., and Scrivener, O. (1999). Pollution wash-off modeling on impervious surface: Calibration, validation, and transposition. *Water Science and Technology*, 39, 17–24.

Part III

Air Quality Monitoring, Land Use/Land Cover Changes, and Environmental Health Concern

14 Using Lidar to Characterize Particles from Point and Diffuse Sources in an Agricultural Field

Michael D. Wojcik, Randal S. Martin, and Jerry L. Hatfield

CONTENTS

14.1 INTRODUCTION

The movement of urban populations into agricultural production areas, combined with the increasing size of these facilities, has elevated the issue of agricultural production emissions to national attention. Accurate measurement of both specific agricultural operations and whole facility aerosol emissions is technically challenging and practically difficult. Currently, air quality regulations pertaining to particulate emission sources are based on data collected by multiple (typically less than five) point-sampled measurements taken near these facilities. These data are then combined with meteorology information and used as input for subsequent numerical models (e.g., AERMOD) to account for wind and time variation (Hoff et al. 2002). This sample-then-model methodology can sometimes make it difficult to determine the effectiveness of specific conservation and management practices due to a number of factors including varying wind direction, turbulence, intermittent emission behavior, or a moving emission source.

The accuracy and cost of emission and management practice studies could be reduced if it were possible to directly measure the flux of emissions from a facility and its components in real or near-real time. While the concept behind the measurement of a physical flux (i.e., the mass transport of material away from a defined surface in a defined time) is intuitively simple in environmental applications, actual flux measurement is practically difficult. Taken in its simplest form, mass flux can be defined as the mean amount of material moving through a defined area per unit time. In a pipe or closed container, flux can be measured as accurately as desired simply by defining the accuracy of the sensors for velocity and movement. There is a tremendous body of work conducted over the past century to extend this concept to uncontained fluxes such as momentum, water vapor, heat, and carbon dioxide from natural and managed surfaces (Friehe 1986).

While initial studies were limited to mean measurements and derived diffusion coefficients by the slow response of available sensors, the general availability of robust, fast-response sensors has made eddy correlation flux measurement the method of choice for flux determination in the atmospheric boundary layer (Eichinger et al. 2005; Chávez et al. 2005). Emission measurements from agricultural sources challenge the assumptions and costs associated with this method. Disruptions to uniform field flow, complex terrain, structures, surface treatments, and mobile sources, when combined with unconfined wind vectors, make mean determination difficult. The goal of this chapter is to discuss a technique of using lidar to virtually surround a facility or operation—equivalent to deploying a vast number of rapid response sensors—to map the emission plume and track its movement in an agricultural field. We utilized a multiwavelength, micropulse lidar and periodically throughout a measurement calibrate the optical return signal *in situ* using standard point sensors. This combination allowed us to build real-time and averaged particulate mass concentration fields, which are combined with the mean wind field to produce the flux measurement.

Techniques for the retrieval of microphysical aerosol parameters using multiwavelength lidar have been reported since the 1980s, with major progress being made in the past 5 years (Heintzenberg et al. 1981; Rajeev and Parameswaran 1998; Müller et al. 1999; Böckmann 2001; Veselovskii et al. 2004; Althausen et al. 2000; Marchant et al. 2010). Unambiguous and stable retrieval of aerosol physical parameters using lidar can require up to 12 empirically derived quantities, most of which are not easily derived optically (Zavyalov et al. 2006). The instrumentation that is required to provide both multiwavelength elastic scatter lidar and Raman information is expensive, complicated to operate, and not transportable to a field test location. A significant database of atmospheric aerosol characteristics has been obtained using a combination of satellite and ground-based observations (Hess et al. 1998; Dubovik et al.

2002). Using this database, it has been shown that the physical properties of assumed aerosol mixtures can be successfully retrieved based on measurements of backscatter coefficients at only three wavelengths (Rajeev and Parameswaran 1998; Sasano and Browell 1989; Del Guasta et al. 1994). Given the complex nature of agricultural aerosols, which most certainly differ from those in the database, direct *in situ* aerosol characterization is the most reliable method for calibrating lidar return signals.

The Aglite lidar system is a robust, agile, and easily operated system that displays emitted particulate distributions in a few seconds under most meteorological and diurnal conditions. Aglite uses three wavelengths combined with information derived from an array of point sensors to distinguish between different types and sizes of particulate emissions. The combination of point samplers and scanning lidar provides near-real-time measurements of facility operations, which is then used to evaluate emission fluxes and operational approaches to minimize aerosol emissions.

To address the problems associated with exposure to high particulate matter levels, we performed a study to determine the control effectiveness of conservation management practices (CMPs) for agricultural tillage using Aglite. Aglite, when coupled with only a few point measurement devices, can map particulate matter (PM) emissions at high spatial and temporal resolutions, allowing for accurate comparisons of various CMPs for a variety of agricultural practices (Bingham 2009). The main operation of this study was to deploy an elastic lidar system, together with a network of air samplers, to measure PM emissions from agricultural operations in order to answer the following questions:

1. What are the magnitude, flux, and transport of PM emissions produced by agricultural practices where CMPs are being implemented versus the magnitude, flux, and transport of PM emissions produced by agricultural practices where CMPs are not being implemented?
2. What are the control efficiencies of equipment being used to implement the CMP? If resources allow assessing additional CMPs, what are the control efficiencies of the equipment?
3. Can these CMPs for a specific crop/practice be quantitatively compared, controlling for soil type, soil moisture, and meteorological conditions?

14.2   LITERATURE REVIEW

There are a handful of published articles pertaining to PM emissions from agricultural tilling, with the majority of the studies being performed in the state of California. There are also several articles that collectively examine impacts of a variety of conservation tillage practices with respect to soil characteristics, fuel consumption, cost of production, and air emissions (Holmén et al. 2001a,b). The use of an elastic lidar system by the University of California at Davis (UC Davis) to examine dust plumes resulting from tillage activities was presented by Holmén et al. (1998). Qualitatively, the constructed system was able to track the plume emitted from the moving source and provide a two-dimensional (2-D) vertical, downwind map of the plume. It was observed that the plume heights were often above the point samplers located at 10 m along the downwind plane. The authors suggested that the best fugitive dust sampling procedures would include a combination of elastic lidar and strategically placed point samplers.

Two papers presented by Holmén et al. in series in 2001 further discuss tillage PM_{10} emission rate investigations by UC Davis using filter-based mass concentration samplers and qualitative measurements from the previously mentioned elastic lidar system (Holmén et al. 2001a,b). The 24 samples listed within the articles were collected from Fall 1996 to Winter 1998 in the San Joaquin Valley during a wide range of environmental (temperature = 7°C–35°C, relative humidity = 20% to 90%, and from prior to the season's first precipitation to periods between winter storms) and soil moisture conditions (1.5%–20%). Tillage operations examined were disking, listing, root cutting, and ripping. Calculated PM_{10} emission rates ranged from 0 to 800 mg·m^{-2} (0 to 6.9 lb·acre^{-1}), the mean ±1 standard deviation was 152 ± 240 mg·m^{-2} (1.4 ± 2.1 lb·acre^{-1}), and the median was 43 mg·m^{-2} (0.4 lb·acre^{-1}). One point made by Holmén et al. (2001b) is that several environmental conditions (temperature profile, relative humidity, soil moisture, etc.) can have very significant effects on PM emissions and should be monitored and accounted for in emission rate measurement and reporting. As a result, the reliability of direct comparisons of emission rates measured under different environmental conditions must be carefully examined.

The studies published by researchers at UC Davis and herein previously discussed were part of a much larger investigation of agricultural PM_{10} emission rates in the San Joaquin Valley as funded by the US Department of Agriculture Special Research Grant Program. Findings of this broad study are published in Flocchini et al. (2001). A summary of emission factors is listed in Table 14.1 for different types of agricultural tillage along with the crop and time of year. As seen in results measured by Holmén et al. (2001b), the emission factors reported by Flocchini et al. for agricultural tillage were significantly influenced by environmental conditions such as the near-ground temperature profile, relative humidity, and soil moisture. The potential variability with the same agricultural implements under opposing extreme environmental conditions may be larger than the variation from the type of crop or equipment used for tilling.

In recent years, effort has been focusing on quantification of aerosol cloud movement and cloud concentration using *in situ* ground truth instrumentation. Using a technique similar to our own, Hiscox et al. (2008) used a nephelometer (MetOne, GT-640A) as a single-point calibration for an elastic lidar system to estimate the three-dimensional (3-D) PM_{10} and TSP plume concentrations and plume dispersion characteristics resulting from a cotton tilling and harvesting operation. The use of a nephelometer leaves the absolute concentration values of these results subject to ambiguity due to the inherent assumptions about the aerosol density and aerosol size distribution that are native to nephelometry. A more recent study carried out by Wang et al. (2009) has compared the outputs from a Lagrangian particle model to raw lidar imagery to examine PM_{10} plume motion and dynamics for a disking operation. The 3-D concentration distributions generated by a Lagrangian model were compared with the lidar-scanned images, and accuracy of the model was calculated for different time and spatial regimes using a single sonic anemometer for collecting

TABLE 14.1

Emission Factors and Uncertainties for Land Preparation as Reported by Flocchini et al. (2001)

Date	Emission Factor (mg/m²)	Uncertainty	Date	Emission Factor (mg/m²)	Uncertainty
		Stubble Disk			
10/27/1995	257.7	NC	11/6/1998	50.0	146%
11/3/1995	49.3	9%	11/6/1998	28.4	145%
11/3/1995	27.4	470%	11/6/1998	35.0	NC
11/3/1995	231.0	4%	11/6/1998	28.0	10%
11/3/1995	136.7	7%	11/6/1998	117.0	18%
11/3/1995	140.8	6%	11/6/1998	32.4	9%
11/3/1995	286.1	5%	11/6/1998	58.9	8%
11/15/1995	537.9	9%	11/6/1998	93.5	9%
11/15/1995	542.2	125%	11/6/1998	74.2	8%
6/24/1997	430.0	17%			
		Finish Disk			
11/26/1996	124.3	3%	12/4/1996	9.2	NC
11/26/1996	142.4	4%	12/4/1996	0.6	NC
11/26/1996	97.5	5%	12/4/1996	3.5	NC
12/2/1996	91.0	9%	12/5/1996	−0.5	NC
		Ripping/Chisel			
6/24/1997	765.0	5%	6/25/1997	331.0	5%
6/26/1997	112.0	5%	6/25/1997	577.0	6%
6/26/1997	776.0	3%			
		Root Cutting			
11/16/1996	30.0	12%	11/16/1996	36.0	8%

Source: Flocchini, R. G. et al., *Interim Report: Sources and Sinks of PM₁₀ in the San Joaquin Valley.* Crocker Nuclear Laboratory, UC-Davis, CA, 2001.

meteorological data. The average spatial offset between the Lagrangian model and lidar images was 6% during the simulation periods. Because the Lagrangian model is driven by meteorology, its near-field accuracy is highest when input time averages are near the turbulent flow time regime and decreases with height due to errors in the input wind field that was only measured at one height.

The US Environmental Protection Agency (US EPA) (2001) used the empirically derived equation shown below to estimate the quantity of PM emitted from all agricultural tilling processes.

$$E = c \, k \, s^{0.6} \, p \, a \qquad (14.1)$$

where E is the PM emission (in lbs), c is a constant (4.8 lb/acre-pass), k is the dimensionless particle size multiplier (TSP = 1.0, PM_{10} = 0.21, $PM_{2.5}$ = 0.042), s is the silt content of surface soil (%), p is the number of passes or tillings in a year, and a is the number of acres of land tilled. The above equation was developed to estimate TSP emissions (k = 1.0) and has since been scaled to estimate PM_{10} and $PM_{2.5}$ emissions by using the respective k value. Average values of s are tabulated in Table 4.8-6 in the literature (US EPA 2001) as a function of soil type on the soil texture classification triangle.

A comparison between standard tilling practices and conservation tilling (strip-till) in dairy forage production on two farms in the San Joaquin Valley is given by Madden et al. (2008). Both strip-till and standard till operations were monitored for PM_{10} emissions over two tillage cycles at both farms. Results show that conservation tillage practices reduce PM_{10} emissions from one farm by 86% and 52% for 2004 and 2005, respectively. At the second farm, conservation tillage emissions were reduced by 85% and 93% for 2004 and 2005, respectively. Derived emission rates are presented in Figure 14.1 and Table 14.2. Madden et al. (2008) attributed these reductions, in part, to a reduced total number of passes (e.g., from three to six passes in standard tillage to one pass in conservation tillage) and the ability for conservation tillage operations to be done under a higher soil moisture content than standard operations. Dust concentrations produced by agricultural implements used at a UC Davis research farm west of Davis, CA, were reported by Clausnitzer and Singer (1996). Personal exposure samplers measuring respirable dust (RD) concentrations, particles that may reach the alveolar region of the lungs when breathed in (i.e., with a 50% cut-point diameter of 4 μm), were mounted on implements in 22 different operations over a 7-month period in 1994; only 18 operations with replicate samples were reported. Average RD concentrations measured on the implement ranged from 0.33 mg·m⁻³ for disking corn stubble to 10.3 mg·m⁻³ for both land planning and ripping operations. While RD concentration was heavily influenced by the type of operation, other factors determined to be significant in dust production were relative humidity, air temperature, soil moisture, wind speed, and tractor speed.

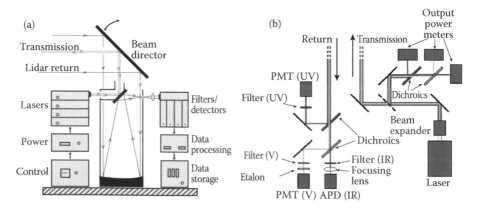

FIGURE 14.1   (a) Aglite lidar conceptual arrangement, and (b) physical layout of the optical components on the optical bench.

TABLE 14.2
Conventional and Conservation Tillage Emission Rates Reported by Madden et al. (2008) for Tillage in a Dairy Forage Crop Rotation

Season/Year	Sweet Haven Dairy		Barcellos Farms	
	Operation	Average Emission Factor (mg/m^2)	Operation	Average Emission Factor (mg/m^2)
Spring 2004	ST: first disking	198	ST: first disking	259
	ST: second disking (w/roller)	1035	ST: second disking	917
	ST: third disking (w/roller)	114	ST: listing	615
	ST: planting	103	ST: bed disking	25
	CT: strip-tilling	181	ST: Bed mulching	89
	CT: planting	26	ST: ring roller	566
			ST: planting	96
			ST: ring roller	104
			CT: planting	394
Spring 2005	ST: first disking	139	ST: first disking	51
	ST: second disking (w/roller)	375	ST: second disking	123
	ST: third disking (w/roller)	404	ST: circle harrow	337
	ST: planting	263	ST: listing	466
	CT: strip-tilling	180	ST: bed disking	109
	CT: planting	385	ST: bed mulching	384
			ST: planting	481
			CT: planting	130

Source: Madden, N. M. et al., *Atmospheric Environment*, 42, 3795–3808, 2008.
Note: ST = standard tillage method; CT = conservation tillage method.

Further investigation of the data set presented by Clausnitzer and Singer (1996) and of another data set collected on a different UC Davis research farm was reported by Clausnitzer and Singer (2000). Both sets of data focus on RD concentrations as measured on the agricultural implement and the analysis examined environmental influences on the measured concentration. Again, soil moisture and air temperature were found to be significant factors in RD production. The RD production with respect to soil moisture was well fitted by a power function, with the curve predicting RD concentrations becoming significantly steeper below 5%. Air temperature was hypothesized to be significant in that it was a surrogate measurement of atmospheric instability (i.e., as temperature increases near the surface, the atmosphere becomes less stable and may carry greater quantities of dust upward).

Baker et al. (2005) examined differences between dust concentrations resulting from standard and conservation tillage practices in the San Joaquin Valley over a 2-year cotton–tomato crop rotation, each under two different cover crop scenarios:

(1) no cover crop and (2) a cover crop forage mixture. Total dust (TD), particles < 100 μm in aerodynamic diameter, and RD samplers were stationed on the implements to collect samples in the plume. For both TD and RD, the presence or lack of a cover crop in the standard till treatment did not seem to affect concentrations. Summed concentrations for conservation tillage without a cover crop were about one-third of standard tillage, and for conservation tillage with a cover crop, they were about two-thirds for both dust fractions measured. Reductions in summed concentrations with conservation tillage were attributed to fewer operations, including the elimination of the dustiest (i.e., disking and power incorporation). When comparing operations common to all four treatments, tomato planting and harvesting in conservation till produced higher concentrations than standard till (i.e., thought to be due to increased organic matter on the surface), and concentrations during cotton harvesting, which does not disturb the soil, were equivalent for all treatments. This study was part of a larger effort to quantify the effects of conservation tillage in California on crop production, soil quality, and time and resources dedicated to production as outlined by Mitchell et al. (2008) and Veenstra et al. (2006).

Upadhyaya et al. (2001) compared the Incorpramaster, a one-pass tillage instrument, against a conventional combination of disking twice and land planning twice based on fuel consumption, timeliness, and effect on soil. Studies on four experimental fields at UC Davis showed no statistical difference between resulting soil conditions (i.e., bulk density changes, soil moisture changes, and aggregate size), but the Incorpramaster (i.e., a new one-pass tillage implement) used between 19% and 81% less fuel with a mean reduction of 50%. The time savings ranged from 67% to 83% with a mean reduction of 72%. In most cases, two passes with the Incorpramaster were required to achieve the same soil conditions as the four passes in the conventional tillage method.

Three conservation tillage methods were compared against the standard tillage method in cotton production and reported in Mitchell et al. (2006) in terms of yield, yield quality, tractor passes, fuel, and production costs. A single field near Fresno, CA, was divided in area among seven tillage treatments: (1) standard, (2) no till/chop, (3) no till, (4) ridge till/chop, (5) ridge till, (6) strip till/chop, and (7) strip till. Prior to both cotton growing seasons examined, small-grain wheat was planted in the field to enhance soil properties; this crop was sprayed with herbicide, and in treatments 2, 4, and 6, it was chopped with a mower prior to tillage activities. In the other treatments (1, 3, 5, and 7), the dead wheat was either incorporated by the tilling or left standing. Yield and yield quality were statistically the same for both years for all treatments, though the standard treatment was numerically higher the second year. Conservation tillage treatments reduced tractor passes by 41% to 53% over the standard method and estimated fuel reduction was 48% to 62% for the conservation practices. The estimated overall production costs of the conservation tillage systems were 14% to 18% lower than the conventional system. Mitchell et al. (2008) estimated by extrapolation from other work that whole-tillage process particulate matter emissions would also be decreased (Table 14.2).

Particulate matter is released during agricultural tillage activities from both the operational activity of the tillage implement as well as the tractor in use. Emissions from the tractor mainly originate from the tires in contact with the soil and the combustion engine. Attempts to quantify the PM emitted in agricultural tractor exhaust have been made by the US EPA and the California Air Resources Board (CARB) in software designed to estimate off-road engine emissions on a county or regional scale (CARB 2003a,b). The US EPA software program was developed by the NONROAD, with the latest version distributed in 2005 (EPA 2008). Emission factors from compression ignition (diesel) engines used in the NONROAD model are calculated by adjusting a zero-hour, steady state measured emission factor (EF_{ss}) for engine deterioration with operation time (DF) and a transient adjustment factor (TAF) that accounts for variations from steady state engine loading and speed, as shown in the following equation (US EPA 2004):

$$EF_{adj(PM)} = EF_{ss} \, TAF \, DF\text{-}S_{PMadj}, \qquad (14.2)$$

where $EF_{adj(PM)}$ is the adjusted PM emission factor, and S_{PMadj} is the emission factor adjustment accounting for the use of a diesel fuel with a sulfur content different than the default concentrations, as fuel sulfur level is known to affect PM emissions. The units for $EF_{adj(PM)}$, EF_{ss}, and S_{PMadj} are g·hp^{-1}·hr^{-1}, where hp stands for horsepower, and TAF and DF are both unitless. All four variables on the right-hand side of Equation 14.2 vary with model year and engine size, expressed in hp, according to measured values and/or the emission standards each model year and engine size was designed to meet. The selection of values for steady state emission factors and all the adjustment variables given in US EPA (2004) was performed using a variety of tests and resources, including the Nonroad Engine and Vehicle Emission Study (NEVES) Report (US EPA 1991), or by setting the values such that the adjusted PM emissions were equal to model year-specific emission standards.

CARB has also developed a model to forecast and backcast daily exhaust emissions from off-road engines, including agricultural tractors, called OFFROAD. Similarly to the US EPA's NONROAD model, emission factors (EF) for each engine size and model year are calculated based on a zero-hour emission rate (ZH) with a deterioration factor (DF) applied to account for engine wear with use, as in Equation 14.3. The derived emission factor is then multiplied by the load factor (LF), the maximum rated average horsepower (HP), and the amount of time the engine is active through the year (Activity) in hours per year,

$$EF = ZH + DR\text{·}CHrs \qquad (14.3)$$

$$P = EF \, HP \, LF \, Activity \, CF, \qquad (14.4)$$

where $CHrs$ is the cumulative engine operation hours. The expression for the amount of pollutant released in tons·day^{-1}, P, is shown in Equation 14.4 where CF is the conversion factor from units of grams per year to tons per day. The values for the EF and DR are derived from measured values or they are calculated based on requirements to meet the proposed emissions limits for future years (CARB 1999).

Kean et al. (2000) estimated off-road diesel engine, locomotive, and marine vehicle emissions of NO_x and PM_{10} for 1996 based on fuel sales. Diesel engine exhaust emission factors were developed based on information provided in the development of the US EPA NONROAD off-road vehicle emissions model with supplemental information in order to calculate emissions based on fuel consumption. A fleetwide average PM_{10} emission factor was determined for farm diesel equipment to be 3.8 $g \cdot kg^{-1}$ of fuel used, at an average mass per volume of 0.85 $kg \cdot L^{-1}$ of diesel fuel. Fuel sales surveys from 1996 were used to calculate regional and national emissions. In the off-road category, which includes farm equipment, the US EPA NONROAD model calculated on average 2.3 times higher emissions, which was attributed to higher engine activity assumed in the US EPA model than represented in the reported fuel sales data.

14.3   MEASUREMENTS AND METHODS

Aglite system description. The primary system requirements of the Aglite lidar were to (1) make the system eye-safe at the operating range, (2) make the system sufficiently robust and portable to be deployed at an agricultural site, and (3) make the scanning and data logging sufficiently fast to capture the dynamics and structure of an entire plume.

The desire to make Aglite as eye-safe as possible drove the decision to employ a micropulse laser that somewhat mitigates eye-safety issues because of the low individual pulse energies in the microwatt range. Aglite uses photon-counting detection due to these low pulse energies. The lidar has a narrow field of view (FOV) and uses narrowband optical filters to limit the solar background. The design concepts for Aglite and its component layout are shown in Figure 14.1a and b, and are based in part on the design by McGill et al. (2002).

The laser is an injection-seeded Nd:YAG equipped with second harmonic and third harmonic generation crystals and operates at 10-kHz repetition rate. The laser simultaneously emits three coboresighted wavelengths: 1064, 532, and 355 nm. Immediately after leaving the laser head, the beam passes first through a 7 × beam expander, at which point the beam far-field full divergence angles are 0.37, 0.30, and 0.27 mrad, respectively. Average output powers employed are 4.35, 0.48, and 0.93 W, respectively, yielding pulse energies of 435, 50, and 93 µJ. After passing through several turning mirror and optional neutral density filters, the beam is directed onto a launch mirror mounted on the central obstruction of the telescope, at which point the laser is coaligned with the receiver telescope. Aglite employs a 28-cm Newtonian telescope with a focal length of 0.91 m. A field stop confines the telescope's full field-of-view to 0.45 mrad. Movement of the large pointing mirror controls the direction the lidar looks, but does not change the relative alignment of the outgoing beam and return signal. The pointing mirror actuators have a pointing knowledge resolution of 1 mrad and pointing repeatability of 5 mrad.

As shown in Figure 14.1b, the signal is collected by the telescope and separated into its component wavelengths using dichroic mirrors. Each channel then passes through several optical filters including a narrowband interference filter. The 1064-nm channel is focused onto a silicon avalanche photodiode (APD), while the remaining two channels pass into photomultiplier tubes (PMTs). The choice of detectors for Aglite was based on the generally superior signal-to-noise ratio (SNR) for PMTs in the visible and ultraviolet wavelengths, given sufficient filtering of the green intensity maximum of sunlight, and the infrared sensitivity of the APD. This conclusion has also been confirmed by Agishev et al. (2006). A commercial high-speed, multichannel photon counting interface was used to log and record the photon counts for each channel. Custom laser drive and signal processing hardware and software used to operate Aglite were described in detail by Cornelsen (2005).

The minimum system range gate is 6 m; however, the range resolution for the data presented here is approximately 12 m, as defined by the detector electronics, the respective laser pulse lengths, and the pulse discrimination times of the detectors. Aglite's maximum range is 15 km. A measurement integration time of 1 s was used for all data presented here.

The laser head, transmission optics, and receiver optics are all mounted on a single optical bench and vertically oriented in a six-point vibration-isolated frame along with the telescope. The optical path is housed under light-tight aluminum covers that also serve to keep the optics clean. The entire frame is covered with removable panels.

The beam steering is accomplished using a flat mirror attached to motorized actuators that allow for 270° motion in the horizontal and +45° to −10° motion in the vertical. Angular scan rates from 0.1°/s to 1°/s are used to develop the 3-D map of the source(s), dependent on the range and concentration of the aerosol (Marchant et al. 2009). The mirror and actuators can be retracted hydraulically inside the trailer for travel. A video surveillance camera looks at the director mirror to provide the operator a 5° FOV along the beam path for safety monitoring. Data from the lidar can be linked by WiFi network to other sites and bring environmental data to the operator. In our study, the Aglite lidar system was deployed at a tillage location (Figure 14.2) with parameters being listed in Table 14.3.

The lidar Equation 14.5 describes the lidar return signal as a function of range z for wavelength λ:

$$P_\lambda(z) = P_0 \cdot L \cdot \frac{c\tau}{2} \cdot A_\lambda(z) \cdot \frac{\beta_\lambda(z)}{z^2} \cdot \exp\left(-2\int_0^z \sigma_\lambda(z')\partial z'\right). \tag{14.5}$$

The term $P_\lambda(z)$ is the measured reflected power for distance z and is measured in photon counts, P_0 is the output power of the lidar, L is the lidar coefficient, which

FIGURE 14.2   Three-wavelength Aglite lidar at dusk, scanning a harvested wheat field.

TABLE 14.3
Aglite Lidar System Parameters

Laser/Telescope		Detectors/Bandwidths		Resolutions	
Wavelengths	1064, 532, 355 nm	Effective filter bandwidth (FWHM)	150 pm at 1064 nm 120 pm at 532 nm 150 pm at 355 nm	Data time resolution	0.1 sec possible; typical 1–3 s
Laser type	Solid-state Nd:YAG				
Laser pulse rate	10 kHz	Filter efficiency	70% at 1064 nm 70% at 532 nm 76% at 355 nm	Data range resolution	18 m at 1064 nm 12 m at 532 nm 12 m at 355 nm
Laser output energy	435 μJ at 1064 nm 50 μJ at 532 nm 93 μJ at 355 nm	Etalon free spectral range	37 pm at 532 nm		
Telescope diameter	28 cm	Etalon reflectivity finesse	6.1 at 532 nm	Steering resolution	5 μrad
Telescope type	Newtonian	Etalon peak transmission	>98.4% at 532 nm		
Telescope FOV	450 μrad, full angle	Detector photon counting efficiency	2% at 1064 nm 14% at 532 nm 27% at 355 nm		

represents system efficiency, c is the speed of light, τ is the pulse width of the lidar, $A_\lambda(z)$ is the effective area function, which includes the geometric form factor (GFF), $\beta_\lambda(z)$ is the atmospheric backscatter coefficient, and $\sigma_\lambda(z)$ is the atmospheric extinction coefficient. The backscatter and extinction coefficients are functions of temperature, pressure, humidity, and the background and emitted aerosols.

Expected SNRs calculated by Marchant et al. (2009) using synthetic data for the Aglite lidar at 20% and 100% emitted power and 1 s, full-range resolution are shown in Figure 14.3. These plots were made using system calibration constants measured in the field. Standard temperature and pressure and 0% humidity were assumed. The background aerosol was assumed to have the same properties as the continental average aerosol from the OPAC aerosol database (Hess 1998). In these plots, SNR is defined as the ratio of the mean aerosol backscatter amplitude over the standard deviation of the aerosol backscatter. Because the noise is not correlated, SNR can be increased by both time and range averaging (Del Guasta et al. 1994).

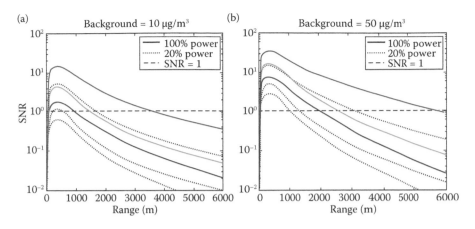

FIGURE 14.3 Aerosol backscatter SNR for the Aglite lidar, corrected for molecular back-scatter with 1 second averaging, as a function of range for 20% and 100% laser power levels at two PM10 background loading levels, (a) is for 10 µg/m³ and (b) is for 50 µg/m³.

14.4 *IN SITU* AEROSOL INFORMATION

An example of an Aglite calibration tower having both MiniVols and OPCs is shown in Figure 14.4a. The solution of Equation 14.5 requires knowledge of the optical parameters of the both the background and source aerosols, which need to be measured at one or more reference points upwind (background) and downwind

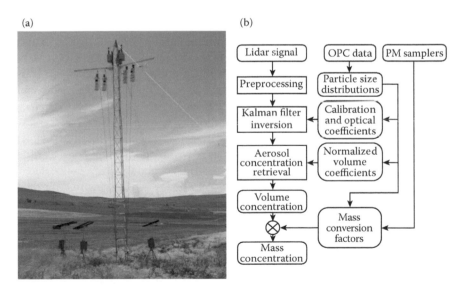

FIGURE 14.4 (a) Example of the array of particulate mass and optical particle counters used in the lidar calibration. (b) Retrieval process used to convert lidar returns to particulate mass fraction. (Modified from Marchant, C. C. et al., *Journal of Applied Remote Sensing*, 3, 033511, 2009, doi:10.1117/1.3097928.)

(background plus source) of the facility. In our approach (Zavyalov et al. 2009), both optical and aerodynamic mass faction sensors were used to develop these parameters (see Figure 14.4b). Aerodynamic mass fraction samples were collected using MiniVols (Airmetrics, OR.), which are portable, self-contained, filter-based impactor particulate samplers. Chow (2006) demonstrated that $PM_{2.5}$ and PM_{10} concentrations measured by MiniVols in California's San Joaquin Valley did not differ statistically from the concentrations measured by the collocated Federal Reference Method (FRM) samplers (Chow et al. 2006).

We have also conducted extensive calibration and intercalibration comparisons of the MiniVols against FRM samplers located at an air quality sampling site in Logan, UT, operated by the Utah Division of Air Quality. *In situ* particle size profiles are collected in parallel with the MiniVol samplers using Aerosol Profilers, Model 9722 (Met One Instruments, Inc., Grants Pass, OR). This optical particle counter (OPC) uses a laser to count and size particles at a sampling rate of 1 L/min with a sheath flow of 2 L/min. The counts are grouped into eight user-specified size bins from 0.3 to >10 μm (0.3–0.5, 0.5–0.6, 0.6–1.0, 1.0–2.0, 2.0–2.5, 2.5–5.0, 5.0–10.0, and ≥10 μm in our studies). The OPCs can be read out at a range of times from 2 to 60 s per sample.

14.4.1 Lidar Retrieval Calibration

The raw data from a lidar system are in units of photons. The units used to describe concentration of airborne particulate matter are often mass per unit volume (i.e., $g \cdot m^{-3}$). While it is conceivable that with exquisite knowledge of both the optical properties of the lidar itself (e.g., mirror reflectivity, filter transmission) and the optical properties of the aerosol under test (n, density, size distribution, concentration), one could calculate from first principles the relationship between photon counts of all three Aglite laser wavelengths and particle concentration; such a problem, however, quickly becomes intractable in any realistic setting. In practice, it is far more reliable to simply calibrate the Aglite laser system periodically against several *in situ* aerosol sensors. This approach avoids the uncertainties associated with changes in individual optical or electronic components of the lidar itself and addresses the well-known problem of the heterogeneous aerosol mixtures. In these experiments, we typically calibrated Aglite every 15 min. A "calibration" event consists of aiming Aglite at a calibration tower upon which MiniVol and OPC are collocated, and collecting approximately 2 min of lidar data. In this way, we knew that for those 2 min the OPC and MiniVol concentration and mass readings define the exact same aerosol that Aglite is interrogating; thus, the calibration station data directly map onto the lidar photon counts. Subsequent changes in either Aglite, local meteorology, or the aerosol itself are actively calibrated into the signal each time a calibration event occurs.

The details of our lidar calibration and aerosol retrieval process are discussed in detail by Marchant and Zavyalov (Marchant 2007; Zavyalov et al. 2006). Our method for the determination of whole facility emissions does not depend on the specific method used for solving the lidar equation; for the purposes of the following discussion, we will use the conventional Klett method for inverting the lidar equation into concentration while simultaneously acknowledging that in many cases more suitable retrieval algorithms exist (Marchant 2010).

The lidar return power from range z for a single scatter is shown in Equation 14.5. In the case of two distinct classes of scatters, $\beta_\lambda(z)$ and $\sigma_\lambda(z)$ represent the total backscatter and extinction from the sum of a background scattering component plus an emission source scattering component. The background scattering component represents homogeneous scatterers, including both background aerosol scattering and molecular scattering, which is expected to be constant over the relatively short ranges near the ground where Aglite is used. These contributions of aerosol scatterers to these coefficients are derived from aerosol sampler data, while the contributions due to molecular scattering are calculated using data from portable meteorology. The algorithm used to retrieve aerosol physical parameters from a raw Aglite lidar signal shown schematically in Figure 14.4b involves four major steps (Zavyalov et al. 2006; Klett 1985) that account for the geometrical form factor of the telescope receiving optics and scattered sunlight background radiation. The routine then calculates the optical parameters (backscatter and extinction coefficients) of the background and source aerosols at three wavelengths. A solution to the lidar equation for two scatterers was shown by Fernald et al. (1972). The algorithm uses Klett's form of the solution, which is mathematically equivalent, but in a more compact form.

An *in situ* OPC instrument is used to provide the reference point values needed for Klett's solution (Klett 1985). This OPC is mounted at the top of a calibration tower and the lidar beam path is directed past it off to the side (Figure 14.4a). The backscatter coefficients of the background aerosols are calculated using Mie theory and the particle size distribution measured by the OPC, while the molecular backscatter coefficients are calculated using the current temperature, pressure, and humidity as measured by meteorological instruments. These measured backscatter coefficients provide the reference values needed by Klett's retrieval equation. These backscatter coefficients are divided by the return power measured at the reference range, resulting in calibration constants for the lidar measurement. When the lidar is not pointed past this calibration tower, these calibration constants are used to determine backscatter values at the reference range.

The main assumptions in the retrieval process are as follows: a bimodal lognormal aerosol size distribution, aerosol particles are spherical, and the aerosol index of refraction, mode radii, and mode geometric standard deviations are constant in time and space. The mode radius describes the peak value of a mode, while geometric standard deviation describes the width of a mode. Once the particle size distribution and number concentration parameters are estimated, the mass concentration of particles with different size ranges (e.g., $PM_{1.0}$, $PM_{2.5}$, PM_{10}) is calculated using aerodynamically sized information of the particulate chemical composition and concentration measured by the MiniVol samplers. Figure 14.4a shows a calibration experiment arrangement, where multiple OPC and PM samplers are arrayed together for estimation of mass conversion factor (MCF) used to convert optical data measured with OPC and lidar to mass concentration units. Twenty-minute lidar stares at the tower under uniform conditions were used in this case to develop error performance data on the lidar and retrieval processes (Zavyalov et al. 2009).

14.4.2 Wind and Environmental Information

Wind profile information is provided by cup anemometers (Met One Instruments, Inc., Grants Pass, OR) located at 2.5, 3.9, 6.2, 9.7, and 15.3 m on two portable towers. These towers also support an array of aerosol samplers. The tower-based wind information is supplemented by tethered balloon profiles collected at 5-min intervals to observe boundary layer structure. The wind, humidity, temperature, and OPC data are transmitted to the lidar- and data-processing trailers for storage and experiment management.

14.4.3 Flux Measurement Method

The lidar's capability to accurately sample 3-D aerosol concentrations entering and leaving an operation in near-real time (1–3 min) makes it possible to measure facility emissions with approximately twice that time resolution. The concept behind our approach, where the facility is treated as one would calculate the source strength in a bioreactor, can be shown in Figure 14.5a. In this approach, the source strength is determined using the mean flow rate through the reactor and the difference in reactive species concentration entering and leaving the vessel. The scanning lidar samples the mass concentration fields entering and leaving the facility, while standard anemometers provide the mean wind speed profile. This same simple relationship applies to defining a box large enough that no source material escapes through the top or side.

In applying this method, the downwind face of the box has to be far enough from the facility that the anemometers provide an accurate wind speed profile. An example of our lidar-derived concentration data can be shown in Figure 14.5b. The location–concentration plot pattern from scanning up one side, across the top, and down the other resembles a common office staple used to clip papers together and is therefore referred to as a "staple scan." The data from the top of

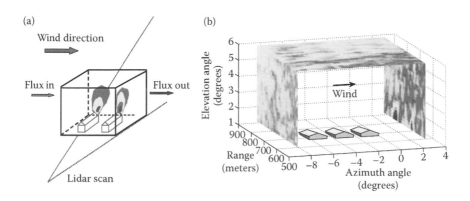

FIGURE 14.5   (a) Conceptual illustration of the scheme for using lidar to generate time-resolved local area particulate fluxes. (b) Example of a "staple" lidar scan over the facility showing aerosol concentrations on the three sides of the box.

the box are monitored such that no source particulate transport passes through the top. The data for the left-side panel of the staple provide the background concentration entering the box, while those on the right provide the background plus facility concentration leaving the box. The short-term flux is calculated by multiplying the area integrated mass concentration difference by the wind speed during the scan.

The flux equation in the integral form for calculating emission is

$$F = \int_r \int_h \bar{v}_\perp(r,h) \cdot \left(C_D(r,h) - C_{\bar{U}}\right) dr \, dh, \tag{14.6}$$

where $\bar{v}_\perp$ is the average wind speed component, the direction of which defines the long axis of the box, and $C_D - C_{\bar{U}}$ forms the mass concentration difference upwind and downwind, integrated over the range (width) and height of the exit plume. In our routine, F is conceptually calculated as

$$F = \sum_{i=R_0}^{R} \sum_{j=H_0}^{H} \bar{v}_{\perp ij} \left(C_{Dij} - C_{\bar{U}}\right) \cdot \Delta r \cdot \Delta h, \tag{14.7}$$

where R_0 and R are the near and far along beam edges of the box, and H_0 and H form the top and bottom of the box, respectively. In many cases, H_0 is set above eye level and concentration is extrapolated to the ground to avoid illuminating personnel and animals. The $\Delta r \cdot \Delta h$ term is the individual area element for which each flux component is calculated by each step in the double summation.

14.5   FLUX MEASUREMENT ERROR ESTIMATION

Since the flux measurement (f) is a function of several variables, $f(x_1, x_2,...,x_n)$, and the uncertainty of each variable, x_i, is known, the uncertainty of f can be calculated. If the variables are assumed to be independent, the flux error can be expressed as the square root of the sum of the squares of the uncertainty induced by each individual variable (Met One Instruments, Inc., Grants Pass, OR), that is

$$\Delta f = \sqrt{\sum_{i=1}^{n} \left(\frac{\partial f}{\partial x_i} \Delta x_i\right)^2}, \tag{14.8}$$

which for our flux calculations breaks out as

$$\Delta F^2 = \sum_i \sum_j \left(\left(\frac{\partial F}{\partial \bar{v}_{\perp ij}} \Delta \bar{v}_{\perp ij}\right)^2 + \left(\frac{\partial F}{\partial C_{Dij}} \Delta C_{Dij}\right)^2 + \left(\frac{\partial F}{\partial C_{\bar{U}}} \Delta C_{\bar{U}}\right)^2\right) \cdot \Delta A_{ij}^2 \tag{14.9}$$

or

$$\Delta F = \left(\sum_i \sum_j \left(\left(\left(C_{Dij} - C_{\bar{U}} \right) \Delta \bar{v}_{\perp ij} \right)^2 + \left(\bar{v}_{\perp ij} \Delta C_{Dij} \right)^2 + \left(\bar{v}_{\perp ij} \Delta C_{\bar{U}} \right)^2 \right) \right)^{1/2} \cdot \Delta A_{ij}, \quad (14.10)$$

where $v_\perp$ is the wind speed in the direction perpendicular to the lidar scan (cos θ corrected). Components that are crosswise to the box do not contribute to the flux error because the box is defined large enough that none of the source material leaves through the sides or top. Specific terms in our error analysis include $v_\perp$, the concentration terms, and the MCF, which is included in Equation 14.10 as a constant in the concentration calculation. Additional understanding of the flux error estimate can be seen from further rearrangement. Substituting averaged parameters over the inferred area from Equations 14.6 to 14.9, a simplified equation can be obtained:

$$\left(\frac{\Delta_F}{F} \right)^2 = \left(\frac{\Delta_v}{v_0} \right)^2 + \frac{\Delta_{C_D}^2 + \Delta_{C_U}^2}{\left(C_D - C_U \right)^2} + \left(\frac{\Delta_A}{A} \right)^2 \approx \left(\frac{\Delta_v}{v_0} \right)^2 + 2 \frac{\Delta_{C_{D,U}}^2}{\left(C_D - C_U \right)^2} + \left(\frac{\Delta_A}{A} \right)^2. \quad (14.11)$$

Typically, the mass concentration errors are the same from the downwind and upwind sides. In this case, the right-hand side approximation is valid. As shown by Zavyalov et al. (2009), the errors are now shown in fractional form. In this form, two things become obvious that are not as easily seen in the earlier form. First, the center term includes a 2, which enters in because the flux requires both upwind and downwind calculations. The second is the relationship between the values of C_D and C_U. When the two terms are of similar size, their difference in the denominator quickly dominates over the other terms.

Wind speed and direction errors are dominated by sampling issues and not by the instrument calibration as we are estimating the wind field for a parameter averaged over a 200 to 300 m area at a particular time. For this error calculation, we estimate the wind errors using the standard deviation of the direction and velocity over a 20-min sampling period. For our field campaigns described in this analysis, these errors are typically 10% to 15% of the wind value. We set our wind dataloggers to collect 1-min averages and standard deviations to provide a quality control value for the flux data.

Short-term error calculations after Marchant et al. (2009) for the Aglite system and the Zavyalov et al. (2009) particulate volume concentration retrieval calibration as applied in the flux calculation are shown in Table 14.4. To understand the flux error, we consider that the sides of the flux box include data collected over about 1 min and ranging from 600 to 1000 m. In the flux calculation, the range and scan data are rolled into the single flux number. Flux error analysis data were collected during a calibration stare past the OPC-MiniVol array with the beam horizontal to the ground and an upwind scan taken after the stare. The upper section shows the SNR ($\Delta C_v/C_v$) for a (1-s) stare without OPC noise, while the lower section shows the measured system (lidar plus OPC) volume concentration SNR for the lidar measurement during the 1-min scan time typically used for flux measurements, as in Chow et al. (2006).

TABLE 14.4
Aglite Lidar 1- and 60-s Aerosol SNR Measured at 5% Power under Uniform Conditions at the Ranges Normally Used in Flux Measurements

	Lidar SNR (Average/Standard Deviation) during a Horizontal Stare					
Range	PM	600 m	700 m	800 m	900 m	1000 m
PM$_{2.5}$	9.0	69	66	63	60	57
PM$_{10}$	25.4	6.3	6.0	5.7	5.4	5.2
TSP	50.2	4.4	4.2	4.0	3.8	3.7
	Concentration Calibration SNR (Lidar and OPC) during a 60-Second Stare					
PM$_{2.5}$	9.0	239	229	218	208	197
PM$_{10}$	25.4	22	21	20	19	18
TSP	50.2	15	15	14	13	13

Source: Marchant, C. C. et al., *Journal of Applied Remote Sensing*, 3, 033511, 2009, doi:10.1117/1.3097928; Zavyalov, V. V. et al., *Journal of Applied Remote Sensing*, 3, 033522, 2009, doi:10.1117/12.833.

The data in Table 14.4 were collected in system performance experiments under stable aerosol conditions with continental aerosol. For these measurements, the system was operating at approximately 5% power, and OPC measured particulate background concentrations were 9.0, 25.4, and 50.2 μg/m³ (MCF = 1).

These values can be compared with the overall SNR calculated for the system in Figure 14.3, where SNR is given in terms of backscatter. The SNR for increasing particle diameter decreases as the particle-size-to-wavelength ratio increases.

A flux calculation example, with the error estimates and magnitude data demonstrating the primary terms of Equation 14.10, and the resultant hourly average flux or emission strength estimate obtained using the Aglite system are shown in Table 14.5. These data are typical values from our system precision experiments and are designed to show the experienced precision.

TABLE 14.5
Example Aglite Lidar System Derived Aerosol Average Emission Flux Component Measurement Error and Estimated Flux Error Determined for Uniform Conditions

PM Bin	Wind Direction (°)	C_D–C_U (μg/m³) @ MCF = 1	MCF	C_D–C_U (μg/m³)	Wind (m/s)	Area (m²)	Mass Flux (g/s)
PM$_{2.5}$	330 ± 9.2	0.70 ± 0.002	3.04 ± 0.3	2.19 ± 0.21	4.7 ± 0.69	1000	0.01 ± 0.005
PM$_{10}$	330 ± 9.2	21.0 ± 0.58	2.10 ± 0.3	44.10 ± 0.64	4.7 ± 0.69	1000	0.21 ± 0.043
TSP	330 ± 9.2	57.4 ± 2.24	1.88 ± 0.3	107.91 ± 17.7	4.7 ± 0.69	1000	5.07 ± 0.11

14.6 FLUX MEASUREMENT EXAMPLES AND RESULTS

In this section, we show three examples where the Aglite system is applied to agricultural system analysis, and we compare those examples with traditional sampler/model results. Each example illustrates the system's effectiveness for long- and short-term measurements and shows lessons learned as the system has evolved.

14.6.1 SWINE FINISHING FACILITY MEASUREMENTS

The Aglite measurement system was applied to an Iowa swine feeding emissions experiment August 24 through September 8, 2005. The swine farm data provided a uniform, fixed, nearly constant flux demonstration (except for periods when road dust plumes from a nearby county road occurred). The fairly steady wind and steady operations during this experiment provided ideal conditions for demonstrating the flux calculation method. The facility consisted of three separate parallel barns, each housing around 1250 pigs, with 1.4-m-tall screen-vented windows running along nearly the length of the north and south sides of the barns. The areas of the facility not used for barns, feed-bin footprints, or access roads were covered with maintained grass. Cultivated fields surrounded the facility with corn to the north, south, and west and soybeans to the east. The barns were located approximately 650 m from the lidar. A 20-m tower was sited between two of the barns to support the aerosol and micrometeorological instruments at three logarithmically spaced heights. A particulate diagnostic trailer was located 50 m in the general downwind direction from the barns. Other instrument support towers were located around the facility.

Example single scans of the upwind (C_U, PM_{10} only) and downwind (C_D) staple face show the mass concentration values for PM_{10}, $PM_{2.5}$ between 400 and 900 m from the lidar (Figure 14.6). The figure shows the typical structure observable in a uniform background and typical plume profile. Each vertical scan was collected in approximately 1.25 min.

The vertical profile of horizontal mass concentration difference of the downwind minus upwind layers can be easily obtained from these data (Figure 14.6c). Single scan differences, of course, do not account for accumulation or depletion in the measurement box due to wind speed variation during a scan, for input background variation, or for storage in or flushing of the box due to the existing large-scale wind eddy structure (i.e., we do not attempt to measure the same air mass at the upwind and downwind scans). Negative features can be observed in the individual profiles. Several scans are required to achieve a meaningful mean estimate of the facility emission. For calculation efficiency, we calculate flux through the downwind surface first and then the upwind flux, differencing the flux rather than concentration. Choosing an area that fully includes the source plume but not a lot of extra area eliminates the need to spend resources calculating for pixels with difference equal to zero. The C_D and C_U area average measurements provide aerosol mass concentration ($\mu g \cdot m^{-3}$), which, when multiplied by the wind speed ($m \cdot s^{-1}$), provides the area average flux (F_D and F_U, $\mu g \cdot m^{-2} s^{-1}$). Differencing ($F_D - F_U$) and integrating over the plume area provide the facility emission estimate (F_S, $\mu g \cdot s^{-1}$) for that staple.

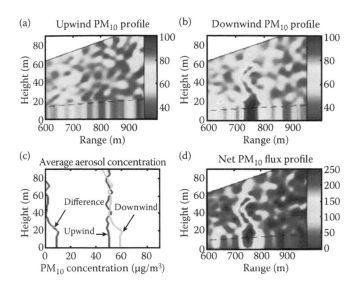

FIGURE 14.6 Single upwind (a) and downwind (b) PM_{10} scans of the swine facility airmass showing the distribution of the background and facility leaving plume concentrations ($\mu g/m^3$), the horizontally averaged PM_{10} concentrations and their difference (c), and the PM_{10} particulate flux ($\mu g \cdot m^{-2} \ s^{-1}$) distribution (d) calculated when the difference is multiplied by the wind speed profile.

Emission summary data collected by various methods during the Iowa measurement sequence are shown in Table 14.6. Martin (2006) provided the emissions calculated from the sampler data and modeled facility flux during this experiment. Emission rates were also estimated from lidar-measured fluxes dividing total day fluxes by the number of pigs inside the flux box. This site has a gravel road on the

---

TABLE 14.6

Comparison of the Ambient (Background) and Facility Mass Concentration and Emission Rate (g/Pig-Day) Data Measured by Filters (23-h Base), OPC (24-h Base), and the Lidar (2-h Base)

	PM Samplers		OPC Data		Lidar Data	
C_M–$\mu g/m^3$	Ambient	Plume	Ambient	Plume	Ambient	Plume
E_M–g/pig/day	(C_U)	(C_D)	(C_U)	(C_D)	(C_U)	(C_D)
C_M–PM_{10}	38.7 ± 5.4	49.4 ± 8.3	34.4 ± 24	42.2 ± 28	37.1 ± 18	52.8 ± 21
(with dust)			28.6 ± 7.8	38.7 ± 7.8	30.2 ± 2.5	46.4 ± 6.5
Without dust						
C_M–$PM_{2.5}$	13.3 ± 3.2	14.7 ± 3.3	14.3 ± 9.0	17.2 ± 9.7	11.2 ± 7.2	12.8 ± 6.5
(with dust)			13.7 ± 4.7	16.7 ± 6.6	9.5 ± 0.8	11.6 ± 1.4
Without dust,						
E_M–PM_{10}		0.83 ± 0.44				0.42 ± 0.13
E_M–$PM_{2.5}$		0.09 ± 0.03				0.09 ± 0.04

upwind (south) side that had traffic at a rate of one to two cars per hour. The road dust could not be separated in the impactor particulate sampler data, but was identifiable and could be processed separately in the OPC and lidar data (see PM concentrations "without dust" in Table 14.6). The modeled PM sampler data are similar to PM_{10} and $PM_{2.5}$ emission rate values from the lidar. The PM and OPC data were collected over an 8-day period, while the lidar data were calculated hourly.

This was an early deployment of the Aglite system, and consistent flux measurement scans were collected only for a 2-h period for each particulate class. Considering the large difference in the sample periods and that fugitive dust was not excluded from the filter data, we conclude that these data show the Aglite system's capability to characterize a stationary facility with fairly uniform emissions in a relatively short period of time. While the magnitude of the $PM_{2.5}$ emission rate shows excellent agreement, the lidar PM_{10} emission rate was roughly half of what was indicated by the long-term filter data. The $PM_{2.5}$ emission was relatively constant over the entire sampling period, while significant structure was observed in the PM_{10} data dominated by road dust and feeding operations. Since the filter data incorporated emissions from both road and facility operations, this difference is consistent with the lower level of road traffic that occurred during the lidar flux measurement period.

14.6.2 COTTON GIN MEASUREMENTS

Measurements at a working cotton gin facility were made from December 11 to 14, 2006. For this deployment, 13 MiniVol impactors (total) were distributed in clusters with PM_1, $PM_{2.5}$, PM_{10}, and TSP heads, and five MetOne OPCs were used for particulate characterization. Facility emission rates were not calculated from inverse modeling using measured PM concentrations as explained by Martin et al. (2006) due to the facility layout (emissions mainly coming from the elevated cyclone outlets). The lidar was located 800 m SW of the gin, with reference towers directly north and east of the lidar. A wind profile tower supporting the five levels of anemometers and temperature sensors, a wind direction and rain gauge, and two levels of OPC and MiniVol clusters was located near the gin. A second, similarly instrumented tower located south of the gin was used to provide ambient conditions. The season presented a nighttime fog challenge that limited lidar operation to daylight hours, and occasional gin operation interruptions were observed as the operators performed maintenance and mechanical adjustments. The site provided a diurnally rotating wind condition that made emission aerosol measurements more challenging.

Lidar operations showed two continuous plumes in scans above the facility, one from gin stack effluents not captured by the cyclones, and a smaller plume originating at the seed pile, which we assume resulted from the wind picking up aerosols from the falling seed stream. Other activities such as vehicle movement and dumping and transferring the cyclone trash were intermittent sources captured by the lidar. Sequential lidar measurements taken during two days of fairly uniform conditions are shown in Figure 14.7. Of the 111 scans, 62 were taken on December 12 and the remainder on December 14. These data show relatively consistent gin operation, with

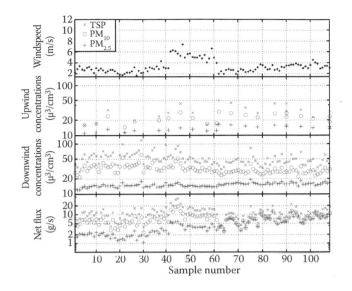

FIGURE 14.7 Wind speed, upwind and downwind volume concentrations, and mass flux calculated using the Aglite data collected on December 12 and 14, 2006. December 14 data begin at point 62.

both days punctuated by downwind concentration spikes associated with noncontinuous activities on the site. There is an increase in fine particulates the second day, with nearly equal $PM_{2.5}$ and PM_{10} flux. The top plot shows the wind speed value used in the flux calculation, with the two middle plots showing the volume concentration of C_U and C_D in $\mu m^3 \cdot cm^{-3}$. While this is a somewhat unusual unit, it is the last step before converting to mass/m^3 and is equivalent to $\mu g \cdot m^{-3}$ if the volume to MCF is 1 (the particulates had the density of water). The net flux in the bottom plot is the product of the $C_D - C_U$ difference multiplied by the MCF and the wind speed. The MCF values are presented in Table 14.7. An error in the MCF calculation (Zavyalov et al. 2009) could explain the increase in PM2.5 value approximately doubling on the second day.

At the time these data were collected, it was assumed that the upwind aerosol concentrations were uniform and would be sampled at a significantly lower rate that the downwind values. This experiment and the almond experiment discussed in the following section show that uniform upwind conditions cannot be assumed and that even distant upwind activity can add pulses of upwind particulates. Measurement variability for the combined period and for each of the individual days is shown in Table 14.7.

14.6.3 Almond Harvesting Measurements

The Aglite system was applied to a mobile source emitter in an almond harvest at the Nickels Soils Laboratory Research Farm near Woodland, CA, from September 26 to October 11, 2006. This experiment compared lidar-based measurements to

TABLE 14.7

Cotton Gin Aerosol Mass Concentration, MCF, and Flux Statistics for Figure 14.7

Concentration (μg/m³)	Whole Period			First Day			Second Day		
Flux (g/s)	$PM_{2.5}$	PM_{10}	TSP	$PM_{2.5}$	PM_{10}	TSP	$PM_{2.5}$	PM_{10}	TSP
C_U-ave	40.2 ± 1.28	42.4 ± 3.1	52.2 ± 7.1	38.4 ± 1.9	43.0 ± 6.2	46.1 ± 11.9	41.7 ± 1.2	41.8 ± 2.7	57.1 ± 7.9
C_U-σ	2.92	7.06	16.27	2.89	9.52	18.18	2.07	4.63	13.36
C_D-ave	45.1 ± 0.7	48.3 ± 3.5	79.5 ± 7.1	44.7 ± 0.7	72.6 ± 3.5	105.9 ± 7.1	44.1 ± 0.7	46.1 ± 1.5	72.7 ± 4.6
C_D-σ	2.95	13.78	28.11	2.95	13.78	28.11	2.51	5.34	16.58
MCF				3.04	2.10	1.88	2.69	1.63	1.59
F-ave	4.5 ± 0.5	7.1 ± 0.6	11.7 ± 1.0	2.7 ± 0.3	7.2 ± 0.8	12.6 ± 1.5	6.8 ± 0.7	7.0 ± 0.8	10.7 ± 1.3
F-σ	2.82	3.06	5.52	1.24	3.34	6.04	2.57	2.70	4.65

Note: Concentrations and flux means are shown with 95% confidence intervals.

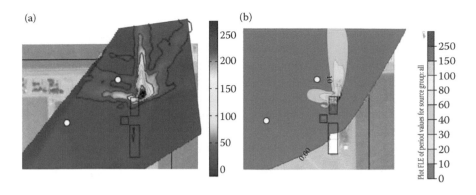

FIGURE 14.8 A comparison of lidar (a) and ISCST3 model (b) derived $PM_{2.5}$ concentrations for a cotton gin under variable wind conditions. The model does not include the background aerosol. White circles are the wind and sampler tower locations.

sampler-based model results, for a mobile source, tree-obstructed harvest (Figure 14.8). This orchard was a working orchard in an almond-producing area surrounded by orchards under a variety of management schemes. The orchard has two varieties in rows orientated north/south with an average tree height of about 7 m. The soil surface condition was bare ground with a slight crust, and the trees were irrigated by drip system. For this deployment, 20 MiniVol samplers (total) distributed in clusters with PM_1, $PM_{2.5}$, PM_{10}, and TSP heads were used along with 10 MetOne OPCs.

Figure 14.9 shows wind speed, volume-based concentrations, and flux data for 138 scans collected over a 4-h period of orchard-sweeping operations. The operation

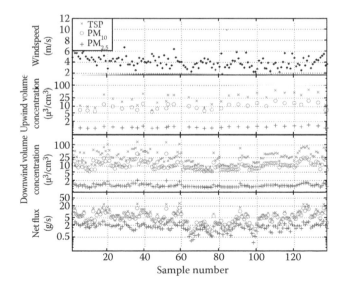

FIGURE 14.9 Wind speed, upwind and downwind concentrations, and flux for 138 scans during a 4-h period of mock almond sweeping operations on October 11, 2006.

was a "mock" or "simulated sweeping" activity that was conducted the day after the nut harvest was completed. While measurements were made during actual sweeping and harvest, these data were unusable because the light and variable winds on those days caused plume-mixing with surrounding field operations. The same equipment and procedures were used for the "mock" sweep as for the actual sweeping operation.

The top plot in Figure 14.9 is the wind speed observed for each scan, showing a fairly consistent northerly wind between 2 and 6 m·s^{-1}, averaging 4 m·s^{-1}. At sample 60, the sweeping method was changed. In the first 59 samples, the sweeper fan operated, but it did not operate with the higher-scan numbers. Since the source was mobile, the plume from the upwind side is filtered by the trees, and some areas of the orchard contributed more emission than others (see Figure 14.8). As expected for this mechanical operation, the PM fractions show significantly higher concentrations of larger particulates, with little contribution from PM$_{2.5}$ for the operation (with and without the fan). The upwind concentration during the measurement period was relatively smooth with some variations in TSP during the operation. The downwind concentrations were much more variable, and the higher concentrations were associated with the sweeper locations near the downwind end of the orchard, especially during turns at the downwind end of the rows. We attributed the area emission differences—without validation data—to local differences in the soil surface being swept and to reduced orchard filtering. The data show a significant difference in TSP emissions with and without the fan. The PM$_{10}$ data remained relatively constant throughout the entire operation, and the TSP emission increased significantly when the fan operated. Fan-generated dust was heaviest during end-of-row turns, where vehicle traffic disturbed the surface. Some operation variability occurred because the operator stopped occasionally for short periods that were not coordinated with the lidar scan sequence. Both location in the orchard (surface effect) and distance into the orchard affected the concentrations, especially for larger particles.

As with the cotton gin data shown previously, the two middle plots in Figure 14.9 show the volume concentration of C_U and C_D in μm^3·cm^{-3}. Volume concentration is shown to avoid confusion caused by uncertainty in the MCF. The net flux, in the bottom plot, is the C_D–C_U difference multiplied by the MCF and the wind speed. The MCF values are presented in Table 14.8. As Zavyalov et al. (2009) pointed out, the MCF determination was the largest measurement error in our system during these operations. Specific values can be summarized for the operation (Table 14.8). As with the cotton data, two things are notable. First, as particle size increases, the particulate density (MCF) decreases, and the error associated with the largest particulates increases significantly. This correlation was discussed by Zavyalov et al. (2009) and is expected with this optical system. Because the particle sizes are larger than the lidar and OPC wavelengths being used for the measurement, accuracy decreases as particle size increases.

Comparison of the lidar-based concentrations with the output of the ISCST3 model is shown in Figure 14.10. Final emission rates were determined with inverse modeling techniques using observed aerosol concentrations at five 2-m height sampling sites and one 9-m height sampling site along the downwind side of the orchard. The area emission estimates (Figure 14.10) were obtained using hourly average wind speeds and directions taken at 5 m and an average source emission rate of

TABLE 14.8

Mock Almond Sweeping Aerosol Upwind and Downwind Mass Concentration and Mass Flux Statistics Associated with Figure 14.9, Showing the Values for the Entire Operation, Sweeping without the Fan (<#60) and with the Fan Operating (>#59)

$C_{U,D}$ = ($\mu g/m^3$) Flux = (g/s)	Whole Operation			With Fan			Without Fan		
	$PM_{2.5}$	PM_{10}	TSP	$PM_{2.5}$	PM_{10}	TSP	$PM_{2.5}$	PM_{10}	TSP
C_U-Ave	9.5 ± 0.4	28.6 ± 3.5	33.1 ± 6.7	9.2 ± 0.4	25.0 ± 4.0	26.2 ± 7.5	9.8 ± 0.5	31.3 ± 5.1	38.2 ± 9.7
C_U-σ	1.063	10.376	19.509	0.786	7.660	14.399	1.172	11.446	21.524
C_D-Ave	10.0 ± 0.3	33.1 ± 2.8	41.6 ± 5.2	10.4 ± 0.5	37.7 ± 4.9	50.4 ± 9.3	9.7 ± 0.3	29.6 ± 2.9	34.9 ± 5.5
C_D-σ	1.71	16.69	31.39	2.02	19.53	36.67	1.34	13.24	24.92
MCF	8.369	2.739	1.263						
F-Ave	1.7 ± 0.1	5.7 ± 0.6	7.1 ± 1.0	2.0 ± 0.15	6.8 ± 0.8	8.7 ± 1.5	1.5 ± 0.2	4.8 ± 0.8	5.9 ± 1.2
F-σ	0.68	3.50	5.91	0.57	3.34	5.99	0.69	3.41	5.60

Note: Averages are shown with 95% confidence intervals.

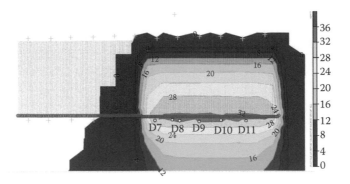

FIGURE 14.10 Modeled PM$_{10}$ aerosol concentrations (μg·m^{-3}) for the 4-h October 11, 2006 almond sweeping exercise, based on the five aerosol sampling stations (D72-D11) on the downwind side of the orchard using the ISCST3 model with an average emission rate of 5.53 μg m^{-2}·s^{-1}.

5.53 μg m^{-2}·s^{-1} to achieve a "best fit" to the measured sampler profile. Sampler-to-model-concentration ratio for the operation averaged 1.00, with a variation ranging from 0.46 to 1.57 (Table 14.9). The data confirm the significant variation in surface emission that the lidar observed. Period average concentrations measured by the samplers varied from 15.3 to 49.5 μg·m^{-3} (Table 14.9), while observed individual lidar scan plume concentrations (averaged over the active plume) ranged from 15 to 106 μg·m^{-3}.

Some discussion is warranted about the potential for sampling error due to the relatively slow vertical scans used in the above analysis. Eddy covariance flux measurement frequencies in the 20-Hz range are often specified. However, eddy covariance sampling volumes are very small and the eddy frequency correspondingly high. In our case, the sampling volume is typically 1 to 200 m long and is being used to

TABLE 14.9

Model and Particulate Sampler Measurements for the Total Sweeping Period Shown in Figure 14.9, Showing the Area Variability Not Picked up by the Model

Sampler Location	Measured Concentration μg·m^{-3}	Modeled Concentration μg·m^{-3}	Measured/Model Average = 1.00
D7 (2m)	49.50	31.44	1.574
D8 (2m)	20.49	32.65	0.628
D8 (9m)	20.66	12.20	1.693
D9 (2m)	24.35	32.28	0.754
D10 (2m)	29.31	32.93	0.890
D11 (2m)	15.31	33.18	0.461

derive an estimate of the mean concentration crossing the plane of the scan, not the variation. In addition, we are sampling at a maximum rate of 10 Hz.

The difference was that the beam was vertically scanned, not held at a single height. These differences in sampling suggest that the scales involved should be carefully analyzed. The scan rate must be considered for sources where the emission stability is not understood.

For the applications presented here, we make the following observations: first, we are measuring relatively continuous processes, with emission variation time periods longer than a single staple scan. While turbulent transfer-driven emission puffs observed at a point may not be captured, emissions from the operation are effectively sampled. The comparisons we show against the ISCST3 model appear to validate the approach. Second, this system is ideal for mobile source operations such as a plowing or harvesting operation. Since the lidar data allow the flux box to move with the source, each staple may be rotated and adjusted when position and general wind direction change. This contrasts against static sampling systems where the plume moves on and off the sampler at random.

If transport from the process is interrupted by a sudden wind shift or a pause in the operation, that transmission period can be easily identified and deleted from the sample sequence. The Aglite data processing system monitors these factors, and these periods are excluded by the quality control process. Unlike a fixed sampler where the plume wanders on and off, the lidar measures the entire plume with each scan.

The data presented here demonstrate how the Aglite system effectively measures emissions under a wide range of conditions. The swine production facility was a stable, consistent source—except for the road traffic. The strong peaks in the lidar data from the road dust allowed the affected scans to be eliminated (Marchant et al. 2009). The flux data reported in this paper include these spikes because they could not be eliminated from the filter data that are shown for comparison. The cotton gin data provided a different challenge, with significantly different aerosol types between the background and facility emissions. The almond data set shows the Aglite's capability to track a mobile source and determine the difference in the equipment's operation method in a short period of time. Aglite not only followed the mobile system but also quantified the difference. Variations resulting from local surface conditions were also quantified.

14.7   CONCLUSIONS

In this chapter, we demonstrated a process for measuring near-real-time whole facility and agricultural operation particulate fluxes. The three-wavelength lidar allows aerosol emissions to be determined and calibrated using characteristics from fixed-point measurements. Aerosol concentration and distribution entering and leaving the facility are differenced and multiplied by the mean wind speed to develop the mass flux. We also described a method that provides unambiguous measurement and characterization of the particulate emissions from agricultural production operations in near-real time. Such an approach assumes only conservation of mass, eliminating reliance on boundary layer theory, reliance on knowledge about the absolute radiometric performance of the lidar instrument and detailed optical properties of the aerosol itself. Point-sampled data are used to provide the aerosol characterization needed for the particle concentration and size fraction calibration, while the lidar provides 3-D mapping of particulate concentrations entering, moving around, and leaving the facility. The 3-D concentration image, collected by the lidar, allows the plume profile to be identified and tracked so that a virtual box can be built around the facility plume. Differences between downwind and upwind measurements provide an integrated aerosol concentration profile, which, when multiplied by the wind speed profile, produces the facility source flux.

This field practice has shown the details of the flux calculation process and has provided a detailed error analysis. The examples given have shown that under relatively constant wind conditions, fluxes with errors in the 10% to 25% range can be developed with sampling periods in the 30-min to 1-h time frame for both fixed and mobile sources. System performance at a CAFO, a mobile harvester, and a fixed product processing plant are used as examples. While the Aglite system is flexible and agile under light and variable wind conditions, it does not allow flux measurement, as the internal mixing and random transport of materials from the source do not allow accurate measurements by any means during such conditions. Also, when emissions from upwind fields or facilities provide a significantly variable aerosol input, longer sample periods are required to achieve the same flux accuracy.

While not the purpose of this chapter, these examples show that to be of value for either regulatory, conservation practice, or improved method development efforts, the flux and fast response emission measurements must be combined with accurate and insightful data and understanding of the processes being evaluated. When these are combined, significant advances in agricultural emission mitigation should be possible.

ACKNOWLEDGMENTS

The development of the Aglite system was performed under USDA Agreement number 58-3625-4-121 and the National Laboratory for Agriculture and the Environment in Ames, IA. This work benefitted from the invaluable guidance and assistance provided by Dr. J. Preuger and R. Pfeiffer of the USDA, and Dr. G. Bingham and K. Moore from Energy Dynamics Laboratory. Any opinions, findings, conclusions, or recommendations expressed in this publication are those of the authors and do not necessarily reflect the view of the USDA.

REFERENCES

Agishev, R., Gross, B., Moshary, F., Gilerson, A., and Ahmed, S. (2006). Simple approach to predict APD/PMT lidar detector performance under sky background using dimensionless parametrization. *Optics and Lasers in Engineering*, 44(8), 779–796.

Althausen, D., Müller, D., Ansmann, A., Wandinger, U., Hube, H., Clauder, E., and Zörner, S. (2000). Scanning 6-wavelength 11-channel aerosol lidar. *Journal of Atmospheric and Oceanic Technology*, 17(11), 1469–1482.

Baker, J. B., Southard, R. J., and Mitchell, J. P. (2005). Agricultural dust production in standard and conservation tillage systems in the San Joaquin Valley. *Journal of Environmental Quality*, 34, 1260–1269.

Bingham, G. E., Marchant, C. C., Zavyalov, V. V., Ahlstrom, D. J., Moore, K. D., Jones, D. S., Wilkerson, T. D. et al. (2009). Lidar based emissions measurement at the whole facility scale: Method and error analysis. *Journal of Applied Remote Sensing*, 3, 033510, doi: 10.1117/1.3097919.

Böckmann, C. (2001). Hybrid regularization method for the ill-posed inversion of multiwavelength lidar data in the retrieval of aerosol size distributions. *Applied Optics*, 40(9), 1329–1342.

California Air Resources Board (CARB) (1999). *Emissions Inventory of Off-road Large Compression-ignited Engines (≥25 hp) Using the New OFFROAD Emissions Model.* MSC99-32. December 1999.

CARB (2003a). *Area Source Methods Manual*, Section 7.4: Agricultural Land Preparation.

CARB (2003b). *Area Source Methods Manual*, Section 7.5: Agricultural Harvest Operations.

Chávez, J. L., Neale, C. M. U., Hipps, L. E., Prueger, J. H., and Kustas, W. P. (2005). Comparing aircraft-based remotely sensed energy balance fluxes with eddy covariance tower data using heat flux source area functions. *Journal of Hydrometeorology*, 6(6), 923–40.

Chow, J. C., Watson, J. G., Lowenthal, D. H., Chen, L.-W A., Tropp, R. J., Park, K., and Magliano, K. A. (2006). $PM_{2.5}$ and PM_{10} mass measurements in California's San Joaquin Valley. *Aerosol Science and Technology*, 40(10), 796–810.

Clausnitzer, H. and Singer, M. J. (1996). Respirable-dust production from agricultural operations in the Sacramento Valley, California. *Journal of Environmental Quality*, 25, 877–884.

Clausnitzer, H. and Singer, M. J. (2000). Environmental influences on respirable dust production from agricultural operations in California. *Atmospheric Environment*, 34, 1739–1745.

Cornelsen, S. (2005). *Electronics Design of the AGLITE Instrument.* M.S. thesis, Utah State University, Logan, UT.

Del Guasta, M., Morandi, M., Stefanutti, L., Stein, B., and Wolf, J. P. (1994). Derivation of Mount Pinatubo stratospheric aerosol mean size distribution by means of a multiwavelength lidar. *Applied Optics*, 33, 5690–5697.

Dubovik, O., Holben, B., Eck, T. F., Smirnov, A., Kaufman, Y. J., King, M. D., Kino, D., Tanre, D., and Slutsker, I. (2002). Variability of absorption and optical properties of key aerosol types observed in worldwide locations. *Journal of the Atmospheric Sciences*, 59(3), 590–608.

Eichinger, W. E., Holder, H. F., Knight, R., Nichols, J., Cooper, D. I., Hipps, L. E., Kusta, W. P., and Prueger, J. H. (2005). Lidar measurement of boundary layer evolution to determine sensible heat fluxes. *Journal of Hydrometeorology*, 6(6), 840–53.

Fernald, F. G., Herman, B. M., and Reagan, J. A. (1972). Determination of aerosol height distributions by lidar. *Journal of Applied Meteorology*, 11(3), 482–489.

Flocchini, R. G., James, T. A., Ashbaugh, L. L., Brown, M. S., Carvacho, O. F., Holmén, B. A., Matsumura, R. T., Trezpla-Nabalgo, K., and Tsubamoto, C. (2001). *Interim Report: Sources and Sinks of PM_{10} in the San Joaquin Valley.* Crocker Nuclear Laboratory, UC-Davis, CA.

Friehe, C. A. (1986). Fine-scale measurements of velocity, temperature and humidity in the atmospheric boundary layer. *Probing the Atmospheric Boundary Layer*, D. Lenschow (ed.), American Meteorological Society, Boston, pp. 29–38.

Heintzenberg, J., Muller, H., Quenzel, H., and Thomalla, E. (1981). Information content of optical data with respect to aerosol properties: Numerical studies with a randomized minimization-search-technique inversion algorithm. *Applied Optics*, 20(8), 1308–15.

Hess, M., Koepke, P., and Schult, I. (1998). Optical properties of aerosols and clouds: The software package OPAC. *Bulletin of the American Meteorological Society*, 79(5), 831–844.

Hiscox, A. L., Miller, D. R., Holmén, B. A., Yang, W., and Wang, J. (2008). Near-field dust exposure from cotton field tilling and harvesting. *Journal of Environmental Quality*, 37 (2), 551–556.

Hoff, S. J., Hornbuckle, K. C., Thorne, P. S., Bundy, D. S., and O'Shaughnessy, P. T. (2002). Emissions and community exposures from CAFOs. In *Iowa Concentrated Animal Feeding Operations Air Quality Study*. Iowa State University and the University of Iowa Study Group, pp. 45–85.

Holmén, B. A., Eichinger, W. E., and Flocchini, R. G. (1998). Application of elastic LIDAR to PM_{10} emissions from agricultural nonpoint sources. *Environmental Science and Technology*, 32, 3068–3076.

Holmén, B. A., James, T. A., Ashbaugh, J. L., and Flocchini, R. G. (2001a). LIDAR-assisted measurement of PM_{10} emissions from agricultural tilling in California's San Joaquin Valley—Part I. LIDAR. *Atmospheric Environment*, 35, 3251–2364.

Holmén, B. A., James, T. A., Ashbaugh, J. L., and Flocchini, R. G. (2001b). LIDAR-assisted measurement of PM10 emissions from agricultural tilling in California's San Joaquin Valley—Part II: Emission factors. *Atmospheric Environment*, 35, 3265–3277.

Kean, A. J., Sawyer, R. F., and Harley, R. A. (2000). A fuel-based assessment of off-road diesel engine emissions. *Journal of the Air and Waste Management Association*, 50, 1929–1939.

Klett, J. D. (1985). Lidar inversion with variable backscatter/extinction ratio. *Applied Optics*, 24, 1638–1683.

Madden, N. M., Southard, R. J., and Mitchell, J. P. (2008). Conservation tillage reduces PM10 emissions in dairy forage rotations. *Atmospheric Environment*, 42, 3795–3808.

Marchant, C. C. (2007). Algorithm development of the Aglite-lidar instrument. M.S. thesis, Utah State University, <http://digitalcommons.usu.edu/etd/107/>.

Marchant, C. C. (2010). Retrieval of Aerosol Mass Concentration from Elastic Lidar Data. Ph.D. dissertation, Electrical and Computer Engineering, Utah State University, Logan, UT.

Marchant, C. C., Moon, T. K., and Gunther, J. H. (2010). An iterative least-squares approach to elastic lidar retrievals for well characterized aerosols. *IEEE Transactions on Geoscience and Remote Sensing*, 48, 2430–2444.

Marchant, C. C., Wilkerson, T. D., Bingham, G. E., Zavyalov, V. V., Andersen, J. M., Wright, C. B., Cornelsen, S. S., Martin, R. S., Silva, P. J., and Hatfield, J. L. (2009). Aglite lidar: A portable elastic lidar system for investigating aerosol and wind motions at or around agricultural production facilities. *Journal of Applied Remote Sensing*, 3, 033511, doi:10.1117/1.3097928.

Martin, R. S., Moore, K. D., and Doshi, V. S. (2006). Determination of particle (PM_{10} and $PM_{2.5}$) and gas-phase ammonia (NH_3) emissions from a deep-pit swine operation using arrayed field measurements and inverse Gaussian plume modeling. *Proceedings of the Workshop on Agricultural Air Quality: State of the Science*, V. Aneja and W. Schlesinger (eds.), Potomac, MD.

McGill, M., Hlavka, D., Hart, W., Scott, V. S., Spinhirne, J., and Schmid, B. (2002). Cloud physics lidar: Instrument description and initial measurement results. *Applied Optics*, 41(18), 3725–3734.

Mitchell, J. P., Munk, D. S., Prys, B., Klonsky, K. K., Wroble, J. F., and De Moura, R. L. (2006). Conservation tillage production systems compared in San Joaquin Valley cotton. *Journal of California Agriculture*, 60(3), 140–145.

Mitchell, J. P., Southard, R. J., Madden, N. M., Klonsky, K. M., Baker, J. B., De Moura, R. L., Horwath, W. R. et al. (2008). Transition to conservation tillage evaluated in San Joaquin Valley cotton and tomato rotations. *Journal of California Agriculture*, 62(2), 74–79.

Müller, D., Wandinger, U., and Ansmann, A. (1999). Microphysical particle parameters from extinction and backscatter lidar data by inversion with regularization: Theory. *Applied Optics*, 38(12), 2346–57.

Rajeev, K. and Parameswaran, K. (1998). Iterative method for the inversion of multiwavelength lidar signals to determine aerosol size distribution. *Applied Optics*, 37(21), 4690–4700.

Sasano, Y. and Browell, E. V. (1989). Light scattering characteristics of various aerosol types derived from multiple wavelength lidar observations. *Applied Optics*, 28, 1670–1679.

Upadhyaya, S. K., Lancas, K. P., Santos-Filno, A. G., and Raghuwanshi, N. S. (2001). One-pass tillage equipment outstrips conventional tillage method. *Journal of California Agriculture*, 55(5), 44–47.

U.S. EPA (1991). *Nonroad Engine and Vehicle Emission Study*—Report. EPA 460/3-91-02, Washington, DC.

U.S. EPA (2001). *Procedures Document for National Emission Inventory*. Criteria Air Pollutants 1985–1999. EPA-454/R-01-006, Washington, DC.

U.S. EPA (2004). *Exhaust and Crankcase Emission Factors for Nonroad Engine Modeling*—Compression-Ignition. EPA420-P-04-009. Washington, DC.

U.S. EPA (2008). *NONROAD Model (nonroad engines, equipment, and vehicles).*, http://www.cpa.gov/otaq/nonrdmdl.htm#techrept (accessed August 21, 2008).

Veenstra, J. J., Horwath, W. R., Mitchell, J. P., and Munk, D. S. (2006). Conservation tillage and cover cropping influence soil properties in San Joaquin Valley cotton-tomato crop. *Journal of California Agriculture*, 60(3), 146–153.

Veselovskii, I., Kolgotin, A., Griaznov, V., Muller, D., Franke, K., and Whiteman, D. N. (2004). Inversion of multiwavelength Raman lidar data for retrieval of bimodal aerosol size distribution. *Applied Optics*, 43(5), 1180–1195.

Wang, J., Hiscox, A. L., Miller, D. R., Meyer, T. H., and Sammis, T. W. (2009). A comparison of Lagrangian model estimates to light detection and ranging (LIDAR) measurements of dust plumes from field tilling. *Journal of the Air & Waste Management Association*, 59, 1370–1378.

Zavyalov, V. V., Marchant, C. C., Bingham, G. E., Wilkerson, T. D., Swasey, J., Rogers, C., Ahlstrom, D., and Timothy, P. (2006). Retrieval of physical properties of particulate emission from animal feeding operations using three wavelength elastic lidar measurements. *Proceedings of SPIE*, 6299, 62990S, doi:10.1117/12.680967.

Zavyalov, V. V., Marchant, C. C., Bingham, G. E., Wilkerson, T. D., Hatfield, J. L., Martin, R. S., Silva, P. J. et al. (2009). Aglite lidar: calibration and retrievals of well characterized aerosols from agricultural operations using a three-wavelength elastic lidar. *Journal of Applied Remote Sensing*, 3, 033522, doi:10.1117/12.833.

15 Measurement of Aerosol Properties over Urban Environments from Satellite Remote Sensing

Min M. Oo, Lakshimi Madhavan Bomidi, and Barry M. Gross

CONTENTS

15.1 INTRODUCTION

It is well known that accurate global characterization of aerosol optical depth (AOD) is essential in accurately determining the energy balance for climate change studies (Charlson et al. 1992) as well as quantifying fine particle pollutants and subsequent health risks (Wilson and Suh 1997). Regarding climate forcings, aerosol interactions can affect the climate both directly and indirectly. For example, direct forcing impacts occur through modification of the earth's albedo due to highly scattering aerosols that result in a net cooling by scattering solar radiation back into space (Twomey 1974, 1977). At the same time, aerosols can modify cloud processes through so-called indirect mechanisms (Twomey 1980; Ackerman et al. 2004). Since observations of clouds and aerosols cannot be done at the same time from space, the main validation of this effect relies on statistical metrics (Nakajima et al. 2001) where individual aerosol and cloud properties are averaged in time and space.

All of these forcing mechanisms should be contrasted against greenhouse gas (GHG) emissions and associated warming mechanisms. In particular, GHG emissions are based on accurate determination of these gases. Fortunately, such measurements are easier from space since they rely mainly on infrared absorption, which can be well measured by polar orbiting spectrometers such as the National Aeronautics and Space Administration (NASA) Atmospheric Infrared Sounder (AIRS) and near-infrared (NIR) spectrometers such as Scanning Imaging Absorption Spectrometer for Atmospheric Cartography (SCIAMACHY) on board the European Space Agency (ESA) Environmental Satellite (ENVISAT) (Crevoisier et al. 2003; Rublev and Uspenskii 2006). For these reasons, the uncertainty of climate forcing due to GHG is much less than those due to direct and indirect aerosol mechanisms. This is clearly illustrated in the radiative forcing error bars depicted in Figure 15.1 taken from the Forth Assessment Report (FAR) (Forster et al. 2007) produced by Intergovernmental Panel on Climate Change (IPCC) (2007) (http://www.ipcc.ch). Complicating issues even more is that trace gases are rather homogenous in the atmosphere whereas aerosols are much more variable in space and time.

On the other hand, aerosols can have significant effects on human health with exposure linked to various health risks including heart- and lung-related distress such as asthma. In fact, the United States Environmental Protection Agency (US EPA) has strict particulate matter (PM) National Ambient Air Quality Standards (NAAQS) setting 24-h exposure at 35 $\mu g \cdot m^{-3}$ (http://www.dec.ny.gov/). In particular, fine mode particulates (PM less than 2.5 μm in diameter, PM2.5) are most

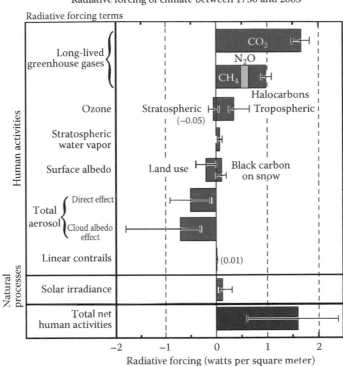

FIGURE 15.1 Climate forcings due to atmospheric constituents. (From IPCC Assessment Report, 2007.)

important since these small particles penetrate deep into the lung tissue and can result in significant lung and cardiac distress (US EPA 1996).

One major reason that aerosol measurements over urban areas are important is that extended urban centers such as New York City, Mexico City, and Shanghai can have significant aerosol loadings with air quality levels that are above US EPA standards. Monitoring air pollution levels from space is based on the observation that the column AOD under certain conditions is strongly related to PM2.5 (Al-Saadi et al. 2005; Engel-Cox et al. 2006). Such correlations are strongest in the Eastern United States where dark vegetative surfaces make the retrieval of aerosols more accurate. Unfortunately, other mechanisms make it very difficult to connect satellite-derived AOD to surface particle mass. These include the variability of the planetary boundary layer height (PBL height), the existence of aloft aerosol plumes, as well as the natural variability of aerosols as manifested in the ratio of scattering to particle volume. These difficulties in connecting air quality parameters to satellite-derived AOD and possible strategies in accounting for these effects based on climatologic and/or other ancillary data are presented in detail in the previous work (Al-Saadi et al. 2005; Engel-Cox et al. 2006).

Based on the above discussion, it is important that remote sensing methods be developed that measure AOD over urban areas both for air quality as well as climatology to address scientific issues such as aerosol–cloud interactions (Wu et al. 2009). Such interactions due to heavy urban aerosol loadings are expected to provide the best opportunities in observing some climate modifying mechanisms that are hard to quantify (Wu et al. 2009). In addition, the need for some level of aerosol speciation is clearly of great importance, but unfortunately can only be partially assisted by satellite observations. In fact, the best that can be expected for aerosol observations beyond column optical depth is some degree of separation between fine and coarse modes, which are of obvious help both for air quality as well as environmental management. Such separation can help identify anthropogenic aerosols, which is fine mode dominated, in relation to natural aerosols as well as absorbing aerosols as measured by their single scattering albedo (SSA). Therefore, in this chapter, we focus on the present and future observational capabilities for urban aerosol monitoring and, when possible, assess the capabilities of retrieving not only AOD but also fine-coarse mode separation and SSA. A case study using those satellite observations that provide operational and validated retrievals is deemed most crucial for climate and air-quality applications.

15.1.1 OPTICAL PROPERTIES OF AEROSOLS

The optical properties of atmospheric aerosols are mainly determined by their size distribution and chemical composition as well as overall particle concentration. Unlike trace gases whose optical properties are well characterized and have stable spectral features suitable for identification and quantification, retrieving a complete description of aerosol microphysical and chemical properties from optical data is not feasible due to the lack of sharp identifying spectral and/or angular features in the aerosol scattering process. Historically, with single-channel observations providing a single piece of information (i.e., top of atmosphere [TOA] reflection at a single wavelength), only the most basic aerosol quantity could be retrieved (i.e., AOD). With the advent of multispectral and/or multiangle capacity in the latest observing sensors, decomposition of aerosols into fine and coarse mode (fine/coarse mode fraction) and identification of aerosol absorption properties (i.e., SSA) have made their way into current retrieval algorithms. Extracting this additional information, however, requires that aerosol models be developed that capture the essential properties of aerosols as they exist naturally in ambient conditions.

15.1.2 AEROSOL MODELS

The simplest assumption used for most aerosol models is that aerosols are spherical and are externally mixed (Hess et al. 1998; Seinfeld and Pandis 1998). In this important case, aerosols are described by a small number of parameters that define the size distribution or modes and the complex refractive index. Most commonly, the size distribution is assumed to be of lognormal type with both fine and coarse modes included. In fact, the bimodal lognormal distribution parameterization is used in almost all satellite remote sensing efforts and is given by Equation 15.1, which is consistent with measurement retrievals taken from Aerosol Robotic Network (AERONET) (Holben et al. 2001).

$$\frac{dV}{d\ln(r)} = \frac{V_f}{\sqrt{2\pi}\ln(\sigma_f)}\exp\left(-\frac{\left[\ln r - \ln \bar{r}_f\right]^2}{2\left(\ln\sigma_f\right)^2}\right) + \frac{V_c}{\sqrt{2\pi}\ln(\sigma_c)}\exp\left(-\frac{\left[\ln r - \ln \bar{r}_c\right]^2}{2\left(\ln\sigma_c\right)^2}\right),$$

(15.1)

where the lognormal parameters V_i, $\bar{r}_i$, and σ_i ($i = f,c$) describe the total volume in each mode, the mean mode radius, and mode width, respectively. AERONET is an extensive global scanning sky radiometer network capable of retrieving not only AOD but also bimodal size distribution parameters (Dubovik and King 2000; Dubovik and Holben 2002a,b) and complex refractive index (or SSA) and functions as the main validation network in which satellite remote sensing aerosol retrievals are compared.

The characteristics of aerosols from AERONET for different regions worldwide show significant variability not only in AOD but also in the ratio of fine/coarse mode and SSA (Dubovik and Holben 2002b). The main urban feature in contrast with other cases (desert, oceanic) is the higher fine mode aerosol contribution with angstrom coefficients ranging from $1.2 \le \alpha_{440-870} \le 2.5$ (Schuster et al. 2006). Aerosol absorption in urban areas such as the northeast (NE) coast of the United States has high SSA at 440 nm (e.g., $0.92 \le \omega_0 \le 0.98$), but cases such as Mexico City with heavy biomass burning contributions have SSA much smaller within ranges (e.g., $0.60 \le \omega_0 \le 0.92$). This is consistent with the high degree of variability expected since the albedo strongly depends on the age of the burning aerosol from the combustion source.

In addition, it is possible that the effective mode radii ($\bar{r}_f$ and $\bar{r}_c$) can depend on optical depth. This dynamic feature occurs for hygroscopic aerosols, which, by definition, undergo changes in both physical (effective radius) and optical (refractive index) when subjected to high relative humidity (RH) (Hanel 1976) conditions. Examples of such aerosols include ammonium sulfate ($(NH_4)_2SO_4$), sea salt, and ammonium nitrate (NH_4NO_3), which, for example, are the dominant modes in many urban coastal areas (i.e., northeast United States). However, not all aerosols are hygroscopic (Vlasenko et al. 2005). Most prominent among these are desert (mineral) dusts and most organic aerosol species. Therefore, aerosol models must account for the high probability of these aerosol classes in different areas of the world. For example, aerosol models are best modeled as static over the West Coast of the United States and in many other world regions where dust and organic aerosol components are significant while dynamic models are best suited to hygroscopic aerosols. Further complications arise with mineral dust species since these particles are far from spherical. To model these aerosols, nonspherical models including the most popular model of spheroids (Dubovik et al. 2006) are used in operational models. These models may be considered within the general particle size distribution (PSD) framework since they are characterized by a size distribution and refractive index. However, realistic assumptions on the statistics of the aspect ratio (semimajor/semiminor axis) of constituent ellipsoids must be assumed *a priori*.

15.2 SATELLITE REMOTE SENSING OF AEROSOL OVER LAND

15.2.1 General Considerations of Current Aerosol over Land Algorithms

The algorithms for aerosol retrieval depend strongly on the different characteristics of the sensor, but all algorithms must in some way quantify the land surface contribution as well as obtain some estimate of aerosol type in order to retrieve AOD (Table 15.1). The reason for this is that any solar reflection signal over a land pixel will consist of photons that have either directly interacted with the ground or interact only with the atmosphere. These photons, upon collection, are in essence indistinguishable unless a more sophisticated approach that can estimate the surface contribution or reduce its contaminating effects can be implemented. This is referred to as surface correction, which is an integral part of any aerosol retrieval over land scheme.

In addition, all aerosol retrieval algorithms require accurate and robust cloud screening procedures and different algorithms and sensors with different spatial resolutions have different ways to deal with it. Therefore, a significant uncertainty that is hard to be quantified in many retrieval algorithms could be linked to partial cloud contaminations (Kaufman et al. 2005). Yet the cloud screening procedure associated with each retrieval algorithm is not included since it is beyond the focus of this chapter. In describing the different aerosol algorithms in what follows, the most useful means of partitioning these algorithms is to discuss how they handle the surface correction. Possible strategies include the following:

1. Estimates of ground reflection are made using multiple temporal observations of a given pixel over time periods where the surface properties are not expected to vary significantly. In particular, it is expected that if sufficient measurements are made, it is likely that one or more observations are made under relatively clear-sky conditions, and that these "clear-sky" conditions can be identified by monitoring the TOA reflectance and extracting the weakest cloud cleared signal since, in general, significant aerosol loading tends to increase the reflectance signal. Once the "clear-sky case" is identified, surface estimates can be obtained by assuming the clear-sky atmosphere is composed of a standard Rayleigh component and a small background aerosol component. Once the surface properties are obtained, it is assumed that they are stable in time and can be used for arbitrary aerosol loading. Examples of this approach include long-term advanced very high resolution radiometer (AVHRR) aerosol retrievals, where multiyear data records are mined to provide estimates of surfaces (Hauser et al. 2005a,b), and the Geostationary Operational Environmental Satellite (GOES) Aerosol and Smoke Product (Prados et al. 2007), where continuous geostationary observations are made over 28-day periods to extract a surface mosaic.

2. Estimation of visible (VIS) surface reflectance from a long wavelength channel where the aerosol reflection is negligible. For example, the MODIS 2.1-μm channel reflectance can be assumed in most cases to be aerosol-free, allowing for direct observation of the surface reflection at 2.1 μm.

TABLE 15.1
General Characteristics of Current and Near Future Aerosol over Land Sensors

	AVHRR2/3 (Multiple)	MISR (Terra)	MODIS (Terra/Aqua)	POLDER (PARASOL)	GOES 11, 12, 13, 14	GOES-R	APS (Glory)[a]
Launch	May 1981	December 1999	December 1999/May 2002	December 2004	2000, 2001, 2006, 2009	2015	2011
Resolution at nadir (km)	1.1	0.275	0.25 (B1-B2) 0.50 (B3-B7) 1.00 (B8-B36)	6×7	1 (at 0.6 μm band)	0.5 (at 0.6 μm band)	6
Swath width (km)	2399	400	2330	2400	Full disk	Full disk	6
Multiangles observation	No	Yes	No	Yes	No	No	Yes
Average revisit time	1 day	9 days	1 to 2 days	1 day	30 min	5 min	16 days
Polarization	No	No	No	Yes	No	No	Yes
Advanced Retrieval Parameters							
Fine/coarse AOD	No	Yes	Yes	Yes	No	Yes	Yes
SSA	No	Yes	No	No	No	No	Yes
Microphysical parameters	No	Yes	Yes	Yes	No	Yes	Yes
Product resolution	100 km	16 km	10 km	18.5 km	4 km	2 km	6 km

[a] Even though APS failed to reach orbit on May 2011 due to satellite failure, we believe the inclusion of the capabilities of this instrument for possible future instrument development is instructive to the reader.

For vegetative surfaces, MODIS uses empirical spectral reflection ratios (SRRs) to convert the 2.1-μm reflectance to reflectances in the red (660 nm) and blue (470 nm) channels (Kaufman and Tanré 1998). These relationships were further refined using global observations of simultaneous AERONET/MODIS observations to account for land surface type and sun/view geometry (Levy et al. 2007a) and are now in use in the operational MOD04 product. Unfortunately, as we will observe in Section 15.3, significant difficulties remain when applied to urban areas since the SRRs were globally trained and urban signatures are not well represented in the global matchups.

3. Simultaneous estimation of surface–aerosol parameters by combining the best characteristics of approaches 1 and 2. In this case, sophisticated optimization methods are applied to multiangle, multispectral measurements over different times of observation in order to extract simultaneous aerosol and surface spectral/angular properties. An example of this is the multiangle implementation of atmospheric correction (MAIAC) algorithm (Lyapustin et al. 2007b) applied to Moderate Resolution Imaging Spectroradiometer (MODIS) data and the statistical optimal estimation method for Polarization and Directionality of the Earth's Reflectances (POLDER) (Dobovik et al. 2010).

4. Polarized reflectance observation methods, which are less sensitive to ground but sensitive to aerosol signatures. This is, in principle, a very attractive strategy but is hard to implement due to the very strict polarization accuracy needed and complex optical implementations. At present, the only operational spaceborne polarization-sensitive sensor that exists is POLDER. The channel limitations of this particular sensor for urban aerosol retrievals are spelled out in detail in Section 15.2.6. In an effort to substantially improve on this sensor, NASA had planned to launch the Advanced Polarimetric Scanner (APS) on board GLORY but tragically, this system failed to reach orbit on May 2011.

15.2.2   Single-Channel/Single-View Polar Sensors (AVHRR)

For AVHRR, only a single view angle and a single VIS channel reflectance at 0.63 μm is available to retrieve AOD. Therefore, aerosol classification must be decided *a priori* based on historical climatology (Knapp and Stowe 2002; Stowe et al. 2002; Hauser et al. 2005a,b). While such an approach is clearly less than optimal, it should be kept in mind that the long-term data (e.g., climatic and environmental data) make the AVHRR processing very useful as the long-term data (CDR or EDR) particularly cover oceans where aerosol variability is less and the ocean is accurately described as dark. However, these data sets are not very suitable for urban land surfaces since the surface reflections trained during the retrievals are based on single-channel observations. In particular, with multiple wavelengths, a significant improvement is possible. This can be confirmed by the fact that spectral variation of surface angular dependence is weak so the ground reflection signatures are not independent, providing an opportunity to disentangle aerosol and ground signatures.

15.2.3 MULTIANGLE WITHOUT LONG-WAVE CHANNEL (MISR)

Since aerosols have rich angular scattering behavior, it is reasonable that multi-angle observations can improve retrieval. Multi-angle Imaging Spectroradiometer (MISR) flies on the polar-orbiting Terra platform (i.e., on the same platform with the MODIS-Terra). MISR measured the TOA radiance by using nine cameras that are paired (fore/aft) in a symmetrical arrangement and acquire multispectral images with fixed view angles at 0°, 26.1°, 45.6°, 60.0°, and 70.5°. Each camera is fixed at a particular view zenith angle in the along-track direction and has four spectral bands (446, 558, 672, and 866 nm). A cross-track ground spatial resolution is 275 m (in all nine cameras in 672 nm and the nadirs camera in 446, 558, and 866 nm) or 1.1 km (the remaining eight cameras in 446, 558, and 866 nm) in global mode (operational mode). However, in local mode, 275-m spatial resolution in all 36 channels (nine cameras in four wavelengths) can be achieved (Diner 1999).

The objective of the MISR aerosol retrieval algorithm is to take advantage of both the multiband and multiangle viewing capability to separate scattering radiance from atmosphere and underlying surface without *a priori* knowledge of the land surface type. In the current version of the retrieval algorithm, two steps are included. The first step is based on the principal components analysis of multispectral and multiangle TOA MISR data from which empirical orthogonal functions (EOFs) for each wavelength for all camera angles were obtained to describe the directional reflectance properties of the surface (Martonchik et al. 1998, 2009). A second step was implemented into the processing chain after 3 years of operation (Diner et al. 2005) and makes use of the fact that the bulk of the angular reflectance dependence at TOA is dominated by the surface angular reflectance.

However, as we will see in Section 15.3.1, MISR retrievals in urban areas are often significantly underestimated. This, together with the lack of spatial coverage (i.e., return frequency), makes MISR less suited as an optimal urban aerosol monitoring instrument. In the next section, the MODIS sensor, which is, by far, the most commonly used tool due to its combination of high resolution, large swath, and easy access to data retrievals, is discussed in detail.

15.2.4 MULTISPECTRAL WITH LONG WAVELENGTH CHANNEL (MODIS)

As stated in the previous section, MODIS is the current workhorse in the aerosol community. Flying on both Terra (EOS AM) and Aqua (EOS PM), they are combined to view the entire earth's surface every 1 to 2 days with a repeating orbit every 16 days and wide 2330-km swath. MODIS has 36 spectral bands (from 0.4 to 14.4 μm) with nadir on-ground spatial resolution of 1000, 500, and 250 m.

Historically, the MODIS algorithm for AOD retrieval over land required "dark" pixels as described in the Collection 4 (C004) algorithm (Kaufman et al. 1997; Kaufman and Tanré 1998). Since many land covers such as vegetation and some soils have very low surface reflectance in the red (0.66 μm) and blue (0.47 μm) wavelengths, it is reasonable to use the darkest pixels in the image to explore the aerosol optical properties. For example, dense dark vegetation in an image can be considered as "dark" pixel and the aerosol contribution can be isolated. Isolation of dark targets in the aerosol retrieval algorithm (Kaufman and Sendra 1988) was based on the detection of green forests as dark pixels based on a high normalized difference vegetation index (NDVI).

Still, even moderately dark vegetation scenes have a residual reflectance that cannot be realistically ignored unless the aerosol optical depth is high, and this is even more so for urban scenes. For this reason, some estimate of the ground reflectance is essential and this estimate relies heavily on MODIS having a channel in the NIR (2120 nm) that can see the ground without atmospheric contamination. In C004, an empirical relationship derived from Landsat observations of vegetation for the SRRs was used ($\rho_g^{660}/\rho_g^{2120} = 0.5$ and $\rho_g^{470}/\rho_g^{2120} = 0.25$). In C005, significant progress in improving these relations was made based on global matchups between AERONET and MODIS (Levy et. al 2007a). In particular, having AERONET provide local aerosol properties made it possible to directly retrieve surface reflections at the VIS and short-wave infrared (SWIR) bands simultaneously. From this, it was possible to obtain relevant SRRs as a function of geometry (parameterized by the scattering angle Θ) and land surface type defined by a so-called modified vegetation index (MVI). This MVI uses bands 1240 and 2120 nm as opposed to the standard NDVI that uses 670 and 860 nm to avoid atmospheric contamination in land type classification. However, as we will see in Section 15.3.2, these global SRR models as functions of Θ and MVI require significant "tuning" in urban areas.

The operational MODIS aerosol retrieval algorithm over land is based on a lookup table (LUT) approach (i.e., radiative transfer calculations of TOA reflectance are precomputed for a set of aerosol models). These include four fine modes (continental, generic, absorbing smoke, and nonabsorbing urban) and one coarse mode (nonspherical dust). Details of the microphysical properties of these modes can be found in Levy et al. (2007a). The calculations are performed with the RT3 code developed by Evans and Stephens (1991) at four discrete wavelengths (0.466, 0.553, 0.644, and 2.12 μm, representing MODIS channels 3, 4, 1, and 7, respectively) for seven aerosol loadings ($\tau_{0.55\mu m} = 0.0, 0.25, 0.5, 1.0, 2.0\ 3.0$, and 5.0), and nine solar zenith angles ($\mu_0 = 0.0, 6.0, 12.0, 24.0, 35.2, 48.0, 54.0, 60.0$, and 66.0), 16 sensor zenith angles ($\mu = 0.0$ to 87.14 approximate increments of 6.0), and 16 relative azimuth angles ($\phi - \phi_0 = 0.0°$ to $180.0°$ with increments of $12.0°$). Aerosol phase functions are calculated either using MIEV code (Wiscombe 1981) for spherical assumption or a version of the T-matrix code for spheroid dust described in Dubovik et al. (2006). Because the MODIS overland retrieval algorithm employs only three channels, there is insufficient information content to independently choose or retrieve the fine aerosol model (Remer et al. 2005). Therefore, the MODIS aerosol retrieval algorithm must assign the fine aerosol model *a priori* of the retrieval based on season and location from global aerosol models obtained from AERONET (Remer et al. 2005; Levy et al. 2007a,b).

To summarize, the algorithm performs a simultaneous retrieval over the parameter space whose variables are the AOD at 0.55 μm, the surface reflectance at 2.12 μm, and the mixing fraction η using the three spectral channels (0.47, 0.66, and 2.12 μm). However, as we will see in Section 15.3.1, the operational algorithm is also less than optimum over urban areas. Unlike MISR, the aerosol retrievals are often overbiased in selected urban areas, and as we will see, this is a direct consequence of the surface albedo estimates MODIS uses in the VIS channels.

Finally, it is also of interest to note that the same basic algorithm structure is being applied to Visible Infrared Imager Radiometer Suite (VIIRS) as part of the National Polar-Orbiting Operational Environmental Satellite System (NPOESS) with AOD, angstrom coefficient, and effective radius parameters being reported at 6-km nadir resolution. However, an effort has been made to improve on the range of surface types where the SRRs using extensive Landsat atmospherically corrected training sets over diverse terrain. Updated relations between the VIS and SWIR ground reflections as function of the NDVI have been developed and further training to connect the shape of the directional reflectance (i.e., BRDF; see Section 15.3.2.2) to NDVI and thereby account for non-Lambertian surface reflection has been incorporated (VIIRS 2010). However, the surface model is designed only for vegetated and semivegetated surface types, and overly bright pixels based on threshold tests on TOA reflectance of the 1240- and 2250-nm channels will be flagged so VIIRS retrieval into urban domains is not expected within the VIIRS operational retrieval.

15.2.5 MODIS DEEP BLUE ALGORITHM

Algorithms that can work over bright surfaces would seem to be very natural for urban retrieval. One approach that has become quite useful over desert terrain is the Deep Blue algorithm, so named since critical use of the shortest wavelength (412 nm) blue channel is involved. The general idea is that, because the surface reflection is in general lower as the wavelength decreases and the aerosol (and molecular) reflectance is higher, a better contrast and less surface reflected signal will occur if the 412-nm channel is used in conjunction with the standard blue–red channels.

Like other methods we have seen, estimates of the surface reflectance are determined from analyzing clear-sky scenes within the long-term archives database and quantifying a best estimate based on geolocation and seasonal attributes. To reduce fluctuations, the surface resolution is taken as 0.1 deg × 0.1 deg, making it particularly useful for homogeneous desert cases but less useful for heterogeneous urban surfaces as we will see in validation comparisons presented in Section 15.3.1.

For aerosol retrieval, the Deep Blue algorithm for MODIS combines reflectance measurements at 412-, 490-, and 670-nm channels. Choosing among the different aerosol models is similar to the standard MODIS retrieval. In essence, the retrieved mixing ratio (of dust and smoke models), AOD, and Angstrom exponent values are those that minimize the residual between the measured and LUT calculated reflectance using a maximum-likelihood method. More details of the Deep Blue algorithm can be found in Hsu et al. (2004, 2006).

15.2.6 POLDER: ADVANTAGES OF POLARIMETRIC SENSING

POLDER radiometer provides multispectral, polarized measurements at multiple angles (Deschamps et al. 1994). The POLDER 1 instrument was launched on the ADEOS Japanese platform in August 1996 and the third version of POLDER is flying in the A-Train on board the Polarization and Anisotropy of Reflectance for Atmospheric Sciences Coupled with Observations from a Lidar (PARASOL) platform since December 2004. POLDER provides quantitative global measurements of total as well as polarized solar radiation reflected by the earth-atmosphere system including multiangle (up to 16 directions due to the wide FOV CCD imaging plane) data at eight spectral bands in the VIS and NIR spectral domain (0.443 to 1.02 μm).

The value of this approach resides in the fact that the polarization of the reflected signal from the surface is significantly lower than the polarization of the signal scattered by fine mode aerosol in the atmosphere. Therefore, the relative contribution of the polarized ground reflectance to the TOA polarized reflectance is small and stable enough to determine with semiempirical surface BPDF models (Herman et al. 1997 and Deuzé et al. 2001) obtained from fitting clear-sky surface retrievals. On the other hand, for dust aerosols, POLDER cannot separate atmosphere and surface signal very well since the coarse aerosols produce little polarization in atmospheric scattering. Therefore, POLDER retrieval focuses on the fine aerosol mode and the retrieval algorithm assumes only spherical scattering, which is valid for fine aerosols. Even though the POLDER retrieval algorithm is less effective for coarse mode aerosol, the retrieval method is still a very useful method for aerosol over land since most of the aerosol generated by pollution and biomasses burning is fine mode (Deuzé et al. 2001) providing excellent views of anthropogenic emissions.

In the retrieval algorithm, the aerosol optical depth and model are obtained by matching the precomputed LUT to the multidirectional polarized reflectance measurements at 670 and 865 nm. The aerosol retrieval algorithm considers only fine particles to compute LUT sampling the lognormal distribution range from 0.05 to 0.15 μm in 10 modal radii r_m with a fixed standard deviation of 0.403 and a refractive index of $m = 1.47$-$0.01i$. To simplify the model, POLDER retrievals only consider the polarized light corresponding to single scattering by aerosols, single scattering by molecules, and direct reflection on the surface. Similarly, the bidirectional polarized distribution function (BPDF) from the surface is modeled by assuming that the polarized light comes from specular reflection on surface elements. The assessment of polarized reflectance is based on *a priori* values (as a function of observation geometry and surface type) derived from a statistical analysis of POLDER data (Nadal and Bréon 1999). More details can be found at Deuzé et al. (2001).

Given the 2N (two wavelengths, N-viewing directions) measurements, for each aerosol model, the residual error (ε) is defined as

$$\varepsilon = \sqrt{\frac{1}{2N} \sum_{\lambda} \sum_{j} \left[Q_{cal}\left(\lambda, \Theta_j\right) - Q_{meas}\left(\lambda, \Theta_j\right) \right]^2}, \qquad (15.2)$$

where Q_{cal} and Q_{meas} are the calculated and measured polarized intensity, respectively. The retrieved aerosol parameters are the Angstrom exponent, α, and optical thickness, τ_{670nm}, corresponding to the model that gives the smallest residual error, ε.

However, several points must be emphasized that make POLDER measurements quite difficult for aerosol retrieval over urban surfaces.

1. The algorithm uses only single scattering and assumes only a single fine mode. While these assumptions can be relaxed by including full multiple scattering and assuming a coarse mode, the lack of a long wavelength

channel makes it impossible to separate the fine and coarse components, thereby biasing the AOD. This makes aerosol retrieval in urban coastal areas or areas with transported dust aerosols difficult.

2. The requirement of single scattering makes the AOD retrieval for wavelengths < 670 nm problematic and unfortunately these wavelengths are critical for SSA estimates (Waquet et al. 2009).

3. The polarized surface models need to be tuned to the specific region. Significant errors are seen using a single global parameterized polarized reflectance and it has not yet been established that unique parameterizations for each land classification are sufficient to retrieve aerosols over diverse urban areas.

4. Therefore, regional tuning of the surface seems to be necessary even for POLDER and this point will be discussed in detail for MODIS in Section 15.3.1.

5. The inclusion of a SWIR channel into the polarized measurement system would provide a very significant improvement and allow a separation of fine and coarse modes as well as providing an accurate retrieval of SSA. This point will be explored in detail in Section 15.4 when we discuss the potential of the ill fated APS sensor on board the failed GLORY Launch.

15.2.7 GEOSTATIONARY AEROSOL RETRIEVAL (GOES AEROSOL AND SMOKE PRODUCT)

It is quite apparent that observations from polar orbiters are not frequent enough to address many air-quality measurement and forecasting issues. Clearly, there is a strong need for near-real-time aerosol information with high temporal resolution for air-quality applications. Therefore, it is not surprising that efforts to retrieve aerosol properties from geostationary platforms have been actively pursued.

Unlike the multispectral polar orbiting sensors (MODIS, MISR, etc.), GOES has a single VIS channel and strong assumptions must be made to retrieve the aerosol optical depth. In analogy with AVHRR, which also has a single VIS channel, the aerosol type must be assumed so that a one-dimensional aerosol LUT parameterized solely by the column AOD at 550 nm can be constructed for VIS TOA reflectance. For simplicity and in an effort to be representative of CONUS retrievals as a whole, a background continental model is used with a single scattering albedo of 0.89 (Vermote et al. 1999). Of course, surface contamination over land is again a fundamental issue that any aerosol retrieval algorithm must address. Unlike the multispectral sensors that can tease out estimates of the surface reflection based either on multiangle observations, long wavelength observations that reduce aerosol effects, or polarization sensors such as APS, the GOES Aerosol/Smoke Product (GASP) makes use of the high temporal resolution to estimate the surface.

In particular, it is possible to isolate the ground albedo contributions simply by making observations over sufficiently long periods of time and expecting that in such a long time period, "clean" atmospheres in which the aerosol loading is negligible can be observed. Then, assuming the same surface albedo holds over the entire period, the aerosol contribution can now be determined for all other cloud cleared observations. This very intuitive idea is the underlying basis for the operational GASP product (Knapp et al. 2005; Prados et al. 2007) and may be obtained at ftp://satepsanone.nesdis.noaa.gov/GASP/AOD/ in the binary format.

The implementation of the surface reflection retrieval involves the following steps: (1) For each GOES pixel and each time slice, generate a 28-day VIS TOA reflectance time series. (2) Determine the second smallest reflectance value. (3) Assume this value to be the result of a "clear-sky" observation where only well-characterized molecular scattering and a small fixed (assumed AOD = 0.02) residual aerosol loading are assumed to obscure the ground. (4) Retrieve pixel reflection by inverting against preconstructed reflectance tables suitably interpolated to the sun-view geometry. (5) Do this for all "clear-sky" pixels and all diurnal time slices, thereby producing an atmospherically cleared surface reflection mosaic that can be used for aerosol retrieval.

The choice of using the second darkest pixel instead of the darkest one is to avoid or reduce cases where cloud shadow contamination might arise (Knapp et al. 2005). Also, although the choice of 28 days is fixed in the operational code, it should be pointed out that such rules may not be optimal and can vary seasonally and geographically seasonally. Generally, the choice was made as a way of balancing between a longer time period, which increases the chances of finding an image with clear-sky conditions, and a shorter time period to increase the probability that the underlying surface reflectivity is stable.

GASP performance. To summarize, the main potential error sources include the assumptions underlying the calculation which are that the surface reflectivity is Lambertian and the use of a fixed aerosol model. In addition, assumptions in the value of the background AOD used in calculating the clear-sky composite image can also lead to significant errors by not taking into account geographical or temporal variability in the background AOD. Estimating these errors has led to theoretical estimates of AOD accuracy between 18% and 34% (Knapp et al. 2005). For comparison, these estimated errors are somewhat higher than those of MODIS where AOD retrievals are generally bounded by 15% with root-mean-square (RMS) AOD errors ~0.05 (Levy et al. 2007a,b). On the other hand, over many urban areas, significant additional biases in AOD are observed in MODIS and MISR retrievals (see Section 15.4.1). For example, significant overestimates in MODIS retrieved AOD of the order of 20% are observed (Oo et al. 2010). On the other hand, as seen in Figure 15.2b, high correlations between GOES-East and AERONET observations at the City College of New York (CCNY) are obtained with RMS errors of the order of 10% (Prados et al. 2007). The matchup procedure was suitably filtered based on complete overlap of the GASP pixel over the AERONET site and setting the temporal coincidence criterion to be less than 30 min.

Looking at the quality of AERONET/GOES correlations over the United States as a whole (Figure 15.2a), what is most evident are the strong correlations between the AODs for the northeast and the quite poor correlations for the west and central United States. While part of the differences between these geographical regions can be attributed to lower overall AOD values in the west, the high surface albedo together with significantly enhanced aerosol variability including strong biomass burning and elevated dust events are important root causes for the higher discrepancies. In addition, the data are obtained using the GOES-East imager whose subsatellite point is

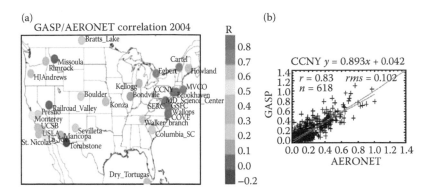

FIGURE 15.2 (a) Correlation between GASP and AERONET AODs. Note the high correlations in the northeast and quite poor performance over the west and central United States. (b) GASP AOD versus AERONET AOD at CCNY. Note the high correlation and lack of significant bias but large AOD RMS errors. (Reproduced from Prados A.I. et al., *Journal of Geophysical Research*, 112, D15201, 2007, doi:10.1029/2006JD007968. With permission.)

centered at 75W and is therefore clearly biased to the eastern portion of the United States.

On the positive side, the performance of the GASP product using the AERONET station at CCNY illustrates the potential for GASP for some urban areas such as those dominated by sulfate aerosol types. However, even in such cases while no bulk bias is seen in general and the correlations are fairly uniform over the diurnal cycle, significant biases as defined by their RMS errors rise dramatically before 1215 UTC or after 2115 UTC (Prados et al. 2007). Reasons for this include larger solar zenith angles magnifying the effect of imprecise atmospheric assumptions, a greater tendency for surface angular effects to be removed, as well as a greater likelihood of cloud shadows (more clouds near sunrise/sunset) affecting the surface composite map. Therefore, optimized performance should be limited between 1215 and 2115 UTC.

In conclusion, the GASP product shows some potential in retrieving AOD over urban areas with high temporal resolution, although RMS errors are significantly larger than MODIS with RMS errors ~0.1 However, it must be pointed out that the GASP product from the current GOES sensor will be superseded with the launch of GOES-R in 2016. In fact, the Advanced Baseline Imager on GOES-R is basically a "MODIS" sensor on a geostationary platform and will make use of multispectral based algorithms bypassing the use of surface mosaics and assumed aerosol models (Kondragunta et al. GOES-R ATBD Document 2010). For these reasons, we can expect performances similar to those of MODIS but for data at 30-min intervals. In addition, further improvements in MODIS retrievals such as the MAIAC approach (see Section 15.4), which makes a concerted effort to simultaneously derive surface and aerosol properties, will naturally be implemented in the GOES-R algorithms.

15.3 CURRENT OPERATIONAL ALGORITHM PERFORMANCE IN URBAN AREAS

In this section, an attempt at concisely presenting the performance of existing satellite retrievals over urban areas is presented. The comparison is not exhaustive and is only meant to illustrate the most significant issues regarding aerosol retrievals over urban areas.

15.3.1 INTERCOMPARISON OF MODIS (STANDARD AND DEEP BLUE) MISR, OMI, AND POLDER WITH AERONET

In order to validate the current operational algorithm of MODIS and MISR, intercomparisons of MODIS, MISR, and AERONET data are performed. Of course, much work on intercomparisons has been reported in the literature (Levy et al. 2007a; Liu et al. 2008; Kahn et al. 2005; Mishchenko et al. 2010) but all such studies tend to have a global focus. In assessing urban sites, it is particularly useful to make use of the MODIS Atmosphere Parameters Subset Statistics (MAPSS) database (Ichoku et al. 2002). The MAPSS database was established to provide a means of intercomparison and cross validation of MODIS Atmosphere products with those of other sensors and instruments (ground-based, airborne, or spaceborne) obtained at different locations around the globe. In particular, relevant statistics for AERONET and satellite products obtained from the original data streams are collected. The process of generating the statistics involves identifying the ground assets (AERONET) on all contributing satellite products extracting the values of the pixel corresponding to each point as well as surrounding pixels falling within a 5 × 5 box centered on the point. For direct comparison, the subset array of pixels for each location is used to compute the following primary statistics:

1. Value of the central pixel ("cval_")
2. Mean of the subset ("mean_")
3. Standard deviation of the subset ("sdev_")

The data files are stored as ASCII table files, in comma-separated-value (.csv) format, with column and header information sufficient to run query-driven searches. For the sunphotometer, the same statistics are obtained for the relevant time series whose duration before and after the overflight time is 30 min. Further supporting satellite/ground validation matchups is a new menu-driven interface built on the Giovanni Online Portal (Prados et al. 2010; Zubko et al. 2010). This interface allows the user to construct overlapping time series data of satellite and ground-based products once a given site is queried. Users can pick any number of data products and can chose the start and end time of the comparisons. Outputs from this product can be extracted in CSV files that are conveniently constructed to allow direct data comparisons across the different products, leaving nulls when data are not available. A first application of the MAPSS-Giovanni tool is shown in Figure 15.3. In all comparisons, the central pixel is used for comparisons rather than the mean since we are

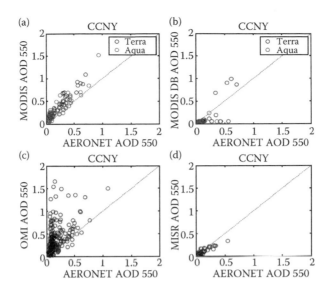

FIGURE 15.3 AERONET AOD versus satellite AOD products from the MAPSS–Giovanni tool over CCNY: (a) MODIS MOD04 (Terra)–MYD04 (Aqua) collection 5.1 product, (b) Deep Blue collection 5.1 (Terra and Aqua), (c) OMI, and (d) MISR.

focused on possible high resolution applications. The time period used is January 1, 2006 to December 31, 2009.

The results illustrate a number of common features in urban aerosol retrieval potential. In panel a, we see a small but clear overestimate of the MODIS operational satellite retrieval for both Aqua and Terra retrievals. This is a feature we will see for other urban sites. In panel b, the MODIS Deep Blue algorithm is seen to result in very poor retrievals both in number and quality. This can be tracked to the algorithm details that require that the bright surfaces be sufficiently homogeneous over the 0.1 × 0.1 degree cells used in the surface construction. This requirement is not met for densely populated urban areas. In panel c, we look at the ozone monitoring instrument (OMI, aboard EOS-Aura) retrievals, which are based primarily on ultraviolet (UV) measurements. Significant difficulties in the OMI algorithm are discussed in Section 15.4.2 and as such cannot be considered a useful sensor for urban aerosol monitoring. In panel d, MISR retrievals are shown to be highly correlated but with a regression slope of the order 0.8. These anomalies have been observed (Khan et al. 2005) at several urban sites and are attributed to the fact that we need to include into the aerosol models urban pollution models with lower SSA, which could raise the AOD and hence better retrieve the particle properties. Even if this anomaly is corrected, we should note that the smaller MISR swath makes collocated measurements much less frequent and hence less useful for air-quality applications. As noted above, MODIS AOD data are significantly overestimated in comparison with AERONET AOD data. This observation is not unique to New York City. For example, similar results were found in the Mexico City urban area during the Megacity Initiative:

Local and Global Research Observation (MILAGRO) project. More details of the MILAGRO project can be found (Castanho et al. 2007) and observed in such urban areas as Beijing and Barcelona. In understanding MODIS overbias, we show (Oo et al. 2010) this to be a manifestation of the underestimation of the algorithm surface VIS/SWIR reflection ratios, which, however, can be regionally corrected as discussed in Section 15.3.2.

In assessing POLDER, it should be recalled that the measurements are sensitive only to fine mode characteristics. Therefore, in comparisons, either dustlike contributions have to be estimated indirectly (Leon et al. 1999) or comparisons should be made on the fine mode AOD such as Fan et al. (2008) who compared POLDER aerosol retrievals with fine mode AOD from AERONET ($r_{eff} < 0.3$ μm). In Fan et al. (2009), the estimates of the surface reflection were based on collecting data with AOD at 865 nm < 0.02 from AERONET and assuming under these conditions that surface polarization is the predominant mechanism. Using the surface model obtained in this manner and reprocessing resulted in correlations > 0.9 and RMS errors < 0.03 for multiple urban areas (Beijing and Xianghe). The performance in matching the fine MODE AOD at 870 nm is seen in Figure 15.4. We note that while much more variability is seen in the center pixel, the mean retrievals are much better behaved. On the other hand, when comparisons were made at other urban sites, retrieval was extremely poor with <20 data matchups in most places and a general underestimation of the fine mode component even after focusing on cases where fine mode AOD is dominant as determined by AERONET retrievals.

In summary, MODIS seems to offer the best overall performance for multiple urban areas including high correlation, fairly consistent overbias signatures, and significantly higher retrieval statistics, which is a combination of the large swath and relatively higher resolution, making it easier for cloud clearing. Still, observed overbiases need to be investigated more carefully on a site-to-site basis to properly interpret the cause as well as mitigate its effects.

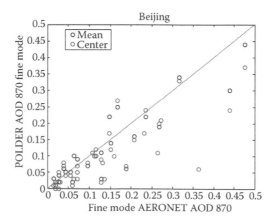

FIGURE 15.4 Comparison of POLDER 870-nm AOD versus AERONET fine mode AOD for Beijing, China.

15.3.2 REGIONAL APPROACHES

15.3.2.1 Regional Training of Urban Surface Models Using Ancillary Data

In the previous section, we discussed the need for regional (ancillary) methods to improve the parameterization of urban surface reflectances to account for observed overbiases. Here, we illustrate these ideas more completely focusing on MODIS where the surface properties are driven by semiempirical models, which calculate the VIS/SWR SRRs. In this section, we show how to generate a per-pixel estimate of the surface VIS/SWIR ratio from high spatial resolution MODIS L1B data (over the period from 2001 to 2007) in combination with *a priori* knowledge of the aerosol properties obtained from AERONET retrievals (Oo et al. 2010). This idea has been further developed and is the basis for the ASRVN product discussed in the next section.

Although the MODIS-defined LUT has four fine aerosol models (continental, generic, smoke, urban) and a dust aerosol model, the C005 Algorithm Theoretical Basis Document (ATBD) shows that, based on extensive AERONET measurements for the northeast region, the fine aerosol model is defined as belonging to the urban aerosol type (Remer et al. 2005; Levy et al. 2007a). Therefore, in obtaining the surface model, we will also assume that the aerosols are of urban type to interpret our comparisons and not have them depend on aerosol classification errors from C005. The physical and optical properties of the five aerosol models are defined in the C005 ATBD (Remer et al. 2005).

During the comparison, we only use atmospheres that we could verify *a priori* as fine mode. In fine mode dominated aerosol conditions, the SWIR reflectance at TOA is totally due to the ground since the SWIR wavelength is transparent to the fine mode aerosol atmosphere (Kaufman et al. 1997). Therefore, we can approximate the surface reflectance at 2.12 µm as the TOA reflectance and not concern ourselves with aerosol contributions at the SWIR during surface retrieval. To identify fine mode dominant cases, we select only comparisons when the angstrom coefficient $\alpha > 1$ (as defined in Equation 15.3), and to further reduce errors due to misclassification of aerosols, we limit the AOD to $\tau_{0.55\mu m} < 0.2$

$$\alpha = \frac{\log\left(\tau_{675nm}/\tau_{1020nm}\right)}{\log(1020/675)}. \tag{15.3}$$

Finally, as stated in the previous section, suitable homogeneity assumptions that were used for intercomparion of the operational product are adopted for regional training. First, we used the MODIS cloud-masking procedure and gas absorption (including water vapor) correction in processing the TOA reflectance. Then, assuming a Lambertian surface albedo, the surface reflectance can be obtained based on Equation 15.4 using cloud clear corrected TOA reflectance and fine mode aerosol atmospheric parameters from LUT:

$$\rho_\lambda^{surf}\left(\theta_0,\theta,\Delta\phi,\tau\right) = \frac{\rho_\lambda^{TOA}\left(\theta_0,\theta,\Delta\phi,\tau\right) - \rho_\lambda^{path}\left(\theta_0,\theta,\Delta\phi,\tau\right)}{s\left(\theta_0,\tau\right)\left(\rho_\lambda^{TOA}\left(\theta_0,\theta,\Delta\phi,\tau\right) - \rho_\lambda^{path}\left(\theta_0,\theta,\Delta\phi,\tau\right)\right) + T_\lambda^{up}\left(\theta_0,\theta,\tau\right)T_\lambda^{dn}\left(\theta_0,\tau\right)},$$

$$\tag{15.4}$$

where ρ_λ^{TOA} is the measured TOA reflectance, $\lambda = 0.47$ and 0.66 µm, ρ_λ^{Path} is the atmospheric path reflectance, s is the atmospheric albedo, T_λ^{dn} is the downward transmission, T_λ^{up} is the upward transmission, and θ_0, θ, and $\Delta\phi$ are the solar zenith angle, sensor (satellite) zenith angle, and relative solar sensor azimuth angle, respectively. More details can be found in Oo et al. (2010).

Therefore, ingesting the AERONET AOD and MODIS urban aerosol phase function is sufficient to obtain the surface reflections for both VIS channels independently from Equation 15.4. Furthermore, since the aerosols are fine mode dominated, the ground reflectance at 2.12 µm is directly obtained allowing us to measure the pixel level SRRs.

The results of these calculations as a function of the scattering angle are shown in Figure 15.5 for three spatial resolutions (10, 3, and 1.5 km) surrounding the CCNY AERONET site. We also masked the inland water area contamination to our derived surface reflectances, since CCNY is so close to the Hudson River (Oo et al. 2010). We first note that the calculated VIS/SWIR reflectance ratios are higher than those implemented from MODIS operational algorithm defined VIS/SWIR reflectance ratios, and these ratios are in good agreement with VIS/SWIR ratios result from Hyperion data for the New York City urban area (Gross et al. 2003). Furthermore, as may be expected, we do not detect any meaningful trend behavior as a function of scattering angle when aggregated over the entire multiyear training period. In fact, any trend observed is significantly smaller than the fluctuations around the trend

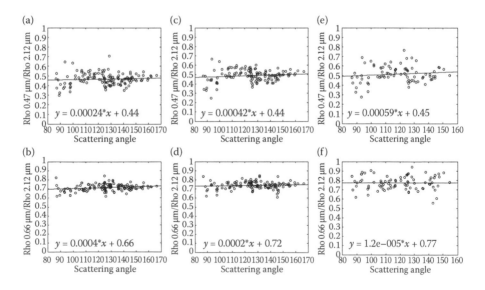

FIGURE 15.5 Surface reflectance ratio of 0.47/2.12 and 0.66/2.12 µm with respect to (x-axis) scattering angle. Mean reflectance ratio for different spatial resolution: 10×10 km (a, b), 3×3 km (c, d), and 1.5×1.5 km (e, f) surrounding to the nearest pixel to AERONET site at CCNY from 2001 to 2007 with fine mode aerosol dominant atmosphere (with inland water areas and rivers masked).

line. It should also be pointed out that the approach was validated against vegetation targets returning values consistent with the C004 and C005 values.

15.3.2.2 Operational Implementation of Surface Retrieval (ASRVN)

The basic ideas we used extensively for New York City were made operational at NASA Goddard Space Flight Center (GSFC). The resultant network has been named the AERONET-based Surface Reflectance Validation Network (ASRVN), resulting in the spectral surface bidirectional reflectance and surface albedo measurements from MODIS data (Lyapustin et al. 2007a; Wang et al. 2009). The algorithm functions on 50×50 km^2 subsets of MODIS L1B data, which uses AERONET aerosol and water vapor information to estimate the atmosphere, allowing direct surface retrievals. The archive data include 6 years of MODIS TERRA and 1 1/2 years of MODIS AQUA data and are available as a standard MODIS product (MODASRVN) at the LAADS Web site http://ladsweb.nascom.nasa.gov/data/search.html for about 100 AERONET sites. Data holdings include data from 2002 until May 4, 2008. Further upgrades to the product are ongoing and new reprocessing is expected by June 2010.

The retrieval product is based on accumulated 16-day TOA MODIS measurements, which, together with AERONET atmosphere measurements, provide optimal LSQ fits for the coefficients of the Li-sparse Ross-thick (LSRT) model of the bidirectional reflectance factor (BRF). In supporting its use as an operational product, the algorithm includes the following quality features:

1. A robust cloud mask algorithm for opaque clouds
2. An aerosol filter to identify residual semitransparent and subpixel clouds
3. Continuity and consistency of updated solution with previously retrieved BRF and albedo
4. Use of the heritage seasonal spectral BRF database to increase data coverage, allowing ASRVN to provide nearly complete coverage for the processing area

The main 1-km products are the BRF coefficients (BRDF/π) and surface albedo. In addition, ASRVN also computes several useful derivative products:

1. *NBRF*, a BRF normalized to nadir view and SZA = 45°, which is identical to the MODIS NBAR product that is part of the MOD43 standard product suite. This product removes the variability due to illumination and view geometry, making the *NBRF* useful for vegetation, surface classification, and so forth.
2. *IBRF*, which is a single observation (i.e., instantaneous) BRF value for the specific viewing geometry of the last day of observations using the latest MODIS measurement assuming that the shape of the BRF, known from previous retrievals, has not changed.

In addition, the MODASRVN product suite includes cloud mask data, the RGB browse images for MODIS TOA reflectance, ASRVN NBRF and IBRF, and the QA flag.

To compare with our previous results, we first look at time series of the IBRF ratios for two different years. The most striking observation is the fairly stable correlation ratio for the 660 (nm)/2120 (nm) ratio with the greatest stability in the urban case as well as the increased value of the IBRF ratio.

This is quite remarkable since the IBRF ratio can be expected to have additional variation due to differences in the angular BRDF that can occur at different wavelengths (Wang et al. 2010). This is to some extent observed in the blue–SWIR ratios where much stronger variability is seen. Additional contributors to the additional fluctuations in the blue would be the increased difficulty in surface retrieval at shorter wavelengths and the smaller surface albedos in the blue. In comparing the ASRVN data in Figure 15.6 with our regional tuning, we note reasonable agreement in the stability and magnitude of the SRRs. However, we find that the ASRVN 660/2120 ratio seems a little higher (0.83) versus our 1.5-km (0.77) result, which we believe may be due to our very conservative water mask. Any water contamination will clearly increase the ratio since water absorption at 2120 nm is so strong that the ratios are significantly increased. Similar results for the increase in SRR are also found in major urban areas, including Mexico City and Beijing, as we discuss in more detail below.

In fact, ASRVN data are particularly helpful in assessing the relationship between MVI (see Section 15.2.4) and the spectral ratio coefficients (SRRs) used by MODIS C005. Results of the dependence of the SRRs with the MVI are given in Figure 15.7 for New York City, Beijing, and Mexico City. The most striking issue is the fact that the general trend connecting MVI to the SRR used within the MODIS C005 is both quantitatively and qualitatively incorrect for all regions. In fact, C005 predicts that the SRR should weakly "increase" with MVI and nonvegetation surfaces (such as bare soils) should have slightly lower SRRs. However, we find in all cases considered that the SRR trend is dramatically different with much higher values at low MVI

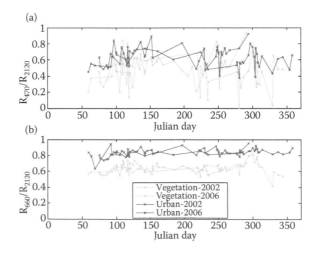

FIGURE 15.6 Time series analysis of the relevant spectral ratios from ASRVN. (a) 470/2120, and (b) 660/2120.

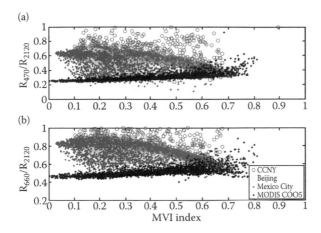

FIGURE 15.7 ASVRN spectral ratios versus MVI with comparison to MODIS C005 selection; (a) 470/2120 and (b) 660/2120.

pixels with values decreasing towards the high vegetation values shared by MODIS C005 (and earlier versions). On the other hand, the New York City data are particularly interesting since this effect is most enhanced. This is going to be a function of possible contamination due to the river together with strong shadowing due to the very high building profiles in the New York City domain. In addition, the regression plots show similar dispersion, allowing us to construct a very simple model based on a least squares quadratic (LSQ) fit using a quadratic model through the data connecting the SRRs with the MVI. For these reasons, we focus on New York City to assess the use of regional trained surface models.

15.3.2.3 Ground Surface Reflectance Ratio Modeling Using End-Member Analysis from ASTER Spectral Libraries

The need to quantify the relationship between surface type and the VIS/SWIR reflection ratios in urban areas is crucial for aerosol retrieval since the current empirical relations being used in MODIS C005 are qualitatively and quantitatively in error, as seen in Figure 15.7. To understand why the SRR versus MVI relationship behaves in this manner, we model the urban environment using a combination of spectra between two distinct classes (called end-members) allowing the classes to be linearly mixed. The two end-members are taken to be vegetation (grass) and a choice between a number of man-made materials indicative of the urban environment. To do this, we use the Advanced Spaceborne Thermal Emission and Reflection Radiometer (ASTER) spectral library (Clark et al. 2003), which has a large database of measured spectra covering from 400 to 2500 nm to construct relevant reflectance spectra. In developing the end-members, we are looking for the simplest model that qualitatively explains the difference in the SRR versus MVI. The relevant man-made spectra classes, together with vegetation (grass/leaves), are given in Figure 15.8a. The grass mixture was taken to be a mixture of healthy (95%) vegetation and dry vegetation (5%). This choice is made to "fit" the ASRVN values at high MVI, which

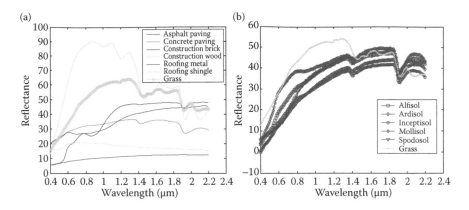

FIGURE 15.8 Spectral reflectances from ASTER Spectral Libraries; (a) man-made materials and (b) soil types. In both cases, grass/leaf spectra are provided as reference.

we take to be composed of pure vegetation. In Figure 15.8b, the soil spectra for different types are given. Unlike the very strong variability observed in urban materials, the soil spectra have less variability than the man-made material spectra. The soil classification groups used can be fully described in USDA (1978). The results of the end-member analysis for variable mixing ratio for each man-made class are given in Figure 15.9.

The most obvious result is that almost all man-made material mixed with vegetation has the requisite behavior where the associated spectral ratios increase as the MVI parameter decreases in contradiction to the MODIS C005 model but in agreement with the ASRVN data sets. The anomalous behavior of construction brick and construction wood should not cause too many problems in an urban environment since these classes are expected to be less frequent than the concrete, asphalt, and roofing metal. It seems that a suitable mixture of these four dominant classes can be used to obtain a suitable end-member that can be mixed with vegetation. In obtaining these results, the spectral response function of the MODIS bands is taken into account.

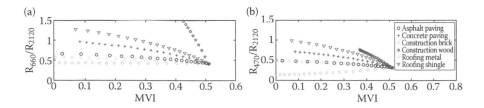

FIGURE 15.9 VIS/SWIR reflectance ratios versus MVI for vegetation/man-made material end-member mixing for different man-made classes. (a) 660/2120 reflectance ratio; (b) 470/2120 reflectance ratio.

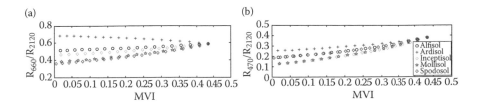

FIGURE 15.10   VIS/SWIR reflectance ratios versus MVI for vegetation/soil end-member mixtures. (a) 660/2120 reflectance ratio; (b) 470/2120 reflectance ratio.

On the other hand, if this is a reasonable method in investigating the SRR versus MVI relationships, we should be able to explain the C005 model, which was globally trained. In the vast majority of AERONET sites, the end-members should be taken to be soils and vegetation. An end-member mixing analysis for five representative soil types is given in Figure 15.10.

This analysis is not meant to be definitive but to illustrate that surface mixtures between man-made and vegetation behave very differently than soil–vegetation mixtures, resulting in quantitative and qualitative differences in the MVI dependence on the SRR values. A more exact approach clearly requires more end-members. For example, it was shown that based on an exhaustive analysis of Landsat ETM+ images of 28 urban centers (Small 2005), three global end-members were able to account for >90% of the scene variability. Therefore, two parameters linear mixing among the three end-members would provide a useful estimator for pixel–pixel estimates of the SRRs. Preliminary application of the three end-member analysis to the SRRs does indeed conform to the general SRR trend observed to the above analysis, but since the surface classification relies on two independent mixing ratios, two independent atmospherically robust surface measurements are needed, and it is not clear at present which measurement parameters are best for this task.

15.3.2.4   Assessment of Regional Training

Using the regionally trained SRR as a function of the MVI (ignoring very small trends in angle) in the operational AOD retrieval processing stream, a significant improvement is found as shown in Figure 15.11. In particular, the significant over-estimation of AOD seen in the operational processing due to an underestimation in the surface albedo in the VIS channels is removed. In fact, the uncertainty even at higher resolution is quite small. In both these figures, the operational error bounds of ($\pm.05 \pm.15\tau$) for AOD retrieval over land were added and show that, except for very low AOD values where even small errors in reflection compensation can lead to significant AOD errors, the AOD retrievals are well within these uncertainty bounds.

To evaluate this method further, we performed an additional regional processing over Mexico City. Unlike the New York City area, Mexico City is located approximately 2 km above sea level and most of the fine mode aerosols fall within the

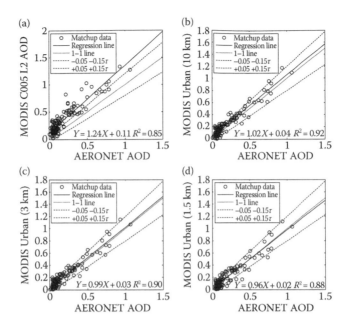

FIGURE 15.11 (a) MODIS L2 aerosol optical depth at 0.55 μm compared with 4-h (~2 h before and 2 h after MODIS [Terra] satellite overpass time) average of AERONET aerosol optical thickness. (b), (c), and (d) are retrieved AOD with new surface reflectance VIS/SWIR ratio plot with average of AERONET AOD, inland water areas, and rivers masked.

smoke aerosol model category. These differences allow us to illustrate that surface improvements may be applied to multiple aerosol conditions. Here we make use of the Mexico City AERONET station data acquired during 2000 to 2007 and duplicate in detail the procedures of the surface reflectance retrieval that were applied to the New York City region (Oo et al. 2010) except that, due to the altitude of Mexico City, a pressure correction was implemented. The resultant VIS/SWIR surface ratios are very similar to the New York City urban area's outcome. The averages and standard deviation of the 0.47/2.12 and 0.66/2.12 μm surface reflectance ratios are approximately (mean = 0.43, std = 0.10) and (mean = 0.70, std = 0.04), respectively, at 10-km resolution and (mean = 0.44, std = 0.12) and (mean = 0.71, std = 0.04), respectively, at 3-km resolution. Of particular interest are the small uncertainties in these ratios at the higher wavelengths, which is reasonable considering that the atmospheric correction uncertainties are expected to be less important. Using these refined regional VIS/SWIR surface reflectance ratios, the significant improvement in AOD retrieval can be observed in Figure 15.12. However, since Mexico City aerosol climatology has a smaller percentage of fine mode aerosol cases, the number of training measurements when compared to New York is much smaller. At 1.5-km resolution, too few cases of fine mode clear-sky cases were available, causing analysis to be limited to 3-km resolution.

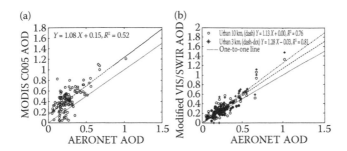

FIGURE 15.12 (a) MODIS L2 aerosol optical depth at 0.55 μm compared with 4-h (~2 h before and 2 h after MODIS [Terra] satellite overpass time) average of AERONET aerosol optical thickness. The dotted line is the one-to-one line and the dashed line is the linear fit line (b) retrieved AOD with new surface reflectance VIS/SWIR ratio versus AERONET AOD; the 'o' is 10 × 10 km resolution and the * is 3 × 3 km resolution.

On the other hand, care must be taken in applying this correction to other urban sites. In fact, we have seen many cases where clear overbias in the AOD retrieval is present. This is illustrated in Figure 15.13 where Kampur and Rome matchups are made. It is clear that for these two cases (as well as others not shown here), even urbanlike cases may not exhibit the telltale overbiases that we observed in New York City, Mexico City, and Beijing.

The reason for these differences can be seen most directly in the statistics of the SRRs for the different urban areas as seen in Figure 15.14. These statistics are the result of 2000 to 2008 aggregate data for the nearest pixel to the associated AERONET site. The most important observation is that the SRR statistics for the cases where overbiased AOD retrievals occur is significantly higher as a group than those for sites where negligible bias is observed. This is consistent with the interpretation that the overbiases are a function of underestimated surface reflections due to

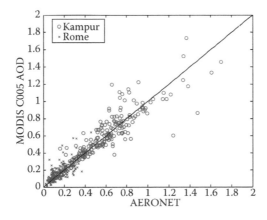

FIGURE 15.13 MODIS L2 AOD comparisons for Kampur and Rome, Italy.

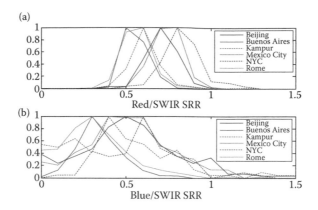

FIGURE 15.14 (a) Red/SWIR SRR frequency statistics illustrating differences between sites with overbiased AOD retrievals (blue) and sites with no bias (red). (b) Same as (a) for the Blue/SWIR SRR. NYC = New York City.

a too small SRR used in the operational processing. Explaining why different urban areas have these discrepancies is, of course, a complex study of the surface behavior of each site in question.

Unfortunately, the regional approach based on custom analysis of a specific region and using ancillary data that are not readily available in most urban centers is far from optimal and approaches that can extract surface properties and aerosol properties simultaneously are needed for complex urban surfaces. One such approach, which we will discuss in more detail in the next section, is the multiangle implementation of atmospheric correction (MAIAC).

15.3.2.5 Modifications to Future MODIS Retrievals (C006)

The use of more AERONET/MODIS data presents opportunities for further improvement by improving aerosol model climatology for different regions as well as refining the SRR factors inherent in the model. For C006, aerosol map climatology is being further refined using AERONET data accumulated since 2005. In addition, modest modifications are being made in C006 to account for biases from the global parameterization, although no unique parameterization for operational processing has been agreed (MODIS C006).

15.4 NEXT-GENERATION APPROACHES

In analyzing the approaches we have explored until now, we see that no single existing platform or algorithm is optimized for aerosol retrieval over urban areas and the best results using existing methods require additional information (aerosol type) or local surface models to improve retrieval performance. However, more robust algorithms and/or instrument systems may help fill the void. We would like to focus on three such cases to illustrate potential improvements.

1. MAIAC
2. Full polarized radiative retrieval from APS on Glory
3. High spatial resolution retrieval (Landsat)

15.4.1 MULTIANGLE IMPLEMENTATION OF ATMOSPHERIC CORRECTION (MAIAC)

The goal for the MAIAC algorithm is to simultaneously retrieve the AOD, surface BRF, and the SRRs based solely on satellite observations. This is in contrast to the operational MODIS C005 code approach, which uses the SRRs that were developed based on global MODIS–AERONET matchups (Levy et al. 2007a). The critical assumption needed is that the time scale over which the surface varies temporally is sufficiently long to accommodate multiple observations. In this way, multiple coregistered images over the same scene can be used to retrieve both aerosol and surface properties simultaneously. To do this, MAIAC first groups the MODIS level 1B pixels into blocks of size $N \times N$ where the aerosol is assumed (at first estimate) to be homogeneous. Then, if K temporal observations are available, the number of retrieval unknowns is $K + 3 \cdot N \cdot N$, where 3 denotes the number of kernel amplitudes that make up the LSRT BRF model (i.e., the 3 LRST coefficients) and K represents the single AOD measured at block scale for each observation time. However, if the BRF function is independent of wavelength, the number of unknowns is now $K + N \cdot N$. Since there are $K \cdot N \cdot N$ observations, the system is overdetermined for $K \geq 3$ and a joint retrieval is possible.

15.4.1.1 Retrieval Algorithm

The general framework of the algorithm is illustrated in the flowchart in Figure 15.15. Although the algorithm details are numerous, the core of the general algorithm starts at block 4 where the SRR of band 3 at pixel level and the AOD at block level ($N = 25$) are retrieved. At the end of this step, the optimal $\left[b_{ij}^3 \right]$ (i.e., the SRR

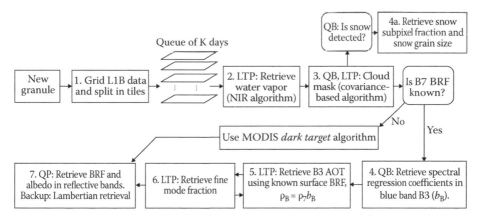

FIGURE 15.15 Flow chart of MAIAC algorithm. (Reproduced from Lyapustin, A. et al., *Journal of Geophysical Research*, 116, D03211, 2011, doi:10.1029/2010JD014986. With permission.)

between band 3 and band 7) for each pixel (i,j) and an optimized estimate of the block level AODs are obtained.

This is done using the following steps:

1. Process the block data for all times within the queue period assuming no aerosol loading (Equation 28 of Lyapustin et al. 2007b). In this way, a first estimate of the SRR for band 3 is made for all days in the queue.
2. Based on the magnitude of the SRRs over the block, the "cleanest" observation in the queue is found by finding the observation where the SRRs are minimized since aerosol turbidity will, in general, increase the SRRs in the VIS channels. This is in the same spirit as the GOES GASP algorithm, which uses a similar minimum reflectance criterion to determine the "clear" sky and produce a pixel level reflectance.
3. Calculate the aerosol optical depth $\{\Delta\tau_k\}$ for each cloud cleared observation (k) where the optical depth difference is referenced to the clear AOD $\Delta\tau_k = \tau_k - \tau_0$ and the SRRs for each day are recalculated assuming different values of optical depth into the forward model. In doing this, the reference value τ_0 is still not fixed.
4. In the final step, a global LSQ minimization (over all days and pixels) is performed to find the correct reference τ_0 and the final SRR's. For each value of τ_0, an LSQ minimization is used to obtain the time invariant SRR and the accompanying global residuals. By minimizing the residuals, the optimal SRRs and reference optical depth are found. Feeding back, the block AODs for each observation are then obtained.

Once the SRRs in band 3 are known (with the assumption that the BRF in band 7 is known), it is possible to retrieve both the optical depth and the fine/coarse mode fraction based on inversion of the full radiative transfer forward model at pixel level. The inversion is similar to the MODIS operational algorithm except that the MAIAC forward model includes the full anisotropy of the LSRT BRF parameters. In summary, the outputs of MAIAC are the same as those of C005, providing optical depth and fine/coarse mode fraction. However, these outputs are retrieved at grid level (i.e., 1-km resolution) in contrast to C005 with retrievals obtained at degraded 10-km resolution. The poor resolution of the MODIS C005 retrievals is a consequence of rejecting pixels whose reflections are too high to be considered sufficiently dark.

15.4.1.2 Performance of MAIAC Retrieval Algorithm (AOD)

The performance of the algorithm is based on intercomparions with AERONET using well-documented matchup procedures (Lyapustin and Wang 2009; Lyapustin et al. 2011). A representative performance is illustrated in Figure 15.16 where MAIAC for both blue (top row) and red (bottom row) channels are compared to the MODIS operational Dark Target algorithm MOD04, C005 (middle row). In the left column, the GSFC matchup, representative of a number of East Coast sites (i.e., Stennis, Walker Branch, and Wallops), shows nearly identical performance with the MOD04 product in both correlation and regression slope. In the right column, comparisons are made at UCLA, which is representative of many West Coast sites (i.e.,

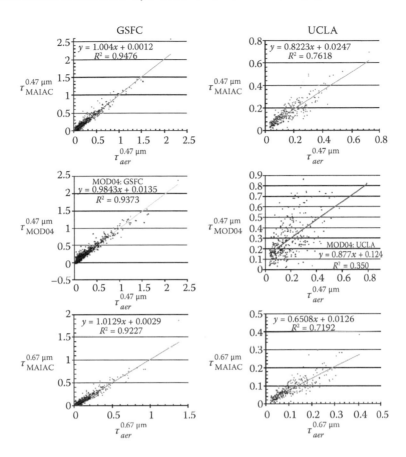

FIGURE 15.16 Left column: East Coast (United States) GSFC AERONET site; right column: West Coast sites (California, United States) UCLA AERONET site. Scatterplots MAIAC-AERONET $AOT_{0.47}$ (top), MOD04-AERONET $AOT_{0.47}$ (middle), and MAIAC-AERONET $AOT_{0.67}$ (bottom) based on MODIS Terra data for 2000 to 2008. (Reproduced from Lyapustin, A. et al., *Journal of Geophysical Research*, 116, D03211, 2011, doi:10.1029/2010JD014986. With permission.)

La Jolla, Fresno, and Table Mountain), having generally brighter surfaces with little or no vegetation. In this case, although the accuracy of MAIAC is lower in comparison to results from the East Coast as may be expected due to the brighter surface, MAIAC shows significant improvement over MOD04 with the MAIAC correlation coefficient approximately 0.77 compared to 0.35 for MOD04.

In analyzing the potential of this method for urban retrievals, we plot the results of the MAIAC retrieval for the CCNY site in Figure 15.17. It is clear that a significant improvement is present in the MAIAC retrieval, which seems to significantly remove the observed overbias. Nevertheless, there are other issues that should be considered in future refinements to MAIAC. For example, it was found (Wang et al. 2010) that in comparison to AERONET-based ASRVN datasets, MAIAC overestimates AOT in the backscattering directions and underestimates it for forward-scattering

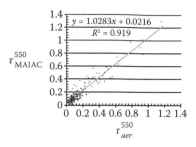

FIGURE 15.17 Urban comparisons of MAIAC and AERONET. (From Lyapustin, personal communication.)

angles with greater anisotropy observed in the blue band than the SWIR band. This implies that the assumption that the BRF angular dependence is constant over all wavelengths is not met and such modifications must be accounted for.

15.4.2 ADVANCED POLARIMETRIC SCANNER (APS) ON GLORY

The APS instrument was meant to be a unique sensor combining the best features of MODIS (spectral coverage to 2120 nm), POLDER (polarization), and MISR (multiangle observations). The instrument design approach was optimized to produce highly accurate polarimetric measurements (e.g., Degree of Linear Polarization (DoLP) < 0.005). The broad spectral range of APS was similar to MODIS 9 bands centered at $\lambda = 410, 443, 555, 670, 865, 910, 1370, 1610,$ and 2200 nm. Furthermore, the critical ability to view a pixel from multiple angles was provided by the scanning the APS IFOV along the spacecraft ground track, providing 250 scattering angles per scene.

15.4.2.1 Retrieval Algorithm Components

The details of the aerosol inversion process are given in Waquet et al. (2009). Here, we briefly discuss only a few elements that are quite unique.

1. Unlike POLDER, which assumes only a fine mode component, both fine and coarse mode aerosols are included. While the coarse mode microphysical parameters are fixed, the fine mode effective radius and the complex refractive index are part of the retrieval process.
2. Unlike all previous inversion methods discussed, the large inversion parameter space makes it extremely advantageous to use more sophisticated inversion techniques. In particular, the optimal estimation technique (Rodgers 2000) is used in the inversion, which in a fully rigorous manner allows for the inclusion of *a priori* information and instrumentation and model uncertainties through explicitly constructed covariance matrices. This allows error budgets to be estimated for each source of uncertainty.
3. Estimation of the surface reflection is related in spirit to the MODIS approach, which estimates the reflectance at the long wavelength channel and uses empirical relations to estimate the VIS channel reflectances. However, APS

uses the polarized TOA observations. In this case, the polarized surface reflection is based on fitting the measurements to a Fresnel-based reflectance model. Unlike unpolarized reflectance, which is not modeled well as a single scattering interaction, the Fresnel model is better suited to the modeling of polarized reflections that are dominated by single reflections. In addition, since the Fresnel reflection is a surface reflection dependent on intrinsic refractive index, the polarized reflectances are not strongly sensitive to wavelength.

4. Most importantly, the use of polarized reflectances reduces the problems associated with unknown surface reflections since heterogeneous surfaces tend to have smaller polarized reflectance signatures. In addition, the increased polarization sensitivity of the fine mode aerosol in contrast to coarse mode underlies the accurate retrieval of fine mode aerosol characteristics important for anthropogenic aerosol identification and monitoring.

5. The addition of the 410-nm channel (in contrast to POLDER) has a profound effect on the ability of APS to obtain an estimate of the SSA (Waquet et al. 2009).

15.4.2.2 Performance of Retrieval Algorithm

As we discussed earlier, the APS instrument was destroyed during launch. Therefore, the future performance using the APS approach is best inferred by exploring the retrieval performance of rotating scanning polarimeter (RSP) instrument, which flies on high altitude aircraft and was designed as a surrogate for the APS instrument. In support of our stated purpose of assessing urban aerosol retrieval, we will use the results obtained from the Aerosol Lidar Validation Experiment (ALIVE) and the MILAGRO field campaign (Waquet et al. 2009). In the comparisons that we made, three cases were examined. Cases a and b were performed on 09/16/05 and 09/19/05 during the ALIVE experiment (Southern Great Plains, United States), while experiment c was performed over Mexico City on 03/15/06 during the MILAGRO experiment.

Most critical to our urban focus were the MILARGO results. RSP retrievals (errors in parenthesis) for the effective radius and the effective variance were 0.15 (0.005) μm and 0.485 (0.01), and the real and imaginary refractive indices are 1.54 (0.01) and 0.027 (0.005), respectively. The small effective radius and high real refractive index together with retrieved imaginary refractive index, which is equivalent to a single-scattering albedo of 0.865 (0.005) at 550 nm, are in good agreement with African savanna biomass burning particles (0.84 < SSA < 0.88) in general and in particular to the fresh biomass burning particles observed in the vicinity of the source (i.e., SSA = 0.86). In addition, the Spectral AOD errors from 410 to 870 nm were less than 5% except near the blue, where retrieval errors approached 8%, which is not unexpected due to additional molecular–aerosol perturbations.

It is useful to consider how APS performance might compare against other satellite observations that might retrieve SSA. In particular, although the main mission of the OMI is ozone, estimates of aerosol properties (only fine mode) are part of the operational retrieval (Torres et al. 1998, 2003, 2007). OMI is a nadir-viewing imaging spectrometer that measures the TOA upwelling radiances in the UV and VIS (270–500 nm) spectral bands. OMI has a wide swath at 2600 km, but with per pixel resolution varying from 13×24 km^2 at nadir to 28×150 km^2 at the swath extremes, the retrieval is particularly susceptible to cloud contamination, making it very difficult to use on a regional (urban) scale. Furthermore, the measurement of the absorbing aerosol optical depth (AAOD) is more accurate and particularly effective when the aerosols have significant absorbing properties (Torres et al. 2007). In this sense, the retrievals are therefore more suited to qualitative determination of absorbing aerosols that are due to extensive biomass burning events. The system has other weaknesses that make it not well suited for urban observations. These include sensitivity of the retrieval to aerosol layer height, trace gas contamination, and surface albedo estimate requirements. Although the product may be improved in combination with other "A-Train" data such as Cloud and Aerosol Lidar and Infrared Pathfinder Spaceborne Observations (CALIPSO) to get information on layer height and/or MODIS that can provide column AOD as a constraint to the OMI retrieval, validation of UV-absorbing AOD is not a simple matter (Satheesh et al. 2009).

In summary, APS will offer the potential of retrieving unique aerosol microphysical properties beyond those possible from conventional intensity-based sensors. This includes accurate fine mode aerosol properties and SSA, which are crucial to monitoring anthropogenic aerosols. In addition, the use of polarized reflectances reduces the problems associated with complex urban surfaces.

Yet there are several difficulties that limit APS from being the ultimate global sensor for aerosols. These include a large footprint (6 km^2) that also makes it susceptible to unresolved clouds. In addition, APS acquires its multiangle coverage by scanning along the platform ground track. The clear weakness of an along-track-scanning (i.e., nadir) instrument is that it does not provide spatial coverage in any way comparable to that of imagers such as MODIS, MISR, and POLDER (Mishchenko et al. 2005). On the other hand, since APS can be flown in combination with a multispectral imager such as MODIS within the A-Train configuration, aerosol information within a narrow swath along the ground track can be used to provide an improved aerosol phase function, which, based on the relatively long correlation scales of aerosol intensive properties (50–100 km), can improve corresponding MODIS retrievals.

Finally, very recent efforts (Dubovik et al. 2010) have been made to utilize the multiple observation approach underlying the MAIAC algorithm to polarization data, thereby combining the benefits of both techniques. In particular, a "statistical" inversion approach that makes use of multiple time observations of POLDER data and develops a consistent optimal estimation-based inversion has been developed and is in the initial validation stages. In addition to utilizing the multiple observations and thus acquiring all the advantages of the MAIAC class of algorithms, the increased measurement space allows for a much more inclusive set of aerosol parameters to be retrieved including a full retrieval of bimodal aerosol size distribution. Of course, in this case, the SSA used in the operational POLDER processing (Deuzé et al. 2001) is no longer valid. To accomplish this, the Dobovik algorithm uses not only the polarized reflectances used in the operational algorithm at 0.67 and 0.78 μm but the unpolarized reflectances as well. This, together with the multitime observations where the surface reflectance properties are assumed stable, makes it possible to use the full spectral range of POLDER measurements, including total and polarized radiances at 0.49, 0.675, and 0.87 μm and total radiances at 0.44, 0.565, and 1.02 μm, and simultaneously retrieve aerosol and surface properties. However, it should be pointed out that the increased parameter space makes LUT solutions unrealistic and the algorithm utilizes continuous optimal estimation approach based on the AERONET microphysical retrieval heritage (Dubovik et al. 2000). Of course, this makes it difficult to apply in real-time processing, and algorithm costs and feasibility still need to be demonstrated.

15.4.3 High Spatial Resolution Retrieval (Landsat)

In this section, we will discuss the potential of urban aerosol retrieval with high spatial resolution sensors with particular emphasis on Landsat. An interesting aspect of Landsat is that the spectral bands of interest are well matched to MODIS VIS and SWIR bands. In fact, the surface spectral ratios being used in MODIS are based on high resolution imagery from Landsat (Kaufman et al. 1997). This leads to possible algorithm techniques that make use of the high spatial resolution for the classification of images to identify surface classes that can have particular spectral ratios. An algorithm of this type was applied (Kawata et al. 2003) for Landsat ETM images. In fact, the algorithm uses similar spectral ratios as those derived over urban and vegetation pixels for New York City discussed in Section 15.3.2.

$$\rho_{B1}^{s} = .25\rho_{B7}^{s}; \quad \rho_{B3}^{s} = .50\rho_{B7}^{s}; \quad \text{Vegetation}$$

$$\rho_{B1}^{s} = .42\rho_{B7}^{s}; \quad \rho_{B3}^{s} = .69\rho_{B7}^{s}; \quad \text{Urban.}$$

(15.5)

The steps of the algorithm are quite simple:

1. Correct for molecular absorption and georectification.
2. Classify the image before atmosphere aerosol correction using combined solar and thermal channels. In fact, the combined use of thermal and solar bands allows improved coarse classification between vegetation and urban pixels (Lu and Weng 2005).
3. Use the MODIS collection 5 algorithm with Landsat B1, B3, and B7 channels based on modification of LUT to account for specifics of Landsat spectral response functions.
4. Apply atmospheric correction (assuming Lambertian coupling) to refine classification (only necessary for sufficiently turbid scenes).

The algorithm showed reasonable performance in single-point matchups with some potential value in retrieving AOD to sub-kilometer resolution. However, as in all validations of images that are not operationally retrieved, the matchups are very sparse and only case-by-case assessments can be carried out, making statistical evaluations inappropriate.

Another approach is the use of multiple high spatial images to find dark pixels that can be used to estimate atmosphere directly (Hadjimitsis et al. 2002; Hadjimitsis 2008) or in contrast. In particular, it is a simple matter to obtain relative aerosol maps where "clean" atmosphere cases serve as a reference. Given such clean observations, aerosol path reflectance maps or AOD maps (based on aerosol model assumptions) can be made. These maps have been demonstrated to show significant correlations with surface pollution parameters such as PM2.5 with $R^2 > 0.45$, although the number of data points in the comparison is quite sparse ($N < 20$). Ultimately, the 16-day repeat time of the Landsat imagery makes it very difficult to use in any meaningful way, although higher resolution surface classification maps that can be borrowed by lower resolution retrievals should be further explored.

15.4.4 SYNERGISTIC EFFORTS

Finally, synergies between passive and active remote sensing offer some unique advantages. For example, Diner et al. (2007) proposed a method known as the Aerosol Global Interaction Satellites (AEGIS) mission, which used a combination of measured data from a passive (i.e., a multiangle spectropolarimetric imager [MSPI]) and an active sensor (a multiwavelength high spectral resolution lidar [HSRL]) to retrieve the 3-D aerosol distribution of size, shape, absorption, and optical depth. The MSPI concept relies heavily on Terra–MISR heritage and infuses with measurement from MODIS, the Total Ozone Mapping Spectrometer (TOMS), and OMI. The combined MSPI-HSRL approach is aimed at eliminating the uncertainty in aerosol microphysical retrieval as well as monitoring and characterizing aerosols over horizontal, vertical, and temporal scales from space.

15.5 CONCLUSIONS

So far, the most promising approach to operational retrieval in urban areas for simple AOD based on correlation accuracy and retrieval frequency is the MODIS retrieval. Previous biases detected in urban areas that were dealt with using regional corrections seem to be reduced with the MAIAC approach, where surface and aerosol properties are derived in a simultaneous fashion, and also accounts for much of the anisotropy of the surface. Furthermore, this approach is directly translatable to the future GOES-R ABI sensor, which will benefit from these algorithmic improvements while greatly improving time resolution. This rapid observation capability will also allow further classification of aerosol scenes based on their transport properties such as identifying aloft dust and smoke plumes that can ultimately be ingested into operational retrievals.

However, intensity measurements alone are ill-equipped to retrieve complex parameters such as SSA or particle size distributions. Ultimately, retrieving these parameters requires high accuracy multiangular, multispectral polarization measurements. As discussed in Section 15.4.2, POLDER in general has severe limitations in the retrieval of higher-level aerosol properties, although the statistical inversion approach that utilizes both polarized and unpolarized reflectances together with multitime observations and smoothness constraints is expected to significantly improve POLDER retrievals, even though a missing SWIR band still is expected to be a significant barrier for urban retrievals. Since the algorithm utilizes optimal estimation-based approaches instead of LUTs due to an extensive parameter set, it is not clear that the algorithm can function for real-time processing. In contrast, APS which was to be launched on GLORY in early 2011 has some unique advantages over POLDER, including the following:

1. Spectral coverage from 410 to 2250 nm allows accurate polarized surface reflection models to be constructed at the SWIR band (2250 nm) and extrapolated to the VIS and channels. This procedure is easier to implement than the corresponding approach in MODIS since polarized reflectances are better modeled as single-scattering events and are therefore more easily modeled based on Fresnel reflection models.
2. Further improvement in parameterized polarized surface reflections has also been extended to more complex surfaces such as mixtures of bare soil and vegetation based on regressions with POLDER observations. These models are themselves based on useful parameterizations of the NDVI. Although no equivalent urban model exists, it is to be hoped that these models can be obtained in the future.
3. Unlike POLDER, which is forced to look at single scattering and therefore can only retrieve AOD at longer wavelengths, the retrieval at short wavelengths requires multiple scattering and the inclusion of both fine and coarse modes. This makes the SWIR channel crucial and also allows for accurate separation of fine–coarse mode AOD and effective radius.
4. The improved modeling allows use of the 410-nm channel crucial for retrieval of aerosol SSA, in contrast with POLDER (PARASOL), whose shortest wavelength is 443 nm.

Opportunities for validation are rare, but impressive isolated matchups have been performed. The MILARGO campaign (AERONET) in Mexico City revealed very good retrieval of fine-mode aerosol optical depth and reasonable estimates of the SSA with local aerosol measurement data. Furthermore, excellent spectral agreement in the retrieval with AOD errors <10% for all wavelengths was obtained. However, as stated earlier, APS would have acquired its multiangle coverage by scanning along the platform ground track. The obvious weakness of an along-track scanning instrument is that it does not provide spatial coverage in any way comparable to that of imagers such as MODIS, MISR, and POLDER. In addition, APS had a large footprint, which would make cloud clearing more difficult. On the other hand, since APS would have been coordinated with MODIS in the A-Train configuration, aerosol information within a narrow swath along the ground track could have been used to provide an improved aerosol phase function, based on the relatively long correlation scales of aerosol-intensive properties (50–100 km), could have improved the corresponding MODIS retrievals.

Finally, it should be mentioned that the next possible system using the APS approach for detailed aerosol retrievals is the Aerosol-Cloud-Ecosystem (ACE) Mission, which is a tier 2 Decadal Survey mission [O.'C. Starr 2011] focusing on clouds and aerosols as well as ocean ecosystems. This mission with a preliminary launch estimate around 2018 would include, among other features, a multiangular polarimetric sensor at a minimum would include the broad spectral coverage needed including coverage into the deep blue, SWIR, as well as additional coverage into the UV to further improve absorbing aerosols. Furthermore, a cross-track capability (~ 400 km) would help eliminate the poor swath coverage of the original APS design. However, the exact technology of the polarimeter is yet to be determined and is the focus of intense study (Diner et al. 2010). This coupled with an onboard multispectral depolarizing lidar can only further improve aerosol property retrieval over urban areas.

REFERENCES

Ackerman, A. S., Kirkpatrick, M. P., Stevens, D. E., and Toon, O. B. (2004). The impact of humidity above stratiform clouds on indirect aerosol climate forcing. *Nature*, 432, 1014–1017.

Al-Saadi, J., Szykman, J., Pierce, R. B., Kittaka, C., Neil, D., Chu, D. A., Remer, L., Gumley, L., Prins, E., Weinstoc, L., MacDonald, C., Wayland, R., Dimmick, F., and Fishman, J. (2005). Improving national air quality forecasts with satellite aerosol observations. *Bulletin of the American Meteorological Society*, 86(9), 1249–1261.

Castanho, A., de Almeida, D., Prinn, R., Martins, V., Herold, M., Ichoku, C., and Molina, L. T. (2007). Analysis of Visible/SWIR surface reflectance ratios for aerosol retrievals from satellite in Mexico City urban area. *Atmospheric Chemistry and Physics*, 7, 5467–5477.

Charlson, R. J., Schwartz, S. E., Hales, J. M., Cess, R. D., Coakley, Jr., J. A., Hansen, J. E., and Hofmann, D. J. (1992). Climate forcing by anthropogenic aerosols. *Science*, 255 (5043), 423–430.

Clark, R. N., Swayze, G. A., Gallagher, A. J., King, T. V. V., and Calvin, W. M. (1993). The U.S. Geological Survey, Digital Spectral Library: Version 1: 0.2 to 3.0 microns, U.S. Geological Survey Open File Report 93-592, 1340 pages, http://speclab.cr.usgs.gov.

Crevoisier, C., Chedin, A., and Scott, N. A. (2003). AIRS channel selection for CO_2 and other trace gas retrievals. *Quarterly Journal of the Royal Meteorological Society*, 129(593), 2719–2740.

Deschamps, P.-Y., Bréon, F. M., Leroy, M., Podaire, A., Bricaud, A., Buriez, J.-C., and Seze, G. (1994). The POLDER mission: Instrument characteristics and scientific objectives. *IEEE Transactions on Geoscience and Remote Sensing*, 32, 598–615.

Deuzé, J. L., Bréon, F. M., and Devaux, C. (2001). Remote sensing of aerosols over land surfaces from POLDER-ADEOS-1 polarized measurements. *Journal of Geophysical Research*, 106(D5), 4913–4926.

Diner, D. J. (1999). Multi-angle Imaging SpectroRadiometer (MISR) Level 2 Aerosol Retrieval Algorithm Theoretical Basis document, http://www-misr.jpl.nasa.gov.

Diner, D. J., Martonchik, J. V., Kahn, R. A., Pinty, B., Gobron, N., Nelson, D. L., and Holben, B. N. (2005). Using angular and spectral shape similarity constraints to improve MISR aerosol and surface retrievals over land. *Remote Sensing of Environment*, 94, 155–171.

Diner, D. J., Boland, S. W., Davis, E. S., Kahn, R. A., Hostetler, C. A., Ferrare, R. A., Hair, J. W., Cairns, B., and Torres, O. (2007). Future mission concept for 3-D remote sensing of aerosols from low earth orbit. *IEEE Aerospace Conference*, Big Sky, MT, pp. 1–9.

Diner, D. J., Cairns, B., and Martins, J. V. (2010). Decadal survey tier 2 mission study summative progress report: ACE polarimeter development. http://dsm.gsfc.nasa.gov/ace/documents/ACE_Report11_Polarimeter_v6.pdf.

Dubovik, O. and King, M. D. (2000). A flexible inversion algorithm for retrieval of aerosol optical properties from sun and sky radiance measurements. *Journal of Geophysical Research*, 105, (D16), 20673–20696.

Dubovik, O. and Holben, B. (2002a). Variability of absorption and optical properties of key aerosol types observed in worldwide locations. *Journal of the Atmospheric Sciences*, 59(3), 590–608.

Dubovik, O. and Holben, B. (2002b). Non-spherical aerosol retrieval method employing light scattering by spheroids. *Geophysical Research Letters*, 29(10), GL0141506, doi: 10.1029/2001GL014506.

Dubovik, O., Sinyuk, A., Lapyonok, T., Holben, B. N., Mishchenko, M., Yang, P., Eck, T. F., Volten, H., Munoz, O., Veihelmann, B., van der Zander, W. J., Leon, J.-F., Sorokin, M., and Slutsker, I. (2006). Application of spheroid models to account for aerosol particle nonsphericity in remote sensing of desert dust. *Journal of Geophysical Research*, 111, D11208, doi:10.1029/2005JD006619.

Dubovik, O., Herman, M., Holdak, A., Lapyonok, T., Tanré, D., Deuzé, J. L., Ducos, F., Sinyuk, A., and Lopatin, A. (2010). Statistically optimized inversion algorithm for enhanced retrieval of aerosol properties from spectral multi-angle polarimetric satellite observations. *Atmospheric Measurement Techniques Discussion*, 3, 4967–5077.

Engel-Cox, J. A., Hoff, R. M., Rogers, R., Dimmick, F., Rush, A. C., Szykman, J. J., Al-Saadi, J., Chu, D. A., and Zell, E. R. (2006). Integrating lidar and satellite optical depth with ambient monitoring for 3-dimensional particulate characterization. *Atmospheric Environment*, 40, 8056–8067.

EPA. (1996). *Air Quality Criteria for Particulate Matter*. Office of Research and Development, EPA/600/P-95/001aF, Volumes I–III, Washington, DC.

Evans, F. F. and Stephens, G. L. (1991). A new polarized atmospheric radiative transfer model. *Journal of Quantitative Spectroscopy & Radiative Transfer*, 5, 413–423.

Fan, X., Goloub, P., Deuze, J.-L., Chen, H., Zhang, W., Tanre, D., and Li, Z. (2008). Evaluation of PARASOL aerosol retrieval over North East Asia. *Remote Sensing of Environment*, 112, 697–707.

Fan, X., Chen, H., Lin, L., Han, Z., and Goloub, P. (2009). Retrieval of aerosol optical properties over the Beijing area using POLDER/PARASOL satellite polarization measurements. *Advances in Atmospheric Sciences*, 26(6), 1099–1107.

Forster, P., Ramaswamy, V., Artaxo, P., Berntsen, T., Betts, R., Fahey, D. W., Haywood, J., Lean, J., Lowe, D. C., Myhre, G., Nganga, J., Prinn, R., Raga, G., Schulz, M., and Van Dorland, R. (2007). Changes in atmospheric constituents and in radiative forcing. In *Climate Change 2007: The Physical Science Basis. Contribution of Working Group I to the Fourth Assessment Report of the Intergovernmental Panel on Climate Change*, Solomon, S., Qin, D., Manning, M., Chen, Z., Marquis, M., Averyt, K.B., Tignor, M., and Miller, H.L. (eds.). Cambridge University Press, Cambridge, United Kingdom and New York. Access at http://www.ipcc.ch/pdf/assessment-report/ar4/wg1/ar4-wg1-chapter2.pdf.

Gross, B., Ogunwuyi, O., Moshary, F., Ahmed, S., and Cairns, B. (2005). Aerosol retrieval over urban areas using spatial regression between VIS/NIR and SWIR hyperion channels, *IEEE Workshop of Remote Sensing of Atmospheric Aerosols*, 43–50.

Hadjimitsis, D. G., Retalis, A., and Clayton, C. R. I. (2002). The assessment of atmospheric pollution using satellite remote sensing technology in large cities in the vicinity of airports. *Water, Air and Soil Pollution*, 2(5–6), 631–640.

Hadjimitsis, D. G. (2008). Description of a new method for retrieving the aerosol optical thickness from satellite remotely sensed imagery using the maximum contrast value principle and the darkest pixel approach. *Transaction of GIS Journal*, 12(5), 633–644.

Hanel, G. (1976). The properties of atmospheric aerosol particles as functions of relative humidity at thermodynamic equilibrium with surrounding moist air. *Advances in Geophysics*, 19, 73–188.

Hauser, A., Oesch, D., Foppa N., and Wunderle, S. (2005a). NOAA AVHRR derived aerosol optical depth over land. *Journal of Geophysical Research*, 110, D08204, doi:10.1029/2004JD005439.

Hauser, A., Oesch, D., and Foppa, N. (2005b). Aerosol optical depth over land: Comparing AERONET, AVHRR and MODIS. *Geophysical Research Letters*, 32, L17816, doi:10.1029/2005GL023579.

Herman, M., Deuzé, J. L., Devaux, C., Goloub, P., Breon, F. M., and Tanre, D. (1997). Remote sensing of aerosols over land surfaces including polarization measurements and application to POLDER measurements. *Journal of Geophysical Research*, 102(D14), 17, 039–049.

Hess, M., Koepke, P., and Schult, I. (1998). Optical properties of aerosols and clouds: The software package OPAC. *Bulletin of the American Meteorological Society*, 79, 831–844.

Holben, B. N., Tanre, D., Smirnov, A., Eck, T. F., Slutsker, I., Abuhassan, N., Newcomb, W. W., Schafer, J., Chatenet, B., Lavenue, F., Kaufman, Y. J., Vande Castle, J., Setzer, A., Markham, B., Clark, D., Frouin, R., Halthore, R., Karnieli, A., O'Neill, N. T., Pietras, C., Pinker, R. T., Voss, K., and Zibordi, G. (2001). An emerging ground-based aerosol climatology: Aerosol optical depth from AERONET. *Journal of Geophysical Research*, 106, 12067–12097.

Hsu, N. C., Tsay, S. C., King, M. D., and Herman, J. R. (2004). Aerosol properties over bright-reflecting source region. *IEEE Transactions on Geoscience and Remote Sensing*, 42, 3, 557–569.

Hsu, N. C., Tsay, S. C., King, M. D., and Herman, J. R. (2006). Deep blue retrievals of Asian aerosol properties during ACE-Asia. *IEEE Transactions on Geoscience and Remote Sensing*, 44(11), 3180–3195.

Ichoku, C., Chu, D. A., Mattoo, S., Kaufman, Y. J., Remer, L. A., Tanré, D., Slutsker, I., and Holben, B. (2002). A spatio-temporal approach for global validation and analysis of MODIS aerosol products. *Geophysical Research Letters*, 29(12), 8006, doi:10.1029/2001GL013206.

Kahn, R. A., Gaitley, B. J., Martonchik, J. V., Diner, D. J., Crean, K. A., and Holben, B. (2005). Multiangle Imaging Spectroradiometer (MISR) global aerosol optical depth validation based on 2 years of coincident Aerosol Robotic Network (AERONET) observations. *Journal of Geophysical Research*, 110, D10S04, doi:10.1029/ 2004JD004706.

Kaufman, Y. J. and Sendra, C. (1988). Algorithm for automatic atmospheric corrections to visible and near-IR satellite imagery. *International Journal of Remote Sensing*, 9(8), 1357–1381.

Kaufman, Y. J., Wald, A. E., Remer, L. A., Gao, B.-C., Li, R.-R., and Flynn, L. (1997). The MODIS 2.1µm channel correlation with visible reflectance for use in remote sensing of aerosol. *IEEE Transactions on Geoscience and Remote Sensing*, 35, 1286–1298.

Kaufman, Y. J. and Tanré, D. (1998). Algorithm for Remote Sensing of Tropospheric aerosol from MODIS, Product ID: MOD04, ATBD document.

Kaufman, Y. J., Remer, L. A., Tanré, D., Li, R. R., Kleidman, R., Mattoo, S., Levy, R., Eck, T., Holben, B. N., Ichoku, C., Martins, J., and Koren, I. (2005). A critical examination of the residual cloud contamination and diurnal sampling effects on MODIS estimates of aerosol over ocean. *IEEE Transactions on Geoscience and Remote Sensing*, 43 (12), 2886–2897.

Kawata, Y., Fukui, H., and Takemata, K. (2003). Retrieval of aerosol optical thickness using band correlation method and atmospheric correction for Landsat-7/ETM+ image data. *IEEE Proceedings of IGARSS '03 Geoscience and Remote Sensing Symposium*. 4, 2173– 2175.

Knapp, K. R. and Stowe, L. L. (2002). Evaluating the potential for retrieving aerosol optical depth over land from AVHRR Pathfinder Atmosphere data. *Journal of the Atmospheric Sciences*, 59, 279–293.

Knapp, K. R., Frouin, R., Kondragunta, S., and Prados, A. I. (2005). Towards aerosol optical depth retrievals over land from GOES visible radiances: Determining surface reflectance. *International Journal of Remote Sensing*, 26(18), 4097–4116.

Kondragunta, S., Ackerman, S., and Ciren, P. (2010). Algorithm theoretical basic document of ABI onboard GOES-R aerosol detection product, NOAA/NESDIS/STAR. http://www.goes-r.gov/products/ATBDs/baseline/AAA_AIP_v2.0_no_color.pdf.

Leon, J. F., Chazette, P., and Dulac F. (1999), Retrieval and monitoring of aerosol optical thickness over an urban area by spaceborne and ground-based remote sensing. *Applied Optics*, 38, 6918–6926.

Levy, R. C., Remer, L., Mattoo, S., Vermote, E., and Kaufman, Y. J. (2007a). Second-generation algorithm for retrieving aerosol properties over land from MODIS spectral reflectance. *Journal of Geophysical Research*, 112, D13211, doi:10.1029/2006JD007811.

Levy, R. C., Remer, L. A., and Dubovik, O. (2007b). Global aerosol optical properties and application to MODIS aerosol retrieval over land. *Journal of Geophysical Research*, 112, D13210, doi:10.1029/2006JD007815.

Liu, L. and Mishchenko, M. I. (2008). Toward unified satellite climatology of aerosol properties: Direct comparisons of advanced level 2 aerosol products. *Journal of Quantitative Spectroscopy & Radiative Transfer*, 109, 2376–2385.

Lu, D. and Weng, Q. (2005). Urban classification using full spectral information of Landsat ETM+ imagery in Marion County, Indiana. *Photogrammetric Engineering & Remote Sensing*, 71, 1275–1284.

Lyapustin, A., Wang, Y., Kahn, R., Xiong, J., Ignatov, A., Wolfe, R., Wu, A., and Bruegge, C. (2007a). Analysis of MODIS-MISR calibration differences using surface albedo around AERONET sites and cloud reflectance. *Remote Sensing of Environment*, 107 (1–2), 12–21.

Lyapustin, A. and Wang, Y. (2009). The time series technique for aerosol retrievals over land from MODIS. In *Satellite Aerosol Remote Sensing Over Land*, A. A. Kokhanovsky and G. de Leeuw (eds.), Springer, Berlin, 69–99.

Lyapustin, A., Wang, Y., Laszlo, I., Kahn, R., Korkin, S., Remer, L., Levy, R., and Reid, J. S. (2011). Multiangle implementation of atmospheric correction (MAIAC): 2. Aerosol algorithm. *Journal of Geophysical Research*, 116, D03211, doi:10.1029/2010JD014986.

Martonchik, J. V., Diner, D. J., Kahn, R. A., Ackerman, T. P., Verstraete, M. M., Pinty, B., and Gordon, H. R. (1998). Techniques for the retrieval of aerosol properties over land and ocean using multi-angle imagery. *IEEE Transactions on Geoscience and Remote Sensing*, 36, 1212–1227.

Martonchik, J. V., Kahn, R. A., and Diner, D. J. (2009). Retrieval of aerosol properties over land using MISR observations. In A. A. Kokhanovsky, and de Leeuw, G., ed., *Satellite Aerosol Remote Sensing Over Land*. Springer, Berlin, 267–291.

Mishchenko, M. I., Cairns, B., Chowdhary, J., Geogdzhayev, I. V., Liu, L., and Travis, L. D. (2005). Remote sensing of terrestrial tropospheric aerosols from aircraft and satellites. *Journal of Physics: Conference Series*, 6, 73–89.

Mishchenko, M. I., Liu, L., Travis, L. D., Cairns, B., and Lacis, A. A. (2010). Toward unified satellite climatology of aerosol properties: 3. MODIS versus MISR versus AERONET. *Journal of Quantitative Spectroscopy & Radiative Transfer*, 111, 540–552.

MODIS C006 Proposed MODIS-Atmosphere Collection 006 Changes *Version 24* (4/22/2010), http://modis-atmos.gsfc.nasa.gov/products_C006update.html.

Nadal, F. and Bréon, F. M. (1999). Parameterization of surface polarized reflectance derived from POLDER spaceborne measurements. *IEEE Transactions on Geoscience and Remote Sensing*, 37, 1709–1718.

Nakajima, T., Higurashi, A., Kawamoto, K., and Penner, J. E. (2001). A possible correlation between satellite-derived cloud and aerosol microphysical parameters. *Geophysical Research Letters*, 28, 1171–1173.

Oo, M. M., Jerg, M., Hernandez, E., Picón, A., Gross, B. M., Moshary, F., and Ahmed, S. A. (2010). Improved MODIS aerosol retrieval using modified VIS/SWIR surface albedo ratio over urban scenes. *IEEE Transactions on Geoscience and Remote Sensing*, 48(3), 983–1000.

Powell, R. L., Robert, D. A., Dennsion, P. E., and Hess, L. L. (2007). Sub-pixel mapping of urban land cover using multiple endmember spectral mixture analysis: Manaus, Brazil. *Remote Sensing of Environment*, 106, 253–267.

Prados, A. I., Kondragunta, S., Ciren, P., and Knapp, K. R. (2007). GOES aerosol/smoke product (GASP) over North America: Comparisons to AERONET and MODIS observations. *Journal of Geophysical Research*, 112, D15201, doi:10.1029/2006JD007968.

Prados, A. I., Leptoukh, G., Johnson, J., Rui, H., Lynnes, C., Chen, A., and Husar, R. B. (2010). Access, visualization, and interoperability of air quality remote sensing data sets via the Giovanni Online Tool. *IEEE Journal of Selected Topics in Applied Earth Observation and Remote Sensing*, 3(3), 359–370, 10.1109/ JSTARS.2010.2047940.

Remer, L. A., Tanré, D., Kaufman, Y. J., Levy, R., and Mattoo, S. (2005). Algorithm for Remote Sensing of Tropospheric Aerosol from MODIS: Collection 005. ATBD document.

Rodgers, C. D. (2000). *Inverse Methods for Atmospheric Sounding: Theory and Practice.* World Science, Hackensack, NJ, USA.

Rublev, A. N. and Uspenskii, A. B. (2006). Estimation of tropospheric carbon dioxide from SCIAMACHY spectrometer measurements under cloud conditions, Earth Research from Space, No. 6.

Satheesh, S. K., Torres, O., Remer, L. A., Babu, S. S., Vinoj, V., Eck, T. F., Kleidman, R. G., and Holben, B. N. (2009). Improved assessment of aerosol absorption using OMI-MODIS joint retrieval. *Journal of Geophysical Research*, 114, D05209, doi:10.1029/ 2008JD011024.

Schuster, G. L., Dubovik, O., and Holben, B. N. (2006). Angstrom exponent and bimodal aerosol size distributions. *Journal of Geophysical Research*, 111, D07207, doi:10.1029/ 2005JD006328.

Seinfeld, J. and Pandis, S. (1998). *Atmospheric Chemistry and Physics.* Wiley-Interscience, New York.

Small, C. (2003). High resolution spectral mixture analysis of urban reflectance. *Remote Sensing of Environment*, 88, 170–186.

Small, C. (2005). A global analysis of urban reflectance. *International Journal of Remote Sensing*, 26, 661–681.

Starr, D. O. 'C. (2011) NASA's Aerosol-Cloud-Ecosystems (ACE) mission, in Hyperspectral Imaging and Sounding of the Environment, OSA Technical Digest (CD) (Optical Society of America), paper HMA4. http://www.opticsinfobase.org/abstract .cfm?URI=HISE-2011-HMA4.

Stowe, L., Jacobowitz, H., Ohring, G., Knapp, K. R., and Nalli, N. R. (2002). The advanced very high resolution radiometer pathfinder atmosphere (PATMOS) data set: Initial analyses and evaluations. *Journal of Climate*, 15, 1243–1260.

Torres, O., Bhartia, P. K., Herman, J. R., and Ahmad, Z. (1998). Derivation of aerosol properties from satellite measurements of backscattered ultraviolet radiation: Theoretical basis. *Journal of Geophysical Research*, 103, 17099–17110.

Torres, O., Decae, R., Veefkind, J. P., and de Leeuw, G. (2003). OMI aerosol retrieval algorithm. In *OMI Algorithm Theoretical Basis Document: Clouds, Aerosols, and Surface UV Irradiance* (http://eospso.gsfc.nasa.gov/eos_homepage/for_scientists/atbd/docs/OMI/ ATBD-OMI-03.pdf).

Torres, O., Tanskanen, A., Veihelmann, B., Ahn, C., Braak, R., Bhartia, P. K., Veefkind, P., and Levelt, P. (2007). Aerosols and surface UV products from Ozone Monitoring Instrument observations: An overview. *Journal of Geophysical Research*, 112, D24S47, doi:10.1029/ 2007JD008809.

Twomey, S. (1974). Pollution and the planetary albedo. *Atmospheric Environment*, 8, 125–1256.

Twomey, S. (1977). The Influence of pollution on the shortwave albedo of clouds. *Journal of the Atmospheric Sciences*, 34, 1149–1152.

Twomey, S. (1980). Cloud nucleation in the atmosphere and the influence of nucleus concentration levels in atmospheric physics. *Journal of Physical Chemistry*, 84, 1459–1463.

USDA (1978). *Soil Taxonomy*. Agriculture Handbook No. 436. USDA, Soil Conservation Service, Washington, DC.

Vermote, E., Tanré, D., Deuzé, J. L., Herman, M., and Morcrette, J. J. (1999). Second Simulation of the Satellite Signal in the Solar Spectrum (6S), Software, Version 4.1, Department of Geography, University of Maryland and Laboratoire d'Optique Atmosphérique, U.S.T.L., available from anonymous ftp (kratmos.gsfc.nasa.gov).

VIIRS (2010). AOT ATBD 2010 VIIRS Aerosol Optical Thickness (AOT) and Particle Size Parameter Algorithm Theoretical Basis Document (ATBD), http://jointmission.gsfc.nasa.gov/project/sciencedocuments/072010/D43313_Rev%20F_VIIRS_AOT_Update_Aerosol-ATBD_Final_ECR%20A-184A_Statement%20A_Submission.pdf.

Vlasenko, A., Sjögren, S., Weingartner, E., Gäggeler, H., and Ammann, M. (2005). Generation of submicron Arizona test dust aerosol: Chemical and hygroscopic properties. *Aerosol Science and Technology*, 39(5), 452–460.

Wang, Y., Lyapustin, A. I., Privette, J. L., Morisette, J. T., and Holben, B. (2009). Atmospheric correction at AERONET locations: A new science and validation data set. *IEEE Transactions on Geoscience and Remote Sensing*, 47(8), 2450–2466.

Wang, Y., Lyapustin, A. I., Privette, J. L., Cook, R. B., Vermote, E. F., and Schaaf, C. L. (2010). Assessment of biases in MODIS surface reflectance due to Lambertian approximation. *Remote Sensing of Environment*, 114, 2791–2801.

Waquet, F., Cairns, B., Knobespiesse, K., Chowdhary, J., Travis, L. D., Schmid, B., and Mishchenko, M. I. (2009). Polarimetric remote sensing of aerosols over land. *Journal of Geophysical Research*, 114, doi:1206, doi:1029/2008JD010619.

Wilson, W. E. and Suh, H. H. (1997). Fine particles and coarse particles: Concentration relationships relevant to epidemiologic studies. *Journal of the Air & Waste Management Association*, 47, 1238–1249.

Wiscombe, W. J. (1981). Improved Mie scattering algorithms. *Applied Optics*, 19, 1505–1509.

Wu, Y., Chaw, S., Gross, B., Moshary, F., and Ahmed, S. (2009). Low and optically thin cloud measurements using a Raman–Mie lidar. *Applied Optics*, 48, 1218–1227.

Zubko, V., Leptoukh, G., and Gopalan, A. (2010). Study of data merging and interpolation methods for use in an interactive online analysis system: MODIS Terra and Aqua daily aerosol case. *IEEE Transactions on Geoscience and Remote Sensing*, 48(12), 4219–4235.

16 DOAS Technique
Emission Measurements in Urban and Industrial Regions

Pasquale Avino and Maurizio Manigrasso

CONTENTS

16.1 INTRODUCTION

16.1.1 BASIS OF DOAS MEASUREMENTS

Measurements of ambient gaseous pollutants using long-path length differential optical absorption spectroscopy (DOAS) in the daytime and nighttime have been carried out at many locations worldwide (Platt et al. 1980a; Barrefors 1996; Brocco et al. 1997; Petrakis et al. 2003; Chiu et al. 2005; Lee et al. 2005; Hao et al. 2006; Avino and Manigrasso 2008b; Triantafyllou et al. 2008). The species of interest include both primary and secondary pollutants such as sulfur dioxide (SO_2), nitrogen dioxide (NO_2), ozone (O_3), formaldehyde, nitrous acid, benzene, and toluene. The focus of data analysis was placed on the impact of these species in the context of atmospheric chemistry.

Among the different methods used to analyze gaseous compounds, the technique involving open-path spectroscopy is ideal for monitoring pollutants for their advantages over classical methods and point-source analyzers (Avino et al. 2002). The DOAS is one of the most versatile optical techniques for determining atmospheric pollutants (Platt and Perner 1980b; Platt et al. 1979; Platt 1994) for routine work to monitor air quality and source emissions (Karlsson 1990). The DOAS analytical method is based on the light absorption of species with fine vibrational structures such as SO_2, NO_2, ozone, nitrous acid, formaldehyde, benzene, and toluene, in both UV and VIS bands. Such a technique is applicable for measuring the air pollution concentrations in the atmospheric environment within a range of several kilometers. It offers the possibility of measuring, simultaneously, primary and secondary air pollutant concentrations by averaging measurements along a long optical path (from 100 m to 10 km). The collective measurements for varying air pollutant species over the measurable ranges of concentrations may generate the average values of specific species area-wide that are certainly better than the point measurements obtained by traditional analyzers at the ground level.

The design philosophy of the DOAS lies upon the interaction of the incident light with gases and molecules present within the optical path length (the distance between the light source and the detector; Figure 16.1). Absorption, fluorescence, Rayleigh scattering, and Raman diffusion are due to the influence of molecules, whereas Mie scattering is due to aerosol particles. Species featured with narrowband absorption in the UV–VIS region are detectable, and their concentrations can be determined by the DOAS systems through the Lambert–Beer law:

$$I_\lambda = I_{0\lambda} \cdot \exp(-L \cdot \sigma_\lambda \cdot C) \tag{16.1}$$

where $I_{0\lambda}$ is the initial intensity emitted and I_λ is the intensity of the radiation emerging from a layer of thickness L (cm), where the species measured is present at the concentration C (molecule·cm^{-3}) with an absorption cross section σ_λ (cm^2·molecule^{-1}) at wavelength λ (nm).

The measured spectra are also influenced by other extinction phenomena such as Rayleigh and Mie scattering and by the variation with wavelength of both the light intensity and the light transmission of the measurement system. Taking into account such contributions, Equation 16.1 becomes

$$A = \ln\left(\frac{I_{0\lambda}}{I_\lambda}\right) = L\left(\sigma_\lambda \cdot C + \varepsilon_{R\lambda} + \varepsilon_{M\lambda}\right) + F(\lambda), \tag{16.2}$$

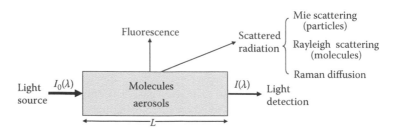

FIGURE 16.1 Light interaction with atmospheric molecules and particles.

where A is the absorbance, $\varepsilon_{R\lambda}$ (molecule·cm^2) and $\varepsilon_{M\lambda}$ (molecule·cm^2) are the Rayleigh and the Mie extinction coefficients, respectively, and $F(\lambda)$ is a function describing the wavelength variation of the light intensity due to the instrument.

Cross-section measurements for a given species can be considered as made of two contributions: one is slowly varying with $\lambda (\sigma_\lambda^s)$ and the other is displaying fast variation (σ_λ^F), called differential cross section (cm^2·molecule^{-1}). Given that the Rayleigh and Mie scattering as well as $F(\lambda)$ display a slow variation with λ (nm), Equation 16.2 can be rewritten as Equation 16.3 to include a fast-varying component due to the light absorption of all the molecular species in the layer of thickness L (cm) and a slow-varying term reflecting contributions due to scattering and instrument (i.e., optics and light source):

$$A = \left[L \sum_i \sigma_{i\lambda}^F \cdot C_i \right] + \left[L \left(\sum_i \sigma_{i\lambda}^S \cdot C_i + \varepsilon_{R\lambda} + \varepsilon_{M\lambda} \right) + F(\lambda) \right]. \tag{16.3}$$

The slow wavelength (nm) component is estimated by a polynomial fitting (e.g., the dotted line in Figure 16.2) and subtracted to the absorbance spectrum (e.g., the continuous line in Figure 16.2) to give the differential absorbance spectrum reporting the differential absorbance (DA) as a function of λ, where DA is expressed as

$$DA = L \sum_i \sigma_{i\lambda}^F \cdot C_i. \tag{16.4}$$

Calibration of the type of instrument may be performed by measuring the spectra of the absorbing gas species in a cell of known optical path length L (cm) and at known concentrations (molecule). Reference spectra for each species are then obtained by subtracting the zero spectrum recorded with measurement performed in the empty cell. The concentrations of each species are calculated by the least-squares fit of Equation 16.4. An estimate of the quality of the fit is represented by the residual area between the fitted and the measured spectra.

The main advantages of the DOAS technique reside in its high sensitivity in association with the long optical path length, which demonstrates the realization of the holistic measurements for a given area. The resultant accuracies should be

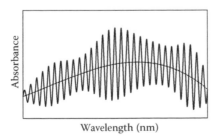

FIGURE 16.2 Polynomial fitting.

higher than the counterparts based on single-point analyzers, given the absence of air sample collection and chemical analyses. The disadvantages are associated with the susceptibility to atmospheric turbulence that may cause intensity variation in the spectra and to meteorological conditions because measurements cannot be carried out in the presence of clouds, fog, and snow due to the strong attenuation of light in the UV–VIS band region.

16.1.2 Interpretation of Pollutant Behavior

The interpretation of atmospheric pollution phenomena is a very complex task because of the simultaneous presence of both emission processes and physiochemical transformations of pollutants associated with meteorological processes (e.g., turbulent diffusion and transport). Even if the pollutants are measured at the ground level, the atmospheric processes are still dependent on the dynamic evolution of the planetary boundary layer (PBL). Both measurements require employing the knowledge of the pollutant behavior and the temporal evolution of the PBL to quantify.

Different techniques such as acoustic experiments, radioanalysis, and analysis of the mechanical turbulence with sonic anemometer could offer valuable information on the evolution of the meteorological conditions but their use is limited due to costs and technical complexity. Yet a very important approach to describe the pollutant evolution is given by the measurement of natural radioactivity due to both radon and its short-lived decay products (Febo et al. 1996; Avino et al. 2003). It is known that the radon ground emission is an almost stationary process considering the time scale for the pollutant observations, and its distribution is spatially homogeneous for about some kilometers (Avino et al. 2000, 2001). Therefore, the temporal evolution of the radon concentration depends only on the PBL dynamics (Allegrini et al. 1994).

The temporal radon evolution can be defined by the following relationship (Allegrini et al. 1994; Avino et al. 2002, 2003):

$$\frac{\partial C_R}{\partial t} = \alpha \left[\Phi_R \right] - \beta \left(C_R \right) + Adv, \tag{16.5}$$

where C is the mixing ratio (molecule) near the ground, α is the stability term (cm), $\Phi(t)$ is the primary emission flux (molecule $cm^{-1} \cdot s^{-1}$), $\beta\{C\}$ is the vertical mass exchange due to eddy diffusion (molecule·s^{-1}), and Adv is the advection term (molecule·s^{-1}).

The term $\Phi(t)$ reflects the temporal variability of source intensity, and the terms α, β, and Adv are linked to the dynamics of the lower layers of the atmosphere. The term Adv can be partially deduced from knowledge of the intensity and direction of the wind at the ground level. The time trend of radon daughter concentration can be used in order to characterize the terms α and β.

From the radon derivative trend, it is possible to define both the stability and instability atmospheric conditions. In fact, high stability conditions at the ground level maximize the contribution of the term $\alpha[\Phi_R]$ as the term $\beta(C_R)$ is negligible, when radon dilution is hindered and the natural radioactivity assumes high values quickly. The transition to instability conditions maximizes the contribution of the term $\beta(C_R)$ as radon accumulation in the low boundary layer results in difficulty so that the poorly modulated natural radioactivity takes low values.

16.2 MEASUREMENT CAMPAIGNS

16.2.1 MEASUREMENT METHOD

The first DOAS system (mod. AR 500, Opsis, Sweden) was installed at the ISPESL monitoring station (via Urbana) located in downtown Rome (near St. Maria Maggiore Cathedral) at about 4 m above the ground level. This is an urban region characterized by high density of traffic flow (Figure 16.3). Some other measurements were performed at a monitoring station inside a green park, Villa Ada Park, located in Rome, from which the data collected can be considered as the background level of this urban area. The second DOAS system was installed on the roof of the ISPESL Mobile Laboratory in an area with high industrial activities. The measurements were carried out at the ground level. The site is actually close to a city with strong industrial emissions.

The DOAS system is made up of the following parts: a light source, a receiver, a spectrophotometer equipped with an optical fiber, and a computer for the system management (e.g., data elaboration and data storage). The light source is a Xenon lamp (i.e., at high pressure and 150 W), in which a transmitter sends the light to the receiver that transmits the beam through the optical fiber into the analyzer (see Figure 16.4).

For both the instruments used in the measurement campaigns, the analytical parameters are almost the same. The distance between the emitter and the receiver ranged between 200 and 250 m (i.e., this term influences the sensitivity of the measures). The absorbance of light from the emitter is continuously measured within the wavelength range 240 to 450 nm to determine several compounds in the atmospheric environment (Table 16.1).

The aromatic hydrocarbons are detected in the wavelength range between 250 and 290 nm, where the major interfering gases are oxygen, ozone, and sulfur dioxide;

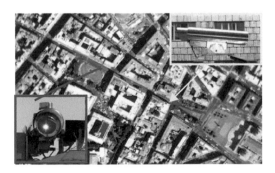

FIGURE 16.3 Sampling site in downtown: DOAS optical path length (250 m) is shown by the dotted line.

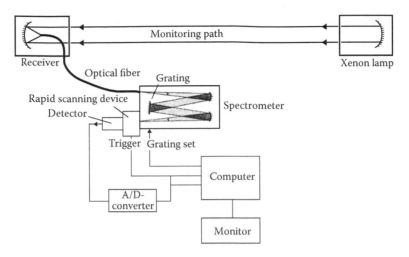

FIGURE 16.4   Scheme of typical DOAS equipment.

around 100 spectra per second are collected in this wavelength range and stored in a register with 1000 channels with a resolution of better than 0.05 nm. The concentrations of air pollutants are automatically calculated from the differential absorbance values according to the Lambert–Beer law (Brocco et al. 1997).

Before starting to discuss the DOAS capability of analyzing primary and/or secondary pollutants, it is worth underlining the basic feature of this methodology by comparing benzene measurements performed by DOAS and GC-PID. A GC-PID analyzer (Syntech-Spectra, The Netherlands) sampling every 15 minutes was used for such comparisons. During comparisons, the DOAS measurement may present "area-wide" average, whereas the GC-PID measurement is based on a single-point measurement. Such difference reflects the fact that two data sets display similar

TABLE 16.1

Operative Conditions of DOAS System

Pollutant	Wavelength (nm)	Time (min)	LOD (ppt)
SO_2	270–380	1	17 (0.2) (Platt 1994)
NO_2	398–450	1	80 (5) (Platt 1994)
O_3	250–320	5	4000 (5) (Platt 1994)
Benzene	240–310	1	200 (2) (Volkamer et al. 1998)
Toluene	240–310	1	500 (0.5) (Balin 1999)
Nitrous acid	320–380	7	40 (5) (Platt 1994)
Formaldehyde	270–340	5	400 (0.2) (Platt 1994)

Note: The limits of detection (LOD) are related to the path lengths (km) reported in brackets.

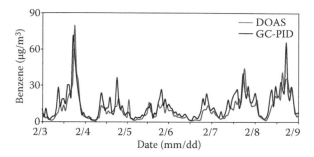

FIGURE 16.5 Comparison benzene trends determined by GC-PID analyzer and DOAS system (February 2009).

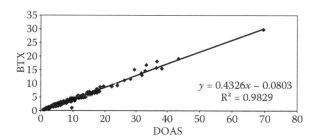

FIGURE 16.6 Correlation curve between GC-PID and DOAS data.

trend of variation (Figure 16.5) and are well correlated (Figure 16.6), but their concentration levels are slightly different under the remixing atmospheric conditions. In any circumstances, lower values are recorded by remote-sensing measurements (i.e., the 4th to the 7th of the month in 2002), whereas similar peak levels appear during stability conditions (Figure 16.5).

16.3 RESULTS AND DISCUSSION

16.3.1 PRIMARY AND SECONDARY POLLUTANTS MEASURED BY A DOAS SYSTEM

The data for nitrous acid, NO_2, O_3, benzene, and toluene concentrations as measured by the two long-path DOAS systems during different seasonal periods in downtown Rome and in the industrial site can be extensively discussed for a side-by-side comparison. Benzene and toluene are primary pollutants that can be associated with traffic and industrial emissions. In the first case, their variation pattern is similar and their ratio is almost constant (a benzene-to-toluene ratio of 3–5) according to those reported values for urban air pollution in the literature (Brocco et al. 1997, 1999). This is exactly the case as what can be observed in downtown Rome (Figure 16.7). Such a ratio, however, is no longer present in the industrial site. As shown in Figure 16.8, the levels of benzene often exceed those of toluene, evidencing additional

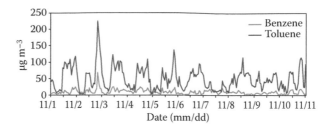

FIGURE 16.7 Typical hourly average concentrations of benzene and toluene measured by the DOAS system in downtown Rome (November 2009).

contribution other than the traffic source, which can be attributed to those sources associated with steel production and oil refinery plants in this area.

The levels of benzene, especially in urban areas, have been changing during the last few decades (Figure 16.9). At the beginning of the 1990s, benzene coming from automobile sources, industrial production, and anthropogenic emissions was present with concentrations between 40 and 80 $\mu g \cdot m^{-3}$ in the atmospheric environment (Avino et al. 2008a; Avino and Manigrasso 2008b). Benzene is carcinogenic for humans (IARC 1998) and chemically stable such that this compound has constituted an element of constant hygienic-sanitary interest for many years. It is interesting to observe that during these almost two recent decades there was a slight but clear decrease in benzene concentration in the atmosphere because of the use of green fuel in Italy. By the same token, the effect on atmospheric concentrations of SO_2 due to the reduction of sulfur content in fuels contributed to cleaner air quality as described in Figure 16.10. It is evidenced that the mean annual-average SO_2 levels decreased from 15 $\mu g \cdot m^{-3}$ in 1991 to 4 $\mu g \cdot m^{-3}$ in 2006.

For sound air-quality management, it is also important to investigate the yearly trend of a secondary pollutant (e.g., ozone; Figure 16.11). In this case, the difference among the trends of primary and secondary pollutants is remarkable: secondary

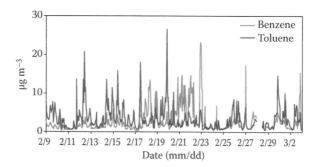

FIGURE 16.8 Typical hourly average concentrations of benzene and toluene measured by the DOAS system in the industrial site (February 2009).

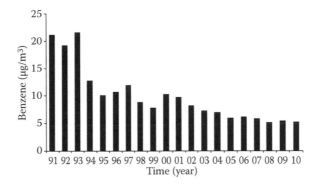

FIGURE 16.9   Yearly benzene trend from 1991 to 2010 in downtown Rome.

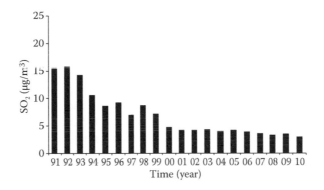

FIGURE 16.10   Yearly SO_2 trend from 1991 to 2010 in downtown Rome.

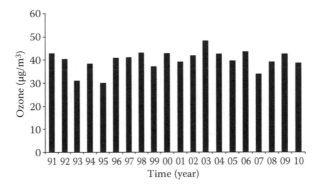

FIGURE 16.11   Yearly ozone trend from 1991 to 2010 in downtown Rome.

pollutants such as ozone were not directly influenced by the changing anthropogenic behavior (i.e., in fact, the concentration levels are similar over the last two decades). Rather, they are intimately tied to the dynamic evolution of PBL and the radical photochemical activity of the atmosphere. The contribution due to the mixing properties of the lower troposphere can be well realized by these measurements based on tropospheric ozone concentrations and natural radioactivity simultaneously, as shown in Figure 16.12.

The patterns of the two data sets in Figure 16.12 are complementary. It describes the fact that the negative derivative values (trends) of natural radioactivity denoting increased PBL dilution potential are closely associated with the positive derivative values (trends) of O_3 concentration, and vice versa. This means that throughout the day, the levels of ozone concentration are basically dominated by the vertical mixing effect in the lower troposphere and by the modulation of PBL mixing height as well. Under particular atmospheric conditions (e.g., high solar radiation, atmospheric stability) and high levels of reactive organic gases (ROGs), a further contribution of O_3 formation is due to photochemical reactions.

Ozone is a secondary pollutant, and it is formed through the reaction of O_2 with atomic oxygen deriving from the photolysis of NO_2. NO_2 is produced through the oxidation of NO by O_3 and the radical RO_2 and HO_2. No ozone accumulation would occur if the formation of NO_2 was due only to the reaction of O_3 with NO. In this case, NO_2 and O_3 show complementary patterns of variation in which the maximum values of O_3 correspond to the minimum values of NO_2, and vice versa, as shown in Figure 16.13. In the case of intense pollution episodes and in the presence of ROGs and OH radicals, the RO_2 and HO_2 radicals may be formed, oxidize NO, and cause O_3 accumulation. The variable O_x, defined as the sum of O_3 and NO_2 concentrations, can be considered as an indicator of the photochemical activity. In fact, its value is approximately constant during periods of low photochemical activity, whereas its trend is characterized by peak values during episodes of photochemical smog (Avino and Russo 2009).

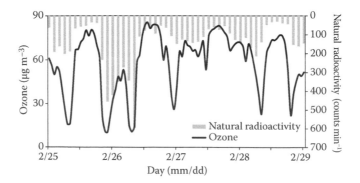

FIGURE 16.12 Trends of hourly averaged ozone concentrations measured by DOAS system and natural radioactivity (February 2010).

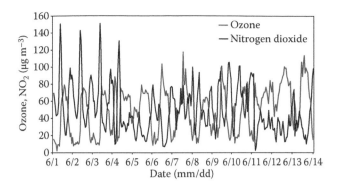

FIGURE 16.13 Daily trends of ozone and nitrogen dioxide measured by a DOAS system in downtown Rome (June 2009).

16.3.2 STUDYING THE DYNAMICS OF POLLUTANT FORMATION: ROLE OF DOAS

A remarkable advantage of the DOAS technique is the possibility of simultaneous measurements of gas pollutants with high time resolution, providing a unique opportunity for studying the dynamics of their formation in atmosphere. Such a feature is particularly valuable for pollutants such as nitrous acid and formaldehyde, which otherwise should be measured with the collection of air samples and chemical analysis with considerably lower time resolution.

The presence of primary formaldehyde in urban areas is mainly due to combustion processes, whereas its secondary origin is due to the photo-oxidation of ROGs. Nitrous acid (HNO_2) has a pivotal role in the formation of OH radical, which is the most important oxidizing species in the daytime atmosphere. Although the mechanism of formation of HNO_2 has not been fully understood yet, it is indicative that this chemical species can be heterogeneously formed with a rate that appears to be the first order in NO_2 and H_2O (Svensson et al. 1987; Jenkin et al. 1988; Harrison et al. 1996), according to the following stoichiometry:

$$NO_2 + NO_2 + H_2O \rightarrow HNO_2 + HNO_3. \qquad (16.6)$$

HNO_2 accumulates in the PBL at night, and it is photolyzed after the sunrise to form NO and the hydroxyl radical OH:

$$HNO_2 \rightarrow NO + OH. \qquad (16.7)$$

The role played by nitrous acid is well described by Figure 16.14a and b, showing a photochemical smog episode triggered characteristically by HNO_2 photolysis (Figure 16.14a).

When nitrous acid concentration drops due to its photolysis reaction with the formation of OH radical, peak concentrations are measured both for O_3 and NO_2. This situation comes along with the photochemical formaldehyde formation as evidenced by the common trend between formaldehyde and of O_x variable (Figure 16.14b).

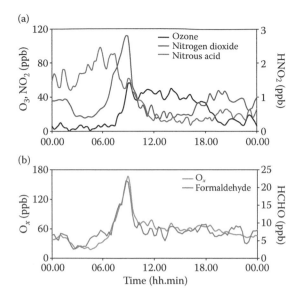

FIGURE 16.14 Photochemical pollution episode: trends of ozone, nitrogen dioxide, and nitrous acid (a) and O_x variable and formaldehyde (b) in April 2010.

Figure 16.15 shows the temporal trends of nitrous acid and of ultrafine particles (UFPs, particles with a diameter less than 100 nm) in the nucleation mode. The nucleation mode refers to particles very close to the sizes of clusters formed from the gas phase. Until about 10 a.m., nitrous acid and particle (3–10 nm) number concentration do not exhibit remarkable variations. Afterward, nitrous acid concentration drops due to the photolysis reaction that releases OH radicals (Equation 16.7). Even if the PBL degree of vertical remixing increases due to thermal convection, the particle number concentration increases, with maximum rate of formation in the same period when the rate of photolysis of nitrous acid is also maximum. Such occurrences

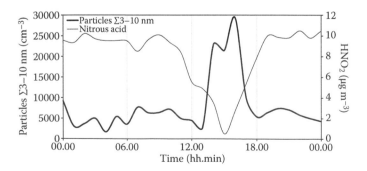

FIGURE 16.15 DOAS concentration of HNO_2 and of ultrafine particles in the nucleation mode measured by a scanning mobility particle sizer (April 2010).

suggest that new particle formation occurred directly from the gas phase following the photo-oxidation of ROGs initiated by reaction with the OH radical.

Among the products of photochemical reactions in atmosphere, peroxyacetil nitrate (PAN) has an important role. PAN [CH3C(O)OONO2] is the principal member of a family of nitrogenous compounds produced by the action of sunlight on NO_x and ROGs (Mills et al. 2007). PAN has very low natural background concentrations (Rubio et al. 2007). While ozone has a relevant source in stratosphere and consequently its levels are relatively high, PAN is deemed a very specific indicator of anthropogenic photochemical air pollution. From a sanitary point of view, PAN is known to be a phytotoxicant and lachrymator. There are concerns with regard to its effects on human health due to the exposure in ambient air, especially in the presence of high levels of ozone. For these reasons, PAN is a suggested agent of skin cancer in photochemically active areas and a possible bacterial mutagen.

Very little data of PAN levels in atmosphere are present in the literature, and it is very difficult to identify guideline values. In Rome, PAN concentrations reach a maximum of ~40 ppb with an average daily maximum of ~6 ppb in summertime and a maximum of ~10 ppb with an average daily minimum of ~3 ppb in wintertime. The variation of PAN pattern is almost regular, depending strictly on the meteorological conditions and both on the ozone (Figure 16.16) and the HCHO levels in atmosphere (i.e., overall on the ROG levels). During smog photochemical episodes, higher levels of HCHO and PAN are often found.

An additional feature of DOAS is its capability of representing the pollution levels of large areas. To emphasize such a valuable feature, Figures 16.17 through 16.19 collectively describe the effort by comparing ozone, benzene, and toluene concentration levels measured at two locations 10 km apart. These two locations are the ISPESL building and the villa Ada green park, both in the Rome area. The comparison of the ozone trends in these sites in springtime can be shown in Figure 16.17. The two patterns of variation are quite similar with maximum peaks recorded at villa Ada station, whereas at the ISPESL site, some differences (circled) can be noted due to photochemical reactions in the presence of hydrocarbons and NO emitted from anthropogenic sources.

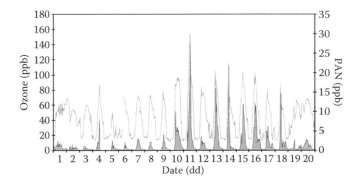

FIGURE 16.16 Typical daily trends of PAN (area) and ozone (line) in summer period (August 2010) in downtown Rome (Villa Ada Park).

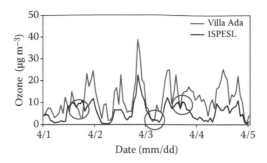

FIGURE 16.17   Comparison of ozone trends determined at via Urbana and Villa Ada park stations in Rome (April 2009).

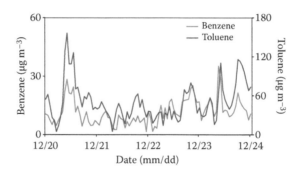

FIGURE 16.18   Typical daily trends of benzene and toluene in downtown Rome (via Urbana) in a cold period (December 2010).

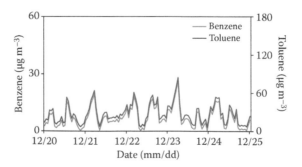

FIGURE 16.19   Typical daily trends of benzene and toluene in downtown Rome (Villa Ada Park) in a cold period (February 2009).

By the same token, Figures 16.18 and 16.19 show the trends of benzene and toluene at via Urbana and villa Ada sites, respectively. A good correlation of the two pollutants is due to their common sources in urban region and industrial site (i.e., basically domestic heating and autovehicular traffic). The different trends observed in the two sites are due to their different levels of anthropogenic emissions. In particular, in downtown Rome, where the emissions are high, benzene reaches up to 40 $\mu g \cdot m^{-3}$ (Figure 16.18), whereas inside the park, benzene reaches up to 25 $\mu g \cdot m^{-3}$ (Figure 16.19), with a similar modulation even if less structured.

16.4 CONCLUSIONS

The DOAS technique is a modern measurement system with high-technology implications useful for integrating information coming from monitoring networks. It allows us to investigate and understand the complex mechanisms of formation and transformation of atmospheric pollutants through the contemporary measures of primary and secondary gaseous species. In this way, the assessment of the air pollution is more representative than using single-point analyzers.

The technique described in this chapter may also allow us to study conventional stable pollutants and nonstable chemical species such as nitrous acid and HCHO, at the same time so as to retrieve precious insight of pollutant formation. The DOAS permits both efforts to generate synergistic information with regard to the dispersion and chemical transformation of the pollutants in order to evaluate very complex pollution phenomena (i.e., photochemical smog episode or atmospheric acidity). Even if the DOAS equipment is very expensive, system operation is easy. Furthermore, the reliability of the analytical measures is very valuable. The results obtained in this study took into account the spatial–temporal variations of the pollutant concentrations due to both the atmospheric heterogeneity and the remote contributions of sources not directly present in the study areas of consideration. This leads to the improvement of environmental management practices with a regional rather than a local approach.

ACKNOWLEDGMENTS

This work was supported under the grant ISPESL/DIPIA/P06/L04 and ISPESL/DIPIA/P06/L06 "Identificazione, analisi e valutazione delle conseguenze delle attività antropiche," 2009–11. The authors wish to thank Luca Lepore, Ida Ventrone (ISPESL), Domenico Brocco, and Antonio Febo (IIA-CNR, Italy) for their kind and helpful suggestions for the manuscript elaboration.

REFERENCES

Allegrini, I., Febo, A., Pasini, A., and Schirini, S. (1994). Monitoring of the nocturnal mixed layer by means of particulate radon progeny measurements. *Journal of Geophysical Research*, 94, 765–777.

Avino, P., Brocco, D., Lepore, L., and Ventrone, I. (2000). Fundamental aspects of carbonaceous particulate measurements in the study of air pollution in urban area. In *Air Pollution 2000,* C. Brebbia (ed.), WIT Press, Wessex, United Kingdom, pp. 301–309.

Avino, P., Brocco, D., and Cecinato, A. (2001). Characterization of carbonaceous particulate matter in the urban area of Rome. In *Urban Transport VII* C. Brebbia and L. J. Sucharov (eds.), WIT Press, Wessex, United Kingdom, pp. 361–370.

Avino, P., Brocco, D., and Scalisi, G. (2002). Criteria for evaluating the urban atmospheric pollution: The results of ten-year monitoring activity in Rome. *Proceedings of the 16th International Clean Air & Environment Conference,* Christchurch, New Zealand, pp. 60–64.

Avino, P., Brocco, D., Lepore, L., and Pareti, S. (2003). Interpretation of atmospheric pollution phenomena in relationship with the vertical atmospheric remixing by means of natural radioactivity measurements (radon) of particulate matter. *Annali di Chimica,* 93, 589–594.

Avino, P., Fanizza, C., and Manigrasso, M. (2008a). 10-years of long-range DOAS measurements in an Italian megacity, Rome. *Proceedings of the 7th International Symposium on Advanced Environmental Monitoring,* Honolulu, HI.

Avino, P. and Manigrasso, M. (2008b). Ten-year measurements of gaseous pollutants in urban air by an open-path analyzer. *Atmospheric Environment,* 42, 4138–4148.

Avino, P. and Russo, M. V. (2009). Peroxyacetil nitrate (PAN) in urban atmosphere: Levels and behavior. *Proceedings of the 2nd International Conference on Environmental Management, Engineering, Planning and Economics (CEMEPE) and SECOTOX Conference,* Mykonos, Greece, pp. 569–574.

Balin, J. (1999). Differential optical absorption spectroscopy for air pollution measurement. Ms.Sc. thesis, Ecole Polytechnique Federale de Lausanne.

Barrefors, G. (1996). Monitoring of benzene, toluene and p-xylene in urban air with differential optical absorption spectroscopy technique. *Science of the Total Environment,* 189, 287–292.

Brocco, D., Fratarcangeli, R., Lepore, L., Petricca, M., and Ventrone, I. (1997). Determination of aromatic hydrocarbons in urban air of Rome. *Atmospheric Environment,* 31, 557–566.

Brocco, D., Febo, A., Lepore, L., and Ventrone, I. (1999). Description of the time-spatial evolution of urban pollution by means long-path measures. *Proceedings of the 2nd International Conference on Urban Air Quality,* Madrid, Spain.

Chiu, K. H., Sree, U., Tseng, S. H., Wu, C. H., and Lo, J. G. (2005). Differential optical absorption spectrometer measurement of NO_2, SO_2, O_3, HCHO and aromatic volatile organics in ambient air of Kaohsiung petroleum refinery in Taiwan. *Atmospheric Environment,* 39, 941–955.

Febo, A., Perrino, C., Giliberti, C., and Allegrini, I. (1996). Use of proper variables to describe some aspects of urban pollution. In *Urban Air Pollution,* I. Allegrini and F. De Santis, (eds.), NATO ASI Series, Springer, Germany, pp. 295–315.

Hao, N., Zhou, B., Chen, D., Sun, Y., Gao, S., and Chen, L. (2006). Measurements of NO_2, SO_2, O_3, benzene and toluene using differential optical absorption spectroscopy (DOAS) in Shanghai, China. *Annali di Chimica,* 96, 365–375.

Harrison, R. M., Peak, J. D., and Collins, G. M. (1996). Tropospheric cycle of nitrous acid. *Journal of Geophysical Research,* 101, 14429–14439.

IARC (1998). Monographs on the evaluation of carcinogenic risks to humans overall evaluations of carcinogenicity: an updating of IARC Monographs. Vol. 1–42, Suppl. 7 (available at http://monographs.iarc.fr/ENG/Monographs/suppl7/suppl7.pdf).

Jenkin, M. I., Cox, R. A., and Williams, D. J. (1988). Laboratory studies of the kinetic of formation of nitrous acid from the thermal reaction of nitrogen dioxide and water vapour. *Atmospheric Environment,* 22, 487–498.

Karlsson, K. (1990). Environmental control using long path measurements. Proceedings of the 1990 EPA/AWMA—International Symposium Research, Triangle Park, NC, 1–7.

Lee, C., Kim, Y. J., Hong, S. B., Lee, H., Jung, J., Choi, Y. J., Park, J., et al. (2005). Measurement of atmospheric formaldehyde and monoaromatic hydrocarbons using differential optical absorption spectroscopy during winter and summer intensive periods in Seoul, Korea. *Water, Air, & Soil Pollution*, 166, 181–195.

Mills, G. P., Sturges, W. T., Salmon, R. A., Bauguitte, S. J.-B., Read, K. A., and Bandy, B. J. (2007). Seasonal variation of peroxyacetylnitrate (PAN) in coastal Antarctica measured with a new instrument for the detection of sub-part per trillion mixing ratios of PAN. *Atmospheric Chemistry and Physics*, 7, 4589–4599.

Petrakis, M., Psiloglou, B., Kassomenos, P. A., and Cartalis, C. (2003). Summertime measurements of benzene and toluene in Athens using a differential optical absorption spectroscopy system. *Journal of the Air & Waste Management Association*, 53, 1052–1064.

Platt, U., Perner, D., and Patz, H. W. (1979). Simultaneous measurement of atmospheric CH_2O, O_3 and NO_2 by differential optical absorption. *Journal of Geophysical Research*, 84, 6329–6335.

Platt, U., Perner, D., Winer, A. M., Harris, G. W., and Pitts, Jr., J. N. (1980a). Detection of NO_3 in the polluted troposphere by differential optical absorption. *Geophysical Research Letters*, 7, 89–92.

Platt, U. and Perner, D. (1980b). Direct measurement of atmospheric CH_2O, HNO_2, O_3 and SO_2 by differential optical absorption. *Journal of Geophysical Research*, 85, 7435–7458.

Platt, U. (1994) *Differential Optical Absorption Spectroscopy (DOAS)*. In *Air Monitoring by Spectroscopic Techniques*, M. W. Sigrist (ed.), Chemical Analysis Series, 127, John Wiley & Sons, Inc., New York, pp. 27–84.

Rubio, M. A., Gramsch, E., Lissi, E., and Villana, G. (2007). Seasonal dependence of peroxyacetylnitrate (PAN) concentrations in downtown Santiago, Chile. *Atmosfera*, 20, 319–328.

Svensson, R., Ljungström, E., and Lindqvist, O. (1987). Kinetics of the reaction between nitrogen dioxide and water vapour. *Atmospheric Environment*, 21, 1529–1539.

Triantafyllou, A. G., Zoras, S., Evagelopoulos, V., Garas, S., and Diamantopoulos, C. (2008). DOAS measurements above an urban street canyon in a medium sized city. *Global NEST Journal*, 10, 161–168.

Volkamer, R., Etzkorn, T., Geyer, A., and Platt, U. (1998). Correction of the oxygen interference with UV spectroscopic (DOAS) measurements of monocyclic aromatic hydrocarbons in the atmosphere. *Atmospheric Environment*, 32, 3731–3747.

17 Interactions between Ultraviolet-B and Total Ozone Concentrations in the Continental United States

Zhiqiang Gao, Wei Gao, and Ni-Bin Chang

CONTENTS

17.1 INTRODUCTION

17.1.1 BACKGROUND

Changes in the earth's atmosphere caused by anthropogenic and natural pollutants have led to a well-documented decline in stratospheric ozone (O_3) and a corresponding increase in ultraviolet (UV) irradiance at higher latitudes (Kerr and Fioletov 2008). At lower latitudes (30.8°S to 30.8°N), however, UV radiation at the earth's surface is highest during the summer season because smaller solar zenith angles minimize the atmospheric path length. In addition to the total ozone content that is interrelated with changes of solar radiation in the near ultraviolet wavelengths, the second largest cause of temporal and geographic variability of UV irradiance is clouds. UVA irradiances (320–400 nm) are not significantly affected by total ozone levels in the stratosphere because they are not strongly absorbed by ozone (Herman et al. 1997). However, the amount of UV radiation penetrating to the earth's surface with wavelengths shorter than 320 nm (UVB) can be significantly reduced by stratospheric ozone absorption, aerosols, clouds, ground albedo, altitude, and Rayleigh scattering in the atmosphere (Herman et al. 1997).

The ozone layer is essential for human life and ecosystem health. If the stratospheric ozone depletion occurred, an increase in UV doses will lead to a variety of adverse health and environmental effects, including human health problems, decreased crop yields, lower ocean productivity, loss of biodiversity, and degradation of materials. Many of these effects are related to UV doses accumulated over the course of a lifetime of human life and ecosystem; therefore, knowledge of the changes in ground-level UV radiation over prolonged periods is required to support relevant environmental and health risk assessments. For example, the thinning of the atmospheric ozone layer has led to elevated levels of UVB at the earth's surface, resulting in an increase in health risks due to deoxyribonucleic acid (DNA) damage in living organisms and possible increase in melanomas. Concern over the harmful effects of increased solar UV radiation on the biosphere has prompted extensive efforts to characterize it at the earth's surface. This chapter focuses on analyzing the long-term covariation of UVB and total ozone based on an integrated database of the National Aeronautical and Space Administration (NASA) total ozone mapping spectrometer (TOMS) satellite images and the U.S. Department of Agriculture (USDA) ground-based sensors and sensor network. This synergistic integration helps identify the varying UVB impacts over differing latitudes across the last three decades. The combination of TOMS data and ground-based measurements of total ozone and UVB collectively addresses the spatiotemporal variations of total ozone concentrations and UVB.

17.1.2 WHAT IS ULTRAVIOLET RADIATION?

UV radiation is part of the electromagnetic (light) spectrum that reaches the earth from the sun. It has wavelengths shorter than visible light, making it invisible to the naked eye. These wavelengths are classified as UVA, UVB, or UVC (Figure 17.1), with UVA the longest of the three at 320 to 400 nm (billionths of a meter). UVA is further divided into two wave ranges: UVA I, which measures 340 to 400 nm, and UVA II, which measures 320 to 400 nm. UVB ranges from 290 to 320 nm. With

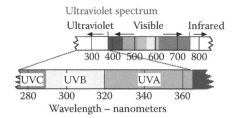

FIGURE 17.1   Diagram showing ultraviolet portion of solar spectrum. (Courtesy of USDA, http://uvb.nrel.colostate.edu/UVB/publications/uvb_primer.pdf [accessed June 2011].)

even shorter rays, most UVC is absorbed by the ozone layer and does not reach the earth. Both UVA and UVB, however, penetrate the atmosphere and play an important role in conditions such as premature skin aging, eye damage (including cataracts), and skin cancers (Skin Cancer Foundation 2011). They also suppress the immune system, reducing the body's ability to fight off these and other maladies (Skin Cancer Foundation 2011).

17.1.3   Ozone Layer and Ozone Hole

The atmosphere is divided into five layers: troposphere, stratosphere, mesosphere, thermosphere, and exosphere. Although ozone is present in small concentrations throughout the atmosphere, most ozone (about 90%) exists in the stratosphere, in a layer between 10 and 50 km above the earth's surface. This ozone layer performs the essential task of filtering out most of the sun's biologically harmful ultraviolet (UVB) radiation. Ozone concentrations in the atmosphere vary according to latitude and altitude, temperature, and weather conditions. Furthermore, aerosols and other particles ejected by man-made and natural events such as volcanic eruptions can have measurable impacts on total ozone levels.

The ozone found in our atmosphere is formed due to an interaction between oxygen molecules and ultraviolet light. When ultraviolet light hits these oxygen molecules (O_2), the reaction causes the molecules to break apart into single atoms of oxygen (UV light + O_2 → O + O). These single atoms of oxygen are very reactive, and a single atom combines with a molecule of oxygen to form ozone, composed of three atoms of oxygen ($2O + 2O_2$ → $2O_3$). This reaction enables the ozone layer to act as a UV filter, absorbing much of the harmful UV radiation before it reaches the surface of the earth.

In 1985, scientists identified a thinning of the ozone layer over the Antarctic during the spring months, which became known as the "ozone hole" (Figure 17.2) (European Commission 2011a). Scientific evidence shows that anthropogenic chemicals are responsible for the creation of the Antarctic ozone hole and likely play a role in global ozone losses (European Commission 2011a). Ozone-depleting substances (ODSs) such as chlorofluorocarbons (CFCs) used as aerosol propellants and refrigerants are used in many commercial products. Regardless of natural seasonal variations of ozone levels, stratospheric ozone levels have been decreasing annually since the 1970s (Tolba et al. 1992), with midlatitudes experiencing greater losses than equatorial regions. In October 1997, the Antarctic ozone hole covered 24 M

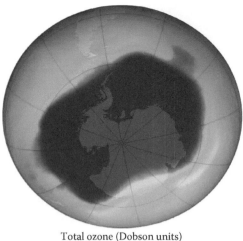

Total ozone (Dobson units)
110   220   330   440   550

FIGURE 17.2   The representation of ozone thickness over Labrador, Canada, measured in Dobson units (DU). (Courtesy of NASA, http://jwocky.gsfc.nasa.gov/teacher/basics/dobson .html [accessed June 2011].)

km^2 with an average of 40% ozone depletion, and ozone levels in Scandinavia, Greenland, and Siberia reached an unprecedented 45% depletion in 1996 (European Commission 2011b).

Scientists set up a unit of measure called the Dobson unit (DU) (Dobson 1957), the most common unit for measuring ozone concentration, to evaluate how much ozone is in the layer. One DU is the number of molecules of ozone required to create a layer of pure ozone 0.01 mm thick at a temperature of 0°C and a pressure of 1 atmosphere (the air pressure at the earth's surface). Expressed another way, a column of air with an ozone concentration of 1 DU would contain about 2.69×10^{16} ozone molecules for every 1 cm^2 of area at the base of the column. For example, a column of air over Labrador, Canada, has a 3-mm-thick ozone layer, so the measurement of the ozone over Labrador is 300 DU (Figure 17.3).

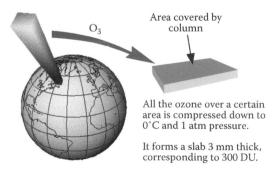

FIGURE 17.3   The location of ozone hole. (Courtesy of NASA, http://ozonewatch.gsfc.nasa .gov/facts/images/ozone_hole.jpg.)

17.2   UVB AND TOTAL OZONE MONITORING

Ground-based measurements permit relatively precise local studies on UV irradiance and its relationship with atmospheric components. Long-term decreases in summertime ozone over Lauder, New Zealand (45°S), have led to substantial increases in peak UV radiation intensities (McKenzie et al. 1999). The increasing trend of UV irradiance has been confirmed by several projects in recent years. For instance, Chubarova (2000) calculated the variability in EW (i.e., erythemally weighted) irradiance for 30 years of UV data (UVEry^2, in W·m^{-2}) in Moscow, Russia. The analysis of variability in UV380 and EW irradiance showed a slight increase in UV values since the middle of the 1980s. den Outer et al. (2005) studied the variability and long-term changes of UV in the Netherlands in relation to ozone and clouds. Linear regression analyses showed that the annual erythemal UV dose received at the ground for all weather conditions increased by 5.5 ± 2% per decade from 1979 to 2003.

To date, spatial coverage of UV radiation monitoring sites is still too sparse and the historical record is still too short to determine global-scale UV patterns and trends using UV radiation measurements alone. UV retrievals have been developed using TOMS onboard the NASA Nimbus 7 satellite to provide continuous spatial coverage over the planet; however, the ground-based network is better suited to monitor local conditions than the current large-pixel (100-km) satellite estimates. Spaceborne satellite observations can provide information of total ozone amount and cloud transmittance or reflectivity. Consequently, TOMS, which has provided both total ozone and cloud reflectivity measurements since the late 1970s, is an important source of derived UV data (Lapeta et al. 2000; Fioletov et al. 2002; Gao et al. 2001). Hence, approaches using TOMS data should be deemed critical alternatives to ground-based measurements for assessments of long-term total ozone and UVB and to include the impacts of past and present UV climate in environmental and health risk assessment (Gao et al. 2010). As a prelude to a full-scale causal investigation, an advanced intercomparison between total ozone and UVB at differing temporal and spatial scales across the continental United States can explore any abnormal trends and unusual correlations (Gao et al. 2010).

The international network of measuring stations structured in the Global Ozone Observation System (GOOS) under the auspices of the World Meteorological Organization (WMO) in Geneva, Switzerland, provides average changes in the concentration of ozone for specific areas (Komhyr 1980). The GOOS product is based on the existing network of ozone measuring stations operating Dobson spectrophotometers in connection to the World Ozone Data Center in Toronto, Canada. The Dobson instrument detects irradiance in eight wavebands from 305.5 to 339.8 nm (Komhyr et al. 1993).

Brewer spectrophotometers, alternative instruments for estimating ozone content, sample eight wavebands between 306.3 and 329.5 nm (McElroy and Kerr 1995). Of the approximately 90 Dobson stations worldwide, 46 have been used recently for total ozone trend studies (Bojkov et al. 1995). Bojkov and Fioletov (1995) applied all available total ozone data collected from more than 150 past and present GOOS stations and found that the ozone decline in percent per decade from 1980 was almost twice as large as the decline calculated from 1970. The cumulative decline of year-round global ozone from 1980 is 4.8%; however, the cumulative year-round decline over middle and polar latitudes from 1980 is more than 7% (Bojkov and Fioletov 1995).

Analysis of the time series of ozone amount has shown that the major persistent sources of ozone variability are the annual cycle, the 2.3-year quasi-biennial oscillation cycle (QBO), and the 11.5-year solar cycle (Herman et al. 2000). Umkehr observations are a commonly used index of ozone calculated from both Dobson (Staehelin et al. 1995) and Brewer (Kerr et al. 1988) spectrophotometers. This metric is the ratio of diffusely transmitted zenith–sky radiance at a wavelength pair in the ultraviolet, in which one wavelength is strongly absorbed by ozone, and the other weakly. Miller (1996) compared the ozone trends and responses to the 11-year solar cycle derived from two data sets they collected from June 1977 to June 1991. They considered data at northern midlatitudes (30°N–50°N) at altitudes between 25 and 45 km. In particular, they investigated the effects of the differences in spatial sampling between the data sets on the derived signals. Findings indicated that the trends derived from the two independent data sets are nearly identical at all levels except 35 km, where the Umkehr data indicate a somewhat more negative trend.

Various seasonal trend analyses were investigated to help retrieve spatiotemporal patterns. Reinsel et al. (1994) performed a seasonal trend analysis of zonal averages of TOMS satellite total ozone data for the comparable period from 1978 to 1991 and confirmed moderately close agreement between trends in Dobson and TOMS data over this period. Trends for the period of 1979 to 1994 are about 1.5% per decade stronger in northern and southern middle and high latitudes than the trends since 1970, highlighting a substantial acceleration of the rate of ozone decline (Bojkov et al. 1995). Allen and Reck (1997) analyzed the spatial and seasonal distributions of daily fluctuations in total ozone from 60°N to 60°S using 14.5-year TOMS data on a 50° latitude by 150° longitude grid. The contributions to these fluctuations from planetary- and medium-scale waves were analyzed using sinusoidal zonal wave filtering. They found that tropical total ozone fluctuations due to planetary-scale waves are slightly larger than those due to medium-scale waves in all seasons. Tourpali et al. (1997) examined the rate of change in the total ozone decline observed during past years over both the northern and the southern hemisphere. In their analysis, the zonal mean total ozone decreased during the last two decades relative to the corresponding mean ozone levels during the undisturbed period of 1964 to 1973.

It is also important to compare the ground-based total ozone measurements against the satellite data, and this has become possible as data became available over the past two decades. McPeters et al. (1994) compared National Oceanic and Atmospheric Administration (NOAA) Solar Backscattered Ultraviolet (SBUV) ozone profiles with Stratospheric Aerosol and Gas Experiment (SAGE) II profiles over the period October 1984 through June 1990. The best agreement with SAGE was seen in the integrated column ozone (cumulative above 15 km), where SAGE II had a 1% negative trend relative to SBUV over the period. Rusch et al. (1994) summarized measurements of ozone retrieved from Nimbus satellite instruments over the 1979 to 1991 period. Total ozone trends from TOMS and SBUV agreed with each other. Near-zero trends were observed at low latitudes, and larger, negative trends (approximately -0.5%/year) occurred near the poles. The column ozone trends depended on the base level altitude of integration but did not exhibit strong latitude dependence. Hilsenrath et al. (1995) compared the SBUV total ozone profile data with the Nimbus SBUV data during the period of data overlap. Total ozone values agreed about 1%, while ozone profile differences ranged from -4% to $+6\%$, depending on latitude and altitude. Ziemke et al. (1998) compared the horizontal resolution of SBUV/2 total ozone fields with those from the TOMS side-scanning photometers. They showed that the latter instruments provide high resolution, easily resolving the medium-scale waves at fixed latitudes that dominate day-to-day mid-latitude total ozone fluctuations. In contrast, SBUV/2 instruments did not because these devices measure only at nadir (straight downward), yielding daily-resolution measurements at a given latitude. They concluded that high-resolution TOMS-like side-scanning total ozone measurements are useful for ozone and UVB monitoring and prediction, rather than SBUV-type nadir observations. Fioletov et al. (2002) prepared data sets based on NOAA-14 SBUV/2, NASA TOMS, ERS-2 Global Ozone Monitoring Experiment (GOME), and ground-based measurements. They used different approaches to homogenize the records over the period 1979 to 2000. Systematic differences of up to 3% were found between different data sets for zonal and global total ozone area weighted average values.

To gain a more accurate estimate of total ozone, research focus has also been placed on improving data assimilation algorithms for both satellite data and ground-based measurements. McCormack and Hood (1997) used a form of the linearized steady-state ozone continuity equation to model the spatial distribution of total ozone in the northern hemisphere (NH) winter for the period 1979 to 1991. They found a significant decrease in the wave-induced variance in the wintertime total ozone distribution, which is consistent with the observed longitude dependence of total ozone trends in NH winter. Eckman et al. (1996) examined the export of ozone-poor air from the polar region following the breakup of the southern hemisphere polar vortex with a three-dimensional chemistry transport model. These simulations revealed that deeper denitrification led to a more persistent column-integrated ozone loss and a slight increase in its progression toward the equator. McPeters et al. (1997) computed total column ozone from a balloon sonde measurement and included an extrapolation of the measured ozone profile to altitudes above the above and beyond the location when the balloon-burst occurred. This total column calculation can be improved by using a monthly average ozone climatology based on ozone profile measurements from the SBUV instrument on the Nimbus 7 satellite. Koelemeijer and Stammes (1999) investigated the influence of errors in cloud parameters on the retrieved ozone column by performing a sensitivity study with a detailed radiative transfer model that included polarization. They showed that clouds influenced the retrieved ozone column via effects of cloud top height, cloud albedo, and cloud fraction. The errors in the retrieved ozone column were within 8%, 6%, and 5% for solar zenith angles of 300°, 600°, and 750°, respectively. Gao et al. (2001) developed a methodology for direct-Sun ozone retrieval using the ultraviolet multifilter rotating shadow-band radiometer (UV-MFRSR). This approach allows the real-time measurement of total vertical column ozone at ground-based stations; three such stations were tracked in this study.

17.3 STUDY AREA

Lapeta et al. (2000) discussed the influence of ozone and temperature profiles on surface UV radiation, and on total ozone column derived from global irradiance measurements. Findings indicated that differences in ozone maximum height as well as in ozone concentration in the upper troposphere have a significant influence on surface UV radiation. Because the downward trends in total ozone began several years before the UV radiation measurements became available, it would be of interest to compare ground-based measurements and TOMS satellite data at differing spatial and temporal scales across the continental United States (Gao et al. 2010).

To address such impacts, this study presents a correlation analysis to examine the trends and correlations of total ozone and UVB in recent years, investigating the temporal and spatial distributions of both over the continental United States. Intercomparisons were made between the daily UV erythemal doses and total column ozone calculated with the NASA Nimbus 7/TOMS UV algorithm and from the USDA ground-based network between 1979 and 2005. Eight of the USDA ground stations, WA02, CA02, CO02, AZ02, NE02, IL02, MD02, and LA02 (Figure 17.4), were selected for detailed investigations of the temporal trends as well as the correlations between the UV index (UVI = 40 × UVEry; WHO 2002; Environment Canada 2008) and the total ozone. This was followed by spatial analyses of multiyear UVI and total zone data computed from TOMS and UV-MFRSR over the continental United States. Such an in-depth spatial and temporal assessment of both UVB and total ozone is significant for future sustainable development in human society.

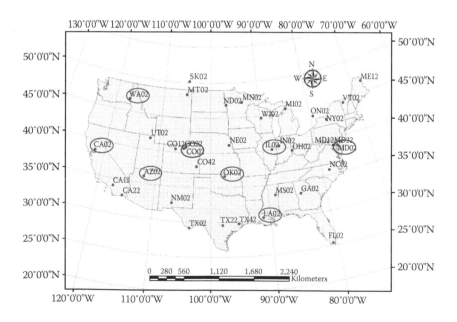

FIGURE 17.4   Locations of 37 USDA stations (selected stations are circled).

17.4 REMOTE SENSING IMAGES AND GROUND-BASED MEASUREMENTS

When compared to USDA ground-based measurements, TOMS data collection is not stable. In 2002, anomalies were noted in TOMS data, and in the following months, Earth Probe TOMS began to experience problems. Because of the data gap that resulted from the anomalies and technical problems after 2002, we could use only the early TOMS data set from 1978 to 2000 in combination with USDA ground-based measurements collected after 2000. Such integration allows us to produce a continuous data set covering the 1978 to 2008 period in this study. The capability of contemporary sensors and sensor networks fosters advanced intercomparisons between the daily UV erythemal doses and total column ozone calculated with any spaceborne data observed at the USDA ground-based stations, as was pursued in the study based on the available data set from 1998 to 2008.

We selected WA02, CA02, CO02, and AZ02 for TOMS data analysis in this study and four additional sites for a detailed statistical analysis based on the USDA data set (Figure 17.4). These eight stations are well distributed geographically, with WA02 and CA02 located in the western coast ranges, CO02 and AZ02 in the Rocky Mountains, NE02 and IL02 in the central lowland, and MD02 and LA02 in the eastern coastal plain. Each station has at least one UV-MFRSR sensor (Figure 17.4).

17.4.1 USDA UVB DATA

The USDA UVB Monitoring and Research Program (UVMRP), a USDA program in the Cooperative State Research, Education and Extension Service (CSREES), began in 1995 and now has 37 sites across the United States. All data from the network are captured by on-site data loggers and downloaded over a phone each evening. Data are made available to the scientific community as well as the general public for next-day retrieval via the network's website (http://uvb.nrel.colostate.edu). These stations use the UV-MFRSR sensor, a seven-channel UV version of the visible multifilter rotating shadow-band radiometer, to measure total and diffuse horizontal irradiances (Figure 17.5). The seven channels are created by ion-assisted deposition filters with a nominal bandwidth of 2 nm at full width at half maximum (FWHM) and nominal band centers at 300, 305, 311, 317, 325, 332, and 368 nm. There is filter-to-filter variation in the nominal wavelength center (± 0.5 nm). The direct beam is obtained by subtraction of the diffuse horizontal from the total horizontal irradiance and includes corrections such as Langley calibration. The measurement is completed in less than 5 seconds at all wavelengths. All three components are recorded every 20 seconds and averaged to 3-minute intervals.

The UV-MFRSR measures the relative intensities of selected pairs of UV wavelengths, as with the Dobson spectrophotometers, to estimate direct-Sun total column ozone. The first pair used in this study consists of 305- and 325-nm wavelengths. Light from both wavelengths is attenuated owing to scattering by air molecules and dust particles in passing through the atmosphere to the instrument. Dobson (1957) developed mathematical equations to calculate total column ozone from direct observation of the sun (also summarized by Komhyr 1998). The total column ozone

FIGURE 17.5   UV-MFRSR sensor at the USDA Fort Collins station, Colorado.

(Ω), expressed in Dobson units (1 DU = 10^{-3} cm pure ozone at standard temperature and pressure), as estimated from a single wavelength pair λ and λ' is (Dobson 1957; Komhyr, 1998)

$$\Omega = \frac{N(\beta_\lambda - \beta_{\lambda'})m(P/P_0) - (\delta_\lambda - \delta_{\lambda'}')\sec Z}{(\lambda_\lambda - \lambda_{\lambda'}')\mu}10^3,$$ (17.1)

where

$$N = \ln\frac{V_{o\lambda}}{V_{o\lambda'}'} - \ln\frac{V_\lambda}{V_{\lambda'}}.$$ (17.2)

In Equations 17.1 and 17.2, V_λ is the measured voltage at the ground at wavelength λ (mV), $V_{o\lambda}$ is the extraterrestrial voltage intercept at zero air mass at wavelength λ (mV), $V_{\lambda'}$ is the measured voltage at the ground at wavelength λ' (mV), $V_{o\lambda'}'$ is the extraterrestrial voltage intercept at zero air mass at wavelength λ' (mV), β_λ is the Rayleigh scattering coefficient (molecular optical depth) at wavelength λ (dimensionless), m is the air mass corresponding to solar zenith at the time of the direct normal irradiance measurement (g), P is the observed station pressure (mbar), P_0 is the mean sea level pressure (mbar), δ_λ is the scattering coefficient (optical depth) of aerosol particles at wavelength λ (dimensionless), and Z is the solar zenith angles (°).

With this methodology, total vertical column ozone between 1998 and 2008 was retrieved from 37 stations over the continental United States (Figure 17.4). The broadband UVB-1 Pyranometer (Yankee Environmental Systems) measured global irradiance in the UVB spectral range of 290 to 320 nm.

17.4.2 TOMS Data

In contrast to ground observations, satellites provide complete global coverage at a moderate resolution with standardized sensors. UV has been observed from space for more than 30 years. Early satellite UV measurements were made by the backscatter ultraviolet (BUV) sensor onboard the Nimbus 4, which was launched in 1970 and continued functioning for several years. Nimbus 7 provided the longest high-quality UV spaceborne observations from 1978 to 2000 with TOMS. This data set can be used for estimating long-term trends in total column ozone. Further, it is useful for investigating seasonal chemical depletions in ozone occurring in both the southern and NH polar springs.

The total column ozone retrieval from satellite data uses a normalized radiance, which is defined as the ratio of the backscattered earth radiance to the incident solar irradiance. Calculating this ratio requires periodic measurements of the solar irradiance. To measure the incident solar irradiance, the TOMS scanner is positioned to view an aluminum diffuser plate that reflects sunlight into the instrument. Retrieval of total column ozone is based on a comparison between the measured normalized radiances and radiances derived by radiative transfer calculations for different ozone amounts and the conditions of the measurement (McPeters et al. 1998).

The erythemal exposure (Exp) data product of TOMS is an estimate of the daily integrated ultraviolet irradiance, calculated using a model of the susceptibility of Caucasian skin to sunburn (erythema). This can be interpreted as an index of the potential for biological damage due to solar irradiation, given the column ozone amount and cloud conditions on each day. Exp is defined by the integral (McPeters et al. 1998)

$$Exp = \frac{1}{d_{es}} \int_{280\,nm}^{400\,nm} S(\lambda(\lambda)\,\lambda)\,d\lambda \int_{t_{ss}}^{t_{sr}} C(\lambda(\vartheta,\tau_{cl})F(\lambda,\vartheta,\Omega\Omega)\varsigma, \qquad (17.3)$$

where d_{es} is the earth–sun distance (au or A.U.), S is the solar irradiance incident on the top of the atmosphere at 1 A.U. (nW·m^{-2}·nm^{-1}), W is the biological action spectrum for erythemal damage, in B.D. m^{-2}, where B.D. = biological damage, t_{sr} and t_{ss} are the time of sunrise and sunset (radians), respectively, C is the cloud attenuation factor (dimensionless), τ_{cl} is the cloud optical thickness (mbar), ϑ is the solar zenith angle (function of time, t; radians), and F is the spectral irradiance at the surface under clear skies, normalized to unit solar spectral irradiance at the top of the atmosphere (dimensionless).

17.4.3 UV Index

For UV-induced erythema (sunburn), the action spectrum adopted by most international organizations is the Commission Internationale de l'Éclairage (International Commission on Illumination CIE) action spectrum (E), using the method described by McKinlay and Diffey (1987). The UVI itself is an irradiance scale computed by multiplying the CIE irradiance (in W m^{-2}) by 40. The clear sky value at sea level in the tropics is normally in the range of 10 to 12 (250–300 mW·m^{-2}), and 10 is an exceptionally high value for northern midlatitudes. This scale has been adopted by WMO and World Health Organization (WHO 2002) and is in use in a number of countries. UV intensity is also described in terms of ranges from low (0–2) to medium (3–5), high (6–7), very high (8–10), and extreme (11+) values. This chapter follows the use of UVI as a representative index to address UVB impact on human society and ecosystems.

17.5 RESULTS AND DISCUSSION

17.5.1 TREND OF UVI AND TOTAL OZONE USING TOMS DATA

Total ozone trend certainly influences UVB (Krzyscin 2000; Malanca et al. 2005). Monthly averages of observations from 1979 to 2000 revealed a consistent trend between stations, thereby supporting the multiyear spatial analyses. Time-series plots of UVI and total ozone associated with each data source for each station (Figures 17.6 through 17.9) confirm the consistency of the seasonal trend as well as identify possible interactions between UVI and total ozone variations. To facilitate interpretation, separate subplots are presented for each of the four seasons at each of the four stations. By comparing sequences associated with both data sources, we can identify the general trends and characterize how the patterns they detect differ geographically, at least at these four stations.

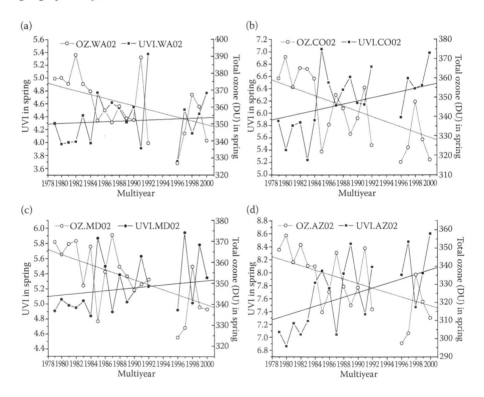

FIGURE 17.6 Seasonal trend analysis of the UVI and total ozone deviations relative to the monthly means associated with ground-based measurements at four stations in spring. ([a] WA02 multiyear seasonal variations, [b] CO02 multiyear seasonal variations, [c] MD02 multiyear seasonal variations, [d] AZ02 multiyear seasonal variations).

The trend analysis for spring (Figure 17.6) UVI measurements showed an increasing long-term trend while the total ozone data showed a decreasing long-term trend from 1979 to 2000. This conflicting variation was moderately evident at both WA02 and MD02 stations, which are at lower elevations than the other stations. At WA02, the fluctuations of the springtime UVI data were mild from 1979 to 1990, yet there were seasonal mean peaks and valleys in 1992 and 1991, respectively. Overall, the UVI measurements reached their minimum in 1996, followed by mild fluctuations of UVI from 1997 to 2000. At this site, the negative correlation between UVI and total ozone in the spring was low, yielding a negative correlation coefficient of 0.46. Stations CO02 and AZ02 are located at high elevations, where contrasting trends of UVI and total ozone are salient. The negative correlation between the springtime UVI and total ozone was strong at these sites, 0.79 and 0.89, respectively, indicating the impact due to the plateau topography. At MD02, however, UVI fluctuated dramatically, and the negative correlation between the UVI and total ozone was also low ($r = -0.58$).

A long-term trend of UVI measurements and total ozone data was exhibited over these four stations in summer (Figure 17.7). Over the 21-year time span, a generally decreasing trend of total ozone was also observed in summer. With the exception of

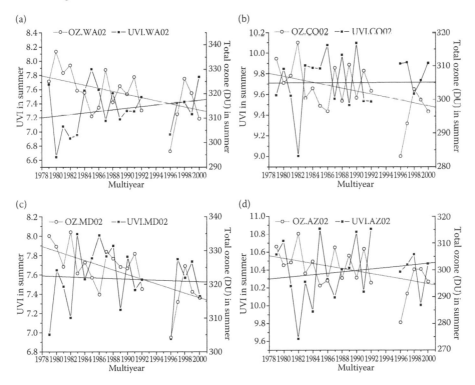

FIGURE 17.7   Seasonal trend analysis of the UVI and total ozone deviations relative to the monthly means associated with ground-based measurements at four stations in summer. ([a] WA02 multiyear seasonal variations, [b] CO02 multiyear seasonal variations, [c] MD02 multiyear seasonal variations, [d] AZ02 multiyear seasonal variations).

the MD02 station, UVI measurements at the other three stations showed a slightly increasing trend over the time period. WA02 uniquely presented an opposite trend between UVI and total ozone with a time lag of about 2 weeks to months. This time lag weakened the negative correlation coefficient between these variables ($r = -0.58$). MD02 had a mild decreasing trend of UVI over the entire time period. When evaluated piecewise, an increasing trend occurred from 1979 to 1985, followed by a decreasing trend from 1985 to 1993. A seasonal mean minimum appeared in 1996, which recovered in 1997 but then continued to decline through 2000. Total ozone seemed to be constantly decreasing over the entire 21 years of the study period, resulting in a poor, negative correlation coefficient with the value of 0.03. At CO02, UVI measurements increased slightly over the 21 years of the study period, revealing an apparent increasing trend from 1979 to 1992. They then stabilized, followed by a slightly decreasing trend in summertime during the remainder of the study period. During the same time period, total ozone was consistently decreasing. The negative correlation coefficient between the UVI measurements and total ozone data at CO02 was 0.64. At AZ02, the UVI measurements were apparently increasing during the entire 21 years, while total ozone was decreasing, resulting in a negative correlation coefficient of 0.5.

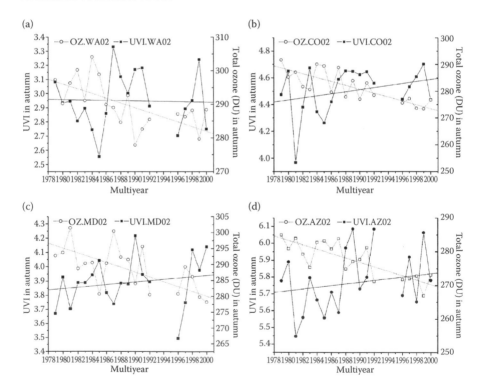

FIGURE 17.8 Seasonal trend analysis of the UVI and total ozone deviations relative to the monthly means associated with ground-based measurements at four stations in autumn. ([a] WA02 multiyear seasonal variations, [b] CO02 multiyear seasonal variations, [c] MD02 multiyear seasonal variations, [d] AZ02 multiyear seasonal variations).

When checking the same record in fall, a long-term trend of the UVI measurements and total ozone data over these four stations was found (Figure 17.8). WA02 showed fluctuations in the UVI measurements, with a slightly decreasing trend from 1979 to 1985 and a mild increasing trend from 1985 to 1987. From 1987 to 1992, the decreasing trend again became obvious, followed by an increasing trend from 1995 to 2000. Such dramatic fluctuations resulted in no net change over the entire time period. Total ozone showed a continuously decreasing trend with no fluctuations. The other three stations revealed an increasing trend in UVI and a decreasing trend in the total ozone. At both CO02 and MD02, the increasing trend of UVI measurements was accompanied by a decreasing trend of total ozone during the 21 years of the study period. At AZ02, although total ozone was observed to be decreasing over the 21 years, the UVI measurements revealed an increasing trend from 1979 to 1992, followed by a decreasing trend from 1994 to 2000. In summary, these four stations collectively exhibited inconsistent fluctuations between the UVI measurements and total ozone data in the fall, with relatively low negative correlation coefficients over the 21 years of the study period. These negative correlations were 0.49 at AZ02, 0.34 at WA02 and MD02, and 0.21 at CO02.

During winter, a long-term trend of UVI measurements and total ozone data persisted over these four stations (Figure 17.9). In general, an increasing trend of UVI

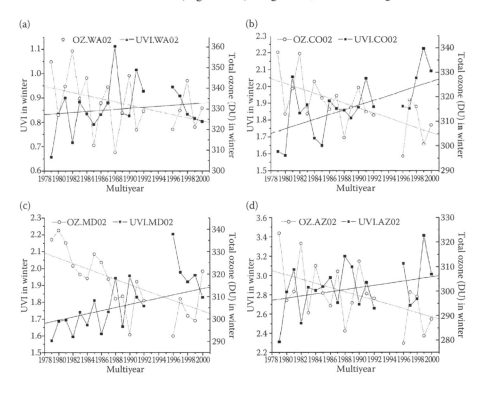

FIGURE 17.9 Seasonal trend analysis of the UVI and total ozone deviations relative to the monthly means associated with ground-based measurements at four stations in winter. ([a] WA02 multiyear seasonal variations, [b] CO02 multiyear seasonal variations, [c] MD02 multiyear seasonal variations, [d] AZ02 multiyear seasonal variations).

measurements occurred with a decreasing trend of total ozone at all four stations over the 21 years of the study period. At the AZ02 station, the UVI measurements seemingly increased from 1979 to 1992, followed by a decreasing trend from 1994 to 2000. The overall trend of total ozone was obviously decreasing during the 21 years of the study period. Within these four stations, WA02 showed a unique pattern. While the total ozone generally decreased during the 21 years of the study period, the UVI measurements increased from 1979 to 1992, followed by a decreasing trend from 1995 to 2000. The remaining three stations exhibited a homogeneous increasing trend of UVI measurements and a homogeneous decreasing trend of total ozone data in the study period. The relatively better consistency of the conflicting trends between the UVI measurements and the total ozone data resulted in larger negative correlation coefficients: 0.76 at AZ02, 0.73 at MD02, 0.50 at CO02, and 0.42 at WA02.

In summary, the UVI measurements were generally increasing and the total ozone data were apparently decreasing from 1979 to 2000, supporting the well-established hypothesis that UV radiation reaching the earth's surface has increased with a concomitant decline in the ozone layer. Site-specific discrepancies did exist, however, leaving room for further investigation. The increasing trend of UVI measurements and the decreasing trend of total ozone were strongest at the CO02 and AZ02 stations, which are both located on plateau topography at middle-low latitude. At WA02 and MD02, located in low-lying areas at high latitude, the increasing trend of UVI measurements was obvious during the 21 years but had dramatic seasonal fluctuations, while total ozone had an apparent decreasing trend with no obvious fluctuations. The negative correlation coefficients between the UVI measurements and total ozone data were relatively high in spring and winter but low in summer and fall. Spatially, they were high at CO02 and AZ02 with high elevations and low latitude, and low at WA02 and MD02 with low elevations and high latitude. This is also confirmed regionally by the spatial analysis in the next section.

17.5.2 Yearly Trend Analysis of UVI and Total Ozone Using USDA Data

The trends of UVB and total ozone over the continental United States were analyzed based on the data set collected between 1998 and 2008 at eight stations (WA02, CA02, CO02, AZ02, NE02, IL02, LA02, and MD02) distributed evenly in different geographical areas (Figure 17.4). These data are derived from the USDA UVB ground-based network. From these UVI and total ozone data sets, the average values of total ozone and UVI can be calculated according to three subperiods of 1998 to 2001, 2002 to 2005, and 2006 to 2008 (Table 17.1). The changes in total ozone and UVI corresponding to each region reveal a clear temporal trend, either increasing or decreasing, with varying degree of difference (see Figures 7.10 and 7.11).

Stations CA02 and WA02, located along the western coast ranges, exhibited a decreasing trend of total ozone. As observed at CA02, there was a decrease in total ozone on average, with values of 319.11 DU in 1998 to 2001, 315.82 DU in 2002 to 2005, and 300.66 DU in 2006 to 2008. However, the changes of UVB revealed an irregular trend at CA02, including an initial decrease, an immediate increase followed by decrease, and an ultimate increase between 1998 and 2008 (Figure 17.10). Yet the UVB trend at CA02 increased during the same time period. The total ozone

TABLE 17.1
Average Values of UVI and Total Ozone (DU) in Three Periods for Eight Stations

Years	OZ.AZ02	UV.AZ02	OZ.CA02	UV.CA02
1998–2001	295.54	6.12	319.11	4.86
2002–2005	292.08	6.37	315.82	4.84
2006–2008	294.66	6.38	300.66	5.06
Years	**OZ.CO02**	**UV.CO02**	**OZ.IL02**	**UV.IL02**
1998–2001	308.99	4.56	314.58	3.77
2002–2005	308.51	4.70	312.61	3.95
2006–2008	301.38	4.69	313.04	4.06
Years	**OZ.MD02**	**UV.MD02**	**OZ.WA02**	**UV.WA02**
1998–2001	318.59	3.89	329.78	3.45
2002–2005	316.73	3.94	326.31	3.47
2006–2008	319.55	4.14	323.38	3.48
Years	**OZ.NE02**	**UV.NE02**	**OZ.LA02**	**UV.LA02**
1998–2001	320.03	3.82	282.61	5.35
2002–2005	316.92	4.07	277.47	5.25
2006–2008	305.87	4.10	270.40	5.45

Note: OZ: ozone; UV: ultraviolet.

at WA02 showed a deceasing trend between 1998 and 2005, with average values of total ozone 329.78 DU in 1998 to 2001, 326.31 DU in 2002 to 2005, and 323.38 DU in 2006 to 2008. The UVB trend at WA02 revealed a mild increase overall. In short, in the coast ranges, the trend of total ozone at CA02 and WA02 was not obvious, while the trend of UVB showed a mild increase.

The change of total ozone observed at station MD02 located at the eastern coastal plain was mild with an increasing trend. But the trend was lower in 1998 to 2005, followed by an increasing trend after 2005, evident from the mean values of total ozone associated with three periods: 318.59 DU in 1998 to 2001, 316.73 DU in 2002 to 2005, and 319.55 DU in 2006 to 2008. In comparison, the trend of UVB at MD02 was clearly increasing. The mean values of UVI were 3.89 DU in 1998 to 2001, 3.94 DU in 2002 to 2005, and 4.14 DU in 2006 to 2008. The fluctuations of total ozone changes at station LA02 were particularly large in 1998 to 2001, with a generally decreasing trend of total ozone in 1998 to 2008. The mean values of total ozone at LA02 were 282.61 DU in 1998– to 2001, 277.47 DU in 2002 to 2005, and 270.4 DU in 2006 to 2008. The overall trend of total ozone was clearly decreasing (Figure 17.11). In contrast, the change of UVB observed at LA02 was mild with an overall increasing trend from 1998 to 2008. In detail, the mean values of UVI were 5.35 DU in 1998 to 2001, 5.25 DU in 2002 to 2005, and 5.45 DU in 2006 to 2008. In summary, in the coastal plain, the trends of total ozone observed at LA02 and MD02 were slightly decreasing, and the changes in total ozone showed more fluctuations; however, this coincided with evident increasing trends of UVB.

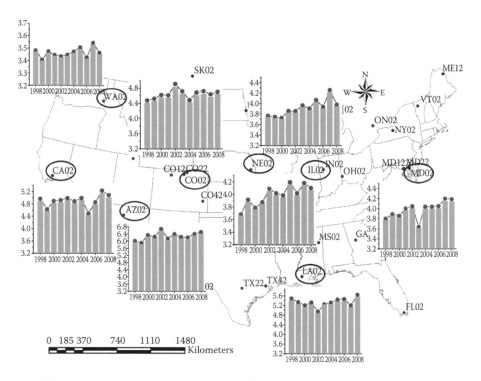

FIGURE 17.10   UVB diagrams of the eight USDA ground stations.

Stations AZ02 and CO02 located at the Rocky Mountains areas represent unique features of a changing landscape and show similar variations of UVI. The UVI wave pattern initially increased, followed by a decrease, and then went up again with an overall increasing trend between 1998 and 2001. Whereas the averages of UVI in 2002 to 2005 at AZ02 and CO02 stations are 6.37 and 4.70 DU, respectively, the corresponding values are 6.38 and 4.69 DU, respectively. We can conclude that the difference of average UVI between these two subperiods is small. In contrast, total ozone observed at AZ02 and CO02 showed an overall decreasing trend, and the wave pattern of total ozone initially decreased followed by an increase and ending with a subsequent decrease, which is exactly opposite to the UVI counterpart. In summary, at AZ02 and CO02, trends of total ozone were decreasing while the trends of UVI were increasing slightly.

Stations NE02 and IL02 are located in the central lowland of the continental United States. UVI observed by NE02 and IL02 exhibited more fluctuations with an overall increasing trend from 1998 to 2008 when compared to other stations. In this time period, total ozone concentrations at NE02 were decreasing, with average values of 320.03 DU in 1998 to 2001, 316.92 DU in 2002 to 2005, and 305.87 DU in 2006 to 2008. We can conclude that changes of total ozone were mild for these three subperiods. In comparison, the trend of total ozone concentrations at IL02 was also decreasing slightly. Although there are two peaks of total ozone in 2003 and 2006, the differences of average total ozone among these three subperiods were small. In

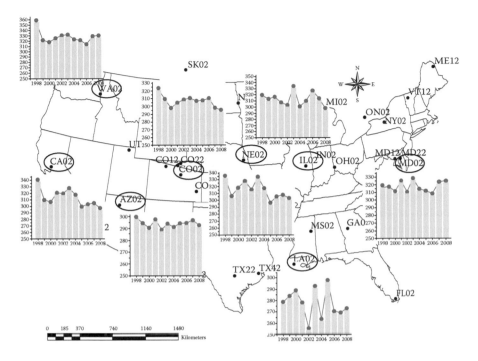

FIGURE 17.11   Total ozone diagrams of the eight USDA ground stations.

summary, the trend of total ozone observed at NE02 and IL02 was mildly decreasing while the trends of UVB were clearly increasing.

17.6   FINAL REMARKS

17.6.1   ARE TOTAL OZONE CONCENTRATIONS REALLY DECREASING?

Many studies have found that the total column ozone in the NH middle latitudes started recovering, maintained, or at least did not decrease since 1997 based on data derived from models, ground-based measurements, and remote sensing. A comparison of previous findings to our observations in this study is interesting. Previous studies, including Reinsel et al. (2002) and Newchurch et al. (2003), examined changes in the ozone profile as well as total column ozone. Some recent analyses collectively support three features—latitudinal, seasonal, and altitudinal dependences of observed changes—as a basis for investigating ozone recovery (WMO 2003; Hofmann et al. 1997; Andersen et al. 2006). Recent data suggest that total column ozone abundances have at least not decreased over the past eight years (1997–2005) for most of the world, including the NH middle latitude (Elizabeth et al. 2007). Reinsel et al. (2005) analyzed total ozone data from satellite and ground-based measurements including the merged TOMS and SBUV merged ozone data (MOD) used during 1978 to 2002. They identified a positive, significant change in trend after 1996 in the NH middle latitudes, even after accounting for some dynamical variables in their statistical model. Hadjinicolaou et al. (2005) found a positive ozone trend for the 1994 to 2003 study period in their dynamically driven model that explains all the observed trends. It suggests that the notion of an early sign of a turnaround due to chemical recovery in total ozone should be treated with caution. Their results are qualitatively consistent with the chemistry climate modeling (CCM) results of Schnadt and Dameris (2002) that predict an accelerated NH ozone recovery from 1990 to 2015 due to tropospheric dynamical changes in contrast with some other CCMs. Andersen et al. (2006) discovered a substantial interest in examining the emerging data, particularly since 1996, for signs of recovery; however, most models suggest a more gradual recovery with an increasing magnitude of recovery rates over time.

Because the continental United States is located in NH middle latitudes, potential errors in the USDA and TOMS methods for estimating total ozone column amounts from the UV measurements may arise from synergistic effects of atmospheric aerosols, cloud distribution, and other particulates. Our analysis, however, shows that total ozone concentrations were decreasing slightly with many fluctuations while the increasing trend of UVI was obvious in continental United States between 1998 and 2008.

17.6.2 SPATIOTEMPORAL UVI DISCREPANCIES BETWEEN TOMS DATA AND USDA GROUND-BASED MEASUREMENTS

The integrated data sets used in this analysis are a combination of TOMS data and USDA ground-based measurements; therefore, a comparison of these two sources of data at some overlapped time point is useful. We selected 2000 for a spatial analysis comparison over four seasons (Figures 17.12 through 17.15).

Spring UVI data comparing USDA ground-based measurements and TOMS data revealed a similar spatial pattern with a disparity in the southwestern United States (Figure 17.12). Higher UVB values appeared in the southwestern United States, including Arizona, New Mexico, and Texas, whereas lower UVB values were present around the Great Lakes area. USDA-based measurements of UVI ranged from 3.36 to 9.72, with a mean of 5.85. In contrast, TOMS UVI values ranged from 3.28 to 10.05 DU, with a mean of 6.52. TOMS UVB values were higher than the corresponding USDA measurement by 0.16 DU, with a larger range of 0.41 between upper and lower bounds, because USDA measurements were taken at the ground level and are affected by topology and cloud impacts whereas TOMS data are not.

Summer UVI data comparing USDA ground-based measurements and TOMS data revealed a similar spatial pattern (Figure 17.13). Higher UVB values appeared in the southwestern United States, including Arizona, New Mexico, and northern Colorado, whereas lower UVB values were present around the Great Lakes area. USDA-based measurements of UVI ranged from 4.88 to 11.04 DU, with a mean of 7.77. In contrast, TOMS UVI ranged from 5.61 to 12.25 DU, with a mean of 6.64. TOMS UVB data were higher than the corresponding USDA measurements by 1.07 because more cloud in summer affects or weakens UVB at the ground level.

Fall UVI data comparing USDA ground-based measurements and TOMS data revealed a similar spatial pattern (Figure 17.14). Higher UVB values appeared in the southwestern United States, including Arizona, New Mexico, and Texas, whereas lower UVB values were present around the Great Lakes area, and the minimum UVB always occurs north of latitude 40°. USDA-based measurements of UVI ranged

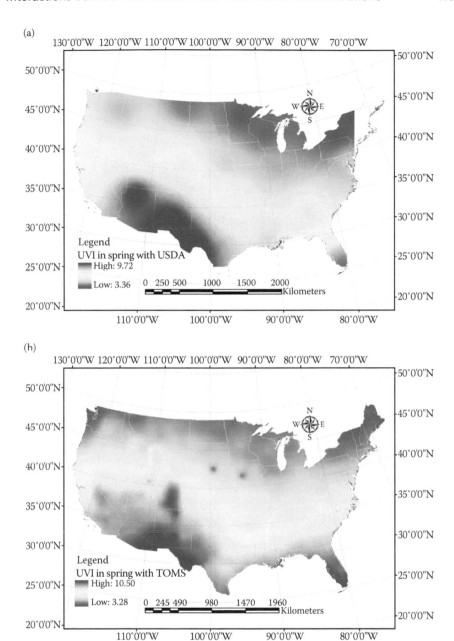

FIGURE 17.12 UVI spatial distributions based on seasonal average in Spring 2000. ([a] UVI spatial distributions with USDA data, [b] UVI spatial distributions with TOMS data.)

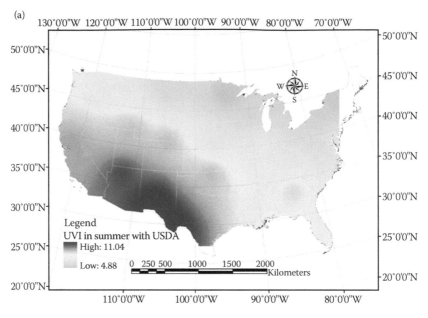

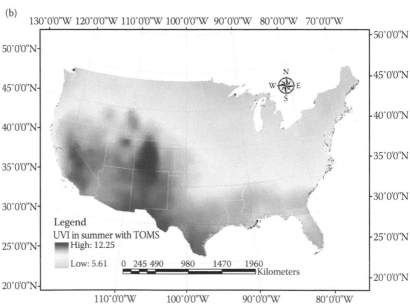

FIGURE 17.13 Maps of UVI spatial distributions based on seasonal average in Summer 2000. ([a] UVI spatial distributions with USDA data, [b] UVI spatial distributions with TOMS data.)

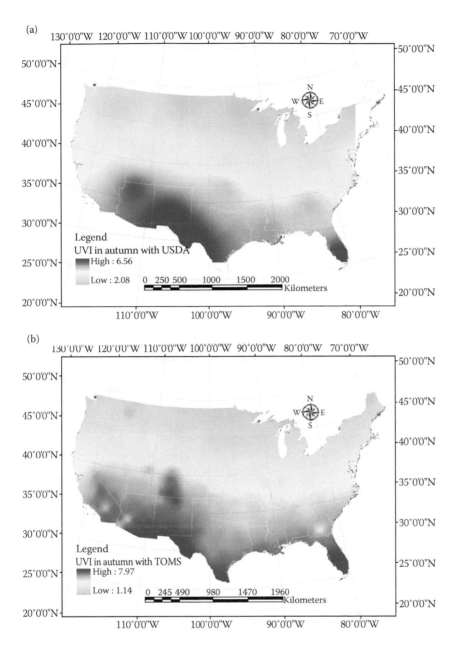

FIGURE 17.14 Maps of UVI spatial distributions based on seasonal average in Fall 2000. ([a] UVI spatial distributions with USDA data, [b] UVI spatial distributions with TOMS data.)

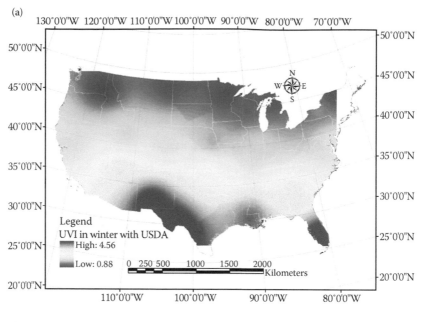

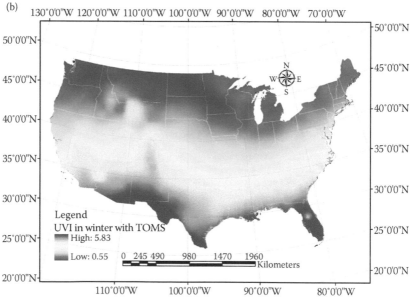

FIGURE 17.15 UVI spatial distributions based on seasonal average in Winter 2000. ([a] UVI spatial distributions with USDA data, [b] UVI spatial distributions with TOMS data.)

from 2.08 to 6.56 DU, with a mean of 3.76. In contrast, TOMS UVI values ranged from 1.14 to 7.97 DU, with a mean of 6.83. TOMS UVB values were higher than the corresponding USDA measurements by 0.65 because more cloud cover in fall affects or weakens the UVB at the ground level. As a consequence, TOMS average UVB was higher than USDA measurements and had a larger range of 2.35 DU between upper and lower bounds.

Winter UVI data comparing USDA ground-based measurements and TOMS data also revealed a similar spatial pattern (Figure 17.15). Higher UVB values appeared in the southwestern United States and coastal region, including Arizona, New Mexico, and Texas, whereas lower UVB values were present around the Great Lakes area. USDA-based measurements of UVI ranged from 0.88 to 4.56 DU, with a mean of 1.97. In contrast, TOMS UVI values ranged from 0.55 to 5.83 DU, with a mean of 2.11. TOMS UVB data were higher than the corresponding USDA measurements by 0.14 due to clouds weakening UVB at the ground level. As a consequence, TOMS UVB values were higher than USDA measurements in general and had a range of 1.60 DU between upper and lower bounds.

Accordingly, consistency of UVI spatial patterns can be confirmed across all four seasons. Both sources of UVB distribution exhibit a south–north gradient in which the maximum always occur in the southern states. The UVB in summer was the strongest, with an average 7.77 to 8.84 DU and maximum 11.04 to 12.25 DU. UVB in winter was weakest, with an average 1.97 to 2.11 DU and maximum 4.56 to 5.83 DU. Simultaneous interpolation among all 37 USDA observation stations cannot cover all corners of the continental United States, which is why the state of Maine was not included in some of the UVI maps. However, TOMS UVB data can reflect spatial variations and do not have such limitations, which is why minute changes of UVB can be retrieved and observed in diagrams based on TOMS data.

17.7 CONCLUSIONS

This study analyzed the changes of UVB and total ozone in an interactive mode in the continental United States based on the USDA ground-based measurements and TOMS data. For the four stations selected in our demonstration, total ozone decreased from 1979 to 2000, yet the UVI apparently increased from 1979 to 1992, remained stable for a time period, and then slightly decreased from 1995 to 2000. The negative correlation coefficients between the UVI measurements and total ozone data were relatively high in spring and winter, but they were low in summer and fall. To continue this investigation without being affected by the TOMS anomalies in 2002, we switched to the USDA ground-based measurements at eight stations from 1998 to 2008. Within this period, total ozone decreased slightly, while UVB clearly increased. Geographically, the UVI multiyear mean gradually decreased northward while the magnitude of total ozone gradually decreased southward. The annual fluctuations of total ozone were relatively high in northern states and had no specific linkage with the elevation; yet the annual fluctuations of UVI had an obvious association with the elevation. This implies that the annual fluctuations of UVI are high at relatively higher altitudes. In all cases, annual UVI and total ozone were negatively correlated, strongest in eastern and western parts of the United States, and weakest in northern states.

Over the past three decades, at a monthly scale, the UVI increased while total ozone decreased in the continental United States at eight selected USDA stations, while spatial distributions of UVI and total ozone showed substantial variations from coastal zones to the midwestern United States; however, the tendency toward recovery of ozone layer in the continental United States cannot be fully confirmed.

ACKNOWLEDGMENT

The authors are grateful for the financial support of the USDA NIFA project (2010-34263-21075) in this study.

REFERENCES

Allen, D. R. and Reck, R. A. (1997). Daily variations in TOMS total ozone data. *Journal of Geophysical Research-Atmospheres*, 102, 13603–13608.

Andersen, S. B., Weatherhead, E. C., Stevermer, A., Austin, J., and Fleming, E. L. (2006). Comparison of recent modeled and observed trends in total column ozone. *Journal of Geophysical Research*, 111, D02303, 1–10.

Bojkov, R. D., Bishop, L., and Fioletov, V. E. (1995). Total ozone trends from quality-controlled ground-based data (1964–1994). *Journal of Geophysical Research-Atmospheres*, 100, 25867–25876.

Bojkov, R. D. and Fioletov, V. E. (1995). Estimating the global ozone characteristics during the last 30 years. *Journal of Geophysical Research-Atmospheres*, 100, 16537–16551.

Chubarova, N. Y. and Nezval, Y. I. (2000). Thirty year variability of UV irradiance in Moscow. *Journal of Geophysical Research-Atmospheres*, 105, 12529–12539.

den Outer, P. N., Slaper, H., and Tax, R. B. (2005). UV radiation in the Netherlands: Assessing long-term variability and trends in relation to ozone and clouds. *Journal of Geophysical Research-Atmospheres*, 110, D02203, doi:10.1029/2004JD004824.

Dobson, G. M. B. (1957). Observers' handbook for the ozone spectrophotometer. In *Annals of the International Geophysical Year, 5 (Part I)*, Pergamon, New York, pp. 46–89.

Environment Canada (2008). http://exp-studies.tor.ec.gc.ca/e/ozone/uv_index_definition.htm.

Eckman, R. S., Grose, W. L., Turner, R. E., and Blackshear, W. T. (1996). Polar ozone depletion: A three-dimensional chemical modeling study of its long-term global impact. *Journal of Geophysical Research-Atmospheres*, 101, 22977–22989.

Elizabeth, C. W. and Signe, B. A. (2005). The search for signs of recovery of the ozone layer. *Nature*, 441(4), 39–45.

European Commission (2011a). http://ec.europa.eu/clima/policies/ozone/layer_en.htm (accessed February 16, 2011).

European Commission (2011b). http://ec.europa.eu/environment/index_en.htm (accessed February 16, 2011).

Fioletov, V. E., Bodeker, G. E., Miller, A. J., McPeters, R. D., and Stolarski, R. (2002). Global and zonal total ozone variations estimated from ground-based and satellite measurements: 1964–2000. *Journal of Geophysical Research-Atmospheres*, 107, 28032–29045.

Gao, W., Slusser, J., Gibson, J., Scott, G., Bigelow, D., Kerr, J., and McArthur, B. (2001). Direct-sun column ozone retrieval by the ultraviolet multifilter rotating shadow-band radiometer and comparison with those from Brewer and Dobson spectrophotometers. *Applied Optics*, 40, 3149–3155.

Gao, Z., Gao, W., and Chang, N. B. (2010). Detection of multidecadal changes in UVB and total ozone concentrations over the continental US with NASA TOMS data and USDA ground-based measurements. *Remote Sensing*, 2(1), 262–277.

Hadjinicolaou, P., Pyle, J. A., and Harris, N. R. P. (2005). The recent turnaround in strato-spheric ozone over northern middle latitudes: A dynamical modeling perspective. *Geophysical Research Letters*, 32, L12821, doi:10.1029/2005GL022476.

Herman, J. R., Piacentini, R. D., Ziemke, J., Celarier, E., and Larko, D. (2000). Interannual variability of ozone and UV-B ultraviolet exposure. *Journal of Geophysical Research-Atmospheres*, 105, 29189–29193.

Herman, J. R., Bhartia, P. K., Torres, O., Hsu, C., Seftor, C., and Celarier, E. (1997). Global dis-tribution of UV-absorbing aerosols from Nimbus 7/TOMS data. *Journal of Geophysical Research-Atmospheres*, 102, 16911–16922.

Hilsenrath, E., Cebula, R. P., Deland, M. T., Lassmann, K., Taylor, S., Wellemeyer, C., and Bhartia, P. K. (1995). Calibration of the NOAA-11 solar backscatter ultraviolet (Sbuv/2) ozone data set from 1989 to 1993 using in-flight calibration data and SSBUV. *Journal of Geophysical Research-Atmospheres*, 100, 1351–1366.

Hofmann, D. J., Oltmans, S. J., Harris, J. M., Johnson, B. J., and Lathrop, J. A. (1997). Ten years of ozonesonde measurements at the south pole: Implications for recovery of springtime Antarctic ozone. *Journal of Geophysical Research,* 102, 8931–8943.

Kerr, J. B. and Fioletov, V. E. (2008). Surface ultraviolet radiation. *Atmosphere-Ocean,* 46, 159–184.

Kerr, J. B., Asbridge, I. A., and Evans, W. F. J. (1988). Intercomparison of total ozone mea-sured by the Brewer and Dobson Spectrophotometers at Toronto. *Journal of Geophysical Research-Atmospheres*, 93, 11129–11140.

Komhyr, W. D., Mateer, C. L., and Hudson, R. D. (1993). Effective Bass-Paur 1985 ozone absorption-coefficients for use with Dobson ozone spectrophotometers. *Journal of Geophysical Research-Atmospheres*, 98, 20451–20465.

Komhyr, W. D. (1980). *Operations Handbook—Ozone Observations with a Dobson Spectrophotometer (WMO Global Ozone Research and Monitoring Project Report 6).* World Meteorological Organization, Geneva, Switzerland.

Komhyr, W. D. (1998). *Operations Handbook—Ozone Observations with a Dobson Spectrophotometer (WMO Global Ozone Research and Monitoring Project Report 6).* World Meteorological Organization, Geneva, Switzerland.

Koelemeijer, R. B. A. and Stammes, P. (1999). Effects of clouds on ozone column retrieval from GOME UV measurements. *Journal of Geophysical Research,* 104(D7), 8281–8294.

Krzyscin, J. W. (2000). Total ozone influence on the surface UV-B radiation in the late spring-summer 1963–1997: An analysis of multiple timescales. *Journal of Geophysical Research-Atmospheres*, 105, 4993–5000.

Lapeta, B., Engelsen, O., Litynska, Z., and Kylling, B. (2000). Sensitivity of surface UV radiation and ozone column retrieval to ozone and temperature profiles. *Journal of Geophysical Research-Atmospheres*, 105, 5001–5007.

Malanca, F. E., Canziani, P. O., and Arguello, G. A. (2005). Trends evolution of ozone between 1980 and 2000 at midlatitudes over the Southern Hemisphere: Decadal differences in trends. *Journal of Geophysical Research-Atmospheres*, 110, 30601–30611.

McPeters, R. D., Bhartia, P. K., Krueger, A. J., and Herman, J. R. (1998). *Earth Probe Total Ozone Mapping Spectrometer (TOMS) Data Products User's Guide.* NASA Technical Publication, 1998-206895, Goddard Space Flight Center Greenbelt, MD.

McPeters, R. D., Labow, G. J., and Johnson, B. J. (1997). A satellite-derived ozone clima-tology for balloonsonde estimation of total column ozone. *Journal of Geophysical Research-Atmospheres*, 102, 8875–8885.

McPeters, R. D., Miles, T., Flynn, L. E., Wellemeyer, C. G., and Zawodny, J. M. (1994). Comparison of SBUV and SAGE-II ozone profiles—Implications for ozone trends. *Journal of Geophysical Research-Atmospheres*, 99, 20513–20524.

McElroy, C. T. and Kerr, J. B. (1995). Table mountain ozone intercomparison—Brewer ozone spectrophotometer Umkehr observations. *Journal of Geophysical Research-Atmospheres*, 100, 9293–9300.

McKinlay, A. F. and Diffey, B. L. (1987). A reference spectrum for ultraviolet induced erythema in human skin. In *Human Exposure to Ultraviolet Radiation: Risks and Regulations.* W. R. Passchler, B. F. M. Bosnajokovic (eds.), Elsevier, Amsterdam, the Netherlands.

McCormack, J. P. and Hood, L. L. (1997). Modeling the spatial distribution of total ozone in northern hemisphere winter: 1979–1991. *Journal of Geophysical Research-Atmospheres,* 102, 13711–13717.

McKenzie, R., Conner, B., and Bodeker, G. (1999). Increased summertime UV radiation in New Zealand in response to ozone loss. *Science,* 285, 1709–1711.

Miller, A. J. (1996). Comparisons of observed ozone trends and solar effects in the stratosphere through examination of ground-based Umkehr and combined solar backscattered ultraviolet (SBUV) and SBUV 2 satellite data. *Journal of Geophysical Research-Atmospheres,* 101, 9017–9021.

National Aeronautics and Space Administration (NASA) (2011) http://ozonewatch.gsfc.nasa.gov/facts/images/ozone_hole.jpg accessed by June 2011.

National Aeronautics and Space Administration (NASA) (2011) http://jwocky.gsfc.nasa.gov/teacher/basics/dobson.html accessed by June 2011.

Newchurch, M. J., Yang, E. S., Cunnold, D. M., and Reinsel, G. C. (2003). Evidence for slowdown in stratospheric ozone loss: First stage of ozone recovery. *Journal of Geophysical Research,* 108, D164507, doi:10.1029/2003JD003471.

Reinsel, G. C. (1994). Seasonal trend analysis of published ground-based and TOMS total ozone data through 1991. *Journal of Geophysical Research-Atmospheres,* 99, 5449–5464.

Reinsel, G. C., Miller, A. J., Elizabeth, C. W., and Lawrence, E. F. (2005). Trend analysis of total ozone data for turnaround and dynamical contributions. *Journal of Geophysical Research,* 110, D16306, doi:10.1029/2004JD004662.

Reinsel, G. C., Elizabeth, C. W., George, C. T., and Alvin, J. M. (2002). On detection of turnaround and recovery in trend for ozone. *Journal of Geophysical Research,* 107, D104078, doi:10.1029/2001JD000500.

Rusch, D. W., Clancy, R. T., and Bhartia, B. K. (1994). Comparison of satellite measurements of ozone and ozone trends. *Journal of Geophysical Research-Atmospheres,* 99, 20501–20511.

Schnadt, C., Dameris, M., Ponater, M., Hein, R., Grewe, V., and Steil, B. (2002). Interaction of atmospheric chemistry and climate and its impact on stratospheric ozone. *Climate Dynamics,* 18, 501–517.

Skin Cancer Foundation (2011). http://www.skincancer.org/ (accessed June 2011).

Staehelin, J., Schill, H., Högger, B., Viatte, P., Levrat, G., and Gamma, A. (1995). Total ozone observation by sun photometry at Arosa Switzerland. *Optical Engineering,* 34, 1977–1986.

Tolba, M. K., El-Kholy, O. A., El-Hinnawi, E., Holdgate, M. W., McMichael, D. F., and Munn, R. E. (1992). Ozone depletion. In *The World Environment 1972–1992: Two Decades of Challenge,* O. El-Kholy (editor), Chapman and Hall, New York.

Tourpali, K., Tie, X. X., Zerefos, C. S., and Brasseur, G. (1997). Decadal evolution of total ozone decline: Observations and model results. *Journal of Geophysical Research-Atmospheres,* 102, 23955–23962.

United States Department of Agriculture (USDA) (2011). UVB radiation: Definition and characteristics. http://uvb.nrel.colostate.edu/UVB/publications/uvb_primer.pdf (accessed June 2011).

WHO (2002). *Global Solar UV Index: A Practical Guide.* World Health Organization (WHO), World Meteorological Organization (WMO), United Nations Environment Program (UNEP), and International Commission on Non-Ionising Radiation Protection (ICNRP), Geneva, Switzerland, 2002, 28.

World Meteorological Organization (2002). Scientific Assessment of Ozone Depletion: 2002. http://ozone.unep.org/Publications/index.aspl (accessed February 16, 2011).

Ziemke, J. R., Herman, J. R., Stanford, J. L., and Bhartia, P. K. (1998). Total ozone UVB monitoring and forecasting: Impact of clouds and the horizontal resolution of satellite retrievals. *Journal of Geophysical Research-Atmospheres,* 103, 3865–3871.

18 Remote Sensing of Asian Dust Storms

Tang-Huang Lin, Gin-Rong Liu, Si-Chee Tsay,
N. Christina Hsu, and Shih-Jen Huang

CONTENTS

18.1 INTRODUCTION

Global environmental changes have become a key issue facing the international community, where variations in the atmospheric environment have received particular attention. The suspended particles in the air, generally called aerosols, are produced either through natural processes or human activities, which deeply affect the air quality. Atmospheric aerosols not only scatter but also absorb significantly the incoming solar radiation within the entire spectrum (Liou 1980). They are also a provider of cloud condensation nuclei that results in the formation of clouds and may thus affect the Earth's radiation budget and rainfalls. Due to the fact that these suspended particles affect the radiation transfer, the aerosol optical depth (AOD, a measure of aerosol loading), which is an important parameter in the radiation transfer equation, can be used to describe the optical properties of the aerosols. Previous research has shown that the aerosol parameters must be taken into account to obtain a more accurate environmental assessment for the simulations of atmospheric circulations and climate patterns (Hansen and Lacis 1990; Hsu et al. 2000; Hansell et al. 2003). Atmospheric aerosol parameters thus play a pivotal role in accessing the global environmental changes (Liu and Shiu 2001; Li et al. 2004; Gao et al. 2006; Yoshioka et al. 2007).

Excluding background particles, the source of the atmospheric aerosols can be essentially sorted into three main categories in Asia. The first category, which is tied to anthropogenic pollutants, originates from cities or industrial areas. The second category, which is dust particles, is brought forth by the outbreaks of dust storms in arid or semiarid areas (e.g., around the Gobi and Taklimakan deserts), usually associated with cold fronts causing the Siberian and Mongolian cyclonic depressions in the springtime (Qian et al. 2002). The third source is the smoke plumes from open combustion (forest fires, land clearing and control burn through fire, and burning of agricultural wastes). Such types of burning produce large amounts of smoke that seriously affect the air quality. Furthermore, the weather pattern could be influenced due to the change of atmospheric aerosol loading in the transfer of solar radiation.

Satellites have always taken the advantage of observing an area from a high vantage point. In monitoring atmospheric aerosols from a wide-area perspective, satellites can provide timely atmospheric observations. On the other hand, growing concerns about environmental protection and climate change have resulted in a keen interest in the development and applications of remote sensing technologies. Since the 1960s, satellite-borne sensors in the visible (VIS) and infrared (IR) regions have been considered an innovative tool in investigating the atmospheric air quality. Durkee et al. (1991) derived the AOD and aerosol size distribution for global analysis by using the advanced very high resolution radiometer (AVHRR) data with band ratios of the red and near-infrared channels. Holben et al. (1992) selected the dark areas of AVHRR data to retrieve land aerosols. Liu et al. (1996) used Système Probatoire d'Observation de la Terre (SPOT) data to retrieve the aerosol characteristic parameters and surface reflectance for local area. The measurement of the AOD over the ocean with AVHRR data has become a routine operation in the National Oceanic and Atmospheric Administration (NOAA) (Rao et al. 1989; Ignatov et al. 1995). In recent years, the data quality of satellite observations has advanced to a level where the monitoring of the air quality, ecosystem, and environment is considered acceptable (Kaufman et al. 1990; King et al. 1999; Tsay et al. 2009).

For meeting the program objectives, there are generally two kinds of orbits, including geostationary and near polar orbits, which satellites may go by. Satellites at very high altitudes, approximately 36,000 km above the Earth's equator, revolve at the same speed of Earth's rotation as geostationary orbits, which allow the satellites to observe and collect information continuously over specific regions. This type of orbit can monitor weather and cloud patterns covering roughly a quarter area of the Earth's surface with high temporal resolution. The observations of geostationary satellites therefore are greatly suitable for monitoring the dust storm outbreaks. Some of these satellite technologies include Geostationary Meteorological Satellite (GMS), Multifunctional Transport Satellite (MTSAT), and Geostationary

Operational Environmental Satellite (GOES). The other popular orbit is near-polar orbit, which travels in approximately a north–south direction. By associating with the Earth's rotation (west–east), the observations of near-polar orbits can provide most of the Earth's surface during a certain time period. Due to the lower altitude of polar orbits (about 800 km), the near-polar orbiting satellites provide better spatial resolution of imaginary than geostationary satellites. The targets are monitored (atmosphere and Earth's surface) in detail when equipping with multiple spectral bands such as AVHRR and Moderate-Resolution Imaging Spectroradiometer (MODIS). For the surveillance of dust storms, Table 18.1 briefly summarizes those satellites/sensors that have been employed to provide the service of dust storm monitoring directly or indirectly.

In this chapter, both geostationary and near-polar satellite data were covered in order to fully monitor the occurrence and intensity of Asian dust storms, respectively. With observational capabilities for wide areas and high temporal resolutions, geostationary satellites seem to be the best choice for dust storm monitoring around the Asian continent from a qualitative perspective. As shown in Figure 18.1, dust plumes can be observed and monitored based on the different characteristics of clouds, terrain, and dust storms exhibited in geostationary satellite data. GMS-5 data can be applied to detect and monitor Asian dust storm events in near real time. As for the intensity and radiometric characteristics of dust storms, the dust particles can be detected distinctly by their colored appearance (reflectivity) on the multiple spectral sensors onboard earth observing system (EOS) satellites (e.g., MODIS), as Figure 18.2 demonstrates. The radiometric characteristics and intensity of dust weather can be evaluated in association with the retrievals of aerosol properties such as AOD,

TABLE 18.1
Research and Instrument of Satellite/Sensor Related to Monitor/Detect Dust Storm Directly or Indirectly

Reference	Satellite/Sensor	Spectral Band	Concept
Chavez et al. (2002)	GOES/I-M Landsat/TM	VIS	Difference in reflectivity
Liu and Lin (2004)	GMS-5/S-VISSR	VIS, IR	Difference in reflectivity and brightness temperature (BT)
Hu et al. (2007)	FY-2C/S-VISSR	VIS, IR	Brightness temperature difference (BTD) with reflectivity
Ackerman (1989)	NOAA/AVHRR	3.7, 11 μm	Correlation between atmospheric turbidity and BTD
Ackerman (1997); Liu and Liu (2006)	MODIS TERRA/MODIS AQUA/MODIS	8.5, 11, 12 μm	Simulated criteria of dust BTs
Tsolmon et al. (2008)	NOAA/AVHRR TERRA/MODIS	11, 12 μm	Brightness temperature difference index (BTDI)
Huang et al. (2007)	AQUA/MODIS, AMSR-E	11, 12 μm 23.8, 89.0 GHz	BTDI and polarized BTs

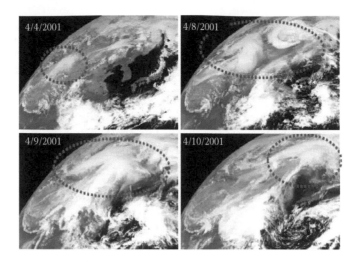

FIGURE 18.1   Visible images from GMS-5 display the dust storm event(s) occurring over Central Asia, which subsequently moved toward the Pacific in early April 2001.

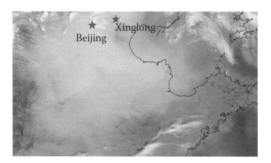

FIGURE 18.2   Dust storm that swept through eastern China was distinctly observed by the color image acquired from Terra/MODIS on March 1, 2009.

Ångström exponent (AE, an indicator of size mode), and single-scattering albedo (SSA, the probability of scattering) (Kauffman et al. 1997; Hsu et al. 2004, 2006; Lin et al. 2011). The optical properties of dust storms retrieved from polar orbiting satellite data are accordingly employed for dust particle identification and dust weather categorization for the quantitative analysis in this chapter.

18.2   GEOSTATIONARY SATELLITE OBSERVATIONS OF DUST STORMS

18.2.1   Overview

In this section, data provided by the Stretched Visible and Infrared Spin Scan Radiometer (GMS-5 S-VISSR; see also Table 18.2) were used to identify dust plumes and monitor the subsequent development of dust storm events. Although other

TABLE 18.2
Bandwidth and Spatial Resolution of GMS-5 S-VISSR Channels

Channels	Bandwidth (μm)	Nadir Spatial Resolution (km)
Visible (VIS)	0.55–0.90	1.25
Water vapor (WV)	6.5–7.0	5.0
Thermal infrared (IR1)	10.5–11.5	5.0
Thermal infrared (IR2)	11.5–12.5	5.0

Source: Japan Meteorological Agency, 1993. Revision of GMS S-VISSR data format. Meteorological Satellite Center, Tokyo, Japan, p. 42.

sensors such as Total Ozone Mapping Spectrometer (TOMS) and ozone monitoring instrument (OMI) provide information regarding air quality and dust coverage over specific locations, their limited orbital observation time makes them hard to provide timely warnings for dust storm occurrences. Since geostationary satellites are capable of providing hourly observations, they offer a higher temporal resolution for dust storm monitoring in time.

The S-VISSR sensor is equipped with one VIS, two thermal infrared (IR1, IR2), and one water vapor (WV) channels. The more the suspended particles become concentrated, the more the observed radiance in the VIS channel is affected when dust storms occur. Based on the different reflecting properties exhibited by clouds, the Earth's surface, and dust storms, we can exploit this characteristic in the detection of dust storms by GMS-5 S-VISSR visible data during their occurrence (Liu et al. 2002). Still, some misdetections might occur around the regions with low cloud cover because of a similar albedo (i.e., apparent reflectance) associated with dust particles and neighborhoods of clouds. All in all, both infrared channels of GMS-5 were used to mitigate misdetections of neighboring dust cloud cluster in regions based on the water vapor content (Liu and Lin 2004). In addition, an automatic operation system associated with satellite receiving system was constructed for dust storm monitoring in near real time with a positive potential for dust storm warning and forecasting.

The albedo (apparent reflectance) observed by the VIS channel was readily obtained when a large concentration of suspended particles were present in the atmosphere. Data from the VIS channel were then applied to detect the occurrence of dust storms from the environmental changes they produced. The characteristics of different geographical environments can be distinguished by employing data from the GMS satellite in the VIS and IR channels, as shown in Figure 18.3 (Liu et al. 2002). The VIS and IR images of different targets, including a clear sea surface, a clear land, and dusty areas along with cloudy areas, may be illuminated together for a better differentiation (Figure 18.3a). The size of each sampling area was 50 × 50 pixels (e.g., about 250 × 250 km). Digital count scatterplots for the VIS and IR channels clearly identified distinct environmental aspects, except when the data were various and vague (Figure 18.3b), indicating that areas covered by dust and cloud plumes can be detected by comparing the clear sky imagery (e.g., background data). Once the background data set (dust storm free) over the region of interest can be

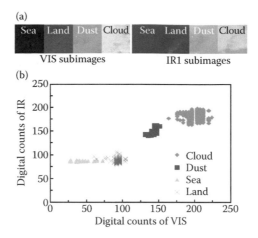

FIGURE 18.3 (a) Images of clear sea surface, clear land, dusty, and cloudy areas acquired by the GMS S-VISSR VIS and IR1 channels, 02:40Z April 7, 2001. (b) Scatterplots of digital counts of the objects in (a). (From Liu, G.-R. et al., 2002, *Asian Journal of Geoinformatics*, 2(4), 3–8.)

constructed before and after the event, identification of the dust storm event would be much easier. In order to avoid the effect of the solar zenith angle, only data that were acquired at the same local time on each date were used for the dust storm detection in this study.

18.2.2   DUST STORM DETECTING WITH GMS-5 DATA

Two sets of GMS-5 S-VISSR images were acquired from January 1, 4, and 10, 2002. The images covered much of Asia and were used to examine a dust storm event. The data set acquired from January 1 to 4 (with no dust storm formation whatsoever) was used to construct the normal background data set (dust-free image). The data set collected on January 10 was applied to monitor the formation of a dust storm and delineate the influential regions with the aid of the background data set. The preliminary results for areas affected by dust particles are shown in Figure 18.4a. In order to delineate the results of the dust storms detected by the GMS-5 data, the sand index was used for the dust intensity, which is defined by the variance of surface albedo in the VIS channel given that the dust-free image is available (see Table 18.3; Liu and Lin 2004). In Figure 18.4a, a green color may be used to depict those areas unaffected by dust storms, indicating areas that the albedo varied under 0.027. Yellow, dark yellow, magenta, red, and purple colors represent variance of albedo in the VIS channel greater than 0.027, indicating areas affected by different dust intensity (sand index). Also, the white regions represent areas with cloud coverage. The preliminary results showed that a lot of purple areas (e.g., dust plumes) appeared over the western Pacific Ocean. January 10 was nearly dust-free, except for regions with yellow color over the Yellow Sea, indicating that the particle concentration was above normal. The neighborhoods of cloud clusters that appeared in the purple regions over

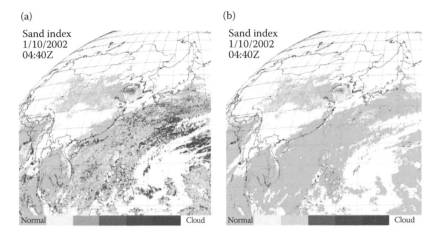

FIGURE 18.4 Comparison of dust plume detection from GMS S-VISSR data without (a) and with (b) the water vapor content, January 10, 2002.

the western Pacific Ocean were mistakenly identified in this dust detection analysis (Figure 18.4a). The main contributors to the misdetection included similar reflection characteristics of dust particles and cloud neighborhoods as well as a lower dynamic range of the S-VISSR in the visible channel (6 bits). Consequently, knowing how to discriminate dust particles from the neighborhoods of cloud clusters plays a key role in error correction. Based on the intrinsic character of dust formation, the water vapor content seems to be the manifest discrepancy between dust and cloud particles, and can thus be used for such correction (Liu and Lin 2004).

Many efforts have been made to estimate the total amount of water vapor over land and ocean surfaces using NOAA/AVHRR thermal infrared and DMSP/SSM/I microwave remote sensing data (Eck and Holben 1994; Sobrino et al. 1999). The estimate was based on a linear relationship exhibited between the total atmospheric water vapor and the brightness temperature difference on split-window channels (e.g., channels of 11 and 12 μm) because the difference of brightness temperatures on split-window channels provides information about the amount of total water vapor.

TABLE 18.3

Definition of Sand Index with Variance of Surface Albedo from Visible Channel of GMS-5

Sand Index	Variance in Surface Albedo	Expressive Color
0	<0.027	Green
1	0.027–0.038	Yellow
2	0.038–0.065	Dark yellow
3	0.065–0.095	Magenta
4	0.095–0.126	Red
5	>0.126	Purple

The split-window channels of GMS-5 data (IR1 and IR2) were used to distinguish between dust objects and neighborhoods of clouds. After examination, the criterion for brightness temperature differences and the characteristics of cloud pixels on VIS and IR channels were used for cloud recognition. Figure 18.4b shows the improvement of results with the information of water vapor content. The original misdetections depicted in purple (Figure 18.4a) have nearly been corrected for areas over the western Pacific Ocean while the actual region affected by dust over the Yellow Sea was still highlighted. These corrections clearly demonstrated that use of both infrared channels of GMS-5 may provide a greater degree of accuracy for the separation of dust pieces and cloud neighborhoods.

To verify the results of dust detection by GMS-5 observations, a comparison was made between the sand index and the AOD retrieved from the MODIS data of the Terra satellite (Tanré et al. 1999; Kaufman et al. 2002). MODIS is a key instrument on board the Terra (EOS AM-1) and Aqua (EOS PM-1) satellites. Terra's orbit around the Earth passes from north to south across the equator in the morning, while Aqua passes from south to north over the equator in the afternoon. Terra MODIS and Aqua MODIS observe the Earth's entire surface every 1 to 2 days, acquiring data in 36 spectral bands (from 0.4 to 14.4 μm). These data can improve our understanding of the global dynamics and processes occurring over land, sea, and the lower atmosphere. MODIS is ready to play a vital role in the development of validated, interactive Earth system models at the global scale.

A dust storm event, which affected a region over eastern Mainland China from March 4 through 9, was observed by GMS-5 and MODIS near the same time on March 7, 2002 and was selected for comparison. The GMS-5 data, which represent the dust storm that affected the region around the East China Sea, may be depicted fully (Figure 18.5a) as the AOD can be observed from MODIS data at about the same time (Figure 18.5b). The black color represents areas with no retrieval because of cloud coverage or a bright surface, such as a desert or beach. The retrieved AOD from MODIS might be affected by particles of sulfates, industrial pollution, biomass

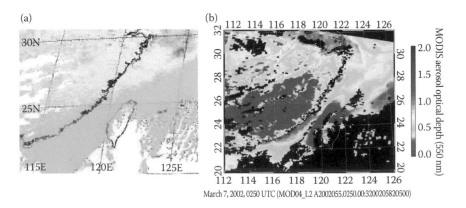

FIGURE 18.5   (a) Dust detection from GMS-5 data at 02:40Z on March 7, 2002. (b) Aerosol optical depth retrieved from Terra/MODIS at 02:50Z on March 7, 2002. (From GSFC/NASA, USA.)

burning, and dust. Still, AOD product from MODIS can provide dust information. In Figure 18.5b, the AOD distribution illustrates the AOD amount that was above normal around the eastern China areas, indicating that they were affected by dust (Liu and Lin 2004). The patterns of dust detected by GMS-5 and AOD retrieved by MODIS data for the study region can be generally correlated well, except for southwestern Taiwan where the higher AOD value might be caused by industrial pollution and/or biomass burning. The GMS-5 and MODIS data acquired on March 17, 2002 provided further verification of the proposed method, with similar distributions demonstrated once again (see also Figure 18.6). It should be noted that different AOD values associated with the represented colors vary twice as much as usual values (see Figure 18.6a and b). The results of this examination demonstrated the strong potential of using GMS-5 data to improve dust detection, as verified by data derived from the MODIS observations shown below.

During March 2002, several strong dust storms occurred over the Mongolian region, seriously affecting northeast China and the Korean peninsula. GMS-5 data were used to detect the progression of the dust storms in the affected regions (Figure 18.7) (Liu and Lin 2004). As usual, the dust-free areas are represented by green color, while the areas covered by clouds are represented by white colors. Yellow, orange, red, and purple colors depict areas affected by the dust with different intensities. Although there were a lot of cloudy areas, the regions affected by the dust storm were evidently detected by referring back to clear sky conditions during this period. On March 22, the dust storm could not be completely monitored at 02:40Z because of cloud coverage, as demonstrated in Figure 18.7a, although detection was significantly enhanced at 06:40Z for areas of eastern China and the Yellow Sea, as demonstrated by Figure 18.7c. On March 24, the dust storm constantly affected these areas, subsequently spreading out to the Korean Peninsula region. This influence is claimed to be the strongest dust storm recorded around the Korean peninsula for the last few years (Figure 18.7f and i). Fortunately, an existing cold front and rainfall around Taiwan significantly decreased the intensity of the dust storm. In this case study, the efficacy of the GMS-5 S-VISSR data for timely dust storm monitoring was clearly confirmed.

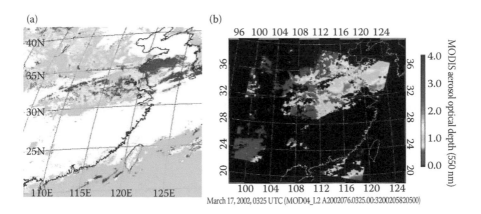

FIGURE 18.6   Same measurements as used in Figure 18.5, but at 04:40Z on March 17, 2002.

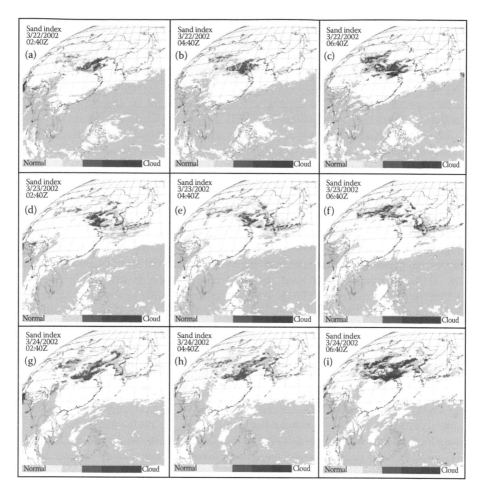

FIGURE 18.7   Results of dust storm detection from GMS-5 data at (a) 02:40Z, (b) 04:40Z, (c) 06:40Z, on March 22, (d) 02:40Z, (e) 04:40Z, (f) 06:40Z, on March 23, and (g) 02:40Z, (h) 04:40Z, (i) 06:40Z, on March 24, 2002.

The dust information in areas where the cold front was passing was not observed because of the constraints of the remote sensing channels. GMS-5 channels do not provide information under cloud layers, and the low dynamic range of S-VISSR in the visible channel provides low sensibility for reflective differences between pure cloud and dust-cloud mixed pixels. For these reasons, information on dust over cloud-covered areas was difficult to discern using the approach developed in this study. In addition, this approach can only be applied for the daytime when the visible channel is applicable.

18.2.3 OPERATIONAL SYSTEM OF DUST STORM MONITORING

Because dust storms possess quick mobility and cause significant damage, providing timely information using a sound operational system (i.e., early warning system) about a potential dust storm requires urgent attention, especially for regions around the source areas of the dust storm (e.g., Central Asia). The speed and impact of dust storms necessitate the construction of such an automatic monitoring system for detecting dust storm formations, their progression, and the areas affected. Based on the analysis in the previous section, the operational mode of dust storm detection and subsequent monitoring may be established with the aid of geostationary satellite receiving systems. With the database of Asian dust storms constructed automatically on a website, the information of dust storm occurrences can be provided in near real time (i.e., an example can be found at http://140.115.111.41/pastel/mtsat/dust.html, which is operated by the Meteorological Satellite Laboratory of Center for Space and Remote Sensing Research, National Central University, Taiwan). Such an automatic system for dust storm detection and monitoring may be even more powerful if it can be expanded into a dust storm forecasting system.

To enhance the current procedures of dust storm detection as well as monitoring air quality assessment, quantitative analysis of dust storms is necessary. The strongest dust storm occurred near the area of Mongolia on April 7, 2001 (Xinhua News Agency 2001b), which was used for the quantitative analysis in this chapter. The aerosol index (AI) (Hsu et al. 1996; Torres et al. 1998; Washington et al. 2003), produced by using the data collected from the Earth Probe TOMS, is operational for global dust storm detection. It has been used frequently through a free website (http://TOMS.gsfc.nasa.gov). The products of TOMS AI were examined for quantitative analysis of a dust storm in this study. This movement enables us to depict the distribution of AI retrieved from TOMS data on April 7, 2001 for areas affected by the dust storm around northeastern China and Korea (Figure 18.8a). A similar dust distribution pattern was also observed from the GMS-5 data on the same date, as illustrated in Figure 18.8b. The dust detection product of GMS-5 was resampled from 5 km to 1° spatial resolution for quantitative comparisons with TOMS AI (1° × 1° in latitude and longitude). Afterward, the data of TOMS and GMS-5 with the same spatial resolution over the areas around northeastern China and the Yellow

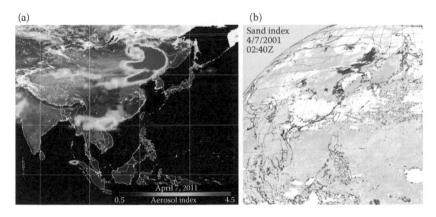

FIGURE 18.8 (a) Spatial distribution of aerosol index from the Earth Probe TOMS data over eastern Asia on April 7, 2001 (from Dr. N. C. Hsu). (b) Sand index of dust storm from GMS-5 data at 02:40Z on April 7, 2001.

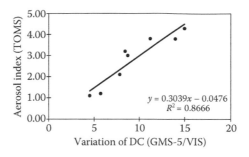

FIGURE 18.9 Relationship between TOMS AI and variation of digital count (DC) in GMS-5 visible channel of the dust plumes on April 7, 2001. (From Liu, G.-R. and Lin, T.-H., 2004, *Terrestrial, Atmospheric and Oceanic Sciences*, 15, 825–837.)

Sea (i.e., affected by the dust storm) were used for analyzing the correlation between AI and variance in albedo over dusty areas. Such a comparison led to a good linear relationship between AI and the digital count variation in the VIS channel of GMS-5 (Figure 18.9). This finding indicates a great potential of geostationary satellite data (GMS-5) that could provide the analysis for the occurrence and intensity of dust storms. This can help with the monitoring and warning in a timely manner. More data sets should be collected and examined for supporting the real-world operational applications.

18.2.4 DISCUSSION

With the optical characteristics of dust plumes (and associated particles) in visible and thermal infrared spectral bands, the capability of the GMS-5 S-VISSR data and its effectiveness for a timely dust storm monitoring was confirmed. This can be assured by the fact that potential errors in dust particle recognition due to cloud clusters were corrected successfully by using both infrared channels of the S-VISSR. As a result of this correction, a dust storm can be detected and monitored more accurately with GMS-5 data. This approach is practical and effective such that it is worthwhile to explore more investigations of dust storm research along this line in the future. Although some satellite data such as TOMS and OMI can monitor the dust storms and the air quality with its specified channels, the orbital observation window and repeat cycle makes them hard to provide a near real-time warning message when a dust storm does occur. On the contrary, the hourly geostationary satellite observations provide a better data set in terms of the temporal resolution.

In addition, to assist policy makers in making accurate decisions concerning the protection of our environment, the next-generation automatic monitoring system may be constructed by combining a geostationary satellite receiving system with TOMS data to expedite the provision of information about dust storm occurrences, progression, and the extent of affected areas. Still, more data of dust storm cases, together with other satellite data, such as MTSAT or MODIS, should be collected and analyzed to enhance the detection of intensity levels for dust storms and facilitate the development for dust storm warnings.

18.3 MONITORING OF DUST STORMS WITH OCI OVER THE OCEAN

The OCI sensor is placed onboard the FORMOSAT-1 (formerly known as ROCSAT-1) satellite, which was launched into a low-earth orbit in January 1999. It has an altitude of 600 km with an inclination angle of 35° and has an orbital period of 97 minutes. It beams downlink telemetry signals to the ground receiving station six times a day. FORMOSAT-1 was designed mainly to carry out the scientific experiments related to the oceanic and atmospheric environment, which include three main payloads: Ocean Color Imager (OCI), Imospheric Plasma Electrodynamics Instrument (IPEI), and Experiment Communication Package (ECP). The primary mission of the OCI sensor is to provide observational data for the analysis of the marine and atmospheric environment (Yang 1995). The OCI sensor provides six channels at 443, 490, 512, 555, 670, and 865 nm, where an additional channel centered at 555 nm is used to evaluate the condition of radiometric decay with time for the former six channels. The OCI is a push-broom scanner with 700-km swath width and 896 pixels in each line. The spatial resolution is 400 m for the 64 pixels located at the center and 800 m for the remaining ones (Table 18.4).

As mentioned before, Asian dust storms oftentimes occur over the Mongolia desert and the arid regions of northwestern China virtually each year (Qiu and Yang 2000). From January through February 2001 alone, three major dust storms were recorded over northwestern China (Xinhua News Agency 2001a). Past records revealed that these events occurred much earlier than normal and were more frequent and powerful than before. The dust storms were often induced by the strong south-moving Siberian cold air mass, leading to significant degeneration

TABLE 18.4
Specifications of FORMOSAT-1/OCI

Parameters	Description
Inclination	35°
Altitude (km)	600
Period (min)	96.6
Spectral bands (nm)	B1 443–453
	B2 480–500
	B3 500–520
	B4 545–565
	B5 660–680
	B6 845–885
Ground resolution (m^2)	800 × 800
Swath width (km)	702
Total pixels per scan line	896
Observation method	Push broom
Acquisition time (local)	9:00–15:00
Tilt	No
Launch date	Jan. 27, 1999

of the air quality over the western Pacific region including China, Korea, Japan, and Taiwan. The residual impact may even affect the air quality over portions of North America. It may be inferred that the occurrence of the dust storms around the arid areas of northwestern China can affect regions on a global scale. Finding ways to monitor and evaluate their influences on the Pacific region, especially through satellite-borne observations, is becoming an increasingly important topic. In this section, the FORMOSAT-1/OCI images were used to estimate the AOD and delineate the time variation of the AOD over the Yellow Sea (Huang et al. 2002).

18.3.1 METHODOLOGY

The observed radiance/reflection of visible bands from the satellite optical sensor is primarily affected by the atmospheric scattering effects of molecular and aerosol particles on the AOD retrievals. The occurrence of a dust storm can thus be detected through the variation of the atmosphere's AOD. Calculations of the AOD were conducted by monitoring the radiometry of the atmosphere from images acquired by the OCI. In general, there are two typical types of atmospheric scattering. These processes are known as Rayleigh and Mie scattering. Mie scattering is caused by the atmospheric aerosols of which the sizes are larger than the wavelength of incident light, such as dust and smoke aerosols suspended in the lower portion of the atmosphere. On the other hand, Rayleigh scattering occurs when the sizes of particles causing the scattering are smaller in size than the incident wavelengths (atmospheric molecules). This type of scattering is wavelength-dependent. In order to quantitatively compare the variations before and after the dust storm event, effects of the Mie scattering and Rayleigh scattering were separated from the total radiance observed by the OCI in this study.

The single-scattering approximation was employed (Buglia 1986) under the consideration of simplifying the mathematical modeling and reducing the calculation time. The total radiance in the single-scattering mode observed by the OCI at the top of the atmosphere in the visible and near-infrared (NIR) regions can be simplified by

$$L_a(\lambda) = \frac{\omega_a(\lambda)\tau_a(\lambda)F_0'(\lambda)P_a}{4\pi\cos\theta},$$ (18.1)

where ω_a is the aerosol single-scattering albedo, λ is the wavelength, τ_a is the aerosol optical depth, and P_a is a parameter that is dependent on the scattering phase function. The Fresnel reflectance θ is the zenith angle. F_0' is the instantaneous extraterrestrial solar irradiance after it is extinguished by the atmosphere, such as

$$F_0'(\lambda) = F_0(\lambda)T(\lambda)$$

$$F_0(\lambda) = \overline{F}_0(\lambda)\left[1 + 0.0167\cos\frac{2\pi(D-3)}{365}\right],$$ (18.2)

where T is the total transmittance, $\overline{F}_0$ is the extraterrestrial average solar irradiance, and F_0 is the extraterrestrial solar irradiance on Julian day D. In our computation, the Henyey–Greenstein aerosol scattering phase function (Gordon and Castano 1987) was used for marine aerosol types. The multiple aerosol scattering mode was further assessed by using a linear transformation (Gordon and Wang 1994). The radiance L was changed to the reflectance ρ_{sa} such as

$$\rho_{sa}(\lambda) = \frac{\pi L_a(\lambda)}{F_0 \cos\theta_0} \tag{18.3}$$

$$\rho_{ma}(\lambda) = I(\lambda) + S(\lambda)\rho_{sa}(\lambda), \tag{18.4}$$

where ρ_{sa} is the reflectance of the single aerosol scattering, ρ_{ma} is the reflectance of multiple aerosol scattering, and θ_0 is the solar zenith angle. $I(\lambda)$ and $S(\lambda)$ are the intercept and the slope, respectively. The parameters were each obtained from the low resolution transmission (LOWTRAN-7) model (Kneizys et al. 1988). As for the estimation of the aerosol optical depth, $\tau_a(\lambda)$, the turbidity formula (Ångström 1961) was employed as such:

$$\tau_a(\lambda) = \beta \cdot \lambda^{-\alpha}, \tag{18.5}$$

where α, Ångström exponent (AE), is commonly used to describe the modes of aerosol size distribution and related to the Junge exponent (AE + 2.0), which describes the slope of the aerosol size distribution, and β is the Ångström turbidity coefficient representing the aerosol concentration, which has a positive correlation with the atmospheric turbidity.

The aerosol albedo of the single scattering, $\omega_a(\lambda)$, was estimated by (Bird and Riordan 1986; Gregg and Carder 1990)

$$\omega_a(0.55\mu m) = (-0.0032AM + 0.972)\exp(3.06 \times 10^{-4}RH) \tag{18.6}$$

$$\omega_a(\lambda) = 0.55\omega_a \cdot \exp\left[-0.095\left[\ln\left(\lambda/0.4\right)\right]^2 + 0.0096342\right], \tag{18.7}$$

where RH (in percentage) is the relative humidity of the atmosphere obtained from the Comprehensive Ocean-Atmosphere Data Set (COADS) of the National Center for Atmospheric Research (NCAR). AM is the air mass character (an integer from 1 to 10). Generally speaking, AM can be used as an indicator for quantifying the aerosol type over the ocean, producing a scale from 1 to 10, where 1 stands for the open ocean (clear air quality) and 10 for coastal areas (rather poor air quality). The AM character is strongly influenced by natural or industrial aerosols originating from neighboring continents (Gathman 1983).

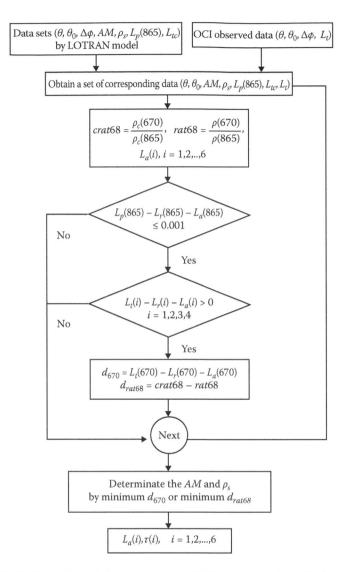

FIGURE 18.10   Procedure of air mass character (AM) and sea surface albedo (ρ_s) determination for retrieving AOD (From Huang, S.-J. et al., 2002, *Journal of Marine Science and Technology*, 10(2), 92–97.)

In determining the AM values during the time of acquisition, this study employed the LOWTRAN-7 model to simulate the total irradiance observation by the six OCI channels under different observation and atmospheric conditions. The observation geometry and times of the observation, sea surface albedo, and different air mass characters were compiled into a database. The channels at 670 and 865 nm were then used to aid the assessment of the aerosol scattering effect and the air mass character parameter in the LOWTRAN model. The total radiance observed by the

FORMOSAT-1 sensor was then compared with the database. The sea surface albedo and *AM* values that matched the database were chosen as the observation conditions during the satellite flyby. Such procedures for the determination of *AM* value and sea surface albedo from OCI observations and the simulations of the LOWTRAN model can be summarized in Figure 18.10, leading to the calculation of AOD associated with the aerosols.

18.3.2 Dust Storm Detection

Generally, the Asian dust storms that originated from northwestern China and Mongolia moved southeastward with the winter–spring circulations, which significantly deteriorated the air quality over the eastern Asian area. Parts of North America were also affected as some of the particles were carried along the easterly stream (Ichikawa and Fujita 1995; Arndt et al. 1998; VanCuren and Cahill 2002; Hsu et al. 2006; Tsay et al. 2009). In this section, a $1° \times 1°$ area in latitude and longitude near the Yellow Sea centered at the location (35.5°N, 122.0°E) was selected as the observation area (see also the region within the white square in Figure 18.11a) to investigate the dust storm effects on the AOD values.

Due to the limited operational time and unique orbit, the OCI was unable to maintain a high revisit rate. In addition, sun-glint and cloudy pixels had to be excluded in advance. As a result, gathering a set of high temporal resolution images of the OCI was not easy. Eventually, a collection of nine qualified OCI images was selected and processed by our atmospheric correction and AOD retrieval algorithm (Huang et al. 2002). Before processing, we also performed a cloud-masking procedure for any cloud-contaminated pixels.

The satellite observed radiance is contributed both by scattering from atmospheric molecules or aerosols and the reflectance from the Earth's surface. The actual radiance component contributed by the dust particles may be enhanced by removing the radiance component of the atmospheric molecules. After the Rayleigh scattering

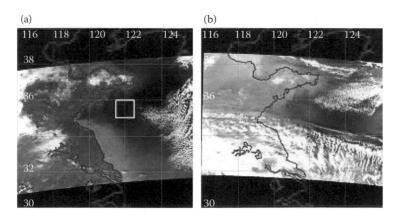

FIGURE 18.11 Comparison of OCI pseudoimages before (a) 07:26Z, January 29, 2001 (before dust storm), and after (b) 06:49Z, February 2, 2001, the dust storm event.

was eliminated, the dust storm effect could be qualitatively demonstrated. The area of extent in Figure 18.11a shows a clear image acquired before a major dust storm was recorded on January 31, 2001. The true-color image was composed by the OCI channels at 443 nm (B), 555 nm (G), and 670 nm (R). For a better visualization, the cloudy pixels were not masked in the image. The coverage of this image included the areas of Shan-Tung Peninsula and the East China Sea. Both the textures of the land topography and the borders of the land/sea border can be clearly identified. The image also shows a darker tone, revealing that the air was clearer because the scattering effect was small over the land and ocean area. Contrary to the former image, the image in Figure 18.11b, which was acquired two days after the dust storm event, indicates a brighter tone in the image that revealed a much stronger scattering effect in the aerosol component. From these Rayleigh corrected images, the effects of the particles from the dust storms on the atmosphere can be clearly demonstrated (Huang et al. 2002).

We further investigated the spatial variation of the radiance along a latitudinal line. In this case, a strip along the 37°N latitude was analyzed. In order to illustrate clearly the radiance variance with the longitude, the mean radiance was obtained from a succession of 0.1° × 0.1° windows along the 37°N latitude between longitude 110°E to 135°E. Figure 18.12 depicted the Mie scattering radiance observed at a 555-nm channel before and after a dust storm event overpassed. Each dot in the figure represents the mean radiance of a 0.1° window area. As mentioned earlier, the radiance rose significantly after the dust storm event. This could be observed over the land or ocean areas under the clear sky circumstances. Although the cloudy pixels were not excluded in this step, the radiance variation could still be seen. The comparison of observed radiances in Figure 18.12 demonstrated that the Mie scattering radiance greatly increased after the dust storm event (Huang et al. 2002). Detailed comparison with cloud-free pixels revealed that the sea surface had a low radiance count because the land surface reflectance was greater than that on the sea surface, around 110°E to 122.4°E (Mainland China) and 126.3°E to 129.4°E (Korea). By comparing the radiance before and after the dust storm event, dust particles could be more clearly distinguished from the ocean area.

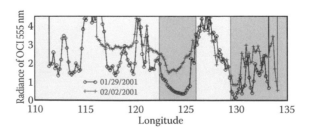

FIGURE 18.12 Comparison of the radiance at 555 nm along the 37°N latitude observed by the OCI, excluding the Rayleigh scattering component before (circle line)/after (plus line) dust storm event. Bright background denotes the land area, while dark background represents the ocean area.

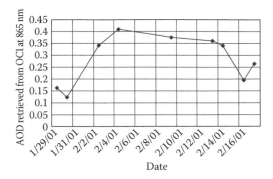

FIGURE 18.13   Variation of the mean AOD over the observational area. (From Huang, S.-J. et al., 2002, *Journal of Marine Science and Technology*, 10(2), 92–97.)

Finally, the temporal variations of the effects produced by the dust storm particles of the study area, as marked in Figure 18.11a, were analyzed. The AOD values retrieved from the OCI observation from the end of January to the middle of February 2001 are shown in Figure 18.13. In this figure, all cloud-contaminated pixels were screened before comparing the actual AOD values contributed by aerosol particles. The results illustrated a drastic AOD increase after January 30, 2001 because a dust storm occurred in Mongolia on January 31, 2001. The AOD value attained its highest level on February 4 and dropped very slowly afterward. During our study period, no major forest fire or volcanic eruptions were reported. This reveals clearly that the dust particles were the main contributor to the AOD changes. In the result, the AOD value rose almost four times higher (from 0.12 up to 0.41). The high AOD indicated high aerosol particle concentration, which was composed mostly of sand or dust particles that were spread into the sky by the strong turbulent winds in Mongolia and then carried out by the Asia winter monsoon.

18.3.3   Discussion

In this section, a scheme for monitoring Asian dust storms over the ocean with the FORMOSAT-1/OCI data was proposed. The air quality can be distinguished from the true-color OCI images in eastern Asia. The true-color images show a deeper tone under clear sky conditions and brighter tones after the eruption of the dust storm. The stronger scattering effect is caused by the suspended particles from the dust storm. It may also be inferred that the dust storm is easier to be distinguished over the ocean when the Rayleigh effect is excluded because the ocean surface is more uniform and shows a lower reflectance than over the land.

The AOD values over the Yellow Sea produced a range from 0.12 to 0.41 during our study period. They rose almost four times higher when the dust storm occurred. The high AOD values exhibit the high aerosol particle concentration suspended in the air. We can therefore deduce that the OCI data can be used to delineate the region under influence and also measure the extent of the Asian dust storms. A suitable algorithm to retrieve the AOD over the land will be developed in the near feature.

18.4   DUST WEATHER CATEGORIZATION WITH MODIS DATA

Aside from the important research in dust storm intensity, extensive investigation with regard to the source areas, frequency, pathway, condition, and trend of Asian dust storms has been performed over the past two decades as well (Tsay et al. 2009). In most previous studies, the intensity of dust weather has often been categorized using standard visibility observations (CCMB 1979) from ground-based weather stations. Still, it is difficult to ensure that the measurement standard is consistent among the hundreds of stations located in this area. Since satellite remote sensing can retrieve relevant atmospheric and/or surface parameters more consistently over large geographic areas, it is a more viable tool in obtaining a more spatially and temporally coherent categorization of dust weather.

Whereas improvements continue to be made in techniques that retrieve dust properties near source regions from satellite measurements, comprehensive analysis of these observations provides scientists with a better understanding of the spatial distribution, frequency, and trend of dust storms (Wang et al. 2000; Hsu et al. 2004, 2006; Remer et al. 2005; Tsay et al. 2009). Given the different observational aspects of the problem (i.e., detailed description can be found in the subsequent subsection), establishing a correlation between the remotely sensed aerosol properties and surface visibility (*VIS*) for dust weather categorization is not an easy task. This has been a subject of many ongoing studies (Qiu and Yang 2000; Qiu 2003). In this section, satellite retrieved aerosol properties such as optical depth, particle size, and turbidity coefficient were analyzed with visibility data observed from ground-based weather stations. Their respective correlation was subsequently used to categorize the corresponding dust weather (Lin et al. 2011).

18.4.1   RETRIEVALS OF AEROSOL PROPERTIES

Due to the complexity in the composition and variability of aerosols, it is difficult to employ a simple and straightforward physical model in calculating the Ångström exponential coefficient (α) with time in the absence of actual data (Cachorro et al. 1987; Lin et al. 2005). However, Equation 18.5 can be expressed as Equation 18.8 when there are two spectral bands available:

$$\alpha = \frac{\ln(\tau_{a\lambda 1}/\tau_{a\lambda 2})}{\ln(\lambda_2/\lambda_1)}, \tag{18.8}$$

where λ_1 and λ_2 represent wavelengths of two spectral bands and $\tau_{a\lambda}$ is their corresponding aerosol optical depth. The variable α can thus be calculated from multi-wavelengths and used for differentiating between coarse and fine particles.

In this section, the newly developed Deep Blue algorithm by Hsu et al. (2004) was employed to characterize the aerosol properties over and near the dust storm source regions. The surface reflectance for a given pixel was determined from a clear-scene database based on its geolocation. The reflectances measured at the 412-, 470-, and 650-nm channels (from MODIS) were subsequently compared to the radiances in a lookup table (i.e., a data table) for different solar zenith, satellite zenith, and relative azimuth angles as well as surface reflectance, AOD, and SSA. Finally, the dust particles can be distinguished from the fine mode pollution particles based on the Ångström exponent (α) derived by the Deep Blue algorithm.

18.4.2 CORRELATION OF AOD, VISIBILITY, AND DUST WEATHER INTENSITY

The AOD is a measure of the amount of photon (sunlight) scattering and/or absorption by airborne particles when passing through a vertical/slant column of the atmosphere. In general, a high AOD value denotes a high concentration of aerosols suspended in the atmosphere, and vice versa. In the field of meteorology, visibility is defined as a measurement of the longest distance at which a dark object can be clearly discerned in the horizontal direction. Visibility hence is a function of aerosols and humidity, including type of aerosols, particle size distribution, concentration, and water vapor content. Despite the different observed directions, the long-term and short-term surface measurements still revealed an inverse relationship between the AOD and visibility (Solis 1999).

The experiential formula of visibility (Middleton 1952) is a function of the horizontal extinction coefficient at 550 nm. Under the assumption that the aerosol extinction coefficient decreases exponentially with the height (Z) and the decreasing rate of the aerosol extinction coefficient (Z_a) is a constant, the AOD (at 750 nm) can be correlated with the visibility (Qiu 2003):

$$\tau_{a750} = \frac{3.912}{VIS}0.733^\alpha Z_a, \qquad (18.9)$$

where Z_a and VIS are in kilometers. Equation 18.9 shows that the AOD is a function of VIS and α. The same result has been derived by Bäumer et al. (2008). Since the particle size distributions usually vary significantly both spatially and temporally (Cachorro et al. 1987), constructing the relationship between AOD and VIS is complex. Furthermore, Equation 18.9 was derived for cases most relevant to urban pollution residing in the boundary layer and may not be suitable for dusty weather conditions. It should also be noted that the Rayleigh extinction is neglected in Equation 18.9. The visibility measurements from the surface are affected not only by aerosol particles but also by atmospheric molecules (McCartney 1976). It would thus be better to also take into account the Rayleigh extinction from the atmospheric molecules.

In another study, the correlation between the visibility and surface aerosol concentration (PM10, particle matter less than 10 μm in aerodynamic diameter) in the Pearl River Delta Region under heavy haze was examined by ground stations (Deng et al. 2008). The correlation coefficient was roughly 0.8 based on an exponential function under a high aerosol concentration that may have been caused by anthropogenic pollutants. Their result suggests that the relationship between the AOD and visibility appears to be exponential rather than linear, but somehow differs with the behavior of dust particles revealed by remotely sensed data. This topic therefore attempts to shed light on the association between the station-measured visibility data and satellite-retrieved AOD during dust weather events. Although many aforementioned studies (Solis 1999; Qiu and Yang 2000; Qiu 2003; Deng et al. 2008) have been made in examining the correlation between the AOD and visibility, they are mostly not under dusty weather conditions.

Four categories of dust weather intensity are specified for the daily reports in the mainland by the China Central Meteorological Bureau. The category of dusty weather is defined based on the reduction in atmospheric visibility shown below (CCMB 1979; Qian et al. 2002).

1. *Dust haze.* Dust particles float up from the ground through winds and are suspended homogeneously in the air. The observed near surface visibility may be about 10 km or less.
2. *Blowing dust.* More dust particles are raised above ground by strong winds compared to those under the dust haze condition in category 1, and the visibility is reduced to 1 to 10 km.
3. *Dust storms.* The dust particles are strongly transported into the air from the surface by storm events or turbulent winds, and the visibility is reduced to 1 km or less.
4. *Dust devils.* The horizontal visibility is less than 0.5 km during such dust storm events. Since dust devil events are rarely observed from satellite, only the first three categories of dust weather are investigated in this chapter. After constructing the relationship between the AOD and visibility, the intensity of dust events will be discerned and categorized by satellite data based on the aforementioned definition.

In order to identify the dust weather characteristics in eastern Asia, a region ranging from 20°N to 60°N and from 70°E to 130°E was selected for study area (see Figure 18.16). Two kinds of data sets are used for the dust weather analysis. The first contains satellite retrieval data from the Aqua/MODIS atmosphere daily level 3 product (available at http://ladsweb.nascom.nasa.gov) covering the regions of interest in this study during springtime from 2003 to 2007. The second is the ground-measured visibility data observed every 3 hours from about 200 ground stations located within the study area (see Figure 18.16) during April 2006. Our analysis also utilized the pertinent dust weather reports, which were mainly used for verifying the dust weather classification result by means of satellite remote sensing.

18.4.3 AEROSOL PROPERTIES DURING SPRINGTIME FROM SATELLITE DATA

The investigation period was focused on springtime, which usually marks the highest level of dust storm activity. As stated in Section 18.3.1, aerosol properties such as the AOD, AE, and SSA can be retrieved from satellite data by using the newly developed Deep Blue approach. Figure 18.14 shows the AOD and AE distributions retrieved from the MODIS/Aqua data during the period from February 1 to May 31, 2007. The AOD data displayed in Figure 18.14a are the composite products of Deep Blue (over land) and Dark Target approaches over ocean (Kaufman et al. 1997). There are two regions with high AOD values in the study area. One is near the northern Taklimakan Desert and the other is around the city of Nanjing. The AOD values over these two regions were quite similar but the AE values were quite different as shown by Figure 18.14b. The distinct size distribution indicates that the particle types are different in these two regions.

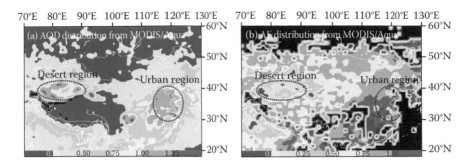

FIGURE 18.14 (a) Spatial distribution of aerosol AOD and (b) AE retrieved from the MODIS/Aqua data during February 1 to May 31, 2007.

In general, the AOD and AE parameters can be used to differentiate between coarse and fine particles (Hsu et al. 2006; Lin et al. 2011). The dust plumes seen in the satellite true color images are commonly associated with increasing AOD and small AE values in the retrieved satellite products, implying an increase in the amount of coarse particles. On the other hand, when both the AOD and AE values increase, there is a higher concentration of fine particles. Accordingly, the AE values can be used to distinguish the varying aerosol types, especially in identifying the coarse particles (dust) from the natural or anthropogenic fine particles (soot or smoke). The results showed in Figure 18.14 completely correspond to the local environment, where dust particles are the main aerosol composition (coarse mode: the average AE is about 0.25) over the Taklimakan Desert, and anthropogenic pollutants (fine mode: the average AE is about 0.75 to 1.0) are heavily concentrated near urban regions (Lin et al. 2011). The AE parameter serves as the principal indicator for dust weather circumstances (i.e., dust particle identification) in the satellite analysis.

According to the previous discussion, the satellite data of a significant dust storm event (March 26–28, 2004) in China were examined to determine the characteristics of the particle size distribution (AE) in areas affected by dust weather. During the ACE-Asia field campaign in April 2001 (Huebert et al. 2003), the monthly AE values from the AERONET measurements in Beijing and Inner Mongolia sites were 0.78 and 0.82, respectively. Accordingly, the average of the AE value (0.8) was selected to identify the dust particles for the dust weather classification.

Dust storm outbreaks occurred over the Taklimakan Desert and Mongolia Plateau on March 25, 2004. The time series AOD variations from the MODIS data illustrated clearly that the dust storms from the Mongolia Plateau passed through Inner Mongolia and moved southeast during March 26 to 28, 2004. The spatial distribution of the AOD (for AE values less than 0.8) showed that the area around the Inner Mongolia Plateau experienced a strong dust storm (Figure 18.15a); meanwhile, the formation of a dust storm at the Hexi Corridor, located to the east of the Taklimakan Desert area, could also be seen. In addition, a heavy aerosol plume over the region near Bohai Bay and Yingkou City was detected by the MODIS data, but was not found in the surface weather reports (Figure 18.15b). The discrepancy may be caused by the combination of dust particles and pollutants in the aerosol plumes. Overall, AOD distribution was quite consistent with the ground-based observations (see also

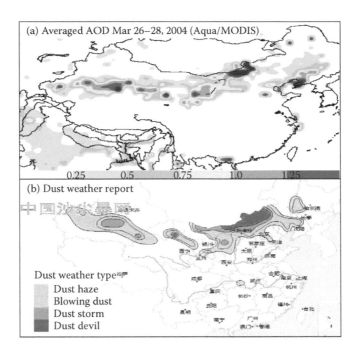

FIGURE 18.15 (a) Spatial distribution of the AOD (for AE values less than 0.8) retrieved from MODIS during a dust storm event in 2004. (b) Dust weather report from surface observations. (From Lin, T.-H. et al., 2011, *International Journal of Remote Sensing*, 32(1), 153–170.)

Figure 18.15). These results indicate that dust weather can indeed be categorized by satellite observed aerosol properties.

18.4.4 Relationship between AOD and Visibility

The visibility observations from 204 ground stations scattered within the study area (see also Figure 18.16) during April 2006 were collected for the correlation of visibility and AOD. The quality of the visibility data from each station was first inspected for the inclusion in our correlation analysis between satellite-retrieved AOD and visibility. After carrying out the data quality assurance (Lin et al. 2011), an exponential relationship exhibits with a high correlation coefficient (about 0.75) between the AOD values and visibility as shown in Figure 18.17a. A similar result was found out by Deng et al. (2008) over the Pearl River Delta Region in southern China. They also pointed out that there was a weak association between the aerosol concentration (PM10) and visibility range during a low-visibility event. In addition to the AOD, the correlation between the turbidity and visibility was also examined and depicted by Figure 18.17b. It is clear that the relationship between the AOD and visibility was stronger than the counterpart between turbidity and visibility from Figure 18.17. The AOD retrieved MODIS data were applied to derive the corresponding visibility in the following dust weather categorization discussion.

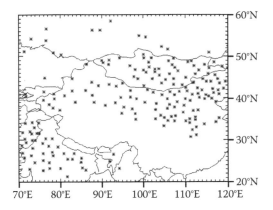

FIGURE 18.16   Location of the 204 ground stations within the study area.

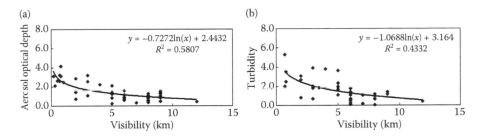

FIGURE 18.17   (a) Respective visibility and AOD scatterplot of the study area during April 2006. (b) Same as (a) but for visibility and turbidity. (From Lin, T.-H. et al., 2011, *International Journal of Remote Sensing*, 32(1), 153–170.)

18.4.5   APPLICATION OF DUST WEATHER CATEGORIZATION

According to the correlation between the AOD and visibility shown in Figure 18.17a, the surface visibility, *VIS*, can be calculated from the satellite-retrieved AOD under the constraint of the AE value (Lin et al. 2011):

$$AOD = -0.727 \cdot ln(VIS) + 2.443. \tag{18.10}$$

The criteria of the dust weather categorization with respect to AOD and AE values and the levels of visibility over China are shown in Table 18.5. Based on the criteria listed in the table, three types of dust weather were categorized for 2003 using the MODIS/Aqua products.

The results of spatial frequency for dust haze, blowing dust, and dust storm events derived from MODIS data are shown in Figures 18.18a, 18.19a, and 18.20a, respectively, with the coincident dust weather reports (Figures 18.18b, 18.19b, and 18.20b). For dust haze, the frequency of spatial distribution from satellite products is quite consistent with the weather reports, excluding the northern Taklimakan Desert (Figure 18.18). For blowing dust weather, discrepancies are seen for the regions around the east Tibet Plateau, the Hexi Corridor to Inner Mongolia Plateau, and the

TABLE 18.5

Criteria Used for Dust Weather Classification in China

Category of Dust Weather	Visibility (km)	AE (Ångström Exponent)	AOD (Aerosol Optical Depth)
Dust haze	>10.0	<0.8	<0.769
Blowing dust	1.0–10.0	<0.8	0.769–2.443
Dust storm	<1.0	<0.8	>2.443

eastern coastal areas from Figure 18.19a and b. The spatial frequency distribution of the dust storm analyzed by the satellite observations matched well with the weather reports, except for the area around the central Tibet Plateau and the Hexi Corridor to Inner Mongolia Plateau (Figure 18.20).

These discrepancies may be caused by the lack of satellite-observed AOD due to the cloudy pixels or absence of ground observations. An analysis of retrieval frequency where aerosol information can be obtained from MODIS/Aqua in 2003 (shown in Figure 18.21) indicated that aerosol products over areas around Tibet Plateau and the Inner Mongolia Plateau were frequently absent in 2003. In particular, fewer than 40% of the days contained retrievals for the region between the Tibet Plateau and Yangtze River Basin. This is possibly caused by the terrain effect (shadows) or frequent cloud cover (Lin et al. 2011). Although some areas like the Tibet Plateau and the Inner Mongolia Plateau exhibited differences, the overall results of dust weather categorization from MODIS aerosol products were very consistent with

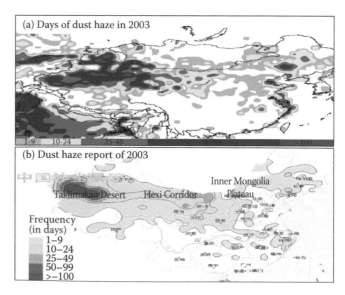

FIGURE 18.18   (a) Spatial frequency of dust haze from MODIS/Aqua data in 2003. (b) Dust weather reports of 2003 from http://www.duststorm.com.cn.

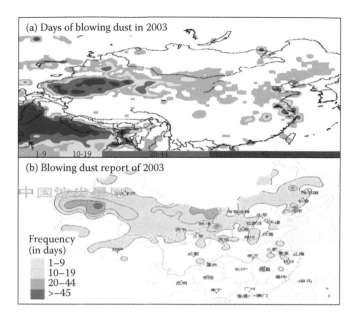

FIGURE 18.19   Same measurements as used in Figure 18.18, but for blowing dust cases.

the dust weather reported by the ground stations and suggested the potential for dust weather monitoring by means of remote sensing.

The spatial frequency distribution of dust weather using MODIS/Aqua products in the springtime from 2004 to 2007 was subsequently investigated to examine the interannual variability of dust weather. The dust haze in central Asia is illustrated

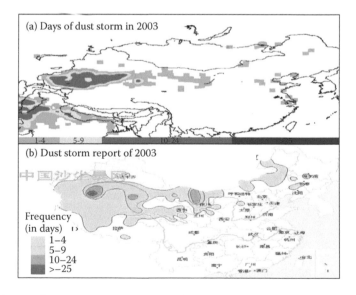

FIGURE 18.20   Same measurements as used in Figure 18.18, but for dust storm cases.

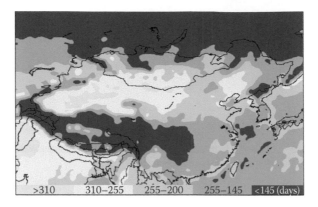

FIGURE 18.21 Number of days where the aerosol optical depth could be provided by MODIS/Aqua products in 2003.

in Figure 18.22. Taklimakan, Gurbantunggut, and Chaidm Deserts witnessed the most frequent dust haze conditions for more than 20 days during the springtime. Areas near the Hexi Corridor and Inner Mongolia Plateau as well as the downwind area of Inner Mongolia also endured dust haze that continued for more than 10 days. For the blowing dust events, the spatial distribution patterns are similar to those for dust haze, excluding the small area coverage and the number of days (Figure 18.23).

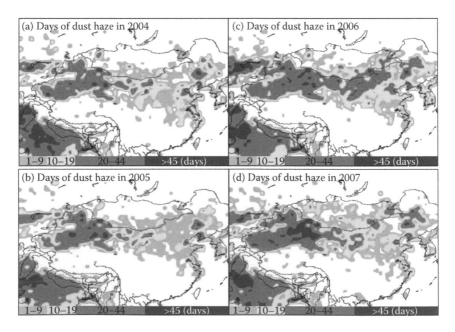

FIGURE 18.22 Results of dust haze spatial frequency distribution from MODIS/Aqua during the springtime in (a) 2004, (b) 2005, (c) 2006, and (d) 2007. (From Lin, T.-H. et al., 2011, *International Journal of Remote Sensing*, 32(1), 153–170.)

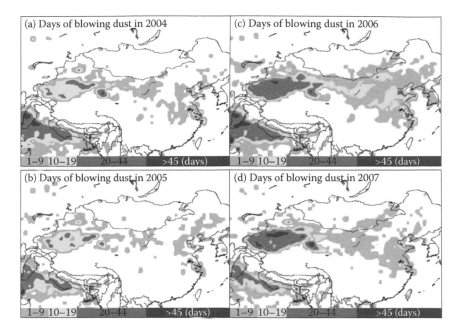

FIGURE 18.23 Same measurements as used in Figure 18.22, but for blowing dust cases. (From Lin, T.-H. et al., 2011, *International Journal of Remote Sensing*, 32(1), 153–170.)

The main active areas of blowing dust were the Taklimakan Desert, Gurbantunggut Desert, and Hexi Corridor. The most notable case was in the Taklimakan Desert. The dust storm active regions are the Taklimakan and Chaidm Deserts, and the highest frequency of dust storms was observed in the western Taklimakan Desert, as shown in Figure 18.24. The results from satellite observation suggest that the highest frequencies of dust weather over China were observed near the Taklimakan Desert. Areas around the Hexi Corridor and Inner Mongolia Plateau also experienced dust haze during springtime (see also Figures 18.22 through 18.24). These results corroborate with the recent dust storm outbreaks based on station data observed from China (Wang et al. 2004), suggesting the high practicability in monitoring dust weather by means of satellite remote sensing.

It is also noted that there is a large interannual variability in the intensity of dust weather during the time period from 2004 to 2007. The number of dust haze days occurring from 2006 to 2007 was almost double in comparison to the period during 2004 to 2005 (Figure 18.22). The spatial distributions and frequencies of blowing dust weather in 2006 and 2007 were more extensive and frequent than 2004 and 2005 as well (Figure 18.23). Furthermore, the occurrences of dust storm events during springtime in 2006 and 2007 appeared to be more active than in 2004 and 2005 (Figure 18.24). The results of dust weather studies using MODIS/Aqua observations indicated the rising trend in frequency and intensity of dust weather in all three categories during the past four years (2004–2007).

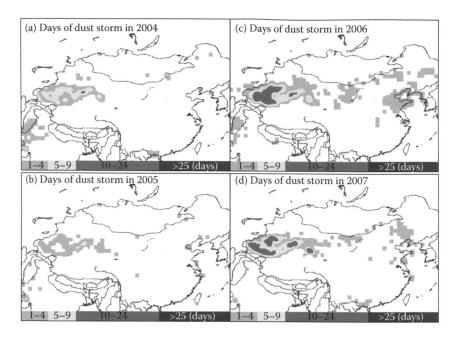

FIGURE 18.24 Same measurements as used in Figure 18.22, but for dust storm cases. (From Lin, T.-H. et al., 2011, *International Journal of Remote Sensing*, 32(1), 153–170.)

18.4.6 DISCUSSION

The ability to use satellite product to categorize dust weather has been verified by comparing remote sensing images to ground truth data (i.e., dust weather reports) in this section. With a wide spatial coverage and a high consistency, satellite remote sensing offers an alternative to spatially limited observations of dust storm events at the ground level. Comparisons between ground station and satellite measurements revealed an exponential relationship with a high correlation coefficient (i.e., greater than 0.75) between the visibility and retrieved AOD under dust weather circumstances. The AOD criteria can thus be used as a substitute for the level of visibility in dust weather categorization provided that the AE value is constrained to be lower than 0.8. The results of dust weather classification generally conformed with the dust storm reports, except for the areas near the Hexi Corridor and Inner Mongolia, which underestimated the spatial distribution of dust storms during 2003. The main reason for the discrepancy was the lack of satellite observations due to some cloudy cases. The effect was significantly mitigated when long-term data were applied. Overall, it can be seen that the dust weather categorization using remote sensing data can be more consistently and efficiently delineated, especially for long-term dust weather investigations.

Although data from ground stations are sometimes difficult to acquire, more data sets (annual) should be analyzed in studying the correlation between the visibility and dust particle properties as well as the criterion of AE value (Ångström exponent coefficient) for dust particle identification. Meanwhile, weather reports regarding the spatial frequency distribution of dust storm events are also very important as they are needed to verify the dust weather categorization accuracy.

18.5 CONCLUSIONS

The results of the previous sections demonstrate the high practicability of dust storm monitoring by means of remote sensing with satellite observations. Still, long-term data sets regarding dust weather events should be collected to characterize the optical properties of dust particles for accurate monitoring of Asian dust storms. In addition, it should be more efficient in dust storm monitoring and warning if polar satellite data (high spatial and spectral resolution) can be combined with geostationary satellite data (high temporal resolution).

The results of this study also indicate that the establishment of a long-term and consistent dust weather database can be conducted effectively by satellite remote sensing. Consequently, the database will benefit our society to issue essential early warnings of Asian dust storm outbreaks and help mitigate further desertification in the future.

REFERENCES

Ångström, A. (1961). Techniques of determining the turbidity of the atmosphere. *Tellus*, 13, 214–223.

Ackerman, S. A. (1989). Using the radiative temperature difference at 3.7 and 11 μm to track dust outbreaks. *Remote Sensing of Environment*, 27, 129–133.

Ackerman, S. A. (1997). Remote sensing aerosols using satellite infrared observations. *Journal of Geophysical Research*, 102(17), 69–79.

Arndt, R. L., Carmichael, G. R., and Roorda, J. M. (1998). Seasonal source-receptor relationships in Asia. *Atmospheric Environment*, 31, 1553–1572.

Bäumer D., Vogel, B., Versick, S., Rinke, R., Möhler, O., and Schnaiter, M. (2008). Relationship of visibility, aerosol optical thickness and aerosol size distribution in an ageing air mass over South-West Germany. *Atmospheric Environment*, 42, 989–998.

Bird, R. E. and Riordan, C. (1986). Simple solar spectral model for direct and diffuse irradiance on horizontal and tilted planes at the Earth's surface for cloudless atmospheres. *Journal of Climate and Applied Meteorology*, 25, 87–97.

Buglia, J. J. (1986). *Introduction to the Theory of Atmospheric Radiative Transfer*. Chapter 5, *NASA reference publication,* 1156, 59–103.

Cachorro, V. E., de Frutos, A. M., and Casanova, J. L. (1987). Determination of the Ångström turbidity parameters. *Applied Optics*, 26(15), 3069–3076.

CCMB (China Central Meteorological Bureau) (1979). *Standard on Weather Observation. Meteorological Press*, Beijing, China, pp. 1–22 (in Chinese).

Chavez, P. S., Mackinnon, D. J., Reynolds, R. L., and Velasco, M. G. (2002). Monitoring dust storms and mapping landscape vulnerability to wind erosion using satellite and ground-based digital images. *Aridlands News Letter*, 51, http//ag.arizona.edu/OALS/ALN/aln51/chavez.html.

Deng X., Tie, X., Wu, D., Zhou, X., Bi, X., Tan, H., Li, F., and Jiang, C. (2008). Long-term trend of visibility and its characterizations in the Pearl River Delta (PRD) region, China. *Atmospheric Environment*, 42, 1424–1435.

Durkee, P. A., Pfeil, F., Frost, E., and Shema, R. (1991). Global analysis of aerosol particle characteristics. *Atmospheric Environment,* 25A, 2457–2471.

Eck, T. F. and Holben, B. N. (1994). AVHRR split window temperature differences and total precipitable water over land surface. *International Journal of Remote Sensing*, 15, 567–582.

Gao, T., Xu, Y., Bo, Y., and Yu, X. (2006). Synoptic characteristics of dust storms observed in Inner Mongolia and their influence on the downwind area (the Beijing-Tianjin region), *Meteorological Applications*, 13, 393–403.

Gathman, S. G. (1983). Optical properties of the marine aerosol as predicted by the Navy model. *Optical Engineering*, 22, 57–62.

Gordon, H. R. and Castano, D. J. (1987). Coastal Zone Color Scanner atmospheric correction algorithm: Multiple scattering effects. *Applied Optics*, 26, 2111–2122.

Gordon, H. R. and Wang, M. (1994). Retrieval of water-leaving radiance and aerosol optical thickness over the oceans with SeaWiFS: A preliminary algorithm. *Applied Optics*, 33, 443–452.

Gregg, W. W. and Carder, K. L. (1990). A simple spectral solar irradiance model for cloudless maritime atmospheres. *Limnology and Oceanography*, 35(8), 1657–1675.

Hansell, R., Tsay, S. C., Ji, Q., Liou, K. N., and Ou, S. (2003). Surface aerosol radiative forcing derived from collocated ground-based radiometric observations during PRIDE, SAFARI, and ACE-Asia. *Applied Optics*, 42, 5533–5544.

Hansen J. E. and Lacis, A. A. (1990). Sun and dust versus greenhouse gases: An assessment of their relative roles in global climate change. *Nature*, 246, 713–719.

Holben, B. N., Vermote, E., Kaufman, Y. J., Tanré, D., and Kalb, V. (1992). Aerosol retrieval over land from AVHRR data application for atmospheric correction. *IEEE Transactions on Geoscience and Remote Sensing*, 30(N2), 212–222.

Hsu, N. C., Herman, J. R., Bhartia, P. K., Seftor, C. J., Torres, O., Thompson, A. M., Gleason, J. F., Eck, T. F., and Holben, B. N. (1996). Detection of biomass burning smoke from TOMS measurements. *Geophysical Research Letters*, 23, 745–748.

Hsu, N. C., Herman, J. R., and Weaver, C. (2000). Determination of radiative forcing of Saharan dust using combined TOMS and ERBE data. *Journal of Geophysical Research*, 105(20), 649–20,661.

Hsu, N. C., Tsay, S. C., King, M. D., and Herman, J. R. (2004). Aerosol properties over bright-reflecting source regions. *IEEE Transactions on Geoscience and Remote Sensing*, 42, 557–569.

Hsu, N. C., Tsay, S. C., King, M. D., and Herman, J. R. (2006). Deep-Blue retrievals of Asian aerosol properties during ACE-Asia. *IEEE Transactions on Geoscience and Remote Sensing*, 44, 3180–3195.

Hu, X. Q., Lu, N. M., Niu, T., and Zhang, P. (2007). Operational retrieval of Asian sand and dust storm from FY-2C geostationary meteorological satellite and its application to real time forecast in Asia. *Atmospheric Chemistry and Physics Discussion*, 7(3), 8395–8421.

Huang, J., Ge, J., and Weng, F. (2007). Detection of Asia dust storms using multisensor satellite measurements. *Remote Sensing of Environment*, 110, 186–191.

Huang, S.-J., Liu, G.-R., Kuo, T.-H., Lin, T.-H., and Liang, C.-K. (2002). Monitoring of Sandstorms with ROCSAT-1 OCI Data. *Journal of Marine Science and Technology*, 10(2), 92–97.

Huebert, B. J., Bates, T., Russell, P. B., Shi, G., Kim, Y. J., Kawamura, K., Carmichael, G., and Nakajima, T. (2003). An overview of ACE-Asia: Strategies for quantifying the relationships between Asian aerosols and their climatic impacts, *Journal of Geophysical Research*, 108(D23), 8633, doi:10.1029/2003JD003550.

Ichikawa, Y. and Fujita, S. (1995). An analysis of wet deposition of sulfate using a trajectory model for East Asia. *Water, Air, and Soil Pollution*, 85, 1921–1926.

Ignatov, A. M., Stowe, L. L., Sakerin, S. M., and Korotaev, G. K. (1995). Validation of the NOAA/NESDIS satellite aerosol product over the North Atlantic in 1989. *Journal of Geophysical Research*, 100(D3), 5123–5132.

Japan Meteorological Agency (1993). Revision of GMS S-VISSR data format. Meteorological Satellite Center, Tokyo, Japan, p. 42.

Kaufman, Y. J., Fraser, R. S., and Ferrare, R. A. (1990). Satellite measurements of large-scale air pollution methods. *Journal of Geophysical Research*, 95(D7), 9895–9909.

Kaufman, Y. J., Wald, A. E., Remer, L. A., Gao, B. C., Li, R. R., and Flynn, L. (1997). The MODIS 2.1 µm channel—Correlation with visible reflectance for use in remote sensing of aerosol. *IEEE Transactions on Geoscience and Remote Sensing*, 35, 1286–1298.

Kaufman, Y. J., Tanré, D., and Boucher, O. (2002). A satellite view of aerosols in the climate system. *Nature*, 419, 215–223.

King, D. M., Kaufman, Y. J., Tanré, D., and Nakajima, T. (1999). Remote sensing of tropospheric aerosols from space: Past, present, and future. *Bulletin of the American Meteorological Society*, 80, 2229–2259.

Kneizys, F. X., Shettle, E. P., Abreu, L. W., Anderson, G. P., Chetwynd, J. H., Gallery, W. O., Selby, J. E. A., and Clough, S. A. (1988). User guide to LOWTRAN 7. *AFGL-TR 880177 Environmental Research Papers*, No. 1010.

Li, F., Vogelmann, A. M., and Ramanathan, V. (2004). Saharan dust aerosol radiative forcing measured from space. *Journal of Climate*, 17, 2558–2571.

Lin, T.-H., Liu, G.-R., and Liang, C.-K. (2005). To construct an effective coefficient of aerosol size distribution for atmospheric turbidity retrieval. *Terrestrial, Atmospheric and Oceanic Sciences*, 16(3), 691–706.

Lin, T.-H., Hsu, N. C., Tsay, S -C., and Huang, S.-J. (2011). Weather categorization of Asian dust storms from surface and satellite observations. *International Journal of Remote Sensing*, 32(1), 153–170.

Liou, K. N. (1980). *An Introduction to Atmospheric Radiation*. Academic Press, New York, 234–292.

Liu, C. H., Chen, A. J., and Liu, G. R. (1996). An image-based retrieval algorithm of aerosol characteristics and surface reflectance for satellite images. *International Journal of Remote Sensing*, 17, 3477–3500.

Liu G.-R., Lin, T.-H., Kuo, T.-H., and Huang, S.-J. (2002). Monitoring of sandstorm with GMS S-VISSR data. *Asian Journal of Geoinformatics*, 2(4), 3–8.

Liu, G.-R. and Lin, T.-H. (2004). Application of geostationary satellite observations for monitoring dust storms of Asia. *Terrestrial, Atmospheric and Oceanic Sciences*, 15, 825–837.

Liu, S. C. and Shiu, C.-J. (2001). Asian dust storms and their impact on the air quality of Taiwan. *Aerosol and Air Quality Research*, 1(1), 1–8.

Liu, S. and Liu, Q. (2006). Detection of dust storms by using daytime and nighttime multispectral MODIS images. *Geoscience and Remote Sensing Symposium*, pp. 294–296, July 31–August 4, 2006, Denver, CO.

Middleton, W. E. K. (1952). *Vision through the Atmosphere*. University of Toronto Press, p. 250.

McCartney, E. J. (1976). *Optics of the Atmosphere, Scattering by Molecules and Particles*. John Wiley & Sons, New York.

Qian, W., Quan, L., and Shi, S. (2002). Variations of the dust storm in China and its climatic control. *Journal of Climate*, 15, 1216–1229.

Qiu, J. (2003). Broadband extinction method to determine aerosol optical depth from accumulated direct solar radiation. *Journal of Applied Meteorology*, 42(11), 1611–1625.

Qiu, J. and Yang, L. (2000). Variation characteristics of atmospheric aerosol optical depths and visibility in North China during 1980–1994. *Atmospheric Environment*, 34, 603–609.

Rao, C. R. N., Stowe, L. L., and McClain, E. P. (1989). Remote sensing of aerosols over the oceans using AVHRR data. Theory, practice and applications. *International Journal of Remote Sensing*, 10(N4), 743–749.

Remer, L. A., Kaufman, Y. J., Tanré, D., Mattoo, S., Chu, D. A., Martins, J. V., Li, R.-R., Ichoku, C., Levy, R. C., Kleidman, R. G., Eck, T. F., Vermote, E., and Holben, B. N. (2005). The MODIS aerosol algorithm, products, and validation. *Journal of the Atmospheric Sciences*, 62, 947–973.

Sobrino, J. A., Raissouni, N., Simarro, J., Nerry, F., and Petitcolin, F. (1999). Atmospheric water vapor content over land surfaces derived from the AVHRR data: Application to the Iberian Peninsula. *IEEE Transactions on Geoscience and Remote Sensing*, 37, 1425–1434.

Solis, J. (1999). *Correlation of Aerosol Optical Depth with Weather Variables*. Institute of Climate and Planets, NASA Goddard Institute of Space Studies, http://icp.giss.nasa.gov/research/ppa/1999/solis/.

Tanré, D., Remer, L. A., Kaufman, Y. J., Mattoo, S., Hobbs, P. V., Livingston, J. M., Russell, P. B., and Smirnov, A. (1999). Retrieval of aerosol optical thickness and size distribution over ocean from the MODIS Airborne Simulator during TARFOX. *Journal of Geophysical Research*, 104, 2261–2278.

Torres, O., Bhartia, P. K., Herman, J. R., and Ahmad, Z. (1998). Derivation of aerosol properties from satellite measurements of backscattered ultraviolet radiation. Theoretical basis. *Journal of Geophysical Research*, 103, 17099–17110.

Tsay, S.-C., Liu, G.-R., Hsu, N. C., and Sun, W.-Y. (2009). Outbreaks of Asian dust storms: An overview from satellite and surface perspectives. *Recent Progress in Atmospheric Sciences: Applications to the Asia Pacific Region, World Scientific*, 373–401.

Tsolmon R., Ochirkhuyag, L., and Sternberg, T. (2008). Monitoring the source of trans-national dust storms in northeast Asia. *International Journal of Digital Earth*, 1, 119–129.

VanCuren, R. and Cahill, T. (2002). Asian aerosols in North America: Frequency and concentration of fine dust. *Journal of Geophysical Research*, 107, doi:10.1029/2002JD002204.

Wang, M. H., Bailey, S., and McClain, C. R. (2000). SeaWiFS provides unique global aerosol optical property data. *EOS, Transactions American Geophysical Union*, 81(18), 197–202.

Wang, X., Dong, Z., Zhang, J., and Liu, L. (2004). Modern dust storms in China: An overview. *Journal of Arid Environments*, 58, 559–574.

Washington, R., Todd, M., Middleton, N. J., and Goudie, A. S. (2003). Dust-storm source areas determined by the Total Ozone Monitoring Spectrometer and Surface observations. *Annals of the Association of American Geographers*, 93, 297–313.

Xinhua News Agency. (2001a). http://www.enviroinfo.org.cn. News release of 2001/01/31 and 2001/02/01, China.

Xinhua News Agency. (2001b). http://www.enviroinfo.org.cn. News release from 2001/04/7 to 2001/04/12, China.

Yang, B. T. (1995). The first ocean remote sensing payload of the ROC: An introduction. *COSPAR Colloquium, Space Remote Sensing of Subtropical Oceans*, Taipei, Taiwan, 13A2-1-13A2-9.

Yoshioka, M., Mahowald, N. M., Conley, A. J., Collins, W. D., Fillmore, D. W., Zender, C. S., and Coleman, D. B. (2007). Impact of desert dust radiative forcing on Sahel precipitation: Relative importance of dust compared to sea surface temperature variations, vegetation changes, and greenhouse gas warming. *Journal of Climate*, 20, 1445–1467.

19 Forest Fire and Air Quality Monitoring from Space

John J. Qu and Xianjun Hao

CONTENTS

19.1 INTRODUCTION

Forest fire is an important natural process of global ecosystems. While wildfires are usually natural hazards impacting environment, air quality, public health, and human properties, prescribed burning is an effective approach for land management to maintain ecosystem health, control invasive species, and reduce the risk of potential large wildfires. According to fire activity statistics from the National Interagency Fire Center (NIFC 2011), in the year 2009 alone, there were 78,792 wildfires, 5,921,786 acres was burned, and 12,429 prescribed fires were reported. A summary of the number of wildfires and area burned during the past 25 years in the United States can be shown in Figure 19.1. Comparing the last 10 years with previous years,

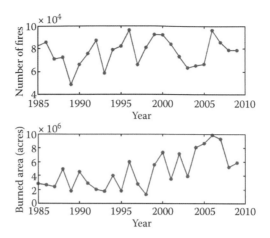

FIGURE 19.1 Statistics of forest fires in the United States. (Data from the National Interagency Fire Center [NIFC 2011].)

a remarkable increase in burned area can be seen, although no significant increase in the number of fires. This implies the increasing trend of fire intensity. Despite the complicated fire–climate interaction, many research works have demonstrated certain linkages between forest fires and climate change (Bevan et al. 2009; Kaufman et al. 2006; Lin et al. 2006; Liu et al. 2010; Martins et al. 2009; Procopio et al. 2004; Vendrasco et al. 2009; Zhang et al. 2008, 2009). Kaufman et al. (2006) found that smoke and pollution aerosols may affect regional cloud cover and have a profound effect on the hydrological cycle and climate. Lin et al. (2006) studied aerosols from forest fires in Amazon and demonstrated important radiative and hydrological effects on the regional climate system. Martins et al. (2009) and Vendrasco et al. (2009) investigated the complicated interaction between precipitation and aerosols from biomass burning. Zhang et al. (2008, 2009) studied smoke aerosol radiative forcing and impacts on moisture transport. Bevan et al. (2009) analyzed time series of aerosol optical depth and found positive feedback between aerosol and drought. Liu et al. (2010) investigated the impacts of climate warming on wildland fire potential and demonstrated that future fire potential may increase significantly in many regions. These results suggest the requirement of more research and management efforts on forest fire. It has become an urgent yet challenging task to develop technical approaches and operational systems to assess fire danger and monitor active fires more timely and accurately, and to investigate the linkages between fire activities and climate change. All these efforts may lead to the provision of effective decision supports to fire managers and government agencies in fire fighting and fuel management.

Systematic satellite-based Earth observation starts from the 1970s with the launch of the Earth Resources Technology (ERTS-1) satellite in 1972 (the name of the series was changed to Landsat in 1975) and the TIROS-N satellite and the Nimbus-7 satellite in 1978. The satellites of the Landsat series (Landsat 1–7) have been providing continuous and consistent measurements of the Earth since the early 1970s. Since then, spaceborne sensors for various missions have been widely used for forest fire monitoring, such as the Advanced Very High Resolution Radiometer (AVHRR), the Tropical Rainfall Measuring Mission (TRMM) Visible and Infrared Scanner (VIRS), the Landsat Multispectral Scanner System (MSS), Thematic Mapper (TM) and Enhanced Thematic Mapper Plus (ETM+), and the Moderate Resolution Imaging Spectroradiometer (MODIS). While active fire detection algorithms have been developed for various satellite sensors (even fire data products have been produced for some sensors), fire danger assessment and fire emission estimation are still in a primitive stage and need further investigation.

This chapter provides a brief review of remote sensing techniques for forest fire and air quality monitoring, including the physical principles, major spaceborne sensors, data products, and operational systems for forest fire detection, burned area mapping, fire danger assessment, smoke plume detection, as well as fire emission estimation. Limitations and potential solutions for future improvements are also discussed.

19.2 ACTIVE FIRE DETECTION

19.2.1 Physical Principle

Active fire detection using spaceborne sensors was first explored in early 1970s (Croft 1973). The theory of active fire detection using thermal sensors was developed in the 1980s by Dozier (1981) and Matson (1981, 1984, 1987) based on the Wien displacement law, which reveals an inverse relationship between the wavelength of the peak emission of a blackbody and its temperature, that is,

$$\lambda_{\max} = \frac{a}{T}, \tag{19.1}$$

where $\lambda_{\max}$ is the peak wavelength, T is the absolute temperature of the blackbody, and a is a constant equaling to $2.8977685 \cdot 10^{-3}$ mK.

The emitted radiance of blackbody at temperatures from 300 to 800 K can be illustrated in Figure 19.2. The peak wavelength of forest fire emission is around 4 μm, which is very sensitive to both flaming fires and smoldering fires, while less sensitive to nonfire background. The peak wavelength of nonfire background emission is around 11 μm, which is much less sensitive to fires compared to the channels around 4 μm. So channels around 4 and 11 μm are usually employed for forest fire detection (Kennedy et al. 1994; Justice et al. 1996). While the temperature of nonfire backgrounds is usually around 300 K, the temperature of flaming forest fires is usually around 800 to 1200 K and can reach as high as 1800 K (Justice et al. 2006). For smoldering fires, the temperature is usually around 450 to 850 K (Justice et al. 2006). Since different sensors have different spectral, spatial, and temporal specifications, forest fire detection algorithms are usually sensor-dependent, using either channels around 4 μm only or a combination of channels around 4 and 11 μm.

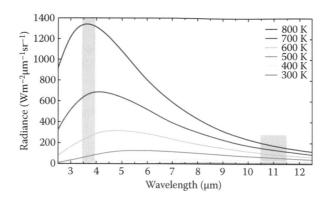

FIGURE 19.2 Radiance of blackbody at temperatures from 300 to 800 K.

The most frequently used sensors for active fire detection include AVHRR and *Geostationary Operational Environmental Satellites* (GOES) Imager managed by the *National Oceanic and Atmospheric Administration* (*NOAA*), as well as Terra/ Aqua, MODIS, and the TRMM VIS, managed by the National Aeronautics and Space Administration (NASA). Operational fire detection algorithms for AVHRR, MODIS, and GOES are introduced here.

19.2.2 MAJOR SENSORS, ALGORITHMS, AND PRODUCTS FOR FOREST FIRE DETECTION

19.2.2.1 AVHRR

AVHRR onboard the NOAA polar-orbiting satellites is the pioneer sensor for forest fire detection from the space. It is a radiometer with four to six channels and daily global coverage at the spatial resolution of 1.1 km (AVHRR/1 had four channels, AVHRR/2 had five channels, and the latest version of AVHRR, i.e., AVHRR/3, has six channels). Spectral characteristics of AVHRR channels used for fire detection are listed in Table 19.1. There are two major categories of AVHRR fire detection algorithms: single-channel algorithms and multichannel algorithms.

Single-channel algorithms are classical fire detection methods developed earlier, using AVHRR channel 3 centered at 3.75 µm with fixed threshold tests based on empirical analysis (Flannigan et al. 1986; Kaufman et al. 1990; Lee and Tag 1990; Setzer 1991; Kennedy et al. 1994; Rauste et al. 1997; Arino and Melinotte 1998; Li et

---

TABLE 19.1

Spectral Characteristics of AVHRR Channels for Forest Fire Detection

Channel/Sensor	AVHRR/1	AVHRR/2	AVHRR/3
3B	3.55–3.93 µm	3.55–3.93 µm	3.55–3.93 µm
4	10.50–11.50 µm	10.30–11.30 µm	10.30–11.30 µm

al. 2000a,b). It is easy to be implemented, but the saturation of channel 3 due to solar reflection over cloud or bright surface may cause false alarms. Since AVHRR channel 3 was not specially designed for fire detection, its saturation temperature is only around 325 K, far below the typical fire temperature and can be easily saturated by forest fires. Although nonfire pixels usually do not reach the saturation temperature, solar reflection can also contribute to channel 3. During nighttime, the contribution from reflected solar irradiance is minimal (Malingreau 1990; Langaas 1992). However, during daytime, solar reflectance from clouds and bright surfaces such as bare soil and rocky surfaces may saturate channel 3 without the existence of fires (Giglio et al. 1999) and thus cause false alarms. Later, multichannel algorithms were developed using either fixed thresholds or contextual thresholds.

The multichannel algorithms can overcome some limitations of the single-spectral algorithm by additional steps to eliminate clouds by using channel 4 and inspecting the difference of brightness temperature at channels 3 and 4 (Li 2001). However, even with multiple channels, fixed thresholds from empirical analysis have limitations for fire detection at the global scale because of the variation of surface features, atmospheric conditions, and fire characteristics (Pu et al. 2004). The contextual algorithm using variable pixel-specific thresholds was introduced by Lee and Tag (1990) and was then developed to various algorithms for regional and global fire detection (e.g., Justice et al. 1996; Eva and Flasse 1996; Flasse and Ceccato 1996; Giglio et al. 1999). Comparing with fixed threshold algorithms, the contextual algorithms are more robust and accurate because of the adaptation to a variety of background properties. Contextual algorithms are usually less affected by the heterogeneous background features and fire characteristics.

Currently, NOAA Satellite Services Division (SSD) provides AVHRR active fire products in operational mode with the Fire Identification, Mapping and Monitoring Algorithm (FIMMA) (FIMMA 2011), which was based on some modifications of the fire detection algorithms proposed by Li et al. (2000, 2001) and Giglio et al. (1999).

19.2.2.2 GOES

GOES operates in a geosynchronous orbit, which allows continuous monitoring of the Earth. GOES Imagers onboard have two infrared channels (3.9 and 10.7 μm) that can be used for forest fire detection. The GOES Automated Biomass Burning Algorithm (ABBA) (Prins and Menzel 1992, 1994) is a contextual multispectral algorithm based on NOAA AVHRR fire detection algorithm developed by Matson and Dozier (1981). It was later redesigned and improved by UW-Madison Cooperative Institute for Meteorological Satellite Studies (CIMSS) in several aspects (Lunetta et al. 1998) and has been used to generate NOAA WF_ABBA fire products (Prins and Menzel 1992, 1994; Prins et al. 1998, 2001a,b). More detailed information can be found from the GOES Biomass Burning Monitoring Program (GOES 2011). Theoretically, geo-stationary satellites are more suitable for forest fire detection because of the continuous measurements at high temporal resolution. However, the current GOES Imager has low spatial resolution at 4 km, which significantly limits the detection of forest fires, especially for small and cool fires. Yet, the next-generation geostationary satellite, GOES-R, has higher spatial resolution and can improve forest fire detection and monitoring effectively.

19.2.2.3 MODIS

The MODIS is a key instrument onboard the Terra and Aqua satellites of NASA Earth Observing System (EOS) mission. Terra MODIS and Aqua MODIS can cover the entire Earth's surface every 1 to 2 days, acquiring data in 36 spectral bands ranging in wavelength from 0.4 to 14.4 μm (MODIS 2011). MODIS has channels specially designed for forest fire detection (i.e., band 21 and band 22 centered at 3.96 μm, with spatial resolution of 1 km at nadir). In the MODIS design, the 3.75 μm channel used in AVHRR was shifted to 3.96 μm to avoid the variable water vapor absorption and to reduce the contributions from solar reflection (Li et al. 2001). Band 21 has saturation brightness temperature as high as 500 K, while band 22 has low signal-to-noise ratio. Moreover, with 36 spectral channels, MODIS can provide more information for excluding cloud pixels and eliminating false alarms more effectively.

Most MODIS fire detection algorithms are the contextual algorithm based on heritage algorithms developed for AVHRR, etc. Kaufman et al. (1998a,b) first proposed the MODIS fire detection algorithm, which selects potential fire pixels by fixed thresholds and confirms fire pixels through contextual tests. The standard MODIS collection 3 fire products were based on the revision by Justice et al. (2002) through adjusting certain threshold values for selecting potential fires and contextual tests. Later, this algorithm was further revised to solve two significant problems in collection 3 fire products: (1) the persistent false detections that occurred in some deserts and sparsely vegetated land surfaces, and (2) frequent omission errors caused by relatively small fires (Giglio et al. 2003). The MODIS version 4 fire product was based on the revision with remarkable improvements to reject false alarms caused by sun glint, desert boundary, and coastal boundary. The current version of the MODIS fire algorithm includes three basic parts: preliminary thresholds to identify potential fire pixels, contextual tests to select tentative fires among the potential fire pixels, and contextual thresholds to reject false alarms.

The MODIS active fire products are produced in operational mode and distributed through the Land Processes Distributed Active Archive Center (LP DAAC) at the U.S. Geological Survey EROS Data Center (EDC). Currently, available MODIS active fire data include the following (Giglio 2010).

1. *MODIS level 2 fire products.* MOD14 (Terra MODIS) and MYD14 (Aqua MODIS) level 2 fire products provide swath data daily at 1-km resolution, including fire mask, algorithm quality, radiative power, and numerous metadata about fire pixel attributes (Giglio 2010).
2. *MODIS level 2G daytime and nighttime fire products.* The MOD14GD/MOD14GN (Terra MODIS) and MYD14GD/MYD14GN (Aqua MODIS) level 2G products provide active fires over daytime and nighttime binned without resampling into an intermediate data format (Giglio 2010).

3. *MODIS level 3 8-day daily composite fire products.* MOD14A1 (Terra MODIS) and MYD14A1 (Aqua MODIS) level 3 products provide 1-km gridded composite of fire pixels in each grid cell over each daily compositing period, and data for 8 days over each compositing period are packed into a single file for convenience (Giglio 2010).

4. *MODIS level 3 8-day summary fire products.* MOD14A2 (Terra MODIS) and MYD14A2 (Aqua MODIS) products provide 1-km gridded composite of fires detected over each 8-day compositing period (Giglio 2010).

These fire data products can be searched and downloaded using the NASA Warehouse Inventory Search Tool (WIST) or the USGS Global Visualization Viewer (GloVis). The Land Processes Distributed Active Archive Center also provides direct FTP access for these products. VIRS onboard TRMM can also be used for forest fire detection (Ji et al. 2002; Giglio et al. 2003). Some other sensors with high spatial resolution such as Landsat TM/ETM+ and the Advanced Spaceborne Thermal Emission and Reflection Radiometer (ASTER) can also be used for fire detection. However, these sensors usually have limited daily spatial coverage, may be not able to capture the fire events, and thus have limitations for operational use. In addition, during the past decades, many satellites for Earth observation have been launched and operated by Europe, Japan, China, India, and so forth. These spaceborne platforms also provide the capability for detecting forest fires.

19.2.3 MAJOR OPERATIONAL SYSTEMS FOR FOREST FIRE DETECTION AND MONITORING

19.2.3.1 NOAA Hazard Mapping System (HMS)

The HMS (HMS 2011) is an interactive processing system operated by the SSD, National Environmental Satellite, Data, and Information Service (NESDIS), NOAA. HMS integrates quality-controlled fire data from various automated fire detection algorithms with GOES, AVHRR, MODIS, and Defense Meteorological Satellite Program (DMSP) Operational Linescan System (OLS). The HMS products can be viewed through the NOAA SSD Fire WEB-GIS page (NOAA SSD 2011) or downloaded as Keyhole Markup Language (KML) files, which can be mapped on Google Earth. A screenshot of the NOAA SSD Fire Web-GIS page, which maps the HMS fire products online with various options, shows some features of this system (Figure 19.3).

19.2.3.2 MODIS Rapid Response System

The MODIS Rapid Response System (LANCE 2011) was developed to provide near real-time MODIS images, as well as a gallery for natural hazards. It is useful for the U.S. Forest Service and the international fire monitoring community to track forest fire events conveniently, for example, a MODIS image from the MODIS Rapid Response System showing wildfires near Los Angeles, CA, on 08/31/2009 (Figure 19.4). The system also provides KML file for visualizing fires on Google Earth.

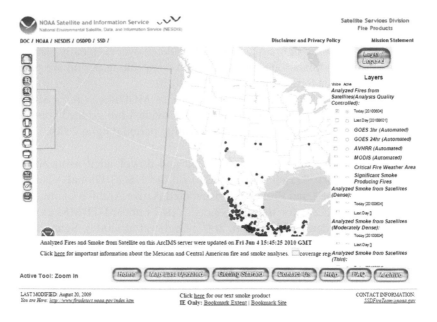

FIGURE 19.3   Screenshot of the SSD Fire Web-GIS page. (NOAA SSD 2011).

FIGURE 19.4   MODIS images of forest fires near Los Angeles, CA, on 08/31/2009. (Courtesy of MODIS Rapid Response System, http://rapidfire.sci.gsfc.nasa.gov/gallery/?2009243-0831/California.A2009243.2105.1km.jpg.)

19.2.3.3   USFS Active Fire Mapping Program

Managed by the USDA Forest Service Remote Sensing Applications Center (RSAC), the USFS Active Fire Mapping Program (RSAC 2011) is an operational satellite-based fire detection and monitoring program. It provides near real-time detection and characterization of wildland fire conditions in a geospatial context for the continental United States, Alaska, Hawaii, and Canada. Currently, MODIS measurement is the primary data source of Active Fire Mapping Program.

19.3 BURNED AREA ESTIMATION

Burned area is one of the most important fire characteristics and critical for evaluating ecosystem disturbance, assessing fire emissions, and studying carbon cycle (Roy et al. 2002, 2005). Forest fire can remove vegetation, change vegetation structure, and leave ash and charcoal over surface. The unique spectral characteristics of ash and charcoal make it possible to discriminate burned area from other surface features in visible (VIS), near-infrared (NIR), and shortwave-infrared (SWIR) channels during a short period after burning (Pereira et al. 1997; Li et al. 2004; Roy et al. 2002, 2005, 2008). Significant vegetation changes caused by fires have relatively longer impacts on surface reflectance. So burned area estimation is usually based on identification of vegetation change or charcoal over surface (Pereira et al. 1997; Roy et al. 2002, 2005, 2008; Li et al. 2004; Barbosa et al. 1999; Fernandez et al. 1997; Fraser et al. 2000). Pereira et al. (1997) reviewed the physical principles and various approaches for burned area mapping and concluded that NIR provides the strongest and less equivocal discriminant ability to identify burned scars, and multitemporal change detection methods are advantageous (Pereira et al. 1997). Many methods for burned area mapping were based on the change of vegetation indices such as Normalized Difference Vegetation Index (NDVI), Enhanced Vegetation Index (EVI), and Normalized Burn Ratio (NBR). Roy et al. (2002, 2005, 2008) developed an approach for burned area mapping using multitemporal MODIS surface reflectance data for the 1.24- and 2.13-μm channels (Roy et al. 2002, 2005, 2008) by exploiting temporal consistency threshold to differentiate between temporary changes and burned areas. Their algorithm has been used to produce burned area data in the MODIS fire products.

Since SWIR bands are less impacted by aerosols, Li et al. (2004) used reflectance of MODIS SWIR (shortwave infrared) bands to separate burned and unburned area directly and provided an approach for near real-time burned area mapping. Middle-infrared (MIR) region has also been investigated for burned area estimation. Pereira et al. (1997) thought that MIR has higher capability to identify burned area than the visible region. However, in the MIR region, radiance is mixed with thermal emission and surface reflectance, which makes it complicated to use the MIR region for burned area mapping. Recently, Libonati et al. (2010) have proposed an approach to map burned area by retrieving MIR reflectance.

Many sensors have been used for burn area mapping, such as TM/ETM+, AVHRR, and MODIS. AVHRR and MODIS can almost provide daily global coverage at spatial resolutions around 1 km at nadir, while Landsat TM and ETM+ can have a high spatial resolution at 30 m. Landsat TM/ETM+ can provide more accurate estimation of burned area when cloud-free measurements are available. The capability of Landsat TM/ETM+ was affected because of the long repeat period (16 days). For MODIS and AVHRR, it is relatively convenient to compose cloud-free measurements close to fire events and estimate burned area through change detection. However, that may get estimates with high uncertainty for small fires.

Currently, MODIS MCD45 burned area product is the major operational data product for burned area mapping (Barbosa et al. 1999; Roy et al. 2002, 2005, 2008). It is a monthly level 3 gridded 500-m product containing per-pixel burning and quality information (Barbosa et al. 1999).

19.4   FIRE DANGER ASSESSMENT

Fire danger assessment is very important for land managers in planning resource allocation facilitating suppression of wildfires and conducting prescribed burning. While remote sensing technology has been successfully used for fire detection and burned area mapping, it is still a challenging task to assess fire danger effectively because fire vulnerability is influenced by diverse factors such as the landscape topography, vegetation type, fuel load, dead and live fuel moisture content, temperature, humidity, wind, and so forth. Most approaches for fire danger assessment rely on selected factors. Vegetation indices such as the NDVI and relative greenness (RG) have traditionally been popular measures of fire risk (Tucker 1977; Paltridge and Barber 1988; Burgan 1996). Temporal decrease in vegetation index has also been used in some studies for fire risk estimation (López et al. 1991; Prosper-Laget et al. 1995; Illera et al. 1996; González et al. 1997). Since SWIR channels are affected by vegetation water content more directly and significantly, vegetation water indices such as the Normalized Difference Water Index (NDWI) (Gao 1996) and the Normalized Difference Infrared Index (NDII) (Hardisky et al. 1983; Hunt and Rock 1989) can also be indicators of potential fire risk. Vegetation temperature can be a measure of fire risk since it increases in drier plants on account of reduced evapotranspiration (Jackson 1986). The combination of vegetation greenness indices and temperature has been found to be related to fuel moisture content and hence can be used for fire risk estimation (Alonso et al. 1996; Chuvieco et al. 1999, 2003b, 2004).

While remote sensing technology can detect the status of live vegetation, especially greenness and water content, it is difficult to estimate the moisture content of understory fuels, especially dead fuels. Dead fuel moisture has a close relationship with meteorological conditions, such as fuel type, fuel temperature, relative humidity, wind, and so forth. Various models have been developed for estimating dead fuel moisture based on meteorological data (Viney 1991). Burgan et al. (1998) developed the Fire Potential Index (FPI) for fire danger potential assessment Wildland Fire Assessment System (WFAS 2011) and that was used in the United States Forest Service (USFS) as an experimental data product. Their approach combined remote sensing measurements with meteorological observations. For example, a WFAS fire danger map for June 6, 2010 indicates fire danger at five levels: low, moderate, high, very high, and extreme (Figure 19.5).

Dasgupta et al. (2006) designed the Fire Susceptibility Index (FSI) based on the physical concept of heat energy of preignition. Heat energy of preignition (Q_{ig}, kJ/kg) can be defined as the heat energy required to bring a fuel from its current temperature to ignition temperature (Bradshaw et al. 1984; Rothermel 1972; Dasgupta et al. 2006) and can thus serve as an indicator of fire risk. FSI can be calculated from live fuel moisture and fuel temperature, both of which can be estimated with

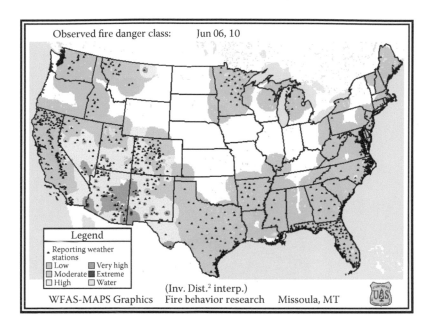

FIGURE 19.5   WFAS fire danger map. (From http://www.wfas.net/.)

remote sensing technology. FSI also has the flexibility to be customized to specific vegetation types or ecoregion for improved performance at local and regional scales.

However, because of the complexity of factors related to fire ignition and fire behavior, simple spectral indices reflecting partial aspects of the factors have limited accuracy. Fire occurrence usually also depends on human activities, and high fire risk may not lead to the occurrence, while fire still can occur in case of low fire risk when ignition is present. It is impractical to validate fire risk assessment effectively.

19.5   SMOKE PLUME DETECTION

Forest fires usually emit a large amount of smoke, which can affect regional air quality and threaten public health. A case study is shown by the MODIS true-color image of smoke plumes from the Georgia Sweat Farm Road Fires/Big Turnaround Fires in May 2007 (Figure 19.6). Although the fires were not very severe, heavy smoke from them significantly affected large regions.

In addition, it has been found that smoke aerosol particles can play important roles in cloud microphysics and radiation balance in the lower atmosphere and smoke plumes may affect cloud formation, and thus have impacts on regional and even global climate (Crutzen et al. 1990; Kaufman et al. 2002; Christopher et al. 1996; Feingold et al. 2001; Koren et al. 2008; Andreae et al. 2004; Breon et al. 2002; Li et al. 2001). Smoke plume detection is also helpful in detecting small and cool fires (Wang et al. 2007; Xie et al. 2007). So it is important to detect and monitor smoke plumes and investigate their impacts on local and regional air quality.

05/02/2007 16:35UTC

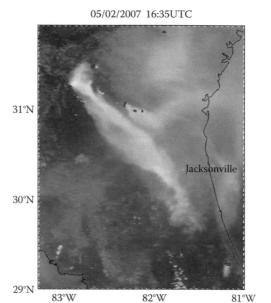

FIGURE 19.6   Georgia Sweat Farm Road fires/Big Turnaround fires.

Since the VIS bands—especially blue bands—are very sensitive to smoke because of Rayleigh scattering, but SWIR bands are less sensitive to smoke, the most direct approaches to detect smoke plumes are based on the combinations of different channels to compose color images that can enhance the impacts of smoke plumes (Chung and Le 1984; Chrysoulakis and Opie 2004; Kaufman et al. 1990; Randriambelo et al. 1998). Multithreshold methods through filtering out nonsmoke pixels by consecutive testing with particular thresholds (Christopher et al. 1996; Chrysoulakis and Cartalis 2003; Li et al. 2001; Xie et al. 2007) are more effective for smoke detection. An important issue is how to determine the thresholds in these methods. Threshold values derived from statistical analysis of regional fire events may have limitations. It is desirable to investigate smoke plumes at regional and global scales so as to develop robust algorithms for more accurate detection of smoke plumes.

For smoke plumes, it is also very important to detect the height and vertical profiles. The Multiangle Imaging Spectroradiometer (MISR) provides a unique view of the Earth at nine angles ranging from nadir to 70.5° in both the forward and afterward directions. Some researchers developed smoke plume detection approaches by exploiting the angular and spectral data of MISR (Mazzoni et al. 2007a,b; Kahn et al. 2007; Patadia et al. 2008; Martin et al. 2010). Kahn et al. (2007) proposed a stereo-matching algorithm to automatically retrieve altitude of aerosol plumes. Mazzoni et al. (2007a) used support vector machines to classify MISR pixels for operational applications and developed automated algorithms for distinguishing smoke from clouds and other aerosols, identifying plumes, and retrieving smoke plume height (Mazzoni et al. 2007b). However, MISR has a very narrow swath and thus may miss many smoke plumes because of limited spatial coverage.

The NASA Cloud-Aerosol Lidar and Infrared Pathfinder Satellite Observation (CALIPSO) can trace vertically through the layers of the atmosphere and thus has the capability to detect vertical profiles of smoke plumes. The CALIPSO lidar cloud and aerosol discrimination algorithm classifies cloud and aerosols at vertical layers (Liu et al. 2009; Omar et al. 2009) and thus provides unique view of smoke plumes. However, CALIPSO also has very narrow swath and can only capture limited smoke plumes.

19.6 FIRE EMISSIONS AND AIR QUALITY

Forest fires release a large amount of particulate matter (PM), carbon monoxide (CO), sulfur dioxide (SO_2), and nitrogen oxides (NO_x), which can have significant impacts on air quality and human health. The Measurements of Pollution in the Troposphere (MOPITT) onboard NASA satellite Terra is an instrument specially designed to observe the distribution, transport, sources, and sinks of carbon monoxide and methane in the troposphere, which can be used to retrieve fire emissions. MOPITT data products include CO mixing ratio profiles, CO total column concentration, and CH_4 total column concentration (Clerbaux et al. 2002, 2004; Deeter et al. 2004, 2010; Edwards et al. 2004; Niu et al. 2004; e.g., monthly CO concentration for May 2004 in the eastern United States [Figure 19.7]).

Currently, it is still a very challenging task to retrieve emissions from space directly. Most approaches are based on integration of spaceborne measurements and modeling. In particular, the BlueSky model can simulate fire emissions from fire characteristics and meteorological data (Figure 19.8). The inputs of BlueSky are fire characteristics and meteorological data, which are critical for fire emission estimation and air quality assessment. Satellite remote sensing can improve the parameterization of BlueSky significantly.

For regional monitoring and analysis of air quality, it is necessary to link BlueSky with air quality models such as the Community Multiscale Air Quality

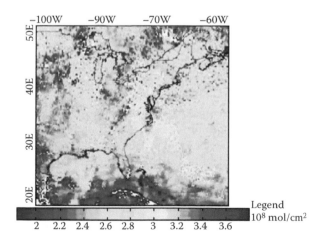

FIGURE 19.7   Monthly CO concentration for May 2004 in the eastern United States.

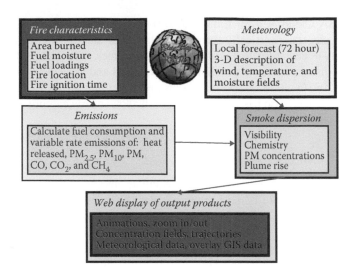

FIGURE 19.8 Components of the BlueSky smoke modeling framework, showing various types of information handled within each component. (From USDA/FS BlueSky Web Page, http://www.fs.fed.us/bluesky.)

(CMAQ) model. CMAQ has been designed to estimate air quality as a whole by including state-of-the-science capabilities for modeling multiple air quality issues, including tropospheric ozone, fine particles, toxics, acid deposition, and visibility degradation (CMAQ 2011). Integration of remote sensing measurements, BlueSky, and CMAQ can provide more accurate estimation of air quality for regional applications.

19.7 DISCUSSION

This study provides a brief overview of the principles, technical approaches, major sensors, and data products for forest fire monitoring from space. For active fire detection, the MODIS, AVHRR, and GOES fire algorithms and products have been developed, improved, and validated over a relatively long period. Because of the relatively low spatial resolution of these sensors and the necessary optimization of thresholds for global applications, small and cool fires may be missed (Wang et al. 2007). Further analysis of fire characteristics over various regions and customization of thresholds may lead to more accurate detection of active fires. In addition, cloud cover is the primary barrier for fire detection from space, and polar orbiting platforms usually have limited daily overpass. Combination of multiple sensors from polar-orbiting observation systems and geostationary systems synergistically may detect active fires more effectively.

For burned area estimation, current MODIS data product is limited by the low spatial resolution. Since the spectral features of burned area may last for a period, it is desirable to integrate burned area estimates from sensors with high spatial resolution whenever high-resolution measurements are available. Besides burned area, burned severity is also an important fire characteristic that can provide more accurate information about fire activity. Though some indices have been developed to assess burned severity, such as Normalized Burned Ratio (NBR), it is necessary to develop the capability for more reliable analysis and validation. Fire danger rating is critical for fire management. Currently, only experimental data products are available. Further analysis and development of algorithms based on physical principles are necessary. For smoke plume detection, most current algorithms are based on classification of scene types. MISR multiple angle measurements and CALIPSO measurements can provide vertical information of smoke plumes, but MISR and CALIPSO have a very narrow swath, which limits them for applications. For fire emission estimation, current spaceborne sensors have limited capability. Although MOPITT can be used to retrieve CO and CH_4, for full assessment for fire emissions, integration of remote sensing measurements and modeling is a promising approach.

19.8 CONCLUSIONS

Forest fire and air quality monitoring from space is an important yet challenging task. While operational remote sensing data products for active fires and burn areas have been produced, much effort is still required toward the effective and efficient monitoring and assessment of fire danger, smoke plume, and fire emissions. Since each sensor has specific spectral and spatial characteristics, most current algorithms for fire monitoring and assessment are sensor-dependent. A multiple sensor approach for fire applications is very promising, although still in the preliminary stage. With more sensors available for different countries and missions, as well as improvement of next-generation sensors in spatial and spectral features, further investigation for integrated solutions will lead to significant improvements in forest fire monitoring and assessment from space.

REFERENCES

Alonso, M., Camarasa, A., Chuvieco, E., Cocero, D., Kyun, I., Martin, M. P., and Salas, F. J. (1996). Estimating temporal dynamics of fuel moisture content of Mediterranean species from NOAA-AVHRR data. *EARSEL Advances in Remote Sensing*, 4, 9–24.

Andreae, M. O., Rosenfeld, D., Artaxo, P., Costa, A. A., Frank, G. P., Longo, K. M., and Silva-Dias, M. A. F. (2004). Smoking rain clouds over the Amazon. *Science*, 303(5662), 1337–1342.

Arino, O. and Melinotte, J. (1998). The 1993 Africa fire map. *International Journal of Remote Sensing*, 19(11), 2019–2023.

Barbosa, P. M., Gregoire, J. M., and Pereira, J. M. C. (1999). An algorithm for extracting burned areas from time-series of AVHRR GAC data applied at a continental scale. *Remote Sensing of Environment*, 69, 253–263.

Bevan, S. L., North, P. R. J., Grey, W. M. F., Los, S. O., and Plummer, S. E. (2009). Impact of atmospheric aerosol from biomass burning on Amazon dry-season drought. *Journal of Geophysical Research*, 114, D09204, doi:10.1029/2008JD011112.

Bradshaw, L. S., Deeming, J. E., Burgan, R. E., and Cohen, J. D. (1984). The 1978 National Fire Danger Rating System: Technical Documentation. General Technical Report INT-169. USDA. Forest Service, Intermountain Forest and Range Experiment Station, Ogden, UT.

Breon, F. M., Tanre, D., and Generoso, S. (2002). Aerosol effect on cloud droplet size monitored from satellite. *Science*, 295(5556), 834–838.

Burgan, R. E. (1996). Use of remotely sensed data for fire danger estimation. *EARSeL Advances in Remote Sensing*, 4(4), 1–8.

Burgan, R. E., Klaver, R. W., and Klaver, J. M. (1998). Fuel Models and Fire Potential from Satellite and Surface Observations. *International Journal of Wildland Fires*, 8(3), 159–170.

Christopher, S. A., Kliche, D. V., Joyce, C., and Welch, R. M. (1996). First estimates of the radiative forcing of aerosols generated from biomass burning using satellite data. *Journal of Geophysical Research*, 101 (D16), 21265–21273.

Chrysoulakis, N. and Cartalis, C. (2003). A new algorithm for the detection of plumes caused by industrial accidents, on the basis of NOAA/AVHRR imagery. *International Journal of Remote Sensing*, 24, 3353–3367.

Chrysoulakis, N. and Opie, C. (2004). Using NOAA and FY imagery to track toxic plumes caused by the 2003 bombing of Baghdad. *International Journal of Remote Sensing*, 25, 5247–5254.

Chung, Y. S. and Le, H. V. (1984). Detection of forest-fire smoke plumes by satellite imagery. *Atmospheric Environment*, 18(10), 2143–2151.

Chuvieco, E., Aguado, I., Cocero, D., and Riano, D. (2003b). Design of an empirical index to estimate fuel moisture content from NOAA-AVHRR images in forest fire danger studies. *International Journal of Remote Sensing*, 24(8), 1621–1637.

Chuvieco, E., Cocero, D., Riano, D., Martin, P., Martinez-Vega, J., de la Riva, J., and Fernando, P. (2004). Combining NDVI and surface temperature for the estimation of live fuel moisture content in forest fire danger rating. *Remote Sensing of Environment*, 92, 322–331.

Chuvieco, E., Deshayes, M., Stach, N., Cocero, D., and Riano, D. (1999). Short-term fire risk: foliage moisture content estimation from satellite data. In *Remote Sensing of Large Wildfires in the European Mediterranean Basin*, E. Chuvieco (ed.), Springer, Berlin, Germany, pp. 17–38.

Clerbaux, C., Hadji-Lazaro, J., Payan, S., Camy-Peyret, C., Wang, J., Edwards, D. P., and Luo, M. (2002). Retrieval of CO from nadir remote-sensing measurements in the infrared by use of four different inversion algorithms. *Applied Optics*, 41(33), 7068–7078.

Clerbaux, C., Gille, J., and Edwards, D. (2004). New Directions: Infrared measurements of atmospheric pollution from space. *Atmospheric Environment*, 38, 4599–4601.

Community Multi-Scale Air Quality (CMAQ) Modeling System (2011). http://www.cmaq-model.org/overview.cfm (accessed March 2011).

Croft, T. A. (1973). Burning waste gas in oil fields. *Nature*, 245, 375–376.

Crutzen, P. J. and Andreae, M. O. (1990). Biomass burning in the Tropics: Impact on atmospheric chemistry and biogeochemical cycles. *Science*, 250(4988), 1669–1678.

Dasgupta, S., Qu, J. J., and Hao X. (2006). Design of a susceptibility index for fire risk monitoring. *IEEE Geoscience and Remote Sensing Letters*, 3(1), 140–144.

Deeter, M. N., Emmons, L. K., Edwards, D. P., Gille, J. C., and Drummond, J. R. (2004). Vertical resolution and information content of CO profiles retrieved by MOPITT. *Geophysical Research Letters*, 31, L15112, doi:10.1029/2004GL020235.

Deeter, M. N., Edwards, D. P., Gille, J. C., Emmons, L. K., Francis, G., Ho, S. P., Mao, D., Masters, D., Worden, H., Drummond, J. R., and Novelli, P. C. (2010). The MOPITT version 4 CO product: Algorithm enhancements, validation, and long-term stability. *Journal of Geophysical Research-Atmospheres*, 115, D07306, doi:10.1029/2009JD013005.

Dozier, J. (1981). A method for satellite identification of surface temperature fields of subpixel resolution, *Remote Sensing of Environment*, 11, 221–229.

Edwards, D. P., Emmons, L. K., Hauglustaine, D. A., Chu, D. A., Gille, J. C., Kaufman, Y. J., Pétron, G., Yurganov, L. N., Giglio, L., Deeter, M. N., Yudin, V., Ziskin, D. C., Warner, J., Lamarque, J.-F., Francis, G. L., Ho, S. P., Mao, D., Chen, J., Grechko, E. I., and

Drummond, J. R. (2004). Observations of carbon monoxide and aerosols from the Terra satellite: Northern Hemisphere variability, *Journal of Geophysical Research*, 109, D24202, doi:10.1029/2004JD004727.

Eva, H. and Flasse, S. (1996). Contextual and multiple-threshold algorithms for regional active fire detection with AVHRR data. *Remote Sensing of Environment*, 14, 333–351.

Feingold, G., Remer, L. A., Ramaprasad, J., and Kaufman, Y. J. (2001). Analysis of smoke impact on clouds in Brazilian biomass burning regions: An extension of Twomey's approach. *Journal of Geophysical Research*, 106, 22,907–22,922.

Fernandez, F., Illera, P., and Casanova, J. L. (1997). Automatic mapping of surfaces affected by forest fires in Spain using AVHRR NDVI composite image data. *Remote Sensing of Environment*, 60, 153–162.

Fire Identification, Mapping and Monitoring Algorithm (FIMMA) (2011). http://www.ssd .noaa.gov/PS/FIRE/Layers/FIMMA/fimma.html (accessed March 2011).

Flannigan, M. D. and Vonder Haar, T. H. (1986). Forest fire monitoring using NOAA satellite AVHRR. *Canadian Journal of Forest Research*, 16, 975–982.

Flasse, S. P. and Ceccato, P. (1996). A contextual algorithm for AVHRR fire detection. *International Journal of Remote Sensing*, 17, 419–424.

Fraser, R. H., Li, Z., and Cihlar, J. (2000). Hotspot and NDVI differencing synergy (HANDS): A new technique for burned area mapping over boreal forest. *Remote Sensing of Environment*, 74, 362–376.

Gao, B. C. (1996). NDWI—A normalized difference water index for remote sensing of vegetation liquid water from space. *Remote Sensing of Environment*, 58, 257–266.

Giglio, L. (2010). *MODIS Collection 5 Active Fire Product User's Guide*, Version 2.4. http:// modis-fire.umd.edu/AF_usermanual.html.

Giglio, L., Kendall, J. D., and Justice, C. O. (1999). Evaluation of global fire detection using simulated AVHRR infrared data. *International Journal of Remote Sensing*, 20, 1947–1985.

Giglio, L., Descloitres, J., Justice, C. O., and Kaufman, Y. J. (2003). An enhanced contextual fire detection algorithm for MODIS. *Remote Sensing of Environment*, 87, 273–282.

GOES Biomass Burning Monitoring Program (GOES) (2011). http://cimss.ssec.wisc.edu/ goes/burn/overview.html (accessed March 2011).

Gonzalez, F. A., Cuevas, J. M., Casanova, J. L., Calle, A., and Illera, P. (1997). A forest fire risk assessment using NOAA-AVHRR images in the Valencia area, eastern Spain. *International Journal of Remote Sensing*, 18, 2201–2207.

Hardisky, M. A., Lemas, V., and Smart, R. M. (1983). The influence of soil salinity, growth form, and leaf moisture on the spectral reflectance of spartina alternifolia canopies. *Photogrammetric Engineering and Remote Sensing*, 49, 77–83.

Hunt, E. R. and Rock, B. N. (1989). Detection of changes in leaf water content using near- and middle infrared reflectances. *Remote Sensing of Environment*, 30, 43–54.

Illera, P., Fernandez, A., and Delgado, J. A. (1996). Temporal evolution of the NDVI as an indicator of forest fire danger. *International Journal of Remote Sensing*, 17, 1093–1105.

Jackson, R. D. (1986). Remote sensing of biotic and abiotic plant stress. *Annual Review of Phytopathology*, 24, 265–287.

Ji, Y. and Stocker, E. (2002). An overview of the TRMM/TSDIS fire algorithm and products. *International Journal of Remote Sensing*, 23, 3285–3303.

Justice, C. O., Kendall, J. D., Dowty, P. R., and Scholes, R. J. (1996). Satellite remote sensing of fires during the SAFARI campaign using NOAA advanced very high resolution radiometer data. *Journal of Geophysical Research*, 101, 23851–23863.

Justice, C. O., Giglio, L., Roy, D., Korontzi, S., Owens, J., Descloitres, J., Alleaume, S., Petitecolin, F., and Kaufman, Y. (2002). The MODIS fire products: Algorithm, preliminary validation and utilization. *Remote Sensing of Environment*, 83, 244–262.

Justice, C., Giglio, L., Boschetti, L., Roy, D., Crirzar, I., Morisette, J., and Kaufman, Y. (2006). MODIS Fire Products Algorithm Technical Background Document. http://modis.gsfc .nasa.gov/data/atbd/atbd_mod14.pdf.

Kahn, R. A., Li, W. H., Moroney, C., Diner, D. J., Martonchik, J. V., and Fishbein, E. (2007). Aerosol source plume physical characteristics from space-based multiangle imaging. *Journal of Geophysical Research-Atmospheres*, 112, D11205, doi:10.1029/2006JD007647.

Kaufman, Y. J., Tucker, C. J., and Fung, I. (1990). Remote sensing of biomass burning in the tropics. *Journal of Geophysical Research*, 95, 9927–9939.

Kaufman, Y. J., Justice, C. O., Flynn, L., Kendall, J. D., Prins, E. M., Giglio, L., Ward, D. E., Menzel, W. P., and Setzer, A. W. (1998a). Potential global fire monitoring from EOS-MODIS. *Journal of Geophysical Research*, 103, 32215–32238.

Kaufman, Y. J., Kleidman, R. G., and King, M. D. (1998b). SCAR-B fires in the tropics: Properties and remote sensing from EOS-MODIS. *Journal of Geophysical Research*, 103, 31955–31968.

Kaufman, Y. J. and Koren, I. (2006). Smoke and Pollution Aerosol Effect on Cloud Cover. *Science*, 313(5787), 655–658.

Kaufman, Y. J., Tanre, D., and Boucher, O. (2002). A satellite view of aerosols in the climate system. *Nature*, 419(6903), 215–223.

Kennedy, P. J., Belward, A. S., and Gregoire, J.-M. (1994). An improved approach to fire monitoring in west Africa using AVHRR data. *International Journal of Remote Sensing*, 15, 2235–2255.

Koren, I., Martins, J. V., Remer, L. A., and Afargan, H. (2008). Smoke invigoration versus inhibition of clouds over the Amazon. *Science*, 321(5891), 946–949.

Langaas, S. (1992). Temporal and spatial distribution of Savannah fires in Senegal and the Gambia, West Africa, 1989–90, derived from multi-temporal AVHRR night images. *International Journal of Wildland Fire*, 2, 21–36.

Lee, T. M. and Tag, P. M. (1990). Improved detection of hotspots using the AVHRR 3.7 um channel. *Bulletin of American Meteorology Society*, 71, 1722–1730.

Li, R.-R., Kaufman, Y. J., Hao, W., Salmon, J. M., and Gao, B. (2004). A technique for detecting burn scars using MODIS data. *IEEE Transactions on Geoscience and Remote Sensing*, 42, 6, 1300–1308.

Li, Z., Nadon, S., Cihlar, J., and Stocks, B. (2000a). Satellite-based detection of Canadian boreal forest fires: Evaluation and comparison of algorithms. *International Journal of Remote Sensing*, 21, 3071–3082.

Li, Z., S. Nadon, and Cihlar, J. (2000b). Satellite detection of Canadian boreal forest fires: Development and application of an algorithm, *International Journal of Remote Sensing*, 21, 3057–3069.

Li, Z., Kaufman, Y. J., Ithoku, C., Fraser, R., Trishchenko, A., Giglio, L., Jin, J., and Yu, X. (2001). A review of AVHRR-based active fire detection algorithms: Principles, limitations, and recommendations. In *Global and Regional Vegetation Fire Monitoring from Space: Planning and Coordinated International Effort,* F. Ahern, J. G. Goldammer, and C. Justice (eds.), SPB Academics Publishing, The Hague, Amsterdam, Netherlands, pp. 199–225.

Libonati, R., DaCamara, C. C., Pereira, J. M. C., and Peres, L. F. (2010). Retrieving middle-infrared reflectance for burned area mapping in tropical environments using MODIS. *Remote Sensing of Environment*, 114(4), 831–843.

Lin, J. C., Matsui, T., Pielke, R. A., and Kummerow, C. (2006). Effects of biomass-burning-derived aerosols on precipitation and clouds in the Amazon Basin: A satellite-based empirical study. *Journal of Geophysical Research*, 111, D19204, doi:10.1029/2005JD006884.

Liu, Y., Stanturf, J., and Goodrick, S. (2010). Trends in global wildfire potential in a changing climate. *Forest Ecology and Management*, 133(1), 40–53.

Liu, Z. Y., Vaughan, M., Winker, D., Kittaka, C., Getzewich, B., Kuehn, R., Omar, A., Powell, K., Trepte, C., and Hostetler, C. (2009). The calipso lidar cloud and aerosol discrimination: Version 2 algorithm and initial assessment of performance. *Journal of Atmospheric and Oceanic Technology*, 26(7), 1198–1213.

Lopez, S., Gonzalzez, F., Llop, R., and Cuevas, J. M. (1991) An evaluation of the utility of NOAA AVHRR images for monitoring forest fire risk in Spain. *International Journal of Remote Sensing*, 12, 1841–1851.

Lunetta, R. S. and Elvidge, C. D. (1998). *Remote Sensing Change Detection: Environmental Monitoring Methods and Applications.* Ann Arbor Press, Ann Arbor, MI, pp. 106–108.

Malingreau, J. P. (1990). The contribution of remote sensing to the global monitoring of fires in tropical and sub-tropical ecosystems. In *Fire in the Tropical Biota*, J. G. Goldammer (ed.), Springer-Verlag, Berlin, Germany, pp. 337–370.

Martin, M. V., Logan, J. A., Kahn, R. A., Leung, F. Y., Nelson, D. L., and Diner, D. J. (2010). Smoke injection heights from fires in North America: Analysis of 5 years of satellite observations. *Atmospheric Chemistry and Physics*, 10(4), 1491–1510.

Martins, J. A., Dias, M. A. F. S., and Goncalves, F. L. T. (2009). Impact of biomass burning aerosols on precipitation in the Amazon: A modeling case study. *Journal of Geophysical Research*, 114, D02207.

Matson, M. and Dozier, J. (1981). Identification of subresolution high temperature sources using a thermal IR sensor. *Photogrammetric Engineering and Remote Sensing*, 47, 1311–1318.

Matson, M., Schneider, S. R., Aldridge, B., and Satchwell, B. (1984). Fire Detection Using the NOAA-series Satellites, NOAA Technical Report NESDIS 7, National Oceanic and Atmospheric Administration, Washington, DC, p. 34.

Matson, M., Stephens, G., and Robinson, J. (1987). Fire detection using data from the NOAA-N satellites. *International Journal of Remote Sensing*, 8(7), 961–970.

Mazzoni, D., Garay, M. J., Davies, R., and Nelson, D. (2007a). An operational MISR pixel classifier using support vector machines. *Remote Sensing of Environment*, 107(1–2), 149–158.

Mazzoni, D., Logan, J. A., Diner, D., Kahn, R., Tong, L. L., and Li, Q. B. (2007b). A data-mining approach to associating MISR smoke plume heights with MODIS fire measurements. *Remote Sensing of Environment*, 107(1–2), 138–148.

Moderate Resolution Imaging Spectroradiometer (MODIS) (2011). http://modis.gsfc.nasa .gov/ (accessed March 2011).

MODIS Rapid Response System (LANCE) (2011). (http://rapidfirc.sci.gsfc.nasa.gov/ (accessed March 2011).

National Interagency Fire Center (NIFC) (2011). http://www.nifc.gov/ (accessed March 2011).

Niu, J., Deeter, M. N., Gille, J. C., Edwards, D. P., Ziskin, D. C., Francis, G. L., Hills, A. J., and Smith, M. W. (2004). Carbon monoxide total column retrievals by use of the measurements of pollution in the troposphere airborne test radiometer. *Applied Optics*, 43(24), 4685–4696.

NOAA Hazard Mapping System (HMS) (2011). http://www.osdpd.noaa.gov/ml/land/hms .html (accessed March 2011).

NOAA Satellite Service Division Fire Products (NOAA SSD) (2011). http://www.firedetect .noaa.gov/viewer.htm (accessed March 2011).

Omar, A. H., Winker, D. M., Kittaka, C., Vaughan, M. A., Liu, Z. Y., Hu, Y. X., Trepte, C. R., Rogers, R. R., Ferrare, R. A., Lee, K. P., Kuehn, R. E., and Hostetler, C. A. (2009). The CALIPSO automated aerosol classification and lidar ratio selection algorithm. *Journal of Atmospheric and Oceanic Technology*, 26(10), 1994–2014.

Paltridge, G. W. and Barber, J. (1988). Monitoring grassland dryness and fire potential in Australia with NOAA/AVHRR data. *Remote Sensing of Environment*, 25, 381–394.

Patadia, F., Gupta, P., Christopher, S. A., and Reid, J. S. (2008). A multisensor satellite-based assessment of biomass burning aerosol radiative impact over Amazonia. *Journal of Geophysical Research-Atmospheres*, 113, D12214, doi:10.1029/2007JD009486.

Pereira, J. M. C., Chuvieco, E., Beaudoin, A., and Desbois, N. (1997). Remote sensing of burned areas: A review. In *A Review of Remote Sensing Methods for the Study of Large Wildland Fires*, E. Chuvieco (ed.), Report of the Megafires Project ENV-CT96-0256, August 1997, Universidad de Alcala, Alcalade Henares, Spain.

Prins, E., Schmetz, J., Flynn, L., Hillger, D., and Feltz, J. (2001a). Overview of current and future diurnal active fire monitoring using a suite of international geostationary satellites. In *Global and Regional Wildfire Monitoring: Current Status and Future Plans*, SPB Academic Publishing, The Hague, Amsterdam, Netherlands, pp. 145–170.

Prins, E. M. and Menzel, W. P. (1992). Geostationary satellite detection of biomass burning in South America. *International Journal of Remote Sensing*, 13, 2783–2799.

Prins, E. M. and Menzel, W. P. (1994). Trends in South American biomass burning detected with the GOES visible infrared spin scan radiometer atmospheric sounder from 1983 to 1991. *Journal of Geophysical Research*, 99, 16719–16735.

Prins, E., Feltz, J., and Schmidt, C. (2001b). An overview of active fire detection and monitoring using meteorological satellites. *Proceedings of 11th Conference on Satellite Meteorology and Oceanography*, Madison, WI, American Meteorology Society, pp. 1–8.

Prins, E. M., Feltz, J. M., Menzel, W. P., and Ward, D. E. (1998). An overview of GOES-8 diurnal fire and smoke results for SCAR-B and the 1995 fire season in South America. *Journal of Geophysical Research*, 103, 31821–31836.

Procopio, A. S., Artaxo, P., Kaufman, Y. J., Remer, L. A., Schafer, J. S., and Holben, B. N. (2004). Multiyear analysis of Amazonian biomass burning smoke radiative forcing of climate. *Geophysical Research Letters*, 31(3), L03108, doi:10.1029/2003GL018646.

Prosper-Laget, V., Douguedroit, A., and Guinot, J. P. (1995). Mapping the risk of forest fire occurrence using NOAA satellite information. *EARSeL Advances in Remote Sensing*, 4, 30–38.

Pu, R., Gong, P., Li, Z., and Scarborough, J. (2004). A dynamic algorithm for wildfire mapping with NOAA/AVHRR data. *International Journal of Wildland Fire*, 13, 275–285.

Rauste, Y., Herland, E., Frelander, H., Soini, K., Kuoremaki, T., and Ruokari, A. (1997). Satellite-based forest fire detection for fire control in boreal forests. *International Journal of Remote Sensing*, 18, 2641–2656.

Rothermel, R. C. (1972). A Mathematical Model for Predicting Fire Spread in Wildland Fuels. Research Paper, INT-115, USDA. Forest Service, Intermountain Forest and Range Experiment Station, Ogden, UT.

Roy, D. P., Boschetti, L., Justice, C. O., and Ju, J. (2008). The collection 5 MODIS burned area product—Global evaluation by comparison with the MODIS active fire product, 2008. *Remote Sensing of Environment*, 112, 3690–3707.

Roy, D. P., Lewis, P. E., and Justice, C. O. (2002). Burned area mapping using multi-temporal moderate spatial resolution data—A bi-directional reflectance model-based expectation approach. *Remote Sensing of Environment*, 83, 263–286.

Roy, D. P., Jin, Y., Lewis, P. E., and Justice, C. O. (2005). Prototyping a global algorithm for systematic fire-affected area mapping using MODIS time series data. *Remote Sensing of Environment*, 97, 137–162.

Setzer, A. W. and Pereira, M. C. (1991). Operational Detection of Fires in Brazil with NOAA-AVHRR. *Proceedings 24th International Symposium on Remote Sensing of the Environment*, Rio de Janeiro, Brazil, pp. 469–482.

Tucker, C. J. (1977). Asymptotic nature of grass canopy spectral reflectance. *Applied Optics*, 16(5), 1151–1156.

USFS Active Fire Mapping Program (RSAC) (2011). http://activefiremaps.fs.fed.us/ (accessed March 2011).

Vendrasco, E. P., Dias, P. L. S., and Freitas, E. D. (2009). A case study of the direct radiative effect of biomass burning aerosols on precipitation in the Eastern Amazon. *Atmospheric Research*, 95(1), 77–87.

Viney, N. R. (1991). A review of fine fuel moisture modeling. *International Journal of Wild Fire*, 1(3), 215–34.

Wang, W., Qu, J. J., Liu, Y., Hao, X., and Sommers, W. (2007). An improved algorithm for small and cool fire detection using MODIS data: A preliminary study in the southeastern United States. *Remote Sensing of Environment*, 108(2), 163–170.

Wildland Fire Assessment System (WFAS) (2011). http://www.wfas.net/ (accessed March 2011).

Xie, Y., Qu, J. J., Xiong, X., Hao, X., Che, N., and Sommers, W. (2007). Smoke plume detection in the eastern United States using MODIS. *International Journal of Remote Sensing*, 28(10), 2367–2374.

Zhang, Z., Fy, R., Yu, H. B., Dickinson, R. E., Juarez, R. N., Chin, M., and Wang, H. (2008). A regional climate model study of how biomass burning aerosol impacts land-atmosphere interactions over the Amazon. *Journal of Geophysical Research*, 113, D14S15, doi:10.1029/2007JD009449.

Zhang, Z., Fy, R., Yu, H. B., Qian, Y., Dickinson, R. E., Dias, M. A. F. S., Dias, P. L. D., and Femandes, K. (2009). Impact of biomass burning aerosol on monsoon circulation transition over Amazonia. *Geophysical Research Letters*, 36, L10814.

20 Satellite Remote Sensing of Global Air Quality

Sundar A. Christopher

CONTENTS

20.1 INTRODUCTION

Urban air quality has become a critical public health concern in many parts of the globe with the increase of urbanization and industrialization during the last few decades. Almost half of the world's population now live in the urban areas, and their number will increase to 4 billion by the end of this decade. Particulate matter (PM) (or aerosols) and ozone are two of the major pollutants affecting the air quality in urban areas throughout the world. PM is a complex mixture of solid and liquid particles that vary in size and composition and remain suspended in the air. Many chemical, physical, and biological components of atmospheric aerosols are identified as being potentially harmful to respiratory and cardiopulmonary human health. Aerosols have many sources from both natural and anthropogenic activities. These include naturally occurring aerosols from windblown dust and episodic activities such as forest fires/agricultural burning (mostly anthropogenic). Increasing human factors such as combustion from automobiles and industries and emissions from power plants also contribute. Apart from direct emissions, PM is also produced by other processes such as gas-to-particle conversion in the atmosphere.

20.2 AEROSOLS AND HUMAN HEALTH

Atmospheric aerosols are one of the most important components of the earth-atmosphere system and play an important role in climate- and weather-related processes (Kaufman et al. 2002). They vary in size from 0.001 to 100 µm. Air pollution has both short- and long-term effects. Short-term impacts include respiratory infections,

479

headaches, nausea, allergic reactions, and irritation to the eyes, nose, and throat. Short-term air pollution can intensify the medical conditions of individuals with asthma and emphysema. Long-term effects include lung cancer, heart disease, chronic respiratory disease, and even damage to the brain, nerves, liver, or kidneys. Various studies have underscored the links between air pollution and human health (e.g., Dockery et al. 1993; Hu et al. 2009; Pope et al. 2009; Wong et al. 2008; Yang et al. 2011). In 1952, London experienced one of the worst smog disasters, which killed more than 4000 people in a few days due to a very high concentration of PM (14 mg·m^{-3}) in the air.

PM with aerodynamic diameters less than 2.5 µm (PM$_{2.5}$) can cause respiratory and lung disease and even premature death. The World Health Organization (WHO) estimates that 4.6 million people die each year from causes directly attributable to air pollution. A medical study (Pope 2000) concludes that fine particles and sulfur oxide–related pollution are associated with causes such as lung cancer and cardio-pulmonary mortality. The same study also states that an increase of 10 µg·m^{-3} in fine particulates can cause approximately a 4%, 6%, and 8% increased risk of all cause, cardiopulmonary and lung cancer mortality, respectively. Indirectly, air pollution significantly affects the economy by increasing medical expenditures and expenditures for preserving the surrounding environment. Therefore monitoring PM air quality is critical.

20.3 MONITORING PARTICULATE MATTER POLLUTION

20.3.1 Ground-Based Monitoring

Various agencies around the world are using ground monitors for measuring air pollution. For example, the United States Environmental Protection Agency (U.S. EPA) monitors air quality by measuring PM and ozone concentration at thousands of ground-based monitoring stations across the country. At the majority of the stations, PM$_{2.5}$ (i.e., fine particle with a diameter of 2.5 µm or less) is measured using a tapered-element oscillating microbalance (TEOM) instrument. A vibrating hollow tube called the tapered element is set in oscillation at resonant frequency and an electronic feedback system maintains the oscillation amplitude. When the ambient air stream enters the mass sensor chamber and particulates are collected at the filter, the oscillation frequency of the tapered element changes, and the corresponding mass change is calculated as the change in measured frequency at time t to the initial frequency at time t_0. The mass concentration is then calculated from dust mass, time, and flow rate. Ideally, only the collection of aerosol mass on the filter should change the tapered element frequency. However, temperature fluctuations, humidity changes, flow pulsation, and change in filter pressure could affect the TEOM performance. Even in best-case scenarios when the operational parameters can be held constant, the heat-induced loss of volatile material could pose serious errors in the PM$_{2.5}$ mass. However, various correction factors are usually applied to adjust for these factors, although the PM$_{2.5}$ mass usually represents the lower limits of a true value (Grover et al. 2005).

The ground monitors have several advantages. The measurement techniques can be standardized and applied across all ground monitors. They can measure pollution throughout the day and can also provide information. They can measure pollution regardless of clouds since these are filter-based measurements that are usually located at the surface. They also have some disadvantages. The obvious one is that they are point measurements and are not representative of pollution over large spatial areas. Costs for installing and maintaining such equipment must also be taken into account.

The U.S. EPA issues National Ambient Air Quality Standards (NAAQS) for six criteria pollutants, namely, ozone, PM, carbon monoxide, sulfur dioxide, lead, and nitrogen oxides. Standards for PM were first issued in 1971 and then revised in 1987. In September 2006, the U.S. EPA revised its 1997 standards to tighten the criteria. The 2006 standards reduced the 24-hour mean $PM_{2.5}$ mass concentration standard from 65 to 35 $\mu g \cdot m^{-3}$ and retained the current annual $PM_{2.5}$ standard at 15 $\mu g \cdot m^{-3}$. The EPA reports an Air Quality Index (AQI) based on the ratio between 24-hour averages of the measured dry particulate mass and NAAQS, and it can range from nearly zero in a very clean atmosphere to 500 in very hazy conditions (see Al Saadi et al. 2005 for further details). In recent years, other countries in Europe and Australia, Japan, and China have also started monitoring $PM_{2.5}$ mass as one measure of air quality conditions.

20.3.2 SATELLITE REMOTE SENSING OF PARTICULAR MATTER POLLUTION

Satellite data have tremendous potential for mapping the global distribution of aerosols and their properties (Chu et al. 2003; Martin 2008). PM air pollution from spaceborne sensors is obtained by largely using passive remote sensing such as reflected solar radiation or emitted thermal radiation. Hoff and Christopher (2009) provided a thorough review of the various satellite sensors available for studying $PM_{2.5}$. In this chapter, we chose the Moderate Resolution Imaging Spectroradiometer (MODIS) onboard the National Aeronautics and Space Administration's (NASA) Terra (morning satellite with equatorial crossing time of 10:30 AM) and Aqua (afternoon satellite with equatorial crossing time of 1:30 PM) satellites that provide systematic retrieval of cloud and aerosol properties over land (King et al. 1992). The MODIS has 36 channels, covers the ultraviolet to the thermal infrared part of the electromagnetic spectrum, and provides near daily global coverage due to its large swath width. Its spatial resolution is from 250 to 1000 m. Aerosol optical depth (AOD) is an important aerosol parameter retrieved from satellite observations, representing columnar loading of aerosols in the atmosphere along with the fraction of fine mode aerosol, which is an indicator of anthropogenic pollution. Several studies conducted over land reveal that MODIS AOD retrievals are within expected uncertainty levels (Remer et al. 2005).

In the absence of clouds, the reflected solar radiation for an aerosol layer from the sun to the Earth atmosphere and back to the satellite is a function of surface reflectance, molecular scattering, and absorption. The top of atmosphere reflectance is a function of sun-satellite viewing geometry and can be related to AOD, which is the columnar value of aerosol extinction (absorption plus scattering). For example, the seasonal distribution of MODIS AOD at 550 nm can be shown in Figure 20.1. In the Northern hemisphere spring and summer, pollution over Asia, Africa, and other parts of the globe is readily seen. Dust is prevalent in Africa during the summer

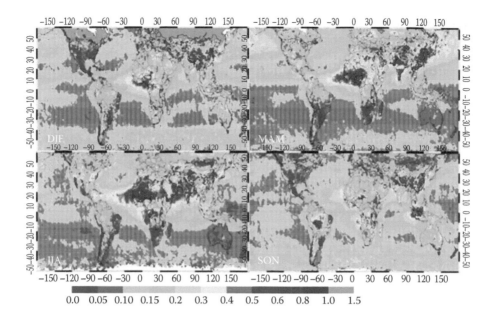

FIGURE 20.1 Seasonal distribution of MODIS mid-visible aerosol optical depth where DJF is December, January, February; MAM is March, April, May; JJA is June, July, August; and SON is September, October, November for 2006.

months, and these dust aerosols can be transported several hundred miles to the Atlantic Ocean. Biomass burning is observed from satellites in the southern hemisphere during August to September and has significant impacts on air quality and climate. This columnar satellite-derived AOD can be related to ground-level $PM_{2.5}$ using the following equation:

$$AOD = PM_{2.5} \cdot H \cdot f(R) \cdot \frac{3Q_{ext.dry}}{4\rho \cdot r_{eff}},\qquad(20.1)$$

where $f(R)$ is the ratio of ambient and dry extinction coefficients, ρ is the aerosol mass density (g·m⁻³), $Q_{ext,dry}$ is the Mie extinction efficiency, r_{eff} is the particle effective radius that is an area weighted mean radius of the particle, and H is the boundary layer height. This equation indicates that if the boundary layer height and other information about the atmosphere and aerosols are known, the satellite-retrieved AOD can be converted to ground-level $PM_{2.5}$. During clear sky conditions and well-mixed boundary layer situations (typical during Terra and Aqua overpasses), AOD can be related to surface $PM_{2.5}$ mass.

This relationship was explored by Wang and Christopher (2003) who correlated ground-level $PM_{2.5}$ with AOD and found an excellent correlation between the two measures. This relationship was then used to calculate air quality indices for the southeastern United States. However, satellite remote sensing of PM air quality is a relatively new area of research. As shown in Table 20.1, many previous studies have

TABLE 20.1

Selected Relevant Literature Survey on Satellite Remote Sensing of Particulate Matter Air Quality

#	Reference	Data and Study Area	Key Conclusions/Remarks
1	Wang and Christopher 2003	MODIS, seven stations, Alabama	Quantitative analysis with space and time collocated hourly $PM_{2.5}$ and MODIS AOD. Demonstrated the potential of satellite data for $PM_{2.5}$ air quality monitoring ($R = 0.7$).
2	Chu et al. 2003	AERONET, MODIS, PM_{10}, 1 station, Italy	Showed relationship between PM_{10} and AOD. More qualitative discussion on satellite capabilities to detect and monitor aerosols globally ($R = 0.82$).
4	Engel-Cox et al. 2004	MODIS, $PM_{2.5}$ continental United States	Correlation analysis over entire United States and discuss difference in relationship over different regions. Qualitative and quantitative analysis over larger area, demonstrated spatial distribution of correlation. Range of R.
5	Hutchinson et al. 2004	MODIS AOD maps, ozone, eastern United States	Used few MODIS AOD maps and discussed the hazy conditions, no correlation analysis, and more emphasis on ozone pollution.
6	Liu et al. 2004	MISR, GEOS-CHEM GOCART, United States	First used MISR data for air quality study and have emphasis on seasonal and annual mean correlation analysis and forecasting ($R = 0.78$).
7	Liu et al. 2005	MISR, GEOS-3 Meteorology, United States	Regression model development and forecasting of $PM_{2.5}$, model generated coarse resolution meteorological fields are used and focused only in eastern United States. 48% explanation of $PM_{2.5}$ variations.
8	Al-Saadi et al. 2005	MODIS, United States	More descriptive paper on IDEA program, which provides online air quality conditions from MODIS and surface measurements over several locations in the United States.
9	Hutchinson et al. 2005	MODIS, Texas	Correlation analysis in Texas. Correlation varies from 0.4 to 0.5 and long time averaging can make correlation greater than 0.9.
10	Engel-Cox et al. 2005	MODIS, United States	Potential of satellite data for monitoring transport of $PM_{2.5}$ over state boundaries and event-specific analysis.

(continued)

TABLE 20.1 (Continued)
Selected Relevant Literature Survey on Satellite Remote Sensing of Particulate Matter Air Quality

#	Reference	Data and Study Area	Key Conclusions/Remarks
11	Gupta et al. 2006	MODIS, meteorology, global 21 locations	Correlation varies from 0.37 to 0.85 over different parts of the world. Cloud fraction, relative humidity, and mixing height information can improve relationship significantly. First study covered several global locations.
12	Engel-Cox et al. 2006	MODIS, lidar, United States	Weak correlation can be significantly improved by using vertical aerosol information from LIDAR measurements.
13	Van Donkelaar et al. 2006	MODIS, MISR, $PM_{2.5}$, GEOS-CHEM, United States and global	Intercomparison between MODIS and MISR over several locations in Canada and United States. R = 0.69 (MODIS) and R = 0.58 (MISR). Different approach used to calculate the fine mass concentration.
14	Koelemeijer et al. 2006	MODIS, $PM_{2.5}$ and PM_{10}, Europe	Mainly focused on Europe. Correlation varies from 0.5 for PM_{10} to 0.6 for $PM_{2.5}$. Use of boundary layer height in analysis improved the relationship.
15	Liu et al. 2007	MODIS, MISR, RUC	Intercomparison between MODIS and MISR in St. Louis area. MISR performed slightly better than MODIS in the region.
16	Hutchinson et al. 2008	MODIS, lidar	An attempt is made to improve AOD-$PM_{2.5}$ relationship by refining MODIS AOD product, optimizing averaging area for MODIS pixels around surface station.
17	Gupta and Christopher 2009a	MODIS, RUC meteorology, southeast United States	Multiregression analysis using model-derived meteorology shows improvement to $PM_{2.5}$–AOD relationships.
18	Gupta and Christopher 2009b	MODIS, southeast United States	A novel neural network method for assessing $PM_{2.5}$ using satellite, ground-based and meteorological information.
19	Hoff and Christopher 2009	Review paper	A comprehensive review of particulate matter air pollution from spaceborne measurements.
20	Van Donkelaar et al. 2010	Global	MODIS, MISR and GEOS-CHEM aerosol vertical profiles.

Note: R = correlation coefficient.

shown the potential of using satellite-derived AOD information as surrogate for air quality conditions. The two main conclusions from Table 20.1 indicate that (1) most of the studies have used MODIS-derived AOD products except for a few studies by Liu et al. (2004, 2005) and Donkelaar et al. (2006), which used AODs from both MISR and MODIS, and (2) the area for most of the studies has been in some part of the United States except studies by Gupta et al. (2006), Koelemeijer et al. (2006), and Donkelaar et al. (2006). One of the reasons is that MODIS gives much better spatial and temporal coverage as compared to MISR, and measurements of $PM_{2.5}$ mass concentration in other parts of the world are limited.

The first study by Wang and Christopher (2003) used $PM_{2.5}$ mass and MODIS AOD data over seven stations in Alabama and presented very good correlation (>0.7) between these two parameters. This study also concluded that, although deriving exact $PM_{2.5}$ mass from satellite could have uncertainties, satellites can provide daily air quality indices with sufficient accuracies. Chu et al. (2003) were more focused on the qualitative analysis of the MODIS product as an alternative for air pollution in the regions where surface measurements are not available. It also shows the potential of satellite monitoring of transport of air pollution from source to near and far urban areas. Hutchison et al. (2004, 2005) mainly focused on air quality over Texas and the eastern United States and on the use of satellite imagery in detecting and tracing the pollution.

A study by Engel-Cox et al. (2004) presented a correlation analysis between MODIS AOD and $PM_{2.5}$ mass over the entire United States. The correlation pattern shows high values in the eastern and midwest portions of the United States, whereas correlations are low in western United States. This study also concludes that high space- and time-resolved observations from satellites can provide synoptic information, visualization of the pollution, and validation of ground-based air quality data and estimations from models. Engel-Cox et al. also published other studies that further emphasize the use of satellite-derived aerosol products in day-to-day air quality monitoring and even in policy-related decision making. One of these papers (Engel-Cox et al. 2006) also presented the application of light detection and ranging (lidar-) derived vertical aerosol profiles to improve $PM_{2.5}$–AOD relationship. MODIS aerosol and cloud data are now being used in the Infusing Satellite Data into Environmental Applications (IDEA, http://www.star.nesdis.noaa.gov/smcd/spb/aq/) program to monitor air quality over the United States. IDEA is a joint effort by various federal agencies including NASA, National Oceanic and Atmospheric Administration (NOAA), and the EPA to improve air quality assessment, management, and prediction by infusing satellite measurements into analysis for public benefit (Al-Saadi et al. 2005).

MISR-derived aerosol products were first used by Liu et al. (2004), which shows similar potential for air quality applications. This study also used chemistry transport models to derive meteorological fields to examine their relative effects on $PM_{2.5}$–AOD relationships. Gupta et al. (2006) compared the $PM_{2.5}$–AOD relationship in different parts of the world such as Europe, Australia, the United States, and Asia. This study shows applications of satellite-derived air quality products at global scales and in the regions where surface $PM_{2.5}$ measurements are not available. Correlations analysis varies in different parts of the world depending on accuracies of MODIS retrieval, cloud contamination in AOD, and height of aerosol layer in the atmosphere. Donkelaar et al. (2006) used GEOS-CHEM- (a chemistry transport model) derived vertical extinction profiles and basic mass formula to calculate mass of fine particles and compare the results over several locations in the United States and Canada. This study also presented a global picture of MODIS- and MISR-derived $PM_{2.5}$ mass concentration, and the results appear promising. More recently, Gupta and Christopher (2009a,b) have used multiple regression and neural network techniques to improve surface $PM_{2.5}$ mass estimations from satellite-derived AOD.

All these studies mainly concluded that the MODIS and MISR AODs are important to assess air quality over large spatial domains and to track and monitor aerosol sources and transport. These studies are based on correlation and linear and multivariant regression among MODIS AOD, ground-based $PM_{2.5}$ mass, and model-derived meteorological parameters. The MODIS-derived AOD, which is a measure of column aerosol loading, cannot be used alone to derive $PM_{2.5}$ mass concentration, which is an indicator of the mass of the dry $PM_{2.5}$ near the surface (Wang and Christopher 2003). Meteorological factors such as surface temperature (T_s), relative humidity (RH), wind speed (WS) and direction (WD), variations in sunlight due to clouds, and seasons are important parameters that affect the relationship between $PM_{2.5}$ and AOD. Changes in these processes, which affect the variability in pollution, are primarily governed by the movement of large-scale high- and low-pressure systems, the diurnal heating and cooling cycle, and local and regional topography. The vertical profile of aerosol mass extinction (β_{ext}) and $f(R)$, which are not always uniform within the boundary layer height (H), are also very important parameters that must be accounted for while deriving relationships between $PM_{2.5}$ and AOD (Wang and Christopher 2003; Gupta et al. 2006).

To show the density of $PM_{2.5}$ ground monitoring networks and how the satellite AOD compares with ground-based $PM_{2.5}$ mass concentrations, the locations of these ground monitors have to be identified first (Figure 20.2). The linear correlation

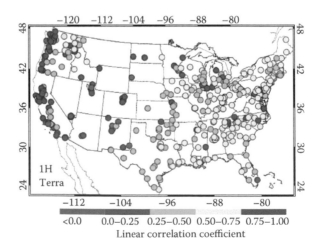

FIGURE 20.2 Location of the $PM_{2.5}$ ground monitors. Color codes represent the correlation coefficient between satellite-AOD and $PM_{2.5}$ mass during the time of the Terra overpass.

coefficients performed using space- and time-matched satellite-PM$_{2.5}$ data sets are also color-coded. This figure indicates that, in general, the correlation coefficients are better in the eastern United States when compared to the west. Difficulties in satellite retrievals over brighter surfaces in the west and selection of aerosol models are some of the major issues for these differences in correlation coefficients. These issues are being addressed currently by various satellite teams.

To forecast air quality, numerical modeling actually requires a system of models and observations including satellite and ground-based data that work together to simulate the emission, transport, diffusion, transformation, and removal of air pollution, and these models include meteorological, emission, and air quality models. Along this direction, various examples of regional pollution events from around the world including agricultural burning in South America/Australia, dust events in China/Africa, and haze events in the eastern United States and Asia can be identified (Figure 20.3). These large-scale pictures provide critical information on episodic events for the air quality community. Furthermore, satellite-derived information can be of tremendous use to air quality modelers. For example, fire locations and emissions derived from satellites are important for forecasting PM$_{2.5}$ mass due to agricultural events (e.g., Christopher et al. 2009). Satellite-based AOT can be used to constrain the fire emissions (Wang et al. 2006) and update the aerosol mass fields through data assimilation (Zhang et al. 2008). However, an algorithm targeted for global retrieval of aerosols may have large errors at regional scale that are of high

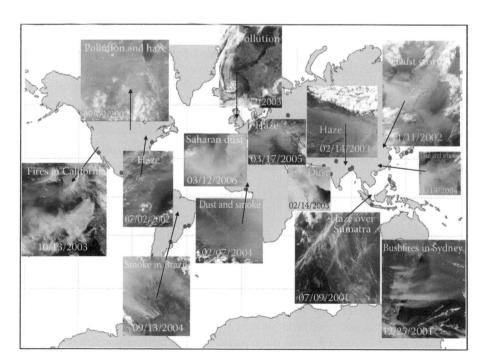

FIGURE 20.3 MODIS satellite imagery of regional pollution events from around the Earth.

interest for air quality applications (Drury et al. 2008). Hence, an integrated use of satellite aerosol product, in particular, the radiance/reflectance data and fire emission data with air quality models, can be designed and calibrated with ground-based data at regional-to-continental scales to improve the air quality forecast and studies (Christopher et al. 2009; Drury et al. 2010).

Finally, tools for visualizing and interpreting air quality information are becoming increasingly important for researchers and decision makers. Prados et al. (2010) described an online tool for accessing and interpreting various air quality data sets. This widely used tool can be used to display satellite data sets coupled with ground-based data sets for visualizing air quality information.

20.4   APPLICATION POTENTIAL

There are numerous examples in the literature indicating the potential of satellite remote sensing for mapping $PM_{2.5}$ concentrations (e.g., Wang and Christopher 2003; Al Saadi et al. 2005; Chu et al. 2003; Wang et al. 2006; von Donkelaar et al. 2010). As an example, we will briefly discuss a 2007 case study. In 2007, massive fires were reported in the Georgia–Florida border in the southeastern United States (Christopher et al. 2009). The total column 550-nm satellite-derived AOD exceeded 1.0 on several days, and ground-based $PM_{2.5}$ mass (i.e., particles less than 2.5 μm in aerodynamic diameter) reached unhealthy levels (>65.5 $\mu g \cdot m^{-3}$). Several hundred miles away from the fire sources, in Birmingham, AL, the impact of the fires was also seen through the high AODs and $PM_{2.5}$ values. Correspondingly, $PM_{2.5}$ mass due to organic carbon obtained from ground-based monitors showed a threefold increase during fire events. Satellite data were especially critical in assessing $PM_{2.5}$ air quality in areas where there were no ground-based monitors. Satellite remote sensing was especially useful for various agencies including the U.S. EPA during this event. Another excellent example of how long-range transport of aerosols from one country affects another country was discussed by Wang et al. (2006). Nearly 1.3 Tg of smoke from biomass burning was transported in Spring 2003 from the Yucatan Peninsula to downwind sources including the southeastern United States. Satellite data sets were used to capture the spatial distribution of these aerosols and assess the magnitude of $PM_{2.5}$ mass concentrations. Satellite-derived fire emissions were also in numerical models to forecast $PM_{2.5}$ mass. As Hoff and Christopher (2009) noted in their review paper, satellite remote sensing has excellent potential for assessing surface $PM_{2.5}$.

20.5   CONCLUSIONS

Satellite remote sensing is the only viable method for monitoring global air pollution. While ground monitors are useful, spaceborne sensors can readily map columnar aerosol concentrations. However, these columnar values can only be used when there are no clouds and bright surfaces such as snow/ice. These columnar values can be related to ground-based PM mass if the vertical distributions of aerosols or boundary layer heights are known. Previous studies have shown the promise and potential for monitoring global pollution from space, and future sensors will continue to improve our capabilities. Satellite data are also being used by the epidemiological community to study health-related issues. The use of satellite data to assess surface $PM_{2.5}$ continues to improve with innovative and improved techniques. Moreover, satellite information (e.g., fire locations and emissions) is being used in numerical models to forecast air quality, which is also an exciting area of research.

REFERENCES

Al-Saadi, J., Szykman, J., Pierce, R. B., Kittaka, C., Neil, D., Chu, D. A., Remer, L., Gumley, L., Prins, E., Weinstock, L., Macdonald, C., Wayland, R., Dimmick, F., and Fishman, J. (2005). Improving national air quality forecasts with satellite aerosol observations. *Bulletin of the American Meteorological Society*, 86(9), 1249–1264.

Christopher, S. A., Gupta, P., Nair, U., Jones, T. A., Kondragunta, S., Wu, Y., Hand, J., and Zhang, X. (2009). Satellite remote sensing and mesoscale modeling of the 2007 Florida/Georgia fires. *IEEE Journal of Selected Topics in Applied Earth Observations and Remote Sensing* (JSTARS-2009-00020), 26, 1–13.

Chu, D. A., Kaufman, Y. J., Zibordi, G., Chern, J. D., Mao, J., Li, C., and Holben, B. N. (2003). Global monitoring of air pollution over land from the Earth Observing System-Terra Moderate Resolution Imaging Spectroradiometer (MODIS), *Journal of Geophysical Research*, 108(D21), 4661, doi:10.1029/2002JD003179.

Dockery, D. W., Pope, C. A., Xu, X., Spengler, J. D., Ware, J. H., and Fay, M. E. (1993). An association between air-pollution and mortality in 6 U.S. cities. *New England Journal of Medicine*, 329(24), 1753–1759.

Drury, E., Jacob, D. J., Wang, J., Spurr, R. J. D., and Chance, K. (2008). Improved algorithm for MODIS satellite retrievals of aerosol optical depths over land. *Journal of Geophysical Research*, 113, D16204, doi:10.1029/2007JD009573.

Drury, E., Jacob, D. J., Spurr, R. J. D., Wang, J., Shinozuka, Y., Anderson, B. E., Clarke, A. D., Dibb, J., McNaughton, C., and Weber, R. (2010). Synthesis of satellite (MODIS), aircraft (ICARTT), and surface (IMPROVE, EPA-AQS, AERONET) aerosol observations over North America to improve MODIS aerosol retrievals and constrain surface aerosol concentrations and sources. *Journal of Geophysical Research*, 115, D14204, doi:10.1029/2009JD012629.

Engel-Cox, J. A., Holloman, C. H., Coutant, B. W., and Hoff, R. M. (2004). Qualitative and quantitative evaluation of MODIS satellite sensor data for regional and urban scale air quality. *Atmospheric Environment*, 38, 2495–2509.

Engel-Cox, J. A., Young, G., and Hoff, R. (2005). Application of satellite remote sensing data for source analysis of fine particulate matter transport events. *Journal of Air and Waste Management*, 55, 9, 1389–1397.

Engel-Cox, J. A., Hoff, R. M., Rogers, R., Dimmick, F., Rush, A. C., Szykman, J. J., Al-Saadi, J., Chu, D. A., and Zell, E. R. (2006). Integrating lidar and satellite optical depth with ambient monitoring for 3-dimensional particulate characterization. *Atmospheric Environment*, 40, 8056–8067.

Grover, B. D., Kleinman, M., Eatough, N. L., Eatough, D. J., Hopke, P. K., Long, R. W., Wilson, W. E., Meyer, M. B., and Ambs, J. L. (2005). Measurement of total $PM_{2.5}$ mass (nonvolatile plus semivolatile) with the filter dynamic measurement system tapered element oscillating microbalance monitor. *Journal of Geophysical Research*, 110, D07S03, doi:10.1029/2004JD004995.

Gupta, P. and Christopher, S. A. (2009a). Particulate matter air quality assessment using integrated surface, satellite, and meteorological products: Multiple regression approach. *Journal of Geophysical Research*, 114, D14205, doi:10.1029/2008JD011496.

Gupta, P. and Christopher, S. A. (2009b). Particulate matter air quality assessment using integrated surface, satellite, and meteorological products: 2. A neural network approach. *Journal of Geophysical Research*, 114, D20205, doi:10.1029/2008JD011497.

Gupta, P., Christopher, S. A., Wang, J., Gehrig, R., Lee, Y. C., and Kumar, N. (2006). Satellite remote sensing of particulate matter and air quality over global cities. *Atmospheric Environment*, 40 (30), 5880–5892.

Hoff, R. and Christopher, S. A. (2009). Remote sensing of particulate matter air pollution from space: Have we reached the Promised Land? *Journal of Air and Waste Management Association*, 59, 642–675.

Hu, Z. (2009). Spatial analysis of MODIS aerosol optical depth, $PM_{2.5}$ and chronic coronary heart disease. *International Journal of Health Geographics*, 8, 27, doi:10.1186/1476-072X-8-27.

Hutchison, K. D., Smith, S., and Faruqui, S. (2004). The use of MODIS data and aerosol products for air quality prediction. *Atmospheric Environment*, 38, 5057–5070.

Hutchison, K. D., Smith, S., and Faruqui, S. (2005). Correlating MODIS aerosol optical depth data with ground-based $PM_{2.5}$ observations across Texas for use in a real-time air quality prediction system. *Atmospheric Environment*, 39(37), 7190–7203.

Hutchison, K. D., Faruqui, S. J., and Smith, S. (2008). Improving correlations between MODIS aerosol optical thickness and ground-based PM2.5 observations through 3D spatial analyses. *Atmospheric Environment*, 42(3), 530–543.

Kaufman, Y., Tanré, D., and Boucher, O. (2002). A satellite view of aerosols in the climate systems. *Nature*, 419, 215–223.

King, M. D., Kaufman, Y. J., Menzel, W. P., and Tanre, D. (1992). Remote sensing of cloud, aerosol, and water vapor properties from the Moderate Resolution Imaging Spectrometer (MODIS). *IEEE Transactions on Geoscience and Remote Sensing*, 30, 2–27.

Koelemeijer, R. B. A., Homan, R. C., and Matthijsen, J. (2006). Comparison of spatial and temporal variations of aerosol optical depth and particulate matter over Europe. *Atmospheric Environment*, 40, 5304–5315.

Liu, Y., Sarnat, J. A., Kilaru, V., Jacob, D. J., and Koutrakis, P. (2005). Estimating ground level $PM_{2.5}$ in the eastern United States using satellite remote sensing. *Environmental Science & Technology*, 39(9), 3269–3278.

Liu, Y., Park, R. J., Jacob, D. J., Li, Q., Kilaru, V., and Sarnat, J. A. (2004). Mapping annual mean ground-level $PM_{2.5}$ concentrations using Multiangle Imaging Spectroradiometer aerosol optical depth over the contiguous United States. *Journal of Geophysical Research*, 109, D22206, doi:10.1029/2004JD005025.

Liu, Y., Franklin, M., Kahn, R., and Koutrakis, P. (2007). Using aerosol optical thickness to predict ground-level $PM_{2.5}$ concentrations in the St. Louis area: A comparison between MISR and MODIS. *Remote Sensing of Environment*, 107(1–2), 33–44.

Martin, R. V. (2008). Satellite remote sensing of surface air quality. *Atmospheric Environment*, 42, 7823–7843.

Pope, C. A. (2000). Epidemiology of fine particulate air pollution and human health: Biologic mechanisms and who's at risk? *Environmental Health Perspectives*, (Suppl. 4), 104, 713–723.

Pope, C. A. III, Ezzati, M., and Dockery, D. W. (2009). Fine-particulate air pollution and life expectancy in the United States. *New England Journal of Medicine*, 360(4), 376–386.

Prados, A. I., Leptoukh, G., Lynnes, C., Johnson, J., Hualan, R., Aijun, C., and Husar, R. B. (2010). Access, visualization, and interoperability of air quality remote sensing data sets via the Giovanni Online Tool. *IEEE Journal of Selected Topics in Applied Earth Observations and Remote Sensing*, 3(3), 359–370.

Remer, L. A., Kaufman, Y. J., Tanré, D., Mattoo, S., Chu, D. A., Martins, J. V., Li, R. R., Ichoku, C., Levy, R. C., Kleidman, R. G., Eck, T. F., Vermote, E., and Holben, B. N. (2005). The MODIS aerosol algorithm, products, and validation. *Journal of Atmospheric Science*, 62, 947–973.

van Donkelaar, A., Martin, R. V., and Park, R. J. (2006). Estimating ground-level $PM_{2.5}$ using aerosol optical depth determined from satellite remote sensing. *Journal of Geophysical Research*, 111(D21), D21201, doi:10.1029/2005JD006996.

van Donkelaar, A., Martin, R. V., Brauer, M., Kahn, R., Levy, R., and Verduzco, C. (2010). Global estimates of ambient fine particulate matter concentrations from satellite-based aerosol optical depth: Development and application. *Environmental Health Perspectives*, 118, 847–855.

Wang, J. and Christopher, S. A. (2003). Intercomparison between satellite-derived aerosol optical depth and $PM_{2.5}$ mass: Implications for air quality studies. *Geophysical Research Letters*, 30(21), 2095, doi:10.1029/2003GL018174.

Wang, J., Christopher, S. A., Nair, U. S., Reid, J. S., Prins, E. M., Szykman, J., and Hand, J. L. (2006). Mesoscale modeling of Central American smoke transport to the United States, part I: "Top-down" assessment of emission strength and diurnal variation impact. *Journal of Geophysical Research*, 111, D05S17, doi:10.1029/2005jd006720.

Wong, C.-M., Vichit-Vadakan, N., Kan, H., Qian, Z., and Teams, P. P. (2008). A multicity study of short-term effects of air pollution on mortality. *Public Health and Air Pollution in Asia (PAPA), Environmental Health Perspectives*, 116, 1195–1202.

Yang, E.-S., Christopher, S. A., Kondragunta, S., and Zhang, X. (2011). Use of hourly Geostationary Operational Environmental Satellite (GOES) fire emissions in a community multiscale air quality (CMAQ) model for improving surface particulate matter predictions. *Journal of Geophysical Research*, 116(D4), D04303.

Zhang, J., Reid, J. S., Westphal, D. L., Baker, N. L., and Hyer, E. J. (2008). A system for operational aerosol optical depth data assimilation over global oceans. *Journal of Geophysical Research*, 113, D10208, doi:10.1029/2007JD009065.

Index

Page numbers followed by f and t indicate figures and tables, respectively.